Study and Solutions Guide for

PRECALCULUS

SECOND EDITION

Larson/Hostetler

Dianna L. Zook

The Pennsylvania State University
The Behrend College

D. C. Heath and Company

Lexington, Massachusetts Toronto

Published simultaneously in Canada.

Printed in the United States of America.

International Standard Book Number: 0–669–17345–2

7 8 9 0

TO THE STUDENT

The *Study and Solutions Guide for Precalculus* is a supplement to the text by Roland E. Larson and Robert P. Hostetler.

As a mathematics instructor, I often have students come to me with questions about the assigned homework. When I ask to see their work, the reply often is "I didn't know where to start." The purpose of the *Study Guide* is to provide brief summaries of the topics covered in the textbook and enough detailed solutions to problems so that you will be able to work the remaining exercises.

A special thanks to Linda M. Bollinger for typing this guide. Also I would like to thank my husband Edward L. Schlindwein for his support during the several months I worked on this project.

If you have any corrections or suggestions for improving this *Study Guide*, I would appreciate hearing from you.

Good luck with your study of precalculus.

Dianna L. Zook
The Pennsylvania State University
Erie, Pennsylvania 16563

CONTENTS

CHAPTER 1

Review of Fundamental Concepts of Algebra

SECTION 1.1

The Real Number System

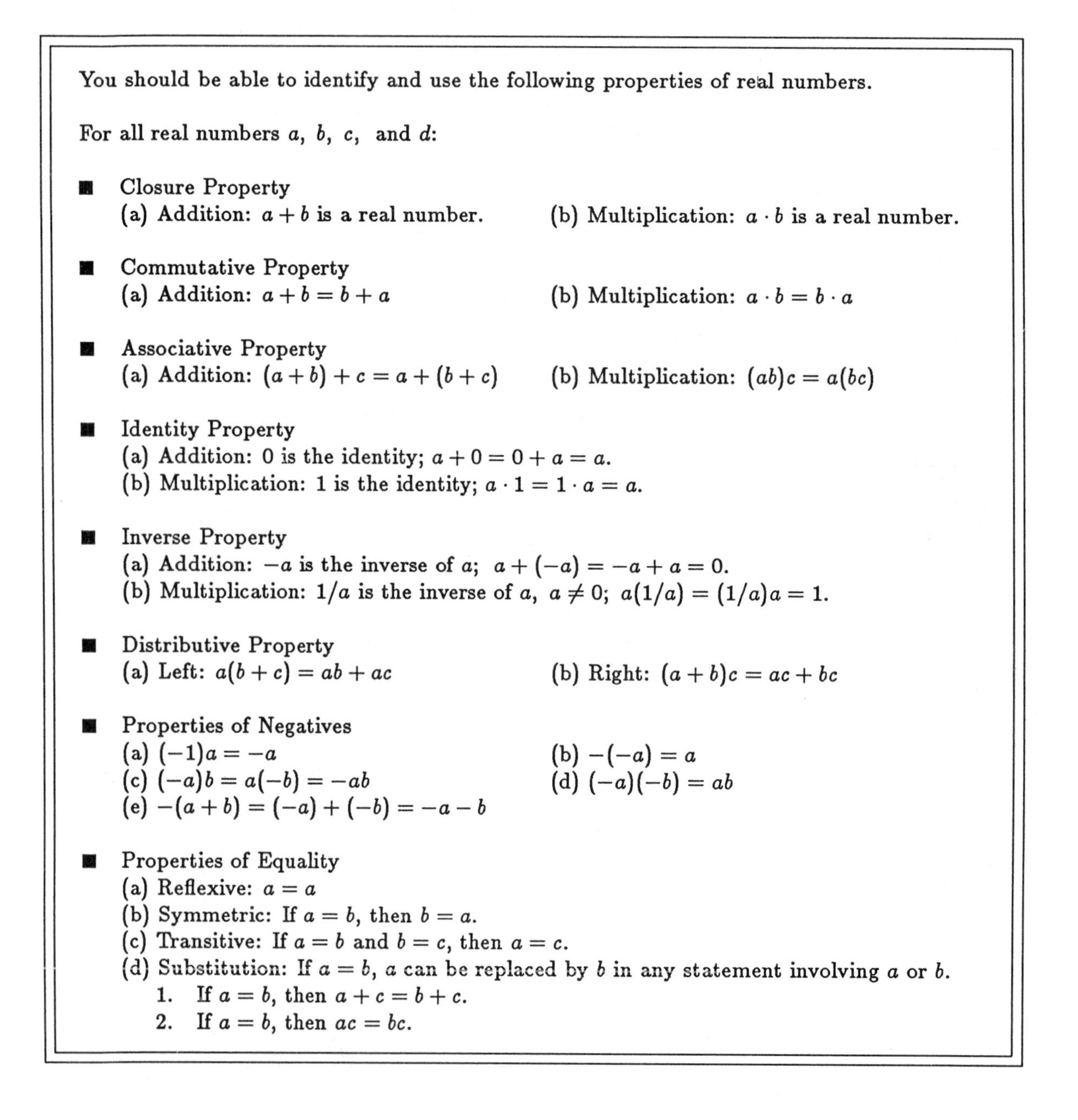

You should be able to identify and use the following properties of real numbers.

For all real numbers a, b, c, and d:

- **Closure Property**
  (a) Addition: $a + b$ is a real number.
  (b) Multiplication: $a \cdot b$ is a real number.

- **Commutative Property**
  (a) Addition: $a + b = b + a$
  (b) Multiplication: $a \cdot b = b \cdot a$

- **Associative Property**
  (a) Addition: $(a + b) + c = a + (b + c)$
  (b) Multiplication: $(ab)c = a(bc)$

- **Identity Property**
  (a) Addition: 0 is the identity; $a + 0 = 0 + a = a$.
  (b) Multiplication: 1 is the identity; $a \cdot 1 = 1 \cdot a = a$.

- **Inverse Property**
  (a) Addition: $-a$ is the inverse of a; $a + (-a) = -a + a = 0$.
  (b) Multiplication: $1/a$ is the inverse of a, $a \neq 0$; $a(1/a) = (1/a)a = 1$.

- **Distributive Property**
  (a) Left: $a(b + c) = ab + ac$
  (b) Right: $(a + b)c = ac + bc$

- **Properties of Negatives**
  (a) $(-1)a = -a$
  (b) $-(-a) = a$
  (c) $(-a)b = a(-b) = -ab$
  (d) $(-a)(-b) = ab$
  (e) $-(a + b) = (-a) + (-b) = -a - b$

- **Properties of Equality**
  (a) Reflexive: $a = a$
  (b) Symmetric: If $a = b$, then $b = a$.
  (c) Transitive: If $a = b$ and $b = c$, then $a = c$.
  (d) Substitution: If $a = b$, a can be replaced by b in any statement involving a or b.
    1. If $a = b$, then $a + c = b + c$.
    2. If $a = b$, then $ac = bc$.

- **Cancellation Laws**
  (a) If $a + c = b + c$, then $a = b$.
  (b) If $ac = bc$, then $a = b$, $c \neq 0$.

- Subtraction: $a - b = a + (-b)$

- Division: $a \div b = a(1/b) = a/b$, $b \neq 0$

- Properties of Fractions ($b \neq 0$, $d \neq 0$)
  (a) Equivalent Fractions: $a/b = c/d$ if and only if $ad = bc$.
  (b) Rule of Signs: $-a/b = a/-b = -(a/b)$ and $-a/-b = a/b$
  (c) Equivalent Fractions: $a/b = ac/bc$, $c \neq 0$
  (d) Addition and Subtraction
    1. Like Denominators: $(a/b) \pm (c/b) = (a \pm c)/b$
    2. Unlike Denominators: $(a/b) \pm (c/d) = (ad \pm bc)/bd$
  (e) Multiplication: $(a/b) \cdot (c/d) = ac/bd$
  (f) Division: $(a/b) \div (c/d) = (a/b) \cdot (d/c) = ad/bc$ if $c \neq 0$.

- Properties of Zero
  (a) $a \pm 0 = a$
  (b) $a \cdot 0 = 0$
  (c) $0 \div a = 0/a = 0$, $a \neq 0$
  (d) If $ab = 0$, then $a = 0$ or $b = 0$.
  (e) $a/0$ is undefined.

- Definition of Absolute Value:

$$|a| = \begin{cases} a, & \text{if } a \geq 0 \\ -a, & \text{if } a < 0 \end{cases}$$

- Properties of Absolute Value
  (a) $|-a| = |a|$
  (b) $|ab| = |a||b|$
  (c) $|a/b| = |a|/|b|$, $b \neq 0$

- Distance Between Two Points on the Real Line
  $d(a,\ b) = |b - a| = |a - b|$

Solutions to Selected Exercises

7. Identify the property illustrated in the equation $2(x+3) = 2x + 6$.

Solution:
By the Distributive Property, we have

$$2(x+3) = 2 \cdot x + 2 \cdot 3 = 2x + 6.$$

15. Identify the properties illustrated in the equation $x(3y) = (x \cdot 3)y = (3x)y$.

Solution:

$$\begin{aligned} x(3y) &= (x \cdot 3)y && \text{by the Associative Property of Multiplication} \\ &= (3x)y && \text{by the Commutative Property of Multiplication} \end{aligned}$$

19. Use the properties of zero to evaluate, if possible, the following expression. If the expression is undefined, state why.

$$\frac{8}{-9 + (6+3)}$$

Solution:

$$\frac{8}{-9 + (6+3)} = \frac{8}{-9+9} = \frac{8}{0}$$

which is undefined since the denominator is zero.

23. Perform the indicated operations: $10 - 6 - 2$.

Solution:

$$10 - 6 - 2 = (10 - 6) - 2 = 4 - 2 = 2$$

27. Perform the indicated operation:

$$2\left(\frac{77}{-11}\right).$$

Solution:

$$2\left(\frac{77}{-11}\right) = 2(-7) = -14$$

35. Perform the indicated operations:

$$\frac{4}{5} \times \frac{1}{2} \times \frac{3}{4}.$$

Solution:

$$\frac{4}{5} \times \frac{1}{2} \times \frac{3}{4} = \frac{1}{5} \times \frac{1}{2} \times \frac{3}{1} = \frac{3}{10}$$

39. Perform the indicated operation: $12 \div \frac{1}{4}$.

Solution:

$$12 \div \frac{1}{4} = 12 \times \frac{4}{1} = 12 \times 4 = 48$$

45. Plot the two real numbers $\frac{5}{6}$ and $\frac{2}{3}$ on the number real line and place the appropriate inequality sign between them.

Solution:

$\frac{2}{3}$ $\frac{5}{6}$

0 1 2 x

$$\frac{5}{6} > \frac{2}{3}$$

47. Use inequality notation to denote the expression "x is negative".

Solution:
"x is negative" can be written as $x < 0$.

51. Use inequality notation to denote the expression "the annual rate of inflation, R, is expected to be at least 3.5% but no more than 6%".

Solution:
The expression "the annual rate of inflation, R, is expected to be at least 3.5% but no more than 6%" can be written

$$3.5\% \leq R \leq 6\%$$

or

$$0.035 \leq R \leq 0.06$$

59. Find the distance between the points $-\frac{5}{2}$ and 0 on the real line.

Solution:

$$d\left(-\frac{5}{2}, 0\right) = \left|0 - \left(-\frac{5}{2}\right)\right| = \left|\frac{5}{2}\right| = \frac{5}{2}$$

63. Find the distance between the points 9.34 and -5.65 on the real line.

Solution:

$$\begin{aligned} d(9.34,\ -5.65) &= |9.34 - (-5.65)| \\ &= |9.34 + 5.65| \\ &= |14.99| = 14.99 \end{aligned}$$

67. Use absolute value notation to describe the expression "the distance between z and $\frac{3}{2}$ is greater than 1".

Solution:
Since

$$d(z,\ \tfrac{3}{2}) = |z - \tfrac{3}{2}| \quad \text{and} \quad d(z,\ \tfrac{3}{2}) > 1$$

we have

$$|z - \tfrac{3}{2}| > 1.$$

71. Let m and n be any two integers. Then $2m$ and $2n$ are even integers and $(2m+1)$ and $(2n+1)$ are odd integers.

(a) Prove that the sum of two even integers is even.
(b) Prove that the sum of two odd integers is even.
(c) Prove that the product of an even integer with *any* integer is even.

Solution:
(a) The sum of two even integers can be written as

$$2m + 2n = 2(m + n)$$

which is even.
(b) The sum of two odd integers can be written as

$$(2m + 1) + (2n + 1) = 2m + 2n + 2 = 2(m + n + 1)$$

which is even.
(c) Let n be any integer. The product of an even integer with n can be written as

$$(2m)n = 2(mn)$$

which is even.

73. One worker can assemble a component in seven days and a second worker can do the same task in five days. If they work together, what fraction of a component can they assemble in two days?

Solution:

One worker assembles 1/7 of a component each day. The other worker assembles 1/5 of a component each day. Together, in two days, they assemble

$$2\left(\frac{1}{7}+\frac{1}{5}\right)=2\left(\frac{5+7}{35}\right)=2\left(\frac{12}{35}\right)=\frac{24}{35}$$

of a component.

77. (a) Use a calculator to order the following real numbers, from smallest to largest.

$$\frac{7071}{5000}, \frac{584}{413}, \sqrt{2}, \frac{47}{33}, \frac{127}{90}$$

(b) Which of the rational numbers in part (a) is closest to $\sqrt{2}$?

Solution:

(a) $\dfrac{7071}{5000}=1.4142$

$\dfrac{584}{413}=1.414043584$

$\sqrt{2}=1.414213562$

$\dfrac{47}{33}=1.42\overline{42}$

$\dfrac{127}{90}=1.41\overline{1}$

$$\frac{127}{90}<\frac{584}{413}<\frac{7071}{5000}<\sqrt{2}<\frac{47}{33}$$

(b) $\dfrac{7071}{5000}$ is closest to $\sqrt{2}$.

SECTION 1.2

Exponents and Radicals

- You should know the following properties of exponents.

  (a) $a^m a^n = a^{m+n}$
  (b) $\dfrac{a^m}{a^n} = a^{m-n}$
  (c) $\dfrac{1}{a^n} = a^{-n}$
  (d) $a^0 = 1,\ a \neq 0$
  (e) $(ab)^m = a^m b^m$
  (f) $(a^m)^n = a^{mn}$
  (g) $\left(\dfrac{a}{b}\right)^m = \dfrac{a^m}{b^m}$
  (h) $|a^2| = |a|^2 = a^2$

- You should know the following properties of radicals.

  (a) $\sqrt[n]{b} = a$ if and only if $a^n = b$
  (b) $\sqrt[n]{a^m} = (\sqrt[n]{a})^m = a^{m/n}$
  (c) $\sqrt[n]{a} \cdot \sqrt[n]{b} = \sqrt[n]{ab}$
  (d) $\dfrac{\sqrt[n]{a}}{\sqrt[n]{b}} = \sqrt[n]{\dfrac{a}{b}},\ b \neq 0$
  (e) $\sqrt[m]{\sqrt[n]{a}} = \sqrt[mn]{a}$
  (f) $(\sqrt[n]{a})^n = a$
  (g) If n is even, $\sqrt[n]{a^n} = |a|$.
  (h) If n is odd, $\sqrt[n]{a^n} = a$.

- You should be able to simplify radicals.

  (a) Remove all possible factors from the radical sign.
  (b) Rationalize the denominator.
  (c) Reduce the index as far as possible.

- You should be able to write numbers in scientific notation.

Solutions to Selected Exercises

5. Evaluate $6x^0 - (6x)^0$ when $x = 10$.

Solution:

$$6(10)^0 - (6 \cdot 10)^0 = 6 \cdot 1 - (60)^0 = 6 - 1 = 5$$

9. Simplify $6y^2(2y^4)^2$.

Solution:

$$6y^2(2y^4)^2 = 6y^2(2)^2(y^4)^2 = 6y^2(4)(y^8) = 24y^{10}$$

13. Simplify

$$\frac{12(x+y)^3}{9(x+y)}.$$

Solution:

$$\frac{12(x+y)^3}{9(x+y)} = \frac{3 \cdot 4(x+y)^{3-1}}{3 \cdot 3} = \frac{4(x+y)^2}{3}$$

15. Simplify $(-2x^2)^3(4x^3)^{-1}$.

Solution:

$$(-2x^2)^3(4x^3)^{-1} = \frac{(-2x^2)^3}{4x^3} = \frac{-8x^6}{4x^3} = -2x^3$$

19. Simplify $(4a^{-2}b^3)^{-3}$.

Solution:

$$(4a^{-2}b^3)^{-3} = (4)^{-3}(a^{-2})^{-3}(b^3)^{-3} = 4^{-3}a^6b^{-9} = \frac{a^6}{4^3b^9} = \frac{a^6}{64b^9}$$

23. Write each number in scientific notation.

(a) 93,000,000
(b) 900,000,000
(c) 0.00000435
(d) 0.000087

Solution:

(a) $93,000,000 = 9.3 \times 10^7$ (decimal moves 7 places to the left)
(b) $900,000,000 = 9 \times 10^8$ (decimal moves 8 places to the left)
(c) $0.00000435 = 4.35 \times 10^{-6}$ (decimal moves 6 places to the right)
(d) $0.000087 = 8.7 \times 10^{-5}$ (decimal moves 5 places to the right)

25. Write each number in decimal form.

(a) 1.91×10^6
(b) 2.345×10^{11}
(c) 6.21×10^0
(d) 9.4675×10^4

Solution:

(a) $1.91 \times 10^6 = 1.91 \times 1,000,000 = 1,910,000$

(b) $2.345 \times 10^{11} = 2.345 \times 100,000,000,000 = 234,500,000,000$

(c) $6.21 \times 10^0 = 6.21 \times 1 = 6.21$

(d) $9.4675 \times 10^4 = 9.4675 \times 10,000 = 94,675$

31. Find the rational exponent form for $\sqrt[3]{-216} = -6$.

Solution:

Rational Exponent Form: $(-216)^{1/3} = -6$

39. Evaluate each expression without using a calculator.

(a) $(\sqrt[6]{326})^6$
(b) $\sqrt[4]{562^4}$

Solution:

(a) $(\sqrt[6]{326})^6 = 326$ (Property 5 of Radicals)

(b) $\sqrt[4]{562^4} = 562$ (Property 6 of Radicals)

41. Evaluate each expression without using a calculator.

(a) $64^{-2/3}$
(b) $\left(\frac{9}{4}\right)^{-1/2}$

Solution:

(a) $64^{-2/3} = \dfrac{1}{64^{2/3}} = \dfrac{1}{(\sqrt[3]{64})^2} = \dfrac{1}{4^2} = \dfrac{1}{16}$

(b) $\left(\dfrac{9}{4}\right)^{-1/2} = \left(\dfrac{4}{9}\right)^{1/2} = \sqrt{\dfrac{4}{9}} = \dfrac{\sqrt{4}}{\sqrt{9}} = \dfrac{2}{3}$

45. Simplify by removing all possible factors from the radical.

(a) $\sqrt{75x^2y^{-4}}$
(b) $\sqrt{5(x-y)^3}$

Solution:

(a) $\sqrt{75x^2y^{-4}} = \sqrt{25x^2y^{-4}(3)} = \sqrt{25}\sqrt{x^2}\sqrt{(y^{-2})^2}\sqrt{3} = 5|x|y^{-2}\sqrt{3} = \dfrac{5|x|\sqrt{3}}{y^2}$

(b) $\sqrt{5(x-y)^3} = \sqrt{(x-y)^2 5(x-y)} = \sqrt{(x-y)^2}\sqrt{5(x-y)} = (x-y)\sqrt{5(x-y)}$

Note: We do not have $|x-y|$ here since $(x-y)$ must be positive for $(x-y)^3$ to be under the radical.

49. Use fractional exponents to verify $\sqrt[6]{(x+1)^4} = \sqrt[3]{(x+1)^2}$.

Solution:

$$\sqrt[6]{(x+1)^4} = (x+1)^{4/6} = (x+1)^{2/3} = \sqrt[3]{(x+1)^2}$$

53. Rewrite each expression by rationalizing the denominator. Simplify your answer.

(a) $\dfrac{5}{\sqrt[3]{(5x)^2}}$ (b) $\dfrac{3}{\sqrt[4]{(3x)^3}}$

Solution:

(a) $\dfrac{5}{\sqrt[3]{(5x)^2}} = \dfrac{5}{\sqrt[3]{(5x)^2}} \cdot \dfrac{\sqrt[3]{5x}}{\sqrt[3]{5x}} = \dfrac{5\sqrt[3]{5x}}{\sqrt[3]{(5x)^3}} = \dfrac{5\sqrt[3]{5x}}{5x} = \dfrac{\sqrt[3]{5x}}{x}$

(b) $\dfrac{3}{\sqrt[4]{(3x)^3}} = \dfrac{3}{\sqrt[4]{(3x)^3}} \cdot \dfrac{\sqrt[4]{3x}}{\sqrt[4]{3x}} = \dfrac{3\sqrt[4]{3x}}{\sqrt[4]{(3x)^4}} = \dfrac{3\sqrt[4]{3x}}{3x} = \dfrac{\sqrt[4]{3x}}{x}$

55. Rewrite each expression by rationalizing the denominator. Simplify your answer.

(a) $\dfrac{3}{\sqrt{5}+\sqrt{6}}$ (b) $\dfrac{8}{\sqrt{2}-2\sqrt{3}}$

Solution:

(a)
$$\begin{aligned}\frac{3}{\sqrt{5}+\sqrt{6}} &= \frac{3}{\sqrt{5}+\sqrt{6}} \cdot \frac{\sqrt{5}-\sqrt{6}}{\sqrt{5}-\sqrt{6}}\\ &= \frac{3(\sqrt{5}-\sqrt{6})}{(\sqrt{5})^2-(\sqrt{6})^2}\\ &= \frac{3(\sqrt{5}-\sqrt{6})}{5-6}\\ &= \frac{3(\sqrt{5}-\sqrt{6})}{-1}\\ &= -3(\sqrt{5}-\sqrt{6})\\ &= 3(\sqrt{6}-\sqrt{5})\end{aligned}$$

(b)
$$\begin{aligned}\frac{8}{\sqrt{2}-2\sqrt{3}} &= \frac{8}{\sqrt{2}-2\sqrt{3}} \cdot \frac{\sqrt{2}+2\sqrt{3}}{\sqrt{2}+2\sqrt{3}}\\ &= \frac{8(\sqrt{2}+2\sqrt{3})}{(\sqrt{2})^2-(2\sqrt{3})^2}\\ &= \frac{8(\sqrt{2}+2\sqrt{3})}{2-12}\\ &= \frac{8(\sqrt{2}+2\sqrt{3})}{-10}\\ &= -\frac{4}{5}(\sqrt{2}+2\sqrt{3})\end{aligned}$$

59. Rewrite each expression by rationalizing the numerator. Simplify your answer.

(a) $\dfrac{\sqrt{3}-\sqrt{2}}{x}$ (b) $\dfrac{\sqrt{15}+3}{12}$

Solution:

(a)
$$\begin{aligned}\frac{\sqrt{3}-\sqrt{2}}{x} &= \frac{\sqrt{3}-\sqrt{2}}{x} \cdot \frac{\sqrt{3}+\sqrt{2}}{\sqrt{3}+\sqrt{2}}\\ &= \frac{(\sqrt{3})^2-(\sqrt{2})^2}{x(\sqrt{3}+\sqrt{2})}\\ &= \frac{3-2}{x(\sqrt{3}+\sqrt{2})}\\ &= \frac{1}{x(\sqrt{3}+\sqrt{2})}\end{aligned}$$

(b)
$$\begin{aligned}\frac{\sqrt{15}+3}{12} &= \frac{\sqrt{15}+3}{12} \cdot \frac{\sqrt{15}-3}{\sqrt{15}-3}\\ &= \frac{(\sqrt{15})^2-(3)^2}{12(\sqrt{15}-3)}\\ &= \frac{15-9}{12(\sqrt{15}-3)}\\ &= \frac{6}{12(\sqrt{15}-3)} = \frac{1}{2(\sqrt{15}-3)}\end{aligned}$$

63. Simplify and/or combine the given radicals.

(a) $2\sqrt{4y} - 2\sqrt{9y} + 10\sqrt{y}$

(b) $6\sqrt{\dfrac{a}{2}} + 5\sqrt{2a}$

Solution:

(a) $$2\sqrt{4y} - 2\sqrt{9y} + 10\sqrt{y} = 2(2\sqrt{y}) - 2(3\sqrt{y}) + 10\sqrt{y}$$
$$= 4\sqrt{y} - 6\sqrt{y} + 10\sqrt{y} = (4 - 6 + 10)\sqrt{y} = 8\sqrt{y}$$

(b) $$6\sqrt{\frac{a}{2}} + 5\sqrt{2a} = \frac{6\sqrt{a}}{\sqrt{2}} + 5\sqrt{2a} = \frac{6\sqrt{a}}{\sqrt{2}} \cdot \frac{\sqrt{2}}{\sqrt{2}} + 5\sqrt{2a}$$
$$= \frac{6\sqrt{2a}}{2} + 5\sqrt{2a} = 3\sqrt{2a} + 5\sqrt{2a} = 8\sqrt{2a}$$

65. Simplify and/or combine the given radicals.

(a) $\sqrt{5x^2y}\sqrt{3y}$

(b) $\dfrac{\sqrt{54a^2}}{\sqrt{2a^4}}$

Solution:

(a) $$\sqrt{5x^2y}\sqrt{3y} = |x|\sqrt{5y}\sqrt{3y} = |x|\sqrt{15y^2} = |x|y\sqrt{15}$$

(b) $$\frac{\sqrt{54a^2}}{\sqrt{2a^4}} = \sqrt{\frac{54a^2}{2a^4}} = \sqrt{\frac{27}{a^2}} = \frac{\sqrt{9 \cdot 3}}{\sqrt{a^2}} = \frac{3\sqrt{3}}{|a|}$$

67. Write $\sqrt{50}\sqrt[3]{2}$ as a single radical.

Solution:

$$\sqrt{50}\sqrt[3]{2} = \sqrt{25 \cdot 2}\sqrt[3]{2} = 5\sqrt{2}\sqrt[3]{2} = 5(2)^{1/2}(2)^{1/3} = 5(2)^{1/2+1/3}$$
$$= 5(2)^{3/6+2/6} = 5(2)^{5/6} = 5\sqrt[6]{2^5}$$

71. Use a calculator to evaluate the given expressions. (Round your answers to three decimal places.)

(a) $2400(1 + 0.06)^{20}$

(b) $750\left(1 + \dfrac{0.11}{365}\right)^{800}$

(c) $\dfrac{(2.414 \times 10^4)^6}{(1.68 \times 10^5)^5}$

(d) $(9.3 \times 10^6)^3(6.1 \times 10^{-4})^4$

Solution:

(a) $$2400(1 + 0.06)^{20} = 2400(1.06)^{20} \approx 7697.125$$

(b) $$750\left(1 + \frac{0.11}{365}\right)^{800} \approx 954.448$$

(c) $$\frac{(2.414 \times 10^4)^6}{(1.68 \times 10^5)^5} = \frac{(2.414)^6 \times 10^{24}}{(1.68)^5 \times 10^{25}} = \frac{(2.414)^6}{(1.68)^5 \times 10} \approx 1.479$$

(d) $$(9.3\times10^6)^3(6.1\times10^{-4})^4 = (9.3)^3\times10^{18}\times(6.1)^4\times10^{-16} = (9.3)^3(6.1)^4(10)^2 \approx 111,369,991.3$$

73. The speed of light is 11,160,000 miles per minute. The distance from the sun to the earth is 93,000,000 miles. Find the time it takes for light to travel from the sun to the earth.

Solution:

$$\frac{93,000,000 \text{ miles}}{11,160,000 \text{ miles/minute}} = 8\tfrac{1}{3} \text{ minutes}$$

75. Use a calculator to approximate the given number. (Round your answer to four decimal places.)

(a) $\sqrt{57}$

(b) $\sqrt[5]{562}$

(c) $\sqrt[3]{45^2}$

(d) $(-10)^{4/5}$

(e) $\sqrt[6]{125}$

(f) $\sqrt[5]{-65}$

Solution:

(a) $\sqrt{57} \approx 7.5498$

(b) $\sqrt[5]{562} = (562)^{1/5} = (562)^{0.2} \approx 3.5477$

(c) $\sqrt[3]{45^2} = (45)^{2/3} = (45)^{0.66666\ldots} \approx (45)^{0.666667} \approx 12.6515$

(d) $(-10)^{4/5} = (-10)^{0.8} = (10)^{0.8} \approx 6.3096$

(e) $\sqrt[6]{125} = \sqrt[6]{5^3} = (5)^{3/6} = (5)^{1/2} = \sqrt{5} \approx 2.2361$

(f) $\sqrt[5]{-65} = (-65)^{1/5} = -(65)^{1/5} = -(65)^{0.2} \approx -2.3045$

77. The amount A after t years in a savings account earning an annual interest rate of r compounded n times per year is $A = P(1 + \frac{r}{n})^{nt}$ where P is the original principal. Complete the following table for \$500 deposited in an account earning 12% compounded daily. [Note that $r = 0.12$ implies an interest rate of 12%.]

[*Hint:* If you have a programmable calculator, try using the programming feature to complete the table.]

t	5	10	20
A	$500\left(1 + \frac{0.12}{365}\right)^{365\times 5} \approx \$910.97$	$500\left(1 + \frac{0.12}{365}\right)^{365\times 10} \approx \$1659.73$	$500\left(1 + \frac{0.12}{365}\right)^{365\times 20} \approx \$5509.41$

t	30	40	50
A	$500\left(1 + \frac{0.12}{365}\right)^{365\times 30} \approx \$18,288.29$	$500\left(1 + \frac{0.12}{365}\right)^{365\times 40} \approx \$60,707.30$	$500\left(1 + \frac{0.12}{365}\right)^{365\times 50} \approx \$201,515.58$

SECTION 1.3

Polynomials: Special Products and Factoring

- You should be able to add, subtract, and multiply polynomials.
- You should know the following special products.
  (a) $(u+v)(u-v) = u^2 - v^2$
  (b) $(u+v)^2 = u^2 + 2uv + v^2$
  (c) $(u-v)^2 = u^2 - 2uv + v^2$
  (d) $(u+v)^3 = u^3 + 3u^2v + 3uv^2 + v^3$
  (e) $(u-v)^3 = u^3 - 3u^2v + 3uv^2 - v^3$
- You should be able to factor polynomials.
  (a) Remove any common factors.
  (b) Know and be able to use the following factoring formulas.
    1. $u^2 - v^2 = (u+v)(u-v)$
    2. $u^2 + 2uv + v^2 = (u+v)^2$
    3. $u^2 - 2uv + v^2 = (u-v)^2$
    4. $u^3 + v^3 = (u+v)(u^2 - uv + v^2)$
    5. $u^3 - v^3 = (u-v)(u^2 + uv + v^2)$
- Be able to factor, if possible, trinomials of the form $ax^2 + bx + c$.
- Be able to factor by grouping.

Solutions to Selected Exercises

1. Simplify $(6x+5)-(8x+15)$.

Solution:

$$(6x+5)-(8x+15) = 6x+5-8x-15 = 6x-8x+5-15 = -2x-10$$

5. Simplify $(15x^2-6)-(-8x^3-14x^2-17)$.

Solution:

$$(15x^2-6)-(-8x^3-14x^2-17) = 15x^2-6+8x^3+14x^2+17 = 8x^3+29x^2+11$$

7. Simplify $5z-[3z-(10z+8)]$.

Solution:

$$5z-[3z-(10z+8)] = 5z-[3z-10z-8] = 5z-[-7z-8] = 5z+7z+8 = 12z+8$$

11. Simplify $(-2x)(-3x)(5x+2)$.

Solution:

$$(-2x)(-3x)(5x+2) = 6x^2(5x+2) = 6x^2(5x) + 6x^2(2) = 30x^3 + 12x^2$$

15. Simplify $(x+3)(x^2-3x+9)$.

Solution:

$$\begin{aligned}(x+3)(x^2-3x+9) &= (x+3)(x^2) + (x+3)(-3x) + (x+3)(9)\\ &= x^3 + 3x^2 - 3x^2 - 9x + 9x + 27\\ &= x^3 + 27\end{aligned}$$

21. Find the product.

$$(x+\sqrt{5})(x-\sqrt{5})(x+4)$$

Solution:

$$\begin{aligned}(x+\sqrt{5})(x-\sqrt{5})(x+4) &= (x^2-5)(x+4) \qquad \text{Special Product}\\ &= x^3 + 4x^2 - 5x - 20\end{aligned}$$

23. Find the product.

$$(3x-5)(2x+1)$$

Solution:

$$(3x-5)(2x+1) = 6x^2 + 3x - 10x - 5 = 6x^2 - 7x - 5$$

25. Find the product.

$$(2x-5y)^2$$

Solution:

$$(2x-5y)^2 = (2x)^2 - 2(2x)(5y) + (5y)^2 = 4x^2 - 20xy + 25y^2$$

27. Find the product.

$$[(x-3)+y]^2$$

Solution:

$$\begin{aligned}[(x-3)+y]^2 &= (x-3)^2 + 2(x-3)y + y^2\\ &= x^2 - 6x + 9 + 2xy - 6y + y^2\\ &= x^2 + 2xy + y^2 - 6x - 6y + 9\end{aligned}$$

31. Find the product.

$$(m-3+n)(m-3-n)$$

Solution:

$$\begin{aligned}(m-3+n)(m-3-n) &= [(m-3)+n][(m-3)-n] \\ &= (m-3)^2-n^2 \\ &= m^2-6m+9-n^2 \\ &= m^2-n^2-6m+9\end{aligned}$$

35. Find the product.

$$(2x-y)^3$$

Solution:

$$(2x-y)^3 = (2x)^3-3(2x)^2y+3(2x)y^2-y^3 = 8x^3-12x^2y+6xy^2-y^3$$

39. Remove the common factor: $(x-1)^2+6(x-1)$.

Solution:

$$(x-1)^2+6(x-1) = (x-1)[(x-1)+6] = (x-1)(x+5)$$

41. Factor $16y^2-9$.

Solution:

$$16y^2-9 = (4y)^2-(3)^2 = (4y+3)(4y-3)$$

43. Factor $(x-1)^2-4$.

Solution:

$$(x-1)^2-4 = [(x-1)+2][(x-1)-2] = (x+1)(x-3)$$

45. Factor x^2-4x+4.

Solution:

$$x^2-4x+4 = x^2-2(2)(x)+(2)^2 = (x-2)^2$$

49. Factor s^2-5s+6.

Solution:

$$s^2-5s+6 = (s-2)(s-3) \text{ since } (-2)(-3)=6 \text{ and } (-2)+(-3)=-5.$$

51. Factor $x^2 - 30x + 200$.

Solution:

$$x^2 - 30x + 200 = (x-10)(x-20) \text{ since } (-10)(-20) = 200 \text{ and } (-10) + (-20) = -30.$$

53. Factor $9z^2 - 3z - 2$.

Solution:

$$9z^2 - 3z - 2 = (3z+1)(3z-2)$$

55. Factor $x^3 - 8$.

Solution:

$$x^3 - 8 = x^3 - 2^3 = (x-2)(x^2 + 2x + (2)^2) = (x-2)(x^2 + 2x + 4)$$

59. Factor $x^3 - x^2 + 2x - 2$ by grouping.

Solution:

$$x^3 - x^2 + 2x - 2 = x^2(x-1) + 2(x-1) = (x-1)(x^2+2)$$

63. Factor $6 + 2x - 3x^3 - x^4$ by grouping.

Solution:

$$6 + 2x - 3x^3 - x^4 = 2(3+x) - x^3(3+x) = (3+x)(2-x^3)$$

65. Completely factor $x^3 - 4x^2$.

Solution:

$$x^3 - 4x^2 = x^2(x-4)$$

69. Completely factor $9x^2 + 10x + 1$.

Solution:

$$9x^2 + 10x + 1 = (9x+1)(x+1)$$

73. Completely factor $2(x+1)(x-3)^2 - 3(x+1)^2(x-3)$.

Solution:

$$\begin{aligned} 2(x+1)(x-3)^2 - 3(x+1)^2(x-3) &= (x+1)(x-3)[2(x-3) - 3(x+1)] \\ &= (x+1)(x-3)[2x - 6 - 3x - 3] \\ &= (x+1)(x-3)(-x-9) \\ &= -(x+1)(x-3)(x+9) \end{aligned}$$

79. Completely factor $2t^3 - 16$.

Solution:

$$2t^3 - 16 = 2(t^3 - 8) = 2(t^3 - 2^3) = 2(t - 2)(t^2 + 2t + 4)$$

81. The probability of three successes and two failures in a certain experiment is given by $10p^3(1-p)^2$. Find this product.

Solution:

$$10p^3(1-p)^2 = 10p^3(1 - 2p + p^2) = 10p^3(p^2 - 2p + 1) = 10p^5 - 20p^4 + 10p^3$$

83. Factor $x^3 + 6x^2 + 12x + 8$ by using the formula

$$(x+a)^3 = x^3 + 3x^2a + 3xa^2 + a^3.$$

Solution:

$$x^3 + 6x^2 + 12x + 8 = x^3 + 3x^2(2) + 3x(2)^2 + (2)^3 = (x+2)^3$$

85. Factor $x^5 - 10x^4 + 40x^3 - 80x^2 + 80x - 32$ by using the formula

$$(x+a)^5 = x^5 + 5x^4a + 10x^3a^2 + 10x^2a^3 + 5xa^4 + a^5.$$

Solution:

$$\begin{aligned} x^5 - 10x^4 + 40x^3 - 80x^2 + 80x - 32 \\ &= x^5 + 5x^4(-2) + 10x^3(-2)^2 + 10x^2(-2)^3 + 5x(-2)^4 + (-2)^5 \\ &= \left(x + (-2)\right)^5 \\ &= (x-2)^5 \end{aligned}$$

SECTION 1.4

Fractional Expressions

- You should know that a rational expression is the quotient of two polynomials.
- You should be able to simplify rational expressions by reducing them to lowest terms. This may involve factoring both the numerator and the denominator.
- You should be able to add, subtract, multiply, and divide rational expressions.
- You should be able to simplify compound fractions by either
  (a) Combining Fractions or
  (b) Rationalizing Fractions.

Solutions to Selected Exercises

3. Fill in the missing numerator so that the two fractions are equivalent.

$$\frac{x+1}{x} = \frac{(\qquad)}{x(x-2)}$$

Solution:

$$\begin{aligned}\frac{x+1}{x} &= \frac{x+1}{x} \cdot \frac{x-2}{x-2} \\ &= \frac{(x+1)(x-2)}{x(x-2)}, \quad x \neq 2\end{aligned}$$

9. Reduce to lowest terms.

$$\frac{3xy}{xy+x}$$

Solution:

$$\frac{3xy}{xy+x} = \frac{3xy}{x(y+1)} = \frac{3y}{y+1}$$

13. Reduce to lowest terms.

$$\frac{x^3 + 5x^2 + 6x}{x^2 - 4}$$

Solution:

$$\frac{x^3 + 5x^2 + 6x}{x^2 - 4} = \frac{x(x+2)(x+3)}{(x+2)(x-2)} = \frac{x(x+3)}{x-2}$$

17. Reduce to lowest terms.

$$\frac{2 - x + 2x^2 - x^3}{x - 2}$$

Solution:

$$\begin{aligned}\frac{2 - x + 2x^2 - x^3}{x - 2} &= \frac{(2-x) + x^2(2-x)}{x-2} \\ &= \frac{(2-x)(1+x^2)}{x-2} \\ &= \frac{-(x-2)(x^2+1)}{x-2} \\ &= -(x^2+1)\end{aligned}$$

19. Reduce to lowest terms.

$$\frac{z^3 - 8}{z^2 + 2z + 4}$$

Solution:

$$\frac{z^3 - 8}{z^2 + 2z + 4} = \frac{(z-2)(z^2+2z+4)}{z^2+2z+4} = z - 2$$

23. Simplify

$$\frac{(x-9)(x+7)}{x+1} \cdot \frac{x}{9-x}.$$

Solution:

$$\frac{(x-9)(x+7)}{x+1} \cdot \frac{x}{9-x} = -\frac{(9-x)(x+7)x}{(x+1)(9-x)} = -\frac{x(x+7)}{x+1}$$

27. Simplify

$$\frac{t^2 - t - 6}{t^2 + 6t + 9} \cdot \frac{t+3}{t^2 - 4}.$$

Solution:

$$\frac{t^2 - t - 6}{t^2 + 6t + 9} \cdot \frac{t+3}{t^2 - 4} = \frac{(t+2)(t-3)}{(t+3)(t+3)} \cdot \frac{t+3}{(t+2)(t-2)} = \frac{t-3}{(t+3)(t-2)}$$

29. Simplify

$$\frac{x^2 + xy - 2y^2}{x^3 + x^2y} \cdot \frac{x}{x^2 + 3xy + 2y^2}.$$

Solution:

$$\frac{x^2 + xy - 2y^2}{x^3 + x^2y} \cdot \frac{x}{x^2 + 3xy + 2y^2} = \frac{(x+2y)(x-y)}{x^2(x+y)} \frac{x}{(x+2y)(x+y)} = \frac{x-y}{x(x+y)^2}$$

33. Simplify

$$\frac{\dfrac{(xy)^2}{(x+y)^2}}{\dfrac{xy}{(x+y)^3}}.$$

Solution:

$$\frac{\dfrac{(xy)^2}{(x+y)^2}}{\dfrac{xy}{(x+y)^3}} = \frac{(xy)^2}{(x+y)^2} \div \frac{xy}{(x+y)^3} = \frac{(xy)^2}{(x+y)^2} \cdot \frac{(x+y)^3}{xy} = xy(x+y)$$

37. Simplify

$$6 - \frac{5}{x+3}.$$

Solution:

$$\begin{aligned} 6 - \frac{5}{x+3} &= \frac{6(x+3)}{x+3} - \frac{5}{x+3} = \frac{6(x+3) - 5}{x+3} \\ &= \frac{6x + 18 - 5}{x+3} = \frac{6x+13}{x+3} \end{aligned}$$

43. Simplify

$$\frac{1}{x^2 - x - 2} - \frac{x}{x^2 - 5x + 6}.$$

Solution:

$$\begin{aligned}\frac{1}{x^2 - x - 2} - \frac{x}{x^2 - 5x + 6} &= \frac{1}{(x-2)(x+1)} - \frac{x}{(x-2)(x-3)} \\ &= \frac{(x-3) - x(x+1)}{(x+1)(x-2)(x-3)} \\ &= \frac{-x^2 - 3}{(x+1)(x-2)(x-3)} = -\frac{x^2+3}{(x+1)(x-2)(x-3)}\end{aligned}$$

49. Simplify

$$\frac{\left(\frac{x+3}{x-3}\right)^2}{\frac{1}{x+3} + \frac{1}{x-3}}.$$

Solution:

(a) Combining method:

$$\begin{aligned}\frac{\left(\frac{x+3}{x-3}\right)^2}{\frac{1}{x+3} + \frac{1}{x-3}} &= \frac{\frac{(x+3)^2}{(x-3)^2}}{\frac{(x-3)+(x+3)}{(x+3)(x-3)}} \\ &= \frac{(x+3)^2}{(x-3)^2} \cdot \frac{(x+3)(x-3)}{2x} = \frac{(x+3)^3}{2x(x-3)}\end{aligned}$$

(b) LCD method:

$$\begin{aligned}\frac{\frac{(x+3)^2}{(x-3)^2}}{\frac{1}{x+3} + \frac{1}{x-3}} \cdot \frac{(x+3)(x-3)^2}{(x+3)(x-3)^2} &= \frac{(x+3)^3}{(x-3)^2 + (x+3)(x-3)} \\ &= \frac{(x+3)^3}{(x^2 - 6x + 9) + (x^2 - 9)} \\ &= \frac{(x+3)^3}{2x^2 - 6x} \\ &= \frac{(x+3)^3}{2x(x-3)}\end{aligned}$$

55. Rationalize the denominator of

$$\frac{3}{\sqrt{x+1}}.$$

Solution:

$$\frac{3}{\sqrt{x+1}} = \frac{3}{\sqrt{x+1}} \cdot \frac{\sqrt{x+1}}{\sqrt{x+1}} = \frac{3\sqrt{x+1}}{x+1}$$

59. Rationalize the numerator of

$$\frac{\sqrt{x+2}-\sqrt{x}}{2}.$$

Solution:

$$\begin{aligned}\frac{\sqrt{x+2}-\sqrt{x}}{2} &= \frac{\sqrt{x+2}-\sqrt{x}}{2} \cdot \frac{\sqrt{x+2}+\sqrt{x}}{\sqrt{x+2}+\sqrt{x}} \\ &= \frac{(x+2)-x}{2(\sqrt{x+2}+\sqrt{x})} = \frac{2}{2(\sqrt{x+2}+\sqrt{x})} \\ &= \frac{1}{\sqrt{x+2}+\sqrt{x}}\end{aligned}$$

63. Simplify

$$\frac{\dfrac{t^2}{\sqrt{t^2+1}} - \sqrt{t^2+1}}{t^2}.$$

Solution:

$$\begin{aligned}\frac{\dfrac{t^2}{\sqrt{t^2+1}} - \sqrt{t^2+1}}{t^2} &= \frac{\dfrac{t^2}{\sqrt{t^2+1}} - \sqrt{t^2+1}}{t^2} \cdot \frac{\sqrt{t^2+1}}{\sqrt{t^2+1}} \\ &= \frac{t^2-(t^2+1)}{t^2\sqrt{t^2+1}} \\ &= -\frac{1}{t^2\sqrt{t^2+1}}\end{aligned}$$

67. Simplify

$$\frac{-x^3(1-x^2)^{-1/2}-2x(1-x^2)^{1/2}}{x^4}.$$

Solution:

$$\begin{aligned}\frac{-x^3(1-x^2)^{-1/2}-2x(1-x^2)^{1/2}}{x^4} &= \frac{-x^3(1-x^2)^{-1/2}-2x(1-x^2)^{1/2}}{x^4}\cdot\frac{(1-x^2)^{1/2}}{(1-x^2)^{1/2}}\\ &= \frac{-x^3-2x(1-x^2)}{x^4(1-x^2)^{1/2}}\\ &= \frac{-x^3-2x+2x^3}{x^4(1-x^2)^{1/2}}\\ &= \frac{x^3-2x}{x^4(1-x^2)^{1/2}}\\ &= \frac{x(x^2-2)}{x^4(1-x^2)^{1/2}}\\ &= \frac{x^2-2}{x^3(1-x^2)^{1/2}}\end{aligned}$$

SECTION 1.5

Linear Equations and Quadratic Equations

- To solve an equation you can:
  (a) Remove symbols of grouping.
  (b) Combine like terms.
  (c) Add or subtract the same quantity to both sides.
  (d) Multiply or divide both sides by the same nonzero quantity.
- If you have multiplied or divided both sides by a variable, check for extraneous solutions.
- To solve quadratic equations:
  (a) Factor, if possible.
  (b) Take the square root of both sides, if possible.
  (c) Complete the square.
  (d) Use the Quadratic Formula

$$x = \frac{-b \pm \sqrt{b^2 - 4ac}}{2a}.$$

- The discriminant of a quadratic equation can be used to determine the type of solutions.
  (a) If $b^2 - 4ac > 0$, then there are two distinct real solutions.
  (b) If $b^2 - 4ac = 0$, then there is one repeated real solution.
  (c) If $b^2 - 4ac < 0$, then there are no real solutions.

Solutions to Selected Exercises

3. Determine whether the given value of x is a solution of the equation $3x^2 + 2x - 5 = 2x^2 - 2$.

(a) $x = -3$
(b) $x = 1$
(c) $x = 4$
(d) $x = -5$

Solution:

(a) $3(-3)^2 + 2(-3) - 5 \stackrel{?}{=} 2(-3)^2 - 2$

$$16 = 16$$

$x = -3$ is a solution.

(b) $3(1)^2 + 2(1) - 5 \stackrel{?}{=} 2(1)^2 - 2$

$$0 = 0$$

$x = 1$ is a solution.

(c) $3(4)^2 + 2(4) - 5 \stackrel{?}{=} 2(4)^2 - 2$

$$51 \neq 30$$

$x = 4$ is not a solution.

(d) $3(-5)^2 + 2(-5) - 5 \stackrel{?}{=} 2(-5)^2 - 2$

$$60 \neq 48$$

$x = -5$ is not a solution.

5. Solve the equation $2(x+5)-7=3(x-2)$.

Solution:

$$\begin{aligned} 2(x+5)-7 &= 3(x-2) \\ 2x+10-7 &= 3x-6 \\ 2x+3 &= 3x-6 \\ -x+3 &= -6 \\ -x &= -9 \\ x &= 9 \end{aligned}$$

7. Solve the equation $\dfrac{5x}{4}+\dfrac{1}{2}=x-\dfrac{1}{2}$.

Solution:

$$\begin{aligned} \frac{5x}{4}+\frac{1}{2} &= x-\frac{1}{2} \\ 4\left(\frac{5x}{4}+\frac{1}{2}\right) &= 4\left(x-\frac{1}{2}\right) \\ 5x+2 &= 4x-2 \\ x+2 &= -2 \\ x &= -4 \end{aligned}$$

9. Solve the equation $0.25x+0.75(10-x)=3$.

Solution:

$$\begin{aligned} 0.25x+0.75(10-x) &= 3 \\ 100[0.25x+0.75(10-x)] &= 100(3) \\ 25x+75(10-x) &= 300 \\ 25x+750-75x &= 300 \\ -50x+750 &= 300 \\ -50x &= -450 \\ x &= 9 \end{aligned}$$

11. Solve the equation $x + 8 = 2(x - 2) - x$, if possible.

Solution:

$$\begin{aligned} x + 8 &= 2(x - 2) - x \\ x + 8 &= 2x - 4 - x \\ x + 8 &= x - 4 \\ 8 &= -4 \qquad \text{Not possible} \end{aligned}$$

Thus, the equation has no solution.

15. Solve the equation

$$\frac{5x - 4}{5x + 4} = \frac{2}{3}.$$

Solution:

$$\begin{aligned} \frac{5x - 4}{5x + 4} &= \frac{2}{3} \\ 3(5x - 4) &= 2(5x + 4) \qquad \text{Cross multiply} \\ 15x - 12 &= 10x + 8 \\ 5x &= 20 \\ x &= 4 \end{aligned}$$

19. Solve the equation

$$\frac{1}{x - 3} + \frac{1}{x + 3} = \frac{10}{x^2 - 9}.$$

Solution:

$$\begin{aligned} \frac{1}{x - 3} + \frac{1}{x + 3} &= \frac{10}{x^2 - 9} \\ \frac{(x + 3) + (x - 3)}{(x - 3)(x + 3)} &= \frac{10}{x^2 - 9} \\ (x^2 - 9)\left(\frac{2x}{x^2 - 9}\right) &= \left(\frac{10}{x^2 - 9}\right)(x^2 - 9) \\ 2x &= 10 \\ x &= 5 \end{aligned}$$

21. Solve the equation

$$\frac{7}{2x+1} - \frac{8x}{2x-1} = -4 .$$

Solution:

$$\begin{aligned}
\frac{7}{2x+1} - \frac{8x}{2x-1} &= -4 \\
(2x+1)(2x-1)\left[\frac{7}{2x+1} - \frac{8x}{2x-1}\right] &= -4(2x+1)(2x-1) \\
7(2x-1) - 8x(2x+1) &= -4(4x^2-1) \\
14x - 7 - 16x^2 - 8x &= -16x^2 + 4 \\
-16x^2 + 6x - 7 &= -16x^2 + 4 \\
6x - 7 &= 4 \\
6x &= 11 \\
x &= \frac{11}{6}
\end{aligned}$$

23. Solve the equation $(x+2)^2 + 5 = (x+3)^2$.

Solution:

$$\begin{aligned}
(x+2)^2 + 5 &= (x+3)^2 \\
x^2 + 4x + 4 + 5 &= x^2 + 6x + 9 \\
4x + 9 &= 6x + 9 \\
4x &= 6x \\
-2x &= 0 \\
x &= 0
\end{aligned}$$

29. Solve $x^2 - 2x - 8 = 0$ by factoring.

Solution:

$$\begin{aligned}
x^2 - 2x - 8 &= 0 \\
(x+2)(x-4) &= 0 \\
x = -2 \quad &\text{or} \quad x = 4
\end{aligned}$$

33. Solve $2x^2 = 19x + 33$ by factoring.

Solution:

$$2x^2 = 19x + 33$$
$$2x^2 - 19x - 33 = 0$$
$$(2x + 3)(x - 11) = 0$$
$$2x + 3 = 0 \quad \text{or} \quad x - 11 = 0$$
$$2x = -3 \quad \text{or} \quad x = 11$$
$$x = -\tfrac{3}{2}$$

35. Solve $3x^2 = 36$ by taking the square root of both sides.

Solution:

$$3x^2 = 36$$
$$x^2 = 12$$
$$x = \pm\sqrt{12}$$
$$x = \pm 2\sqrt{3}$$

41. Solve $x^2 + 4x - 32 = 0$ by completing the square.

Solution:

$$x^2 + 4x - 32 = 0$$
$$x^2 + 4x = 32$$
$$x^2 + 4x + 4 = 32 + 4$$
$$(x + 2)^2 = 36$$
$$x + 2 = \pm 6$$
$$x = -2 \pm 6$$
$$x = 4 \quad \text{or} \quad x = -8$$

43. Solve $9x^2 - 18x + 3 = 0$ by completing the square.

Solution:

$$9x^2 - 18x + 3 = 0$$
$$x^2 - 2x + \tfrac{1}{3} = 0$$
$$x^2 - 2x = -\tfrac{1}{3}$$
$$x^2 - 2x + 1 = -\tfrac{1}{3} + 1$$
$$(x - 1)^2 = \tfrac{2}{3}$$
$$x - 1 = \pm\sqrt{\tfrac{2}{3}}$$
$$x = 1 \pm \sqrt{\tfrac{2}{3}} = 1 \pm \tfrac{\sqrt{6}}{3}$$

47. Use the quadratic formula to solve $16x^2 + 8x - 3 = 0$.

Solution:

$16x^2 + 8x - 3 = 0; \; a = 16, \; b = 8, \; c = -3$

$$x = \frac{-8 \pm \sqrt{8^2 - 4(16)(-3)}}{2(16)} = \frac{-8 \pm \sqrt{256}}{32} = \frac{-8 \pm 16}{32}$$

$$x = \frac{-8 + 16}{32} = \frac{1}{4}$$

$$x = \frac{-8 - 16}{32} = -\frac{3}{4}$$

51. Use the Quadratic Formula to solve $12x - 9x^2 = -3$.

Solution:

$$12x - 9x^2 = -3$$

$$0 = 9x^2 - 12x - 3$$

$$0 = 3(3x^2 - 4x - 1) \qquad \text{Divide both sides by 3}$$

$$0 = 3x^2 - 4x - 1$$

$$a = 3, \; b = -4, \; c = -1$$

$$x = \frac{-(-4) \pm \sqrt{(-4)^2 - 4(3)(-1)}}{2(3)} = \frac{4 \pm \sqrt{16 + 12}}{6}$$

$$= \frac{4 \pm \sqrt{28}}{6} = \frac{4 \pm 2\sqrt{7}}{6}$$

$$= \frac{2(2 \pm \sqrt{7})}{6} = \frac{2 \pm \sqrt{7}}{3}$$

55. Use the Quadratic Formula to solve $(y - 5)^2 = 2y$.

Solution:

$$(y - 5)^2 = 2y$$

$$y^2 - 10y + 25 = 2y$$

$$y^2 - 12y + 25 = 0$$

$$a = 1, \; b = -12, \; c = 25$$

$$y = \frac{-(-12) \pm \sqrt{(-12)^2 - 4(1)(25)}}{2(1)} = \frac{12 \pm \sqrt{144 - 100}}{2}$$

$$= \frac{12 \pm \sqrt{44}}{2} = \frac{12 \pm 2\sqrt{11}}{2}$$

$$= \frac{2(6 \pm \sqrt{11})}{2} = 6 \pm \sqrt{11}$$

61. Complete the square for the quadratic portion of

$$\frac{1}{x^2 - 4x - 12}.$$

Solution:

$$\frac{1}{x^2 - 4x - 12} = \frac{1}{x^2 - 4x + 4 - 4 - 12} = \frac{1}{(x-2)^2 - 16}$$

63. Use a calculator to solve $5.1x^2 - 1.7x - 3.2 = 0$. Round your answer to three decimal places.

Solution:

$5.1x^2 - 1.7x - 3.2 = 0;\ a = 5.1,\ b = -1.7,\ c = -3.2$

$$x = \frac{-(-1.7) \pm \sqrt{(-1.7)^2 - 4(5.1)(-3.2)}}{2(5.1)}$$

$$x = \frac{1.7 \pm \sqrt{68.17}}{10.2} \approx \frac{1.7 \pm 8.2565}{10.2}$$

$x \approx 0.976$ or $x \approx -0.643$

67. Find two numbers whose sum is 100 and whose product is 2500.

Solution:

Let x = one number and $100 - x$ = other number.

$$\begin{aligned} x(100 - x) &= 2500 \\ 100x - x^2 &= 2500 \\ 0 &= x^2 - 100x + 2500 \\ 0 &= (x - 50)^2 \\ x &= 50, \quad 100 - x = 50 \end{aligned}$$

69. Use the cost equation $C = 0.125x^2 + 20x + 5000$ to find the number of units x that a manufacturer can produce for the cost of $C = \$14,000$. (Round your answer to the nearest positive integer.)

Solution:

$$\begin{aligned} C &= 0.125x^2 + 20x + 5000, \quad C = 14000 \\ 14000 &= 0.125x^2 + 20x + 5000 \\ 0 &= 0.125x^2 + 20x - 9000 \qquad \text{Multiply both sides by 8.} \\ 0 &= x^2 + 160x - 72000 \\ 0 &= (x - 200)(x + 360) \end{aligned}$$

Choosing the positive value for x, we have $x = 200$ units.

71. Find the time when the object hits the ground if the object is dropped from a balloon at a height of 1600 feet. (Use the position equation in Example 13.)

Solution:

From Example 13, we have

$$s = -16t^2 + v_0 t + s_0.$$

The initial height s_0 is 1600 feet. Since the object is dropped, v_0 is zero. When the object hits the ground, s is zero feet. Thus, we have the quadratic equation

$$0 = -16t^2 + 0t + 1600$$
$$0 = -16(t^2 - 100)$$
$$0 = -16(t + 10)(t - 10)$$

Choosing the positive value for t, we have $t = 10$ seconds.

75. Two brothers must mow a rectangular lawn 100 feet by 200 feet. Each wants to mow no more than one-half of the lawn. The first starts by mowing around the outside of the lawn. How wide a strip must he mow on each of the four sides? Approximately how many times must he go around the lawn if the mower has a 24-inch cut?

Solution:

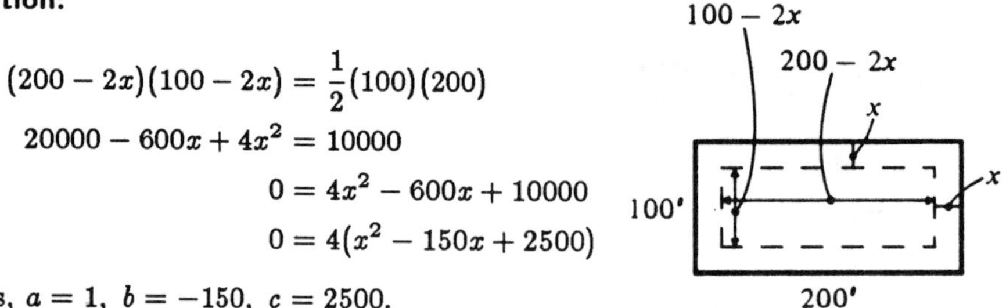

$$(200 - 2x)(100 - 2x) = \frac{1}{2}(100)(200)$$
$$20000 - 600x + 4x^2 = 10000$$
$$0 = 4x^2 - 600x + 10000$$
$$0 = 4(x^2 - 150x + 2500)$$

Thus, $a = 1$, $b = -150$, $c = 2500$.

$$x = \frac{150 \pm \sqrt{(-150)^2 - 4(1)(2500)}}{2(1)} \approx \frac{150 \pm 111.8034}{2}$$

$$x = \frac{150 + 111.8034}{2} \approx 130.902 \text{ ft}$$ Not possible since the lot is only 100 ft wide

$$x = \frac{150 - 111.8034}{2} \approx 19.1 \text{ ft}$$

He must go around the lot

$$\frac{19.098 \text{ ft}}{24 \text{ in}} = \frac{19.098 \text{ ft}}{2 \text{ ft}} = 9.5 \text{ times.}$$

SECTION 1.6

Inequalities

- The procedures for solving a linear inequality are similar to those used for solving linear equations. However, when you multiply or divide both sides of an inequality by a negative number, *reverse* the inequality.
- You should be able to solve inequalities involving absolute value.
  (a) If $|x| < a$, then $-a < x < a$.
  (b) If $|x| > a$, then $x < -a$ OR $x > a$.
- You should be able to solve polynomial inequalities, or inequalities involving fractions.
  (a) Find the critical numbers.
  (b) Check the test intervals.

Solutions to Selected Exercises

3. Determine whether or not the given value of x satisfies the inequality $0 < \frac{x-2}{4} < 2$.

(a) $x = 4$
(b) $x = 10$
(c) $x = 0$
(d) $x = \frac{7}{2}$

Solution:

(a) $x = 4$

$$0 \overset{?}{<} \frac{4-2}{4} \overset{?}{<} 2$$

$0 < \frac{1}{2} < 2$, $\quad x = 4$ is a solution.

(b) $x = 10$

$$0 \overset{?}{<} \frac{10-2}{4} \overset{?}{<} 2$$

$0 < 2 \not< 2$, $\quad x = 10$ is not a solution.

(c) $x = 0$

$$0 \overset{?}{<} \frac{0-2}{4} \overset{?}{<} 2$$

$0 \not< -\frac{1}{2} < 2$, $\quad x = 0$ is not a solution.

(d) $x = \frac{7}{2}$

$$0 \overset{?}{<} \frac{(7/2)-2}{4} \overset{?}{<} 2$$

$0 < \frac{3}{8} < 2$, $\quad x = \frac{7}{2}$ is a solution.

5. Solve $4x < 12$ and sketch the solution on the real number line.

Solution:

$$4x < 12$$
$$x < 3$$

11. Solve $4(x+1) < 2x+3$ and sketch the solution on the real number line.

Solution:

$$\begin{aligned} 4(x+1) &< 2x+3 \\ 4x+4 &< 2x+3 \\ 2x &< -1 \\ x &< -\tfrac{1}{2} \end{aligned}$$

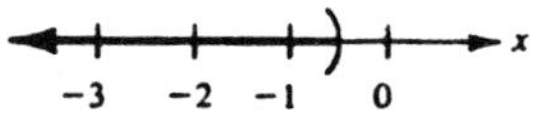

15. Solve

$$-4 < \frac{2x-3}{3} < 4$$

and sketch the solution on the real number line.

Solution:

$$\begin{aligned} -4 < {} & \frac{2x-3}{3} < 4 \\ -12 < {} & 2x-3 < 12 \\ -9 < {} & 2x < 15 \\ -\frac{9}{2} < {} & x < \frac{15}{2} \end{aligned}$$

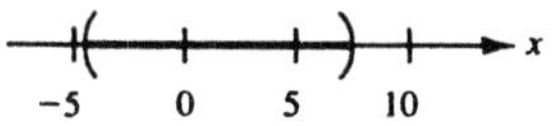

19. Solve $|x/2| > 3$ and sketch the solution on the real number line.

Solution:

$$\left|\frac{x}{2}\right| > 3$$

$$\frac{x}{2} < -3 \quad \text{or} \quad \frac{x}{2} > 3$$

$$x < -6 \quad \text{or} \quad x > 6$$

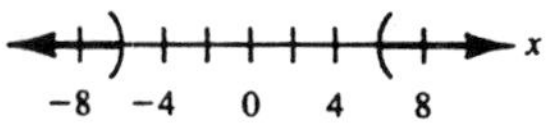

25. Solve

$$\left|\frac{x-3}{2}\right| \geq 5$$

and sketch the solution on the real number line.

Solution:

$$\left|\frac{x-3}{2}\right| \geq 5$$

$$\frac{x-3}{2} \leq -5 \quad \text{or} \quad \frac{x-3}{2} \geq 5$$

$$x-3 \leq -10 \quad \text{or} \quad x-3 \geq 10$$

$$x \leq -7 \quad \text{or} \quad x \geq 13$$

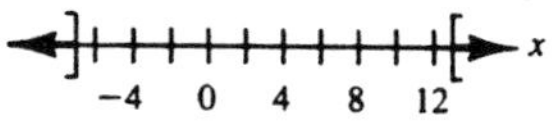

29. Solve $2|x + 10| \geq 9$ and sketch the solution on the real number line.

Solution:

$$
\begin{aligned}
2|x+10| &\geq 9 \\
|x+10| &\geq \tfrac{9}{2} \\
x + 10 \leq -\tfrac{9}{2} \quad &\text{or} \quad x + 10 \geq \tfrac{9}{2} \\
x \leq -\tfrac{29}{2} \quad &\text{or} \quad x \geq -\tfrac{11}{2}
\end{aligned}
$$

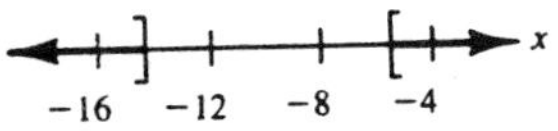

−16 −12 −8 −4 x

33. Find the interval on the real number line for which the radicand in $\sqrt{x-5}$ is nonnegative.

Solution:

The radicand of $\sqrt{x-5}$ is $x - 5$.

$$
\begin{aligned}
x - 5 &\geq 0 \\
x &\geq 5
\end{aligned}
$$

Therefore, the interval is $[5, \infty)$.

39. Use absolute value notation to define the pair of intervals on the real line.

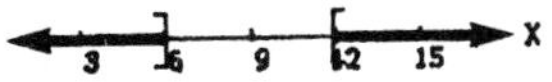

3 6 9 12 15 x

Solution:

$$|x - 9| \geq 3$$

43. Solve the inequality $x^2 > 4$ and graph the solution on the real number line.

Solution:

$$
\begin{aligned}
x^2 &> 4 \\
x < -\sqrt{4} \quad &\text{or} \quad x > \sqrt{4} \\
x < -2 \quad &\text{or} \quad x > 2 \\
(-\infty, -2) \quad &\text{or} \quad (2, \infty)
\end{aligned}
$$

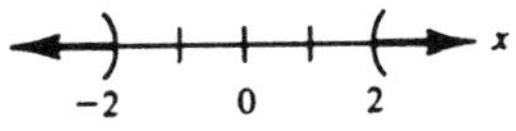

−2 0 2 x

47. Solve the inequality $x^2 + 4x + 4 \geq 9$ and graph the solution on the real number line.

Solution:

$$(x+2)^2 \geq 9$$

$$x + 2 \leq -\sqrt{9} \quad \text{or} \quad x + 2 \geq \sqrt{9}$$

$$x + 2 \leq -3 \quad \text{or} \quad x + 2 \geq 3$$

$$x \leq -5 \quad \text{or} \quad x \geq 1$$

$$(-\infty, -5] \quad \text{or} \quad [1, \infty)$$

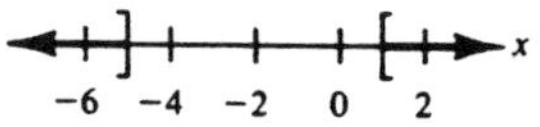

49. Solve the inequality $3(x-1)(x+1) > 0$ and graph the solution on the real number line.

Solution:

$$3(x-1)(x+1) > 0$$

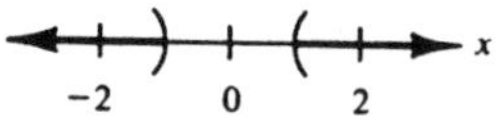

Critical numbers: $x = -1$, $x = 1$
Test intervals: $(-\infty, -1)$, $(-1, 1)$, $(1, \infty)$
Solution intervals: $(-\infty, -1)$, $(1, \infty)$ or $x < -1$, $x > 1$

53. Solve the inequality $4x^3 - 6x^2 < 0$ and graph the solution on the real number line.

Solution:

$$4x^3 - 6x^2 < 0$$

$$2x^2(2x - 3) < 0$$

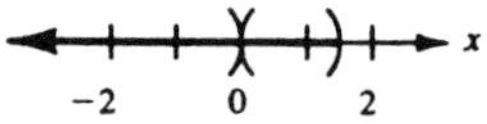

Critical numbers: $x = 0$, $x = \frac{3}{2}$
Test intervals: $(-\infty, 0)$, $(0, \frac{3}{2})$, $(\frac{3}{2}, \infty)$
Solution intervals: $(-\infty, 0)$, $(0, \frac{3}{2})$ or $x < 0$, $0 < x < \frac{3}{2}$
Note: $x = 0$ is *not* a solution.

57. Solve the inequality $1/x > x$ and graph the solution on the real number line.

Solution:

$$\frac{1}{x} > x$$

$$\frac{1}{x} - x > 0$$

$$\frac{1-x^2}{x} > 0$$

$$\frac{(1+x)(1-x)}{x} > 0$$

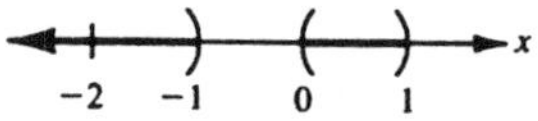

Critical numbers: $x = -1$, $x = 0$, $x = 1$
Test intervals: $(-\infty, -1)$, $(-1, 0)$, $(0, 1)$, $(1, \infty)$
Solution intervals: $(-\infty, -1)$, $(0, 1)$ or $x < -1$, $0 < x < 1$

65. Find the domain of x in the expression $\sqrt{x^2 - 7x + 12}$.

Solution:
The radicand is $x^2 - 7x + 12$.

$$x^2 - 7x + 12 \geq 0$$
$$(x-3)(x-4) \geq 0$$

Critical numbers: $x = 3, \ x = 4$
Test intervals: $(-\infty, 3), \ (3, 4), \ (4, \infty)$
Domain: $(-\infty, 3], \ [4, \infty)$

67. P dollars is invested at a simple interest rate of r. The balance in the account after t years is given by $A = P + Prt$. In order for an investment of \$1000 to grow to *more than* \$1250 in two years, what must the interest rate be?

Solution:
$A = P + Prt$
$t = 2$ years, $P = \$1000, \quad 1250 < A$

$$1250 < 1000 + 1000r(2)$$
$$250 < 2000r$$
$$0.125 < r$$
$$r > 12\tfrac{1}{2}\%$$

71. A rectangle with a perimeter of 100 meters is to have an area of at least 500 square meters. Within what bounds must the length of the rectangle lie?

Solution:

$$2L + 2W = 100$$
$$W = \frac{100 - 2L}{2} = 50 - L$$
$$LW \geq 500$$
$$L(50 - L) \geq 500$$
$$50L - L^2 \geq 500$$
$$0 \geq L^2 - 50L + 500$$

By the quadratic formula the critical numbers are:

$$x = \frac{50 \pm \sqrt{(50)^2 - 4(500)}}{2}, \qquad x = \frac{50 \pm \sqrt{500}}{2} = \frac{50 \pm 10\sqrt{5}}{2} = 25 \pm 5\sqrt{5}$$

Solution interval: $[25 - 5\sqrt{5}, \ 25 + 5\sqrt{5}]$, or $13.8197 \text{ meters} \leq L \leq 36.1803$ meters

SECTION 1.7

Algebraic Errors and Some Algebra of Calculus

- You should be able to recognize and avoid the common algebraic errors listed in this section.
- You should be able to "unsimplify" algebraic expressions by the following methods.
  (a) Unusual Factoring
  (b) Inserting Required Factors
  (c) Rewriting with Negative Exponents
  (d) Writing a Fraction as a Sum of Terms

Solutions to Selected Exercises

3. Find and correct any errors in $5z + 3(x - 2) = 5z + 3x - 2$.

Solution:

$$\begin{aligned} 5z + 3(x-2) &= 5z + 3x - 6 \quad \text{By the Distributive Law} \\ &\neq 5z + 3x - 2 \end{aligned}$$

5. Find and correct any errors in

$$-\frac{x-3}{x-1} = \frac{3-x}{1-x}.$$

Solution:

$$\begin{aligned} -\frac{x-3}{x-1} &= \frac{-(x-3)}{x-1} \\ &= \frac{3-x}{x-1} \quad \text{Only the numerator is multiplied by } (-1). \\ &\neq \frac{3-x}{1-x} \end{aligned}$$

11. Find and correct any errors in $\sqrt{x+9} = \sqrt{x} + 3$.

Solution:

$$\sqrt{x+9} \neq \sqrt{x} + 3$$

The root of a sum does not equal the sum of the roots.

$$\sqrt{x} + 3 = \sqrt{(\sqrt{x}+3)^2} = \sqrt{x + 6\sqrt{x} + 9} \neq \sqrt{x+9}$$

$\sqrt{x+9}$ cannot be simplified.

15. Find and correct any errors in $\dfrac{1}{x+y^{-1}} = \dfrac{y}{x+1}$.

Solution:

$$\frac{1}{x+y^{-1}} = \frac{1}{x+(1/y)} \cdot \frac{y}{y} = \frac{y}{xy+1} \neq \frac{y}{x+1}$$

19. Find and correct any errors in $\sqrt[3]{x^3+7x^2} = x^2\sqrt[3]{x+7}$.

Solution:

$$\sqrt[3]{x^3+7x^2} = \sqrt[3]{x^2(x+7)} = \sqrt[3]{x^2}\sqrt[3]{x+7} \neq x^2\sqrt[3]{x+7}$$

Radicals apply to every factor of the radicand.

23. Find and correct any errors in $\dfrac{1}{2y} = (1/2)y$.

Solution:

$$\frac{1}{2y} = \frac{1}{2} \cdot \frac{1}{y} = \left(\frac{1}{2}\right)\frac{1}{y} \neq \left(\frac{1}{2}\right)y$$

Use the definition for multiplying fractions.

27. Insert the required factor in the parentheses.

$$\frac{2}{3}x^2 + \frac{1}{3}x + 5 = \frac{1}{3}(\quad)$$

Solution:

$$\frac{2}{3}x^2 + \frac{1}{3}x + 5 = \frac{2}{3}x^2 + \frac{1}{3}x + \frac{15}{3} = \frac{1}{3}(2x^2 + x + 15)$$

31. Insert the required factor in the parentheses.

$$x(2x^2+15) = (\quad)(2x^2+15)(2x)$$

Solution:

$$x(2x^2+15) = \left(\frac{1}{2}\right)(2x)(2x^2+15) = \left(\frac{1}{2}\right)(2x^2+15)(2x)$$

35. Insert the required factor in the parentheses.

$$\frac{1}{\sqrt{x}(1+\sqrt{x})^2} = (\quad)\frac{1}{(1+\sqrt{x})^2}\left(\frac{1}{2\sqrt{x}}\right)$$

Solution:

$$\frac{1}{\sqrt{x}(1+\sqrt{x})^2} = \frac{1}{\sqrt{x}}\cdot\frac{1}{(1+\sqrt{x})^2} = (2)\left(\frac{1}{2\sqrt{x}}\right)\frac{1}{(1+\sqrt{x})^2}$$
$$= (2)\frac{1}{(1+\sqrt{x})^2}\left(\frac{1}{2\sqrt{x}}\right)$$

39. Insert the required factor in the parentheses.

$$\frac{3}{x}+\frac{5}{2x^2}-\frac{3}{2}x = (\quad)(6x+5-3x^3)$$

Solution:

$$\frac{3}{x}+\frac{5}{2x^2}-\frac{3}{2}x = \frac{6x}{2x^2}+\frac{5}{2x^2}-\frac{3x^3}{2x^2} = \left(\frac{1}{2x^2}\right)(6x+5-3x^3)$$

43. Insert the required factor in the parentheses.

$$\frac{x^2}{1/12}-\frac{y^2}{2/3} = \frac{12x^2}{(\ )}-\frac{3y^2}{(\ )}$$

Solution:

$$\frac{x^2}{1/12}-\frac{y^2}{2/3} = x^2\left(\frac{12}{1}\right)-y^2\left(\frac{3}{2}\right) = \frac{12x^2}{1}-\frac{3y^2}{2}$$

47. Insert the required factor in the parentheses.

$$3(2x+1)x^{1/2}+4x^{3/2} = x^{1/2}(\quad)$$

Solution:

$$\begin{aligned} 3(2x+1)x^{1/2} + 4x^{3/2} &= 3(2x+1)x^{1/2} + 4xx^{1/2} \\ &= x^{1/2}[3(2x+1) + 4x] \\ &= x^{1/2}(10x+3) \end{aligned}$$

51. Insert the required factor in the parentheses.

$$\frac{1}{10}(2x+1)^{5/2} - \frac{1}{6}(2x+1)^{3/2} = \frac{(2x+1)^{3/2}}{15}(\quad)$$

Solution:

$$\begin{aligned} \frac{1}{10}(2x+1)^{5/2} - \frac{1}{6}(2x+1)^{3/2} &= \frac{3}{30}(2x+1)^{3/2}(2x+1) - \frac{5}{30}(2x+1)^{3/2} \\ &= \frac{1}{30}(2x+1)^{3/2}[3(2x+1) - 5] \\ &= \frac{1}{30}(2x+1)^{3/2}(6x-2) \\ &= \frac{1}{30}(2x+1)^{3/2}2(3x-1) \\ &= \frac{1}{15}(2x+1)^{3/2}(3x-1) \end{aligned}$$

55. Write the following as a sum of two or more terms.

$$\frac{4x^3 - 7x^2 + 1}{x^{1/3}}$$

Solution:

$$\begin{aligned} \frac{4x^3 - 7x^2 + 1}{x^{1/3}} &= \frac{4x^3}{x^{1/3}} - \frac{7x^2}{x^{1/3}} + \frac{1}{x^{1/3}} \\ &= 4x^{3-1/3} - 7x^{2-1/3} + \frac{1}{x^{1/3}} \\ &= 4x^{8/3} - 7x^{5/3} + \frac{1}{x^{1/3}} \end{aligned}$$

59. Write the following as a sum of two or more terms.

$$\frac{x^2 + 4x + 8}{x^4 + 1}$$

Solution:

$$\frac{x^2 + 4x + 8}{x^4 + 1} = \frac{x^2}{x^4+1} + \frac{4x}{x^4+1} + \frac{8}{x^4+1}$$

REVIEW EXERCISES FOR CHAPTER 1

3. Perform the indicated operations and/or simplify $\sqrt[3]{x}(3+4\sqrt[3]{x^2})$.

Solution:

$$\sqrt[3]{x}(3+4\sqrt[3]{x^2}) = x^{1/3}(3+4x^{2/3}) = 3x^{1/3}+4x^{1/3+2/3} = 3\sqrt[3]{x}+4x$$

7. Perform the indicated operations and/or simplify $(x^2-2x+1)(x^3-1)$.

Solution:

$$\begin{aligned}(x^2-2x+1)(x^3-1) &= (x^2-2x+1)x^3-(x^2-2x+1)(1)\\ &= x^5-2x^4+x^3-x^2+2x-1\end{aligned}$$

13. Perform the indicated operations and/or simplify.

$$\frac{x^2-4}{x^4-2x^2-8}\cdot\frac{x^2+2}{x^2}$$

Solution:

$$\frac{x^2-4}{x^4-2x^2-8}\cdot\frac{x^2+2}{x^2} = \frac{x^2-4}{(x^2-4)(x^2+2)}\cdot\frac{x^2+2}{x^2} = \frac{1}{x^2}$$

21. Perform the indicated operations and/or simplify.

$$\frac{1}{x-2}+\frac{1}{(x-2)^2}+\frac{1}{x+2}$$

Solution:

$$\begin{aligned}\frac{1}{x-2}+\frac{1}{(x-2)^2}+\frac{1}{x+2} &= \frac{(x-2)(x+2)+(x+2)+(x-2)^2}{(x+2)(x-2)^2}\\ &= \frac{(x^2-4)+(x+2)+(x^2-4x+4)}{(x+2)(x-2)^2} = \frac{2x^2-3x+2}{(x+2)(x-2)^2}\end{aligned}$$

27. Simplify the compound fraction.

$$\frac{\dfrac{3a}{(a^2/x)-1}}{\dfrac{a}{x}-1}$$

Solution:

$$\frac{\dfrac{3a}{(a^2/x)-1}}{\dfrac{a}{x}-1} = \frac{\left[\dfrac{3a}{(a^2/x)-1}\right]\dfrac{x}{x}}{\dfrac{a-x}{x}} = \frac{3ax}{a^2-x} \cdot \frac{x}{a-x} = \frac{3ax^2}{(a^2-x)(a-x)}$$

31. Insert the missing factor for $x^3 - x^2 + 2x - 2 = (x-1)(\quad)$.

Solution:

$$x^3 - x^2 + 2x - 2 = x^2(x-1) + 2(x-1) = (x-1)(x^2+2)$$

37. Insert the missing factor for $x^4 - 2x^2 + 1 = (x+1)^2(\quad)^2$.

Solution:

$$x^4 - 2x^2 + 1 = (x^2-1)^2 = [(x+1)(x-1)]^2 = (x+1)^2(x-1)^2$$

45. Solve the equation $5x^4 - 12x^3 = 0$.

Solution:

$$\begin{aligned} 5x^4 - 12x^3 &= 0 \\ x^3(5x-12) &= 0 \\ x^3 = 0 \quad &\text{or} \quad 5x - 12 = 0 \\ x = 0 \quad &\text{or} \quad x = \tfrac{12}{5} \end{aligned}$$

47. Solve the equation

$$3\left(1 - \frac{1}{5t}\right) = 0.$$

Solution:

$$\begin{aligned} 3\left(1 - \frac{1}{5t}\right) &= 0 \\ 1 - \frac{1}{5t} &= 0 \\ 1 &= \frac{1}{5t} \\ 5t &= 1 \\ t &= \frac{1}{5} \end{aligned}$$

55. Solve the equation

$$\frac{(x-1)(2x)-x^2}{(x-1)^2}=0.$$

Solution:

$$\frac{(x-1)(2x)-x^2}{(x-1)^2}=0$$
$$(x-1)(2x)-x^2=0$$
$$2x^2-2x-x^2=0$$
$$x^2-2x=0$$
$$x(x-2)=0$$
$$x=0 \quad \text{or} \quad x=2$$

59. Solve the equation $\sqrt{x+4}=3$.

Solution:

$$\sqrt{x+4}=3$$
$$(\sqrt{x+4})^2=(3)^2$$
$$x+4=9$$
$$x=5$$

65. Solve the equation $\sqrt{2x+3}+\sqrt{x-2}=2$.

Solution:

$$\sqrt{2x+3}+\sqrt{x-2}=2$$
$$(\sqrt{2x+3})^2=(2-\sqrt{x-2})^2$$
$$2x+3=4-4\sqrt{x-2}+x-2$$
$$x+1=-4\sqrt{x-2}$$
$$(x+1)^2=(-4\sqrt{x-2})^2$$
$$x^2+2x+1=16(x-2)$$
$$x^2-14x+33=0$$
$$(x-3)(x-11)=0$$
$$x=3,\ \text{extraneous} \quad \text{OR} \quad x=11,\ \text{extraneous}$$

No solution

69. Solve the inequality $x^2 - 4 \leq 0$.

Solution:

$$x^2 - 4 \leq 0$$
$$(x+2)(x-2) \leq 0$$

Critical numbers: $x = 2,\ x = -2$
Test intervals: $x \leq -2,\ -2 \leq x \leq 2,\ x \geq 2$
Solution interval: $-2 \leq x \leq 2$

75. Solve the inequality.

$$\left|x - \tfrac{3}{2}\right| \geq \tfrac{3}{2}$$

Solution:

$$\left|x - \tfrac{3}{2}\right| \geq \tfrac{3}{2}$$

$$x - \tfrac{3}{2} \leq -\tfrac{3}{2} \quad \text{or} \quad x - \tfrac{3}{2} \geq \tfrac{3}{2}$$
$$x \leq 0 \quad \text{or} \quad x \geq 3$$

79. The distance from a spacecraft to the horizon is 1000 miles. Find x, the altitude of the craft, as shown in the figure. Assume that the radius of the earth is 4000 miles.

Solution:

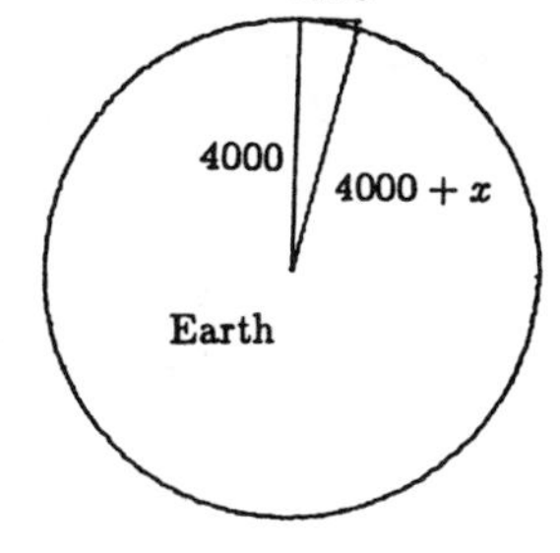

$$1000^2 + 4000^2 = (4000 + x)^2 \qquad \text{Pythagorean Theorem}$$
$$1000^2 + 4000^2 = 4000^2 + 8000x + x^2$$
$$0 = x^2 + 8000x - 1000^2$$
$$x = \frac{-8000 \pm \sqrt{8000^2 - 4(-1000^2)}}{2}$$

Considering only the positive root,

$$x = \frac{-8000 + 2000\sqrt{17}}{2} = -4000 + 1000\sqrt{17}$$
$$x \approx 123.106 \text{ mi}$$

81. Calculate 15^4 in two ways. First, use the exponential key $\boxed{y^x}$. Second, enter 15 and press the square key $\boxed{x^2}$ twice. Why do these two methods give the same result?

Solution:

Since $(15^2)^2 = 15^4$, the two methods will give the same result: $15^4 = 50,625$.

Practice Test for Chapter 1

1. Simplify $4 + 3(18 - 11)$.

2. Simplify $\left(\frac{4}{15} \div 2\right) - \left(5 \times \frac{8}{15}\right)$.

3. Use absolute value notation to describe the expression "the distance between x and -6 is less than 4."

4. Write 0.0000439 in scientific notation.

5. Simplify $(3x^2y^{-1})^2(4x^{-2}y)^{-1}$.

6. Simplify $\sqrt[3]{81x^5y^6}$.

7. Rationalize the denominator of $\dfrac{4}{\sqrt[3]{2}}$ and simplify.

8. Multiply $(x + 3)(x^2 - 4x - 7)$ and simplify.

9. Factor $x^4 - 81$ completely.

10. Factor $x^5 - 4x^3 - x^2 + 4$ completely.

11. Factor $8x^2 + 6x - 9$ completely.

12. Reduce $\dfrac{8x^3 + 8x^2y}{x^2y + xy^2}$ to lowest terms.

13. Simplify $\dfrac{3x}{x^2 - x - 6} - \dfrac{2}{x - 3}$.

14. Simplify $\dfrac{\dfrac{1}{x+1} - \dfrac{1}{x}}{\dfrac{1}{x^2 + x}}$.

15. Simplify $\dfrac{x^2 - 49}{x^2 + 6x - 7} \cdot \dfrac{x^2 - 1}{x}$.

16. Solve $\dfrac{1}{x+2} - \dfrac{3}{x-4} = \dfrac{5}{x^2 - 2x - 8}$.

17. Solve $(x + 12)^2 = 20$ by taking the square root of both sides.

18. Solve $3x^2 + 6x + 2 = 0$ by completing the square.

19. Solve $2x^2 - 3x - 5 = 0$ by the Quadratic Formula.

20. Solve $-3 \le \dfrac{4 - x}{2} \le 5$.

21. Solve $|x - 15| \ge 10$.

22. Solve $x^3 - 9x \le 0$.

23. Solve $\dfrac{4}{x+3} > \dfrac{-1}{x-2}$.

24. Insert the required factor in the parentheses: $x^2(1-2x)^{4/3} - 3x(1-2x)^{1/3} = x(1-2x)^{1/3}(\quad\quad)$.

25. True or False: $\sqrt{36 + x^2} = 6 + x$. Explain.

CHAPTER 2
Functions and Graphs

SECTION 2.1

The Cartesian Plane

- You should be able to plot points.
- You should know that the distance between (x_1, y_1) and (x_2, y_2) in the plane is

  $$d = \sqrt{(x_2 - x_1)^2 + (y_2 - y_1)^2}$$

- You should know that the midpoint of the line segment joining (x_1, y_1) and (x_2, y_2) is

  $$\left(\frac{x_1 + x_2}{2}, \frac{y_1 + y_2}{2}\right)$$

Solutions to Selected Exercises

3. Sketch the square with vertices (2, 4), (5, 1), (2, −2), and (−1, 1).

Solution:

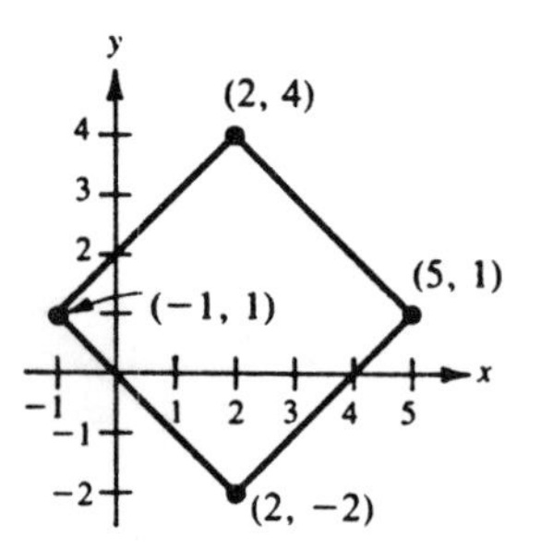

7. Find the distance between the points (−3, −1) and (2, −1).

Solution:

Since the points (−3, −1) and (2, −1) lie on a vertical line, the distance between the points is given by the absolute value of the difference of their x–coordinates.

$$d = |-3 - 2| = 5$$

11. For the indicated triangle (a) find the length of the two sides of the right triangle and use the Pythagorean Theorem to find the length of the hypotenuse, and (b) use the Distance Formula to find the length of the hypotenuse of the triangle.

Solution:

(a) $a = |-3-7| = 10$

$b = |4-1| = 3$

$c = \sqrt{10^2 + 3^2} = \sqrt{109}$

(b) $c = \sqrt{(7-(-3))^2 + (4-1)^2}$

$= \sqrt{10^2 + 3^2}$

$= \sqrt{109}$

15. (a) Plot the points $(-4, 10)$ and $(4, -5)$, (b) find the distance between the points, and (c) find the midpoint of the line segment joining the points.

Solution:

(a)

(b) $d = \sqrt{(-4-4)^2 + (10-(-5))^2}$

$= \sqrt{(-8)^2 + (15)^2}$

$= \sqrt{289}$

$= 17$

(c) $m = \left(\frac{-4+4}{2}, \frac{10+(-5)}{2}\right)$

$= \left(0, \frac{5}{2}\right)$

21. (a) Plot the points $(6.2, 5.4)$, and $(-3.7, 1.8)$, (b) find the distance between the points, and (c) find the midpoint of the line segment joining the points.

Solution:

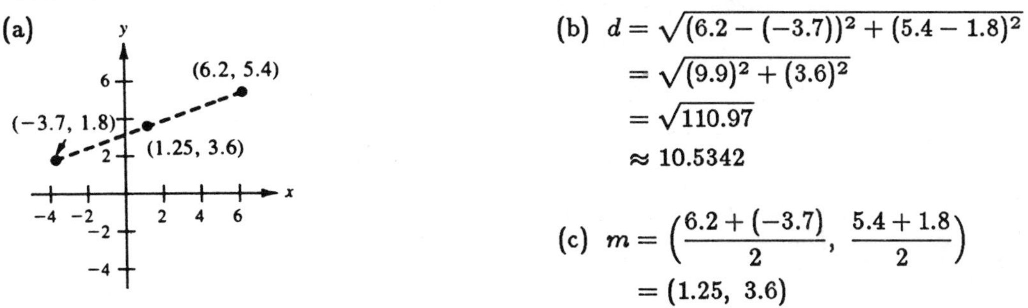

(a)

(b) $d = \sqrt{(6.2-(-3.7))^2 + (5.4-1.8)^2}$

$= \sqrt{(9.9)^2 + (3.6)^2}$

$= \sqrt{110.97}$

≈ 10.5342

(c) $m = \left(\frac{6.2+(-3.7)}{2}, \frac{5.4+1.8}{2}\right)$

$= (1.25, 3.6)$

25. Show that the points $(4, 0)$, $(2, 1)$, and $(-1, -5)$ form the vertices of a right triangle.

Solution:

$$d_1 = \sqrt{(-1-2)^2 + (-5-1)^2} = \sqrt{45}$$
$$d_2 = \sqrt{(2-4)^2 + (1-0)^2} = \sqrt{5}$$
$$d_3 = \sqrt{(4-(-1))^2 + (0-(-5))^2} = \sqrt{50}$$

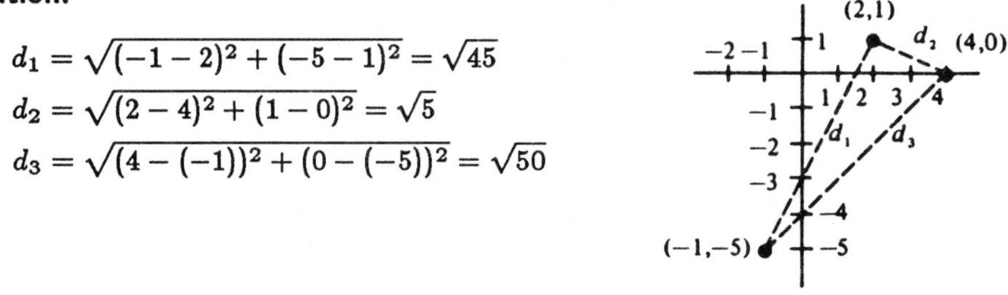

Since ${d_1}^2 + {d_2}^2 = {d_3}^2$, we can conclude by the Pythagorean Theorem that the triangle is a right triangle.

29. Find x so that the distance between $(1, 2)$ and $(x, -10)$ is 13.

Solution:

$$\sqrt{(x-1)^2 + (-10-2)^2} = 13$$
$$\sqrt{x^2 - 2x + 1 + 144} = 13$$
$$x^2 - 2x + 145 = 169$$
$$x^2 - 2x - 24 = 0$$
$$(x+4)(x-6) = 0$$
$$x = -4 \quad \text{or} \quad x = 6$$

33. Find a relationship between x and y so that (x, y) is equidistant from the points $(4, -1)$ and $(-2, 3)$.

Solution:
The distance between $(4, -1)$ and (x, y) is equal to the distance between $(-2, 3)$ and (x, y).

$$\sqrt{(x-4)^2 + (y+1)^2} = \sqrt{(x+2)^2 + (y-3)^2}$$
$$(x-4)^2 + (y+1)^2 = (x+2)^2 + (y-3)^2$$
$$x^2 - 8x + 16 + y^2 + 2y + 1 = x^2 + 4x + 4 + y^2 - 6y + 9$$
$$4 = 12x - 8y$$
$$1 = 3x - 2y$$
$$2y = 3x - 1$$

37. Determine the quadrant in which (x, y) is located so that the conditions $x > 0$ and $y > 0$ are satisfied.

Solution:

$x > 0 \longrightarrow x$ lies in Quadrant I or in Quadrant IV

$y > 0 \longrightarrow y$ lies in Quadrant I or in Quadrant II

$x > 0$ and $y > 0 \longrightarrow (x, y)$ lies in Quadrant I

45. Use the Midpoint Formula twice to find the three points that divide the line segment joining (x_1, y_1) and (x_2, y_2) into four parts.

Solution:

The midpoint of the given line segment is

$$\left(\frac{x_1 + x_2}{2}, \frac{y_1 + y_2}{2}\right).$$

The midpoint between (x_1, y_1) and $\left(\frac{x_1 + x_2}{2}, \frac{y_1 + y_2}{2}\right)$ is

$$\left(\frac{x_1 + \frac{x_1 + x_2}{2}}{2}, \frac{y_1 + \frac{y_1 + y_2}{2}}{2}\right) = \left(\frac{3x_1 + x_2}{4}, \frac{3y_1 + y_2}{4}\right).$$

The midpoint between $\left(\frac{x_1 + x_2}{2}, \frac{y_1 + y_2}{2}\right)$ and (x_2, y_2) is

$$\left(\frac{\frac{x_1 + x_2}{2} + x_2}{2}, \frac{\frac{y_1 + y_2}{2} + y_2}{2}\right) = \left(\frac{x_1 + 3x_2}{4}, \frac{y_1 + 3y_2}{4}\right)$$

Thus, the three points are

$$\left(\frac{3x_1 + x_2}{4}, \frac{3y_1 + y_2}{4}\right), \left(\frac{x_1 + x_2}{2}, \frac{y_1 + y_2}{2}\right), \quad \text{and} \quad \left(\frac{x_1 + 3x_2}{4}, \frac{y_1 + 3y_2}{4}\right).$$

47. Use the Midpoint Formula to estimate the sales of a company for 1983, given the sales in 1980 and 1986. Assume the annual sales followed a linear pattern.

Year	1980	1986
Sales	$520,000	$740,000

Solution:

$$\frac{520,000 + 740,000}{2} = 630,000$$

The estimated sales for 1983 is $630,000.

SECTION 2.2

Graphs of Equations

- You should be able to use the point-plotting method of graphing.
- You should be able to find x- and y-intercepts.
- You should be able to test for symmetry.
- You should know the standard equation of a circle with center (h, k) and radius r:

  $$(x-h)^2 + (y-k)^2 = r^2$$

Solutions to Selected Exercises

5. Determine whether the points (a) $\left(1, \frac{1}{5}\right)$, and (b) $\left(2, \frac{1}{2}\right)$ lie on the graph of the equation $x^2y - x^2 + 4y = 0$.

Solution:

(a) $\left(1, \frac{1}{5}\right)$ lies on the graph since $(1)^2\left(\frac{1}{5}\right) - (1)^2 + 4\left(\frac{1}{5}\right) = \frac{1}{5} - 1 + \frac{4}{5} = 0$.

(b) $\left(2, \frac{1}{2}\right)$ lies on the graph since $(2)^2\left(\frac{1}{2}\right) - (2)^2 + 4\left(\frac{1}{2}\right) = 2 - 4 + 2 = 0$.

7. Find the constant C so that the ordered pair $(2, 6)$ is a solution point of the equation $y = x^2 + C$.

Solution:

$$\begin{aligned} y &= x^2 + C \\ 6 &= (2)^2 + C \\ 6 &= 4 + C \\ C &= 2 \end{aligned}$$

13. Find the x- and y-intercepts of the graph of the equation $y = x^2 + x - 2$.

Solution:

Let $y = 0$. Then $0 = x^2 + x - 2 = (x+2)(x-1)$ and $x = -2$ or $x = 1$.

x-intercepts: $(-2, 0)$ and $(1, 0)$

Let $x = 0$. Then $y = -2$.

y-intercept: $(0, -2)$

17. Find the x- and y-intercepts of the graph of the equation $xy - 2y - x + 1 = 0$.

Solution:

Let $y = 0$. Then $-x + 1 = 0$ and $x = 1$.

x-intercept: $(1, 0)$

Let $x = 0$. Then $-2y + 1 = 0$ and $y = \frac{1}{2}$.

y-intercept: $\left(0, \frac{1}{2}\right)$

21. Check for symmetry with respect to both axes and the origin for $x - y^2 = 0$.

Solution:

By replacing y with $-y$, we have

$$x - (-y)^2 = 0$$
$$x - y^2 = 0$$

which is the original equation. Replacing x with $-x$ or replacing both x and y with $-x$ and$-y$ does not yield equivalent equations. Thus, $x - y^2 = 0$ is symmetric with respect to the x-axis.

25. Check for symmetry with respect to both axes and the origin for

$$y = \frac{x}{x^2 + 1}.$$

Solution:

Replacing x with $-x$ or y with $-y$ does not yield equivalent equations. Replacing x with $-x$ and y with $-y$ yields

$$-y = \frac{-x}{(-x)^2 + 1}$$
$$-y = \frac{-x}{x^2 + 1} \qquad \text{Multiply both sides by } -1$$
$$y = \frac{x}{x^2 + 1}$$

Thus, $y = \dfrac{x}{x^2 + 1}$ is symmetric with respect to the origin.

Note: An equation is symmetric with respect to the origin if it is symmetric with respect to both the x-axis and the y-axis. Also, if an equation is symmetric with respect to the origin, then one of the following is true:

1. The equation has both x-axis and y-axis symmetry or
2. The equation has neither x-axis nor y-axis symmetry.

31. Match $y = x^3 - x$ with its graph.

Solution:

$y = x^3 - x$
x-intercepts: $(-1,\ 0)$, $(0,\ 0)$, $(1,\ 0)$
y-intercept: $(0,\ 0)$
Symmetry: Origin
Matches graph (e)

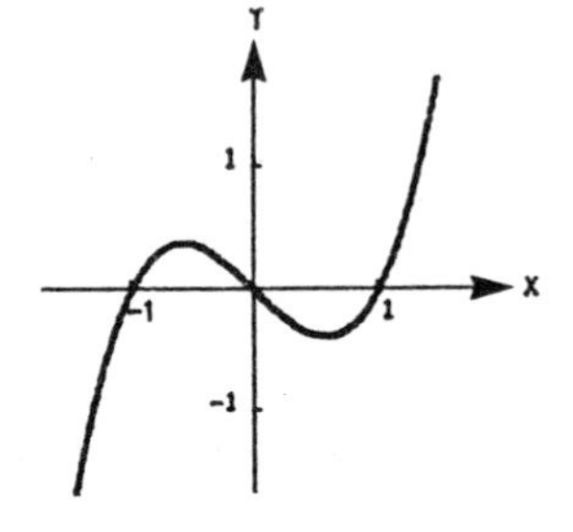

33. Sketch the graph of $y = -3x + 2$. Identify any intercepts and test for symmetry.

Solution:

$y = -3x + 2$
x-intercept: $\left(\frac{2}{3},\ 0\right)$
y-intercept: $(0,\ 2)$
No symmetries

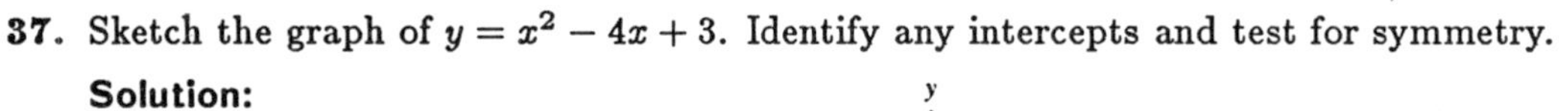

37. Sketch the graph of $y = x^2 - 4x + 3$. Identify any intercepts and test for symmetry.

Solution:

$y = x^2 - 4x + 3 = (x-1)(x-3)$
x-intercepts: $(1,\ 0)$, $(3,\ 0)$
y-intercept: $(0,\ 3)$
No symmetries

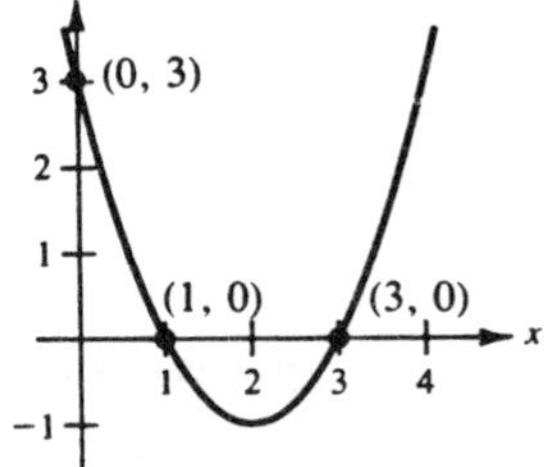

43. Sketch the graph of $y = \sqrt{x-3}$. Identify any intercepts and test for symmetry.

Solution:

$y = \sqrt{x-3}$
x-intercept: $(3,\ 0)$
No y-intercept
No symmetry

x	3	4	7	12
y	0	1	2	3

Note: The domain is $[3,\ \infty)$ and the range is $[0,\ \infty)$.

49. Sketch the graph of $x = y^2 - 1$. Identify any intercepts and test for symmetry.

Solution:

$x = y^2 - 1$
x-intercept: $(-1, 0)$
y-intercepts: $(0, -1)$, $(0, 1)$
x-axis symmetry

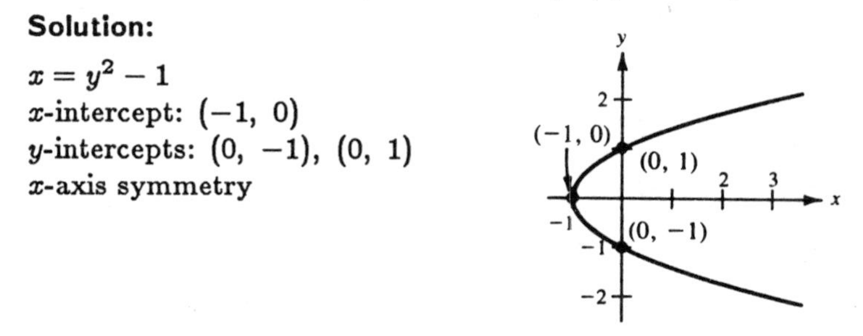

55. Find the standard form of the equation of the circle with center $(2, -1)$ and radius 4.

Solution:

$$(x-2)^2 + (y+1)^2 = 4^2$$
$$x^2 - 4x + 4 + y^2 + 2y + 1 = 16$$
$$x^2 + y^2 - 4x + 2y - 11 = 0$$

57. Find the standard form of the equation of the circle with center $(-1, 2)$ and solution point $(0, 0)$.

Solution:

$$(x+1)^2 + (y-2)^2 = r^2$$
$$(0+1)^2 + (0-2)^2 = r^2 \quad \rightarrow \quad r^2 = 5$$
$$(x+1)^2 + (y-2)^2 = 5$$
$$x^2 + y^2 + 2x - 4y = 0$$

61. Write the following equation of the circle in standard form and sketch its graph.

$$x^2 + y^2 - 2x + 6y + 6 = 0$$

Solution:

$$x^2 + y^2 - 2x + 6y + 6 = 0$$
$$(x^2 - 2x + 1) + (y^2 + 6y + 9) = -6 + 1 + 9$$
$$(x-1)^2 + (y+3)^2 = 4$$

Center: $(1, -3)$
Radius: 2

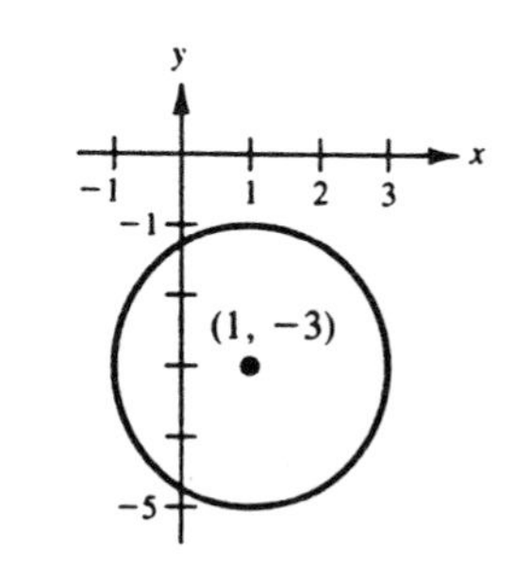

63. Write the following equation of the circle in standard form and sketch its graph.

$$x^2 + y^2 - 2x + 6y + 10 = 0$$

Solution:

$$x^2 + y^2 - 2x + 6y + 10 = 0$$
$$(x^2 - 2x + 1) + (y^2 + 6y + 9) = -10 + 1 + 9$$
$$(x - 1)^2 + (y + 3)^2 = 0$$

Graph is the point $(1, -3)$.

67. Write the following equation of the circle in standard form and sketch its graph.

$$16x^2 + 16y^2 + 16x + 40y - 7 = 0$$

Solution:

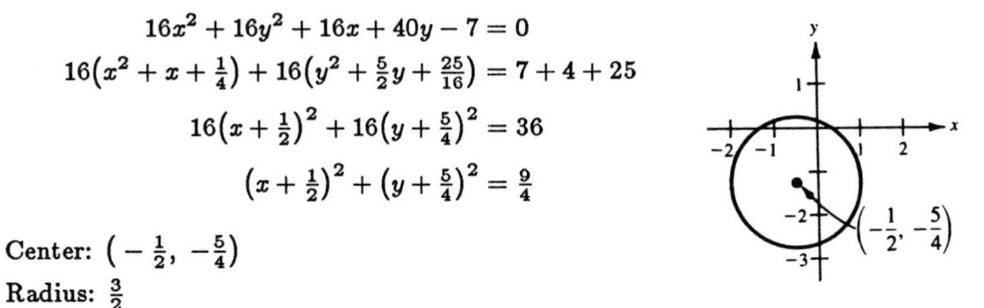

$$16x^2 + 16y^2 + 16x + 40y - 7 = 0$$
$$16\left(x^2 + x + \tfrac{1}{4}\right) + 16\left(y^2 + \tfrac{5}{2}y + \tfrac{25}{16}\right) = 7 + 4 + 25$$
$$16\left(x + \tfrac{1}{2}\right)^2 + 16\left(y + \tfrac{5}{4}\right)^2 = 36$$
$$\left(x + \tfrac{1}{2}\right)^2 + \left(y + \tfrac{5}{4}\right)^2 = \tfrac{9}{4}$$

Center: $\left(-\tfrac{1}{2}, -\tfrac{5}{4}\right)$

Radius: $\tfrac{3}{2}$

SECTION 2.3

Lines in the Plane

You should know the following important facts about lines.

- The slope of the line through (x_1, y_1) and (x_2, y_2) is

$$m = \frac{y_2 - y_1}{x_2 - x_1}.$$

- (a) If $m > 0$, the line rises from left to right.
  (b) If $m = 0$, the line is horizontal.
  (c) If $m < 0$, the line falls from left to right.
  (d) If m is undefined, the line is vertical.

- Equations of Lines
  (a) Point-Slope: $y - y_1 = m(x - x_1)$
  (b) Two-Point: $y - y_1 = \frac{y_2 - y_1}{x_2 - x_1}(x - x_1)$
  (c) Slope-Intercept: $y = mx + b$
  (d) General: $Ax + By + C = 0$
  (d) Vertical: $x = a$
  (e) Horizontal: $y = b$

- Given two distinct nonvertical lines

$$L_1 : y = m_1 x + b_1 \quad \text{and} \quad L_2 : y = m_2 x + b_2$$

  (a) L_1 is parallel to L_2 if and only if $m_1 = m_2$.
  (b) L_1 is perpendicular to L_2 if and only if $m_1 = -1/m_2$.

Solutions to Selected Exercises

9. Plot the points $(-3, -2)$ and $(1, 6)$ and find the slope of the line passing through the points.

Solution:

$$m = \frac{6 - (-2)}{1 - (-3)} = \frac{8}{4} = 2$$

11. Plot the points $(-6,\ -1)$ and $(-6,\ 4)$ and find the slope of the line passing through the points.

Solution:

$$m = \frac{4-(-1)}{-6-(-6)} = \frac{5}{0}$$

The slope is undefined.

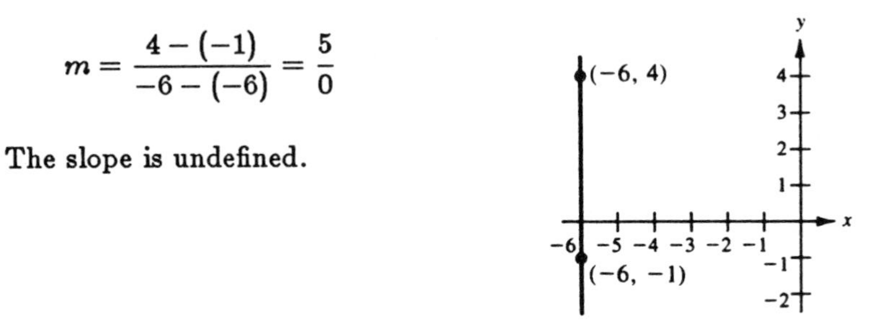

15. Determine if the lines L_1 and L_2 passing through the given pairs of points are parallel, perpendicular, or neither.

$L_1 : (0,\ -1),\ (5,\ 9)$
$L_2 : (0,\ 3),\ (4,\ 1)$

Solution:

The slope of L_1 is $m_1 = \dfrac{9-(-1)}{5-0} = \dfrac{10}{5} = 2.$

The slope of L_2 is $m_2 = \dfrac{1-3}{4-0} = -\dfrac{2}{4} = -\dfrac{1}{2}.$

Since $m_1 \cdot m_2 = 2\left(-\frac{1}{2}\right) = -1$, the lines are perpendicular.

19. Use the point $(2,\ 1)$ on the line and the slope $m = 0$ of the line to find three additional points that the line passes through. (The solution is not unique.)

Solution:

Since $m = 0$, the line is horizontal, and since the line passes through $(2,\ 1)$, all other points on the line will be of the form $(x,\ 1)$. Three additional points are: $(0,\ 1)$, $(1,\ 1)$, $(3,\ 1)$.

25. Find the slope and y-intercept, if possible, of the line specified by $5x - y + 3 = 0$.

Solution:

$$\begin{aligned} 5x - y + 3 &= 0 \\ -y &= -5x - 3 \\ y &= 5x + 3 \end{aligned}$$

Slope: $m = 5$
y-intercept: $(0,\ 3)$

27. Find the slope and y-intercept, if possible, of the line specified by $5x - 2 = 0$.

Solution:

$$5x - 2 = 0$$
$$5x = 2$$
$$x = \tfrac{2}{5} \qquad \text{Vertical line}$$

Slope: Undefined
y-intercept: None

33. Find an equation for the line passing through the points $\left(2, \frac{1}{2}\right)$, $\left(\frac{1}{2}, \frac{5}{4}\right)$ and sketch a graph of the line.

Solution:

$$m = \frac{\frac{5}{4} - \frac{1}{2}}{\frac{1}{2} - 2} = \frac{\frac{3}{4}}{-\frac{3}{2}} = -\frac{1}{2}$$
$$y - \frac{1}{2} = -\frac{1}{2}(x - 2)$$
$$2y - 1 = -(x - 2)$$
$$2y - 1 = -x + 2$$
$$x + 2y - 3 = 0$$

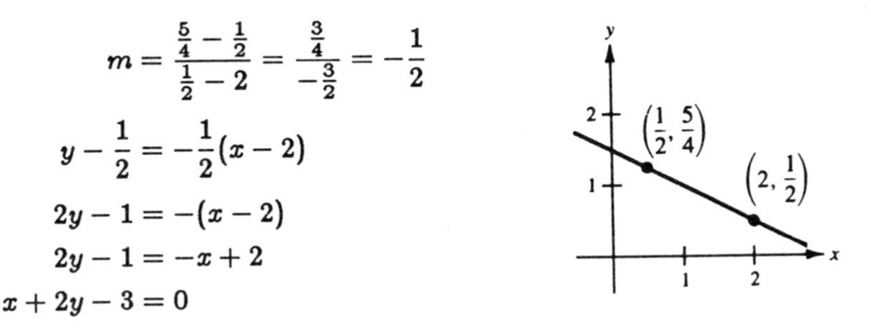

37. Find an equation for the line passing through the points $(1, 0.6)$, and $(-2, -0.6)$ and sketch a graph of the line.

Solution:

$$m = \frac{-0.6 - 0.6}{-2 - 1} = 0.4$$
$$y - 0.6 = 0.4(x - 1)$$
$$y - 0.6 = 0.4x - 0.4$$
$$y = 0.4x + 0.2 \quad \text{or} \quad 2x - 5y + 1 = 0$$

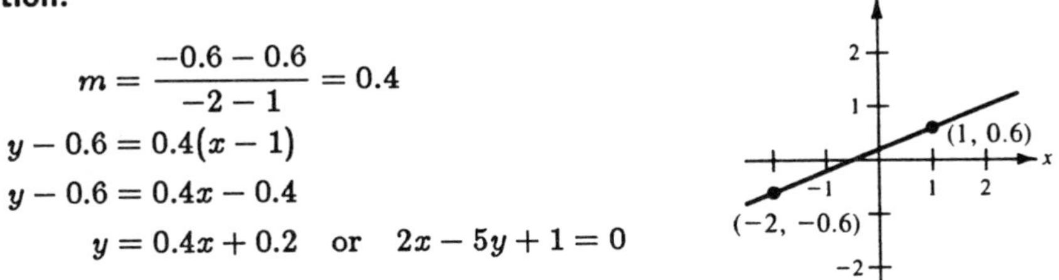

41. Find an equation of the line that passes through the point $(-3, 6)$ and has a slope of $m = -2$. Sketch a graph of the line.

Solution:

$$y - 6 = -2\big(x - (-3)\big)$$
$$y - 6 = -2x - 6$$
$$y = -2x$$

45. Find an equation of the line that passes through the point $(6, -1)$ and has an undefined slope. Sketch a graph of the line.

Solution:

Since the slope is undefined, the line is vertical and since the line passes through $(6, -1)$, its equation is $x = 6$.

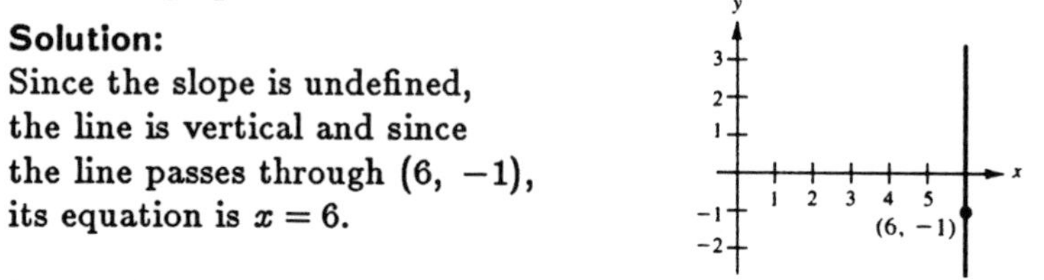

49. Prove that the line with intercepts $(a, 0)$ and $(0, b)$ has the following equation.

$$\frac{x}{a} + \frac{y}{b} = 1, \quad a \neq 0, \; b \neq 0$$

Solution:

Using the points $(a, 0)$ and $(0, b)$ we have

$$m = \frac{b-0}{0-a} = -\frac{b}{a}$$

$$y - 0 = -\frac{b}{a}(x - a)$$

$$y = -\frac{b}{a}x + b$$

$$ay = -bx + ab$$

$$bx + ay = ab$$

$$\frac{bx + ay}{ab} = \frac{ab}{ab}$$

$$\frac{x}{a} + \frac{y}{b} = 1.$$

53. Use the result of Exercise 49 to write an equation of the line with x-intercept $\left(-\frac{1}{6}, 0\right)$ and y-intercept $\left(0, -\frac{2}{3}\right)$.

Solution:

x-intercept: $\left(-\frac{1}{6}, 0\right)$

y-intercept: $\left(0, -\frac{2}{3}\right)$

$$\frac{x}{-1/6} + \frac{y}{-2/3} = 1$$

$$-6x - \frac{3}{2}y = 1$$

$$-12x - 3y = 2$$

$$12x + 3y = -2$$

59. Write the equation of the line through the point $(-6, 4)$ (a) parallel to the line $3x + 4y = 7$ and (b) perpendicular to the line $3x + 4y = 7$.

Solution:

$$\begin{aligned} 3x + 4y &= 7 \\ 4y &= -3x + 7 \\ y &= -\frac{3}{4}x + \frac{7}{4} \end{aligned}$$

The slope of the given line is $m_1 = -3/4$.

(a) The slope of the parallel line is $m_2 = m_1 = -3/4$.

$$\begin{aligned} y - 4 &= -\frac{3}{4}(x - (-6)) \\ y - 4 &= -\frac{3}{4}x - \frac{9}{2} \\ y &= -\frac{3}{4}x - \frac{1}{2} \\ 4y &= -3x - 2 \\ 3x + 4y &= -2 \end{aligned}$$

(b) The slope of the perpendicular line is $m_2 = -1/m_1 = 4/3$.

$$\begin{aligned} y - 4 &= \frac{4}{3}(x - (-6)) \\ y - 4 &= \frac{4}{3}x + 8 \\ y &= \frac{4}{3}x + 12 \\ 3y &= 4x + 36 \\ 4x - 3y &= -36 \end{aligned}$$

63. Find the equation of the line giving the relationship between the temperature in degrees Celsius, C, and degrees Fahrenheit, F. Use the fact that water freezes at 0° Celsius (32° Fahrenheit) and boils at 100° Celsius (212° Fahrenheit).

Solution:

Using the points (0, 32) and (100, 212), we have

$$\begin{aligned} m &= \frac{212 - 32}{100 - 0} = \frac{180}{100} = \frac{9}{5} \\ F - 32 &= \frac{9}{5}(C - 0) \\ F &= \frac{9}{5}C + 32 \end{aligned}$$

67. A store is offering a 15% discount on all items in its inventory. Write a linear equation giving the sale price S, for an item with a list price, L.

Solution:

$$S = L - 0.15L$$
$$S = 0.85L$$

73. Prove that if two distinct lines have equal slope, they must be parallel.

Solution:
In slope-intercept form,

$$L_1 : y = mx + b_1 \quad \text{and} \quad L_2 : y = mx + b_2, \quad b_1 \neq b_2$$

If $m = 0$, then L_1 is a horizontal line through $(0, b_1)$ and L_2 is a horizontal line through $(0, b_2)$. Thus, the lines are parallel. If $m \neq 0$, assume that L_1 and L_2 are not parallel. Then L_1 and L_2 must intersect at some point (a, b). To find this point, set $L_1 = L_2$.

$$mx + b_1 = mx + b_2$$
$$b_1 - b_2 = 0$$

Since $b_1 \neq b_2$, this is not possible. The lines do not intersect; therefore, they must be parallel. If m is undefined, then L_1 and L_2 are vertical lines and since they are distinct, they are parallel.

SECTION 2.4

Functions

- Given an equation, you should be able to determine if it represents a function.
- Given a function, you should be able to do the following.

  (a) Find the domain.
  (b) Find the range.
  (c) Determine if it is one-to-one.
  (d) Evaluate it at specific values.

Solutions to Selected Exercises

5. Evaluate the function at the specified value of the independent variable and simplify the results.

$$f(x) = 2x - 3$$

(a) $f(1)$ (b) $f(-3)$
(c) $f(x-1)$ (d) $f(1/4)$

Solution:

(a) $f(1) = 2(1) - 3 = -1$ (b) $f(-3) = 2(-3) - 3 = -9$
(c) $f(x-1) = 2(x-1) - 3 = 2x - 5$ (d) $f(1/4) = 2(1/4) - 3 = -5/2$

9. Evaluate the function at the specified value of the independent variable and simplify the results.

$$f(y) = 3 - \sqrt{y}$$

(a) $f(4)$ (b) $f(100)$
(c) $f(4x^2)$ (d) $f(0.25)$

Solution:

(a) $f(4) = 3 - \sqrt{4} = 1$ (b) $f(100) = 3 - \sqrt{100} = -7$
(c) $f(4x^2) = 3 - \sqrt{4x^2} = 3 - 2|x|$ (d) $f(0.25) = 3 - \sqrt{0.25} = 2.5$

13. Evaluate the function at the specified value of the independent variable and simplify the results.

$$f(x) = \frac{|x|}{x}$$

(a) $f(2)$ (b) $f(-2)$
(c) $f(x^2)$ (d) $f(x-1)$

Solution:

(a) $f(2) = \dfrac{|2|}{2} = 1$ (b) $f(-2) = \dfrac{|-2|}{-2} = -1$

(c) $f(x^2) = \dfrac{|x^2|}{x^2} = 1$ (d) $f(x-1) = \dfrac{|x-1|}{x-1}$

15. Evaluate the function at the specified value of the independent variable and simplify the results.

$$f(x) = \begin{cases} 2x+1, & x < 0 \\ 2x+2, & x \geq 0 \end{cases}$$

(a) $f(-1)$ (b) $f(0)$
(c) $f(1)$ (d) $f(2)$

Solution:

(a) $f(-1) = 2(-1) + 1 = -1$ (b) $f(0) = 2(0) + 2 = 2$
(c) $f(1) = 2(1) + 2 = 4$ (d) $f(2) = 2(2) + 2 = 6$

17. For $f(x) = x^2 - x + 1$, find

$$\frac{f(2+h) - f(2)}{h}$$

and simplify your answer.

Solution:

$$\begin{aligned} f(x) &= x^2 - x + 1 \\ f(2+h) &= (2+h)^2 - (2+h) + 1 \\ &= 4 + 4h + h^2 - 2 - h + 1 \\ &= h^2 + 3h + 3 \\ f(2) &= (2)^2 - 2 + 1 = 3 \\ f(2+h) - f(2) &= h^2 + 3h \\ \frac{f(2+h) - f(2)}{h} &= h + 3 \end{aligned}$$

25. Find all real values x such that $f(x) = 0$ for $f(x) = x^2 - 9$.

Solution:

$$\begin{aligned} x^2 - 9 &= 0 \\ x^2 &= 9 \\ x &= \pm 3 \end{aligned}$$

27. Find all real values x such that $f(x) = 0$ for

$$f(x) = \frac{3}{x-1} + \frac{4}{x-2}.$$

Solution:

$$\begin{aligned} \frac{3}{x-1} + \frac{4}{x-2} &= 0 \\ 3(x-2) + 4(x-1) &= 0 \\ 7x - 10 &= 0 \\ x &= \frac{10}{7} \end{aligned}$$

31. Find the domain of $h(t) = 4/t$.

Solution:

The domain includes all real numbers except 0, i.e. $t \neq 0$.

35. Find the domain of $f(x) = \sqrt[4]{1 - x^2}$.

Solution:

Choose x-values for which $1 - x^2 \geq 0$. Using methods of Section 2.8, we find that the domain is $-1 \leq x \leq 1$.

39. Determine if y is a function of x for $x^2 + y^2 = 4$.

Solution:

y is not a function of x since some values of x give two values for y. For example, if $x = 0$, then $y = \pm 2$.

43. Determine if y is a function of x for $2x + 3y = 4$.

Solution:

$$\begin{aligned} 2x + 3y &= 4 \\ y &= \tfrac{1}{3}(4 - 2x) \end{aligned}$$

y is a function of x.

47. Determine if y is a function of x for $x^2y - x^2 + 4y = 0$.

Solution:

$$x^2y - x^2 + 4y = 0$$
$$y(x^2 + 4) = x^2$$
$$y = \frac{x^2}{x^2 + 4}$$

y is a function of x.

51. Assume that the domain of $f(x) = x^2$ is the set $A = \{-2, -1, 0, 1, 2\}$. Determine the set of ordered pairs representing the function f.

Solution:

$$\{(-2, f(-2)), (-1, f(-1)), (0, f(0)), (1, f(1)), 2, f(2))\}$$
$$\{(-2, 4), (-1, 1), (0, 0), (1, 1), (2, 4)\}$$

57. Find the value(s) of x for which $f(x) = g(x)$ where $f(x) = \sqrt{3x} + 1$ and $g(x) = x + 1$.

Solution:

$$f(x) = g(x)$$
$$\sqrt{3x} + 1 = x + 1$$
$$\sqrt{3x} = x$$
$$3x = x^2$$
$$0 = x^2 - 3x$$
$$0 = x(x - 3)$$
$$x = 0 \quad \text{or} \quad x = 3$$

61. Express the area A of a circle as a function of its circumference C.

Solution:

$$A = \pi r^2, \quad C = 2\pi r$$
$$r = \frac{C}{2\pi}$$
$$A = \pi\left(\frac{C}{2\pi}\right)^2$$
$$A = \frac{C^2}{4\pi}$$

65. A right triangle is formed in the first quadrant by the x- and y-axes and a line through the point $(1, 2)$, as shown in the figure. Write the area of the triangle as a function of x, and determine the domain of the function.

Solution:

$$A = \frac{1}{2}bh = \frac{1}{2}xy$$

Since $(0,\ y)$, $(1,\ 2)$ and $(x,\ 0)$ all lie on the same line, the slopes between any pair are equal.

$$\frac{2-y}{1-0} = \frac{0-2}{x-1}$$

$$2 - y = -\frac{2}{x-1}$$

$$y = \frac{2}{x-1} + 2$$

$$y = \frac{2x}{x-1}$$

Therefore,

$$A = \frac{1}{2}x\left(\frac{2x}{x-1}\right)$$

$$A = \frac{x^2}{x-1}$$

The domain of A includes x-values such that $x^2/(x-1) > 0$. This results in a domain of $x > 1$.

69. A company produces a product for which the variable cost is \$12.30 per unit and the fixed costs are \$98,000. The product sells for \$17.98. Let x be the number of units produced.

(a) Write the total cost C as a function of the number of units produced.
(b) Write the revenue R as a function of the number of units produced.
(c) Write the profit P as a function of the number of units produced.

(Note: $P = R - C$.)

Solution:

(a) Cost = variable costs + fixed costs
$C = 12.30x + 98,000$

(b) Revenue = price per unit × number of units
$R = 17.98x$

(c) Profit = Revenue − Cost
$P = 17.98x - (12.30x + 98,000)$
$P = 5.68x - 98,000$

SECTION 2.5

Graphs of Functions

- You should be able to determine the domain and range of a function from its graph.
- You should be able to use the vertical line test for functions.
- You should know that the graph of $f(x) = c$ is a horizontal line through $(0,\ c)$.
- You should be able to determine when a function is constant, increasing, or decreasing.
- You should know that f is
  (a) Odd if $f(-x) = -f(x)$.
  (b) Even if $f(-x) = f(x)$.
- You should know the basic types of transformations.

Solutions to Selected Exercises

5. Determine the domain and range of the function $f(x) = \sqrt{25 - x^2}$.

Solution:
From the graph we see that the x-values do not extend beyond $x = -5$ (on the left) and $x = 5$ (on the right). The domain is $[-5,\ 5]$. Similarly, the y-values do not extend beyond $y = 0$ and $y = 5$. The range is $[0,\ 5]$.

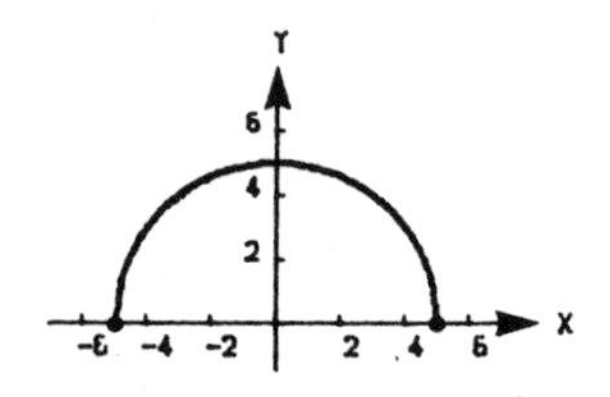

7. Use the vertical line test to determine if y is a function of x where $y = x^2$.

Solution:
Since no vertical line would ever cross the graph more than one time, y *is* a function of x.

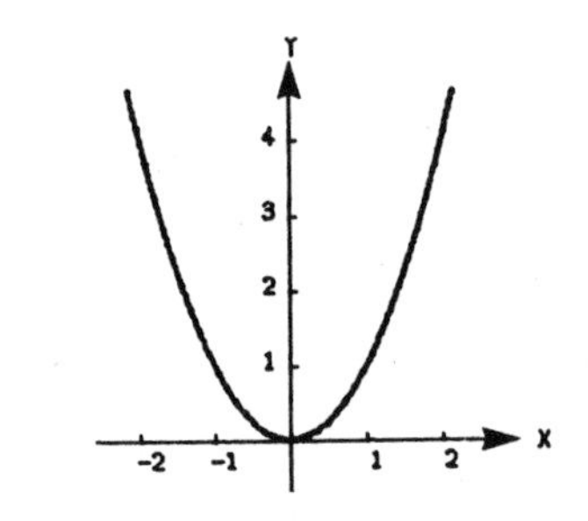

11. Use the vertical line test to determine if y is a function of x where $x^2 = xy - 1$.

Solution:

Since no vertical line would ever cross the graph more than one time, y *is* a function of x.

$$y = \frac{x^2 + 1}{x}$$

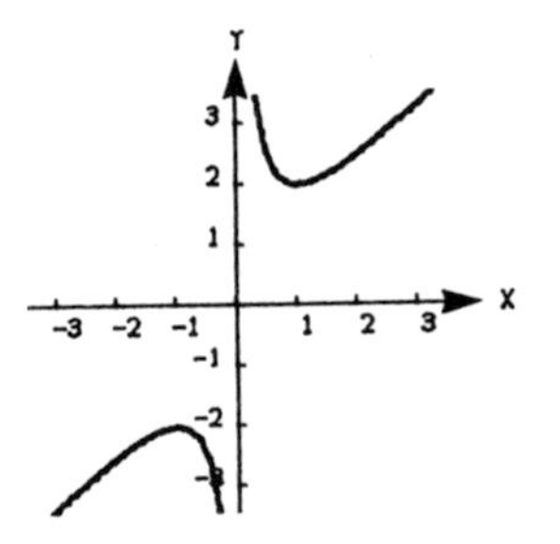

15. (a) Determine the intervals over which the function is increasing, decreasing, or constant, and (b) determine if the function is even, odd, or neither for $f(x) = x^3 - 3x^2$.

Solution:

(a) By its graph we see that f is increasing on $(-\infty,\ 0)$ and $(2,\ \infty)$ and is decreasing on $(0,\ 2)$.

(b)
$$f(-x) = (-x)^3 - 3(-x)^2$$
$$= -x^3 - 3x^2$$

$f(-x) \neq f(x)$ and $f(x) \neq -f(x)$, so the function is neither odd nor even.

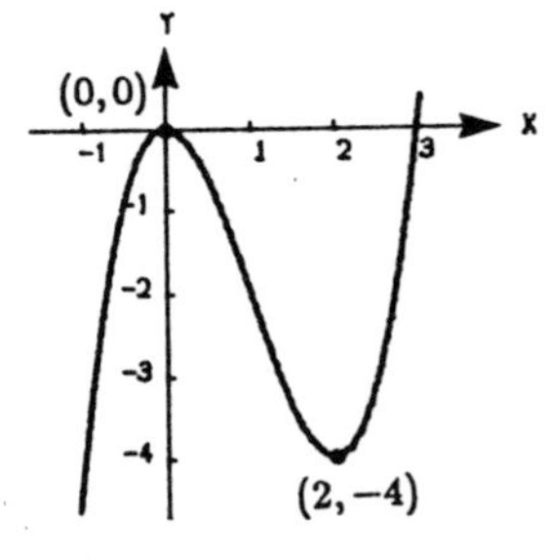

19. (a) Determine the intervals over which the function is increasing, decreasing, or constant, and (b) determine if the function is even, odd, or neither for $f(x) = x\sqrt{x+3}$.

Solution:

(a) By its graph we see that f is increasing on $(-2,\ \infty)$ and decreasing on $(-3,\ -2)$.

(b) $f(-x) = -x\sqrt{-x+3}$

$f(-x) \neq f(x)$ and $f(-x) \neq -f(x)$, so the function is neither odd nor even.

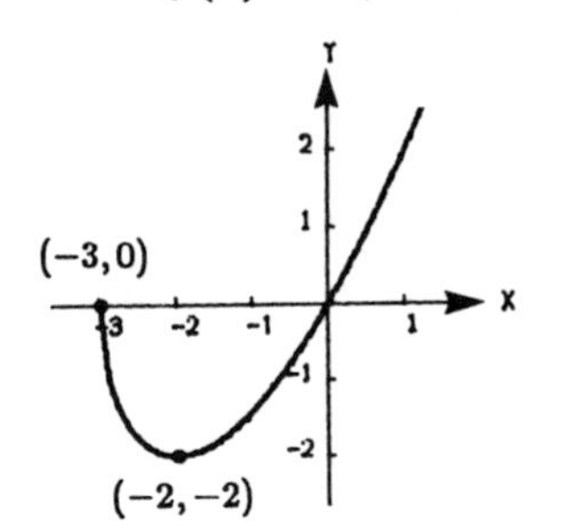

23. Determine whether $g(x) = x^3 - 5x$ is even, odd, or neither.

Solution:

$$g(x) = x^3 - 5x$$
$$g(-x) = (-x)^3 - 5(-x)$$
$$= -x^3 + 5x$$
$$= -(x^3 - 5x)$$
$$= -g(x)$$

Therefore, g is odd.

27. Sketch the graph of $f(x) = 3$ and determine whether the function is odd, even, or neither.

Solution:
$f(x) = 3$
Domain: $(-\infty, \infty)$
Range: $\{3\}$
y-intercept: $(0, 3)$
y-axis symmetry
$f(-x) = 3 = f(x)$
Therefore, f is even.

31. Sketch the graph of $g(s) = s^3/4$ and determine whether the function is odd, even, or neither.

Solution:
$g(s) = s^3/4$
Intercept: $(0, 0)$
Origin symmetry
Domain: $(-\infty, \infty)$
Range: $(-\infty, \infty)$
$g(-s) = -g(s)$
Therefore, g is odd.

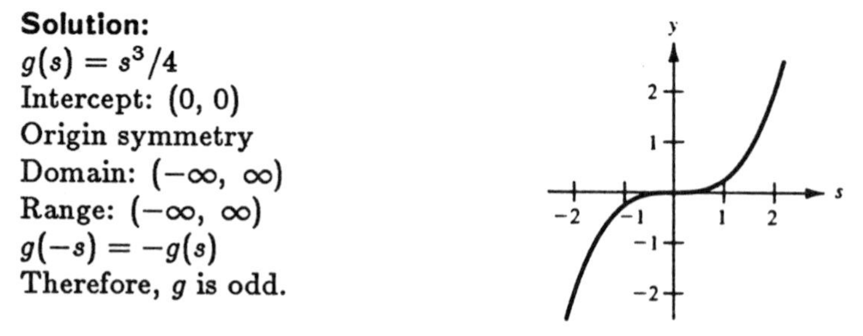

35. Sketch the graph of $g(t) = \sqrt[3]{t-1}$ and determine whether the function is odd, even, or neither.

Solution:
$g(t) = \sqrt[3]{t-1}$
x-intercept: $(1, 0)$
y-intercept: $(0, -1)$
Domain: $(-\infty, \infty)$
Range: $(-\infty, \infty)$

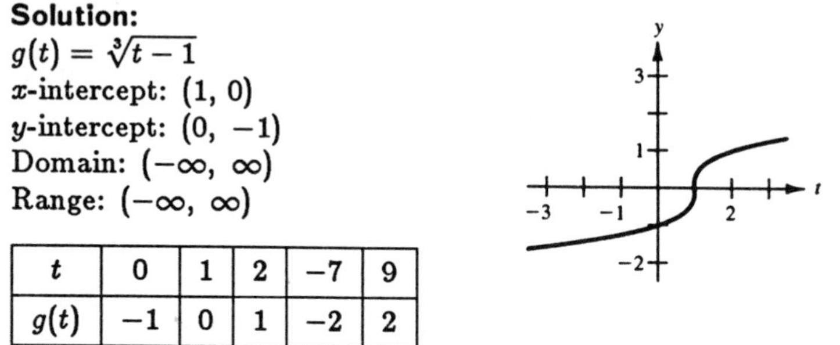

t	0	1	2	-7	9
$g(t)$	-1	0	1	-2	2

$g(-t) = \sqrt[3]{-t-1} \neq g(t)$ and $\neq -g(t)$. Therefore, g is neither odd nor even.

37. Sketch the graph of

$$f(x) = \begin{cases} x+3, & \text{if } x \le 0 \\ 3, & \text{if } 0 < x \le 2 \\ 2x-1, & \text{if } x > 2 \end{cases}$$

and determine whether the function is odd, even, or neither.

Solution:

For $x \le 0$, $f(x) = x + 3$. For $0 < x \le 2$, $f(x) = 3$. For $x > 2$, $f(x) = 2x - 1$. Thus, the graph of f is as shown.

$$f(-x) = \begin{cases} -x+3, & \text{if } x \le 0 \\ 3, & \text{if } 0 < x \le 2 \\ -2x-1, & \text{if } x > 2 \end{cases}$$

So, f is neither odd nor even.

41. Sketch the graph of $f(x) = x^2 - 9$ and determine the interval(s), if any, on the real axis for which $f(x) \ge 0$.

Solution:

$f(x) = x^2 - 9$

x-intercepts: $(-3, 0)$, $(0, 3)$

y-intercept: $(0, -9)$

y-axis symmetry

Domain: $(-\infty, \infty)$

Range: $[-9, \infty)$

$f(x) \ge 0$ on the intervals $(-\infty, -3]$ and $[3, \infty)$.

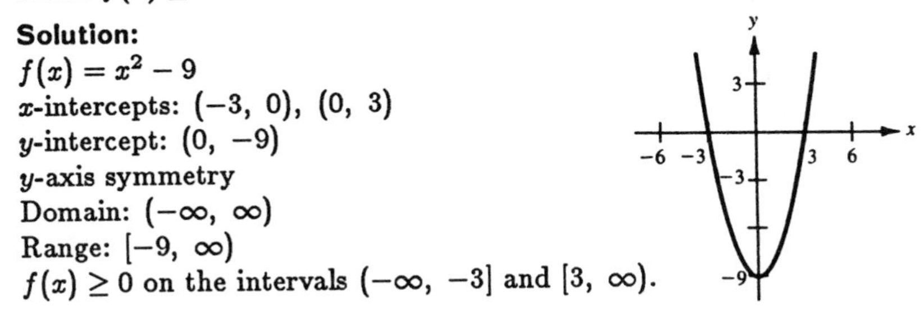

45. Sketch the graph of $f(x) = x^2 + 1$ and determine the interval(s), if any, on the real axis for which $f(x) \ge 0$.

Solution:

$f(x) = x^2 + 1$

x-intercept: None

y-intercept: $(0, 1)$

y-axis symmetry

Domain: $(-\infty, \infty)$

Range: $[1, \infty)$

$f(x) \ge 0$ for all real numbers.

51. Use the graph of $f(x) = \sqrt{x}$ to sketch the graph of each of the following.

(a) $y = \sqrt{x} + 2$

(b) $y = -\sqrt{x}$

(c) $y = \sqrt{x-2}$

(d) $y = \sqrt{x+3}$

(e) $y = 2 - \sqrt{x-4}$

(f) $y = \sqrt{2x}$

Solution:

(a) $y = \sqrt{x} + 2$
Vertical shift 2 units upward

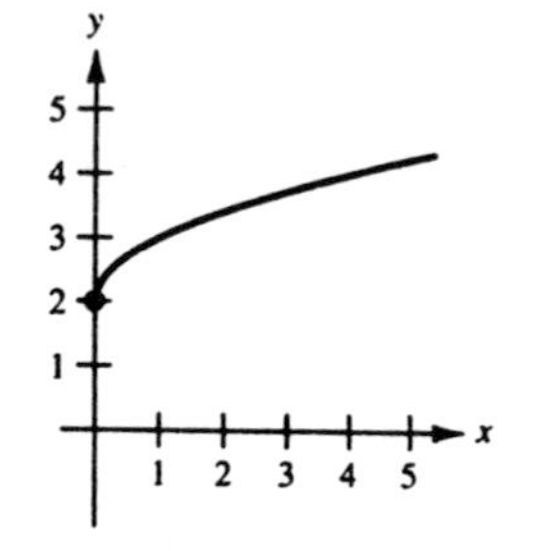

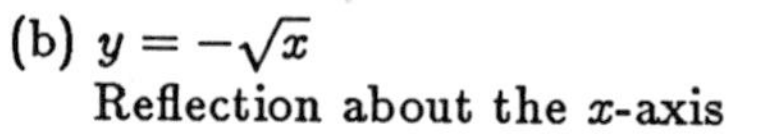
(b) $y = -\sqrt{x}$
Reflection about the x-axis

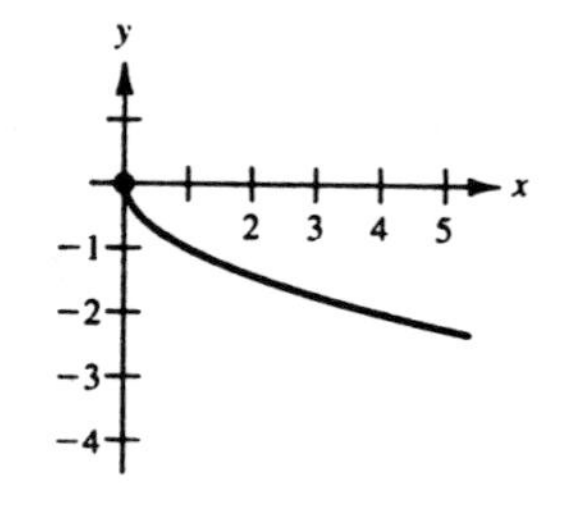

(c) $y = \sqrt{x-2}$
Horizontal shift 2 units to the right

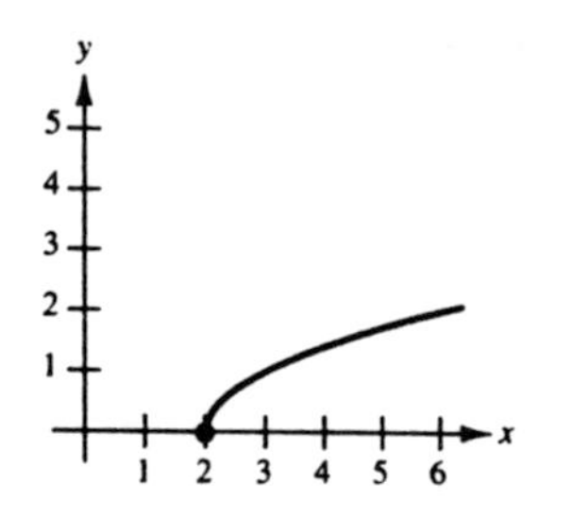

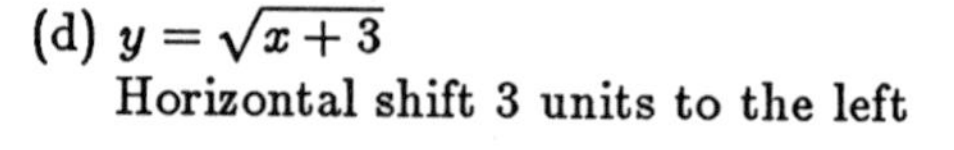
(d) $y = \sqrt{x+3}$
Horizontal shift 3 units to the left

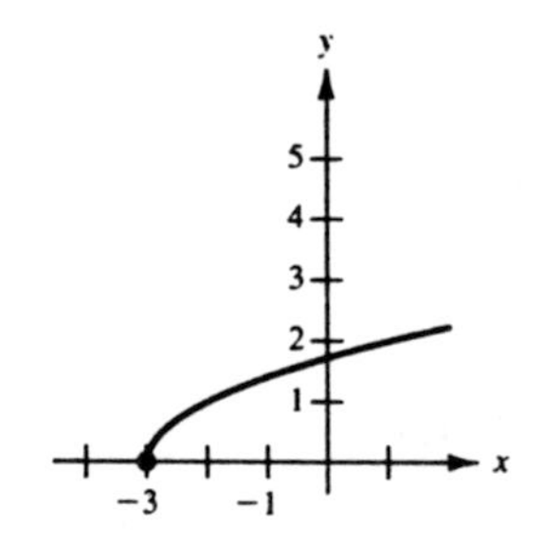

(e) $y = 2 - \sqrt{x-4}$
Reflection about the x-axis, horizontal shift of 4 units to the right and a vertical shift 2 units upward

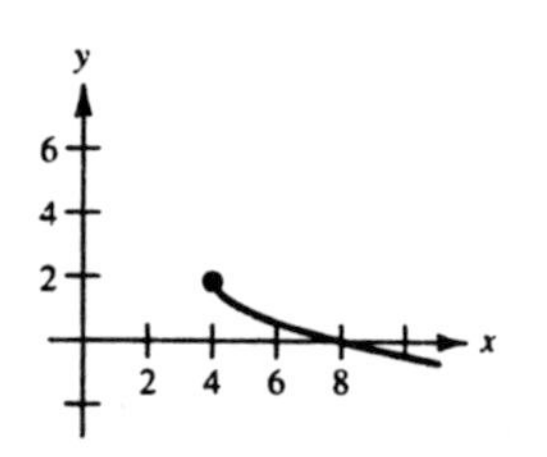

(f) $y = \sqrt{2x}$

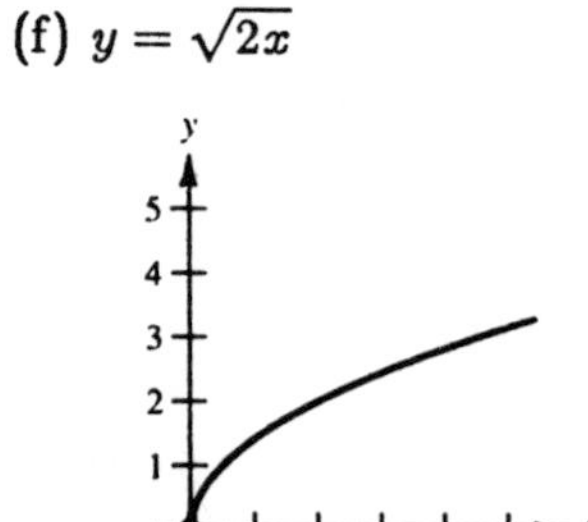

57. Write the height h of the given rectangle as a function of x.

Solution:

$$
\begin{aligned}
h &= \text{top} - \text{bottom} \\
&= (4x - x^2) - x^2 \\
&= 4x - 2x^2
\end{aligned}
$$

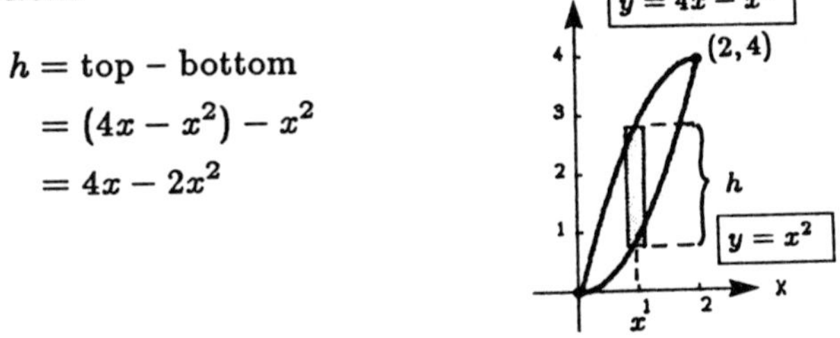

61. Prove that a function of the following form is odd.

$$f(x) = a_{2n+1}x^{2n+1} + a_{2n-1}x^{2n-1} + \ldots + a_3x^3 + a_1x$$

Solution:

$$
\begin{aligned}
f(x) &= a_{2n+1}x^{2n+1} + a_{2n-1}x^{2n-1} + \ldots + a_3x^3 + a_1x \\
f(-x) &= a_{2n+1}(-x)^{2n+1} + a_{2n-1}(-x)^{2n-1} + \ldots + a_3(-x)^3 + a_1(-x) \\
&= -a_{2n+1}x^{2n+1} - a_{2n-1}x^{2n-1} - \ldots - a_3x^3 - a_1x \\
&= -f(x)
\end{aligned}
$$

Therefore, $f(x)$ is odd.

SECTION 2.6

Combinations of Functions

- Given two functions, f and g, you should be able to form the following functions (if defined):
  1. Sum: $(f+g)(x) = f(x) + g(x)$
  2. Difference: $(f-g)(x) = f(x) - g(x)$
  3. Product: $(fg)(x) = f(x)g(x)$
  4. Quotient: $(f/g)(x) = f(x)/g(x),\ g(x) \neq 0$
  5. Composition of f with g: $(f \circ g)(x) = f\big(g(x)\big)$
  6. Composition of g with f: $(g \circ f)(x) = g\big(f(x)\big)$

Solutions to Selected Exercises

5. Find (a) $(f+g)(x)$, (b) $(f-g)(x)$, (c) $(fg)(x)$, and (d) $(f/g)(x)$. What is the domain of f/g?

Solution:

$$f(x) = x^2 + 5, \quad g(x) = \sqrt{1-x}$$

(a) $(f+g)(x) = f(x) + g(x) = x^2 + 5 + \sqrt{1-x}$

(b) $(f-g)(x) = f(x) - g(x) = x^2 + 5 - \sqrt{1-x}$

(c) $(fg)(x) = f(x)g(x) = (x^2+5)\sqrt{1-x}$

(d) $\left(\dfrac{f}{g}\right)(x) = \dfrac{f(x)}{g(x)} = \dfrac{x^2+5}{\sqrt{1-x}}, \ x < 1$

The domain of f/g is $(-\infty, 1)$.

11. Evaluate $(f-g)(2t)$ for $f(x) = x^2 + 1$ and $g(x) = x - 4$.

Solution:

$$\begin{aligned}(f-g)(2t) &= f(2t) - g(2t)\\ &= [(2t)^2 + 1] - [(2t) - 4]\\ &= 4t^2 + 1 - 2t + 4\\ &= 4t^2 - 2t + 5\end{aligned}$$

15. Evaluate $(f/g)(5)$ for $f(x) = x^2 + 1$ and $g(x) = x - 4$.

Solution:

$$\left(\frac{f}{g}\right)(5) = \frac{f(5)}{g(5)} = \frac{(5)^2+1}{5-4} = 26$$

19. Evaluate $(f/g)(-1) - g(3)$ for $f(x) = x^2 + 1$ and $g(x) = x - 4$.

Solution:

$$\left(\frac{f}{g}\right)(-1) - g(3) = \frac{f(-1)}{g(-1)} - g(3) = \frac{(-1)^2 + 1}{-1 - 4} - (3 - 4) = -\frac{2}{5} + 1 = \frac{3}{5}$$

23. Find (a) $f \circ g$, (b) $g \circ f$, and (c) $f \circ f$ for $f(x) = 3x + 5$ and $g(x) = 5 - x$.

Solution:

(a) $$\begin{aligned} f \circ g &= f(g(x)) \\ &= f(5 - x) \\ &= 3(5 - x) + 5 \\ &= 20 - 3x \end{aligned}$$

(b) $$\begin{aligned} g \circ f &= g(f(x)) \\ &= g(3x + 5) \\ &= 5 - (3x + 5) \\ &= -3x \end{aligned}$$

(c) $$\begin{aligned} f \circ f &= f(f(x)) \\ &= f(3x + 5) \\ &= 3(3x + 5) + 5 \\ &= 9x + 20 \end{aligned}$$

25. Find (a) $f \circ g$ and (b) $g \circ f$ for $f(x) = \sqrt{x + 4}$ and $g(x) = x^2$.

Solution:

(a) $$\begin{aligned} f \circ g &= f(g(x)) \\ &= f(x^2) \\ &= \sqrt{x^2 + 4} \end{aligned}$$

(b) $$\begin{aligned} g \circ f &= g(f(x)) \\ &= g(\sqrt{x + 4}) \\ &= (\sqrt{x + 4})^2 \\ &= x + 4 \end{aligned}$$

29. Find (a) $f \circ g$ and (b) $g \circ f$ for $f(x) = \sqrt{x}$ and $g(x) = \sqrt{x}$.

Solution:

(a) $f \circ g = f(g(x)) = f(\sqrt{x}) = \sqrt{\sqrt{x}} = \sqrt[4]{x}$

(b) Same as (a)

31. Find (a) $f \circ g$ and (b) $g \circ f$ for $f(x) = |x|$ and $g(x) = x + 6$.

Solution:

(a) $f \circ g = f(g(x)) = f(x + 6) = |x + 6|$

(b) $g \circ f = g(f(x)) = g(|x|) = |x| + 6$

37. Find functions f and g such that $(f \circ g)(x) = h(x)$ for $h(x) = (2x+1)^2$.

Solution:
Let $f(x) = x^2$ and $g(x) = 2x + 1$, then $(f \circ g)(x) = h(x)$.

Note: This is not a unique solution. For example, if $f(x) = (x+1)^2$ and $g(x) = 2x$, then $(f \circ g)(x) = h(x)$ as well.

41. Find functions f and g such that $(f \circ g)(x) = h(x)$ for $h(x) = 1/(x+2)$.

Solution:
Let $f(x) = 1/x$ and $g(x) = x + 2$, then $(f \circ g)(x) = h(x)$. Again, this is not a unique solution. Other possibilities are:

$$f(x) = \frac{1}{x+2} \quad \text{and} \quad g(x) = x$$

OR

$$f(x) = \frac{1}{x+1} \quad \text{and} \quad g(x) = x + 1$$

OR

$$f(x) = \frac{1}{x^2+2} \quad \text{and} \quad g(x) = \sqrt{x}$$

47. Determine the domain of (a) f, (b) g, and (c) $f \circ g$ for $f(x) = 3/(x^2-1)$ and $g(x) = x + 1$.

Solution:
(a) The domain of $f(x) = 3/(x^2-1)$ includes all real numbers except $x = \pm 1$.
(b) The domain of $g(x) = x + 1$ includes all real numbers.
(c) $f \circ g = f(g(x)) = f(x+1) = \dfrac{3}{(x+1)^2 - 1} = \dfrac{3}{x^2+2x} = \dfrac{3}{x(x+2)}$

The domain of $f \circ g$ includes all real numbers except $x = 0$ and $x = -2$.

53. Prove that the product of two odd functions is an even function.

Solution:
Let $f(x)$ and $g(x)$ be two odd functions and define $h(x) = f(x)g(x)$. Then

$$\begin{aligned} h(-x) &= f(-x)g(-x) \\ &= [-f(x)][-g(x)] \qquad \text{Since } f(x) \text{ and } g(x) \text{ are odd} \\ &= f(x)g(x) \\ &= h(x) \end{aligned}$$

Thus, h is even.

SECTION 2.7

Inverse Functions

- Two functions f and g are inverses of each other if $f(g(x)) = x$ for every x in the domain of g and $g(f(x)) = x$ for every x in the domain of f.
- A function f has an inverse if and only if f is one-to-one.
- Be able to find the inverse of a function, if it exists.

Solutions to Selected Exercises

5. (a) Show that $f(x) = x^3$ and $g(x) = \sqrt[3]{x}$ are inverse functions by showing that $f(g(x)) = x$ and $g(f(x)) = x$, and (b) graph f and g on the same set of coordinate axes.

Solution:

$$f(x) = x^3, \quad g(x) = \sqrt[3]{x}$$

(a) $f(g(x)) = f(\sqrt[3]{x}) = (\sqrt[3]{x})^3 = x$

$g(f(x)) = g(x^3) = \sqrt[3]{x^3} = x$

(b)

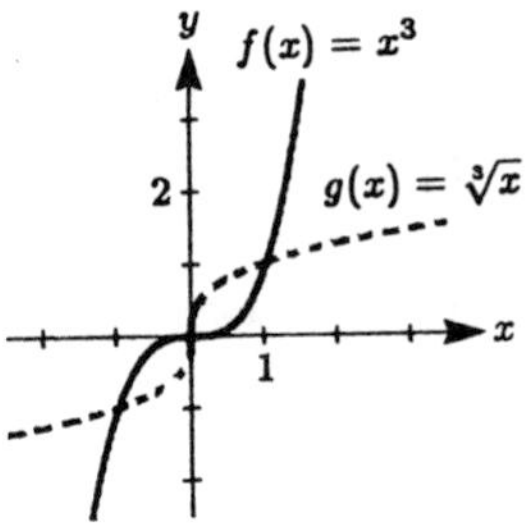

7. (a) Show that $f(x) = \sqrt{x-4}$ and $g(x) = x^2 + 4, \quad x \geq 0$ are inverse functions by showing that $f(g(x)) = x$ and $g(f(x)) = x$, and (b) graph f and g on the same set of coordinate axes.

Solution:

$$f(x) = \sqrt{x-4}, \quad g(x) = x^2 + 4, \quad x \geq 0$$

(a) $f(g(x)) = f(x^2 + 4) = \sqrt{(x^2 + 4) - 4} = \sqrt{x^2} = |x| = x, \quad x \geq 0$

$g(f(x)) = g(\sqrt{x - 4}) = (\sqrt{x - 4})^2 + 4 = (x - 4) + 4 = x$

(b)

$g(x) = x^2 + 4$

$f(x) = \sqrt{x - 4}$

11. Determine whether the function shown is one-to-one.

Solution:

Since the function is decreasing on its entire domain, it is one-to-one.

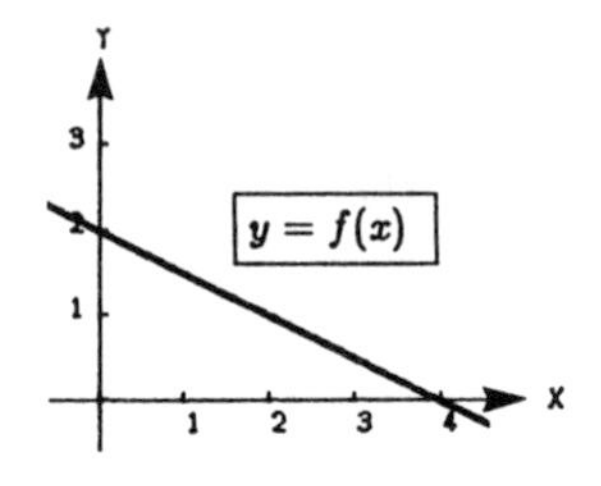

15. Determine whether the function $g(x) = (4 - x)/6$ is one-to-one.

Solution:

Let a and b be real numbers with $g(a) = g(b)$. Then we have

$$\begin{aligned} \frac{4-a}{6} &= \frac{4-b}{6} \\ 4 - a &= 4 - b \\ -a &= -b \\ a &= b \end{aligned}$$

Therefore, $g(x)$ is one-to-one.

19. Determine whether the function $f(x) = -\sqrt{16 - x^2}$ is one-to-one.

Solution:

Since $f(4) = 0$ and $f(-4) = 0$, the function is not one-to-one.

23. Find the inverse of the one-to-one function $f(x) = x^5$. Then graph both f and f^{-1} on the same coordinate plane.

Solution:

$$\begin{aligned} f(x) &= x^5 \\ y &= x^5 \\ x &= \sqrt[5]{y} \\ f^{-1}(x) &= \sqrt[5]{x} \end{aligned}$$

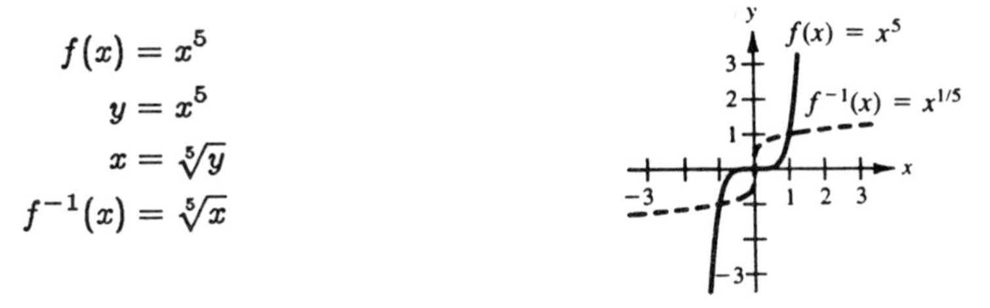

29. Find the inverse of the one-to-one function $f(x) = \sqrt[3]{x-1}$. Then graph both f and f^{-1} on the same coordinate plane.

Solution:

$$\begin{aligned} f(x) &= \sqrt[3]{x-1} \\ y &= \sqrt[3]{x-1} \\ y^3 &= x - 1 \\ x &= y^3 + 1 \\ f^{-1}(x) &= x^3 + 1 \end{aligned}$$

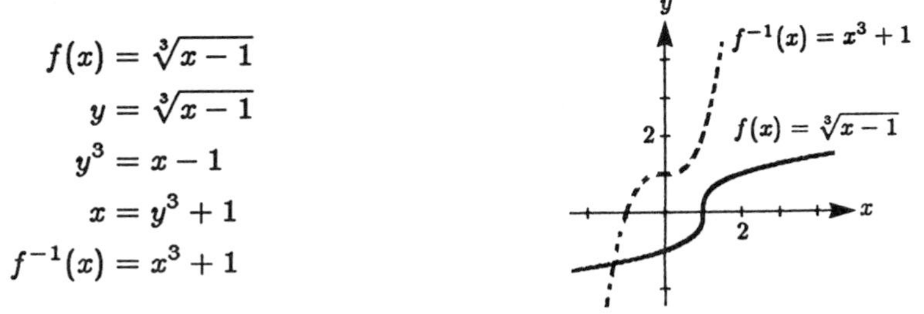

33. Determine whether the function $g(x) = x/8$ is one-to-one. If it is, find its inverse.

Solution:

$$\begin{aligned} g(a) &= g(b) \\ \frac{a}{8} &= \frac{b}{8} \\ a &= b \end{aligned}$$

Therefore, g is one-to-one.

$$\begin{aligned} g(x) &= \frac{x}{8} \\ y &= \frac{x}{8} \\ x &= 8y \\ g^{-1}(x) &= 8x \end{aligned}$$

35. Determine whether the function $p(x) = -4$ is one-to-one. If it is, find its inverse.

Solution:
$p(x) = -4$ for all real numbers x. Therefore, $p(x)$ is not one-to-one.

39. Determine whether the function $h(x) = 1/x$ is one-to-one. If it is, find its inverse.

Solution:

$$h(a) = h(b)$$
$$\frac{1}{a} = \frac{1}{b}$$
$$a = b \qquad \text{Therefore, } h \text{ is one-to-one.}$$
$$h(x) = \frac{1}{x}$$
$$y = \frac{1}{x}$$
$$xy = 1$$
$$x = \frac{1}{y}$$
$$h^{-1}(x) = \frac{1}{x}$$

43. Determine whether the function $g(x) = x^2 - x^4$ is one-to-one. If it is, find its inverse.

Solution:

Since $g(0) = 0$ and $g(1) = 0$, g is not one-to-one.

45. Determine whether the function $f(x) = 25 - x^2$, $x \le 0$ is one-to-one. If it is, find its inverse.

Solution:

$$f(a) = f(b)$$
$$25 - a^2 = 25 - b^2$$
$$a^2 = b^2 \qquad \text{Since } a,\ b \le 0, \text{ we have } a = b \text{ and } f \text{ is one-to-one.}$$
$$f(x) = 25 - x^2, \quad x \le 0$$
$$y = 25 - x^2$$
$$x^2 = 25 - y$$
$$x = -\sqrt{25 - y} \quad \text{Since } x \le 0$$
$$f^{-1}(x) = -\sqrt{25 - x}$$

49. Use the functions $f(x) = (1/8)x - 3$ and $g(x) = x^3$ to find $(f^{-1} \circ g^{-1})(1)$.

Solution:

$$f(x) = (1/8)x - 3 \quad \Longrightarrow \quad f^{-1}(x) = 8(x + 3)$$
$$g(x) = x^3 \quad \Longrightarrow \quad g^{-1}(x) = \sqrt[3]{x}$$

$$(f^{-1} \circ g^{-1})(1) = f^{-1}(g^{-1}(1))$$
$$= f^{-1}(\sqrt[3]{1}) = f^{-1}(1) = 8(1 + 3) = 32$$

53. Use the functions $f(x) = x + 4$ and $g(x) = 2x - 5$ to find $g^{-1} \circ f^{-1}$.

Solution:

$$f(x) = x + 4 \implies f^{-1}(x) = x - 4$$

$$g(x) = 2x - 5 \implies g^{-1}(x) = \frac{x+5}{2}$$

$$\begin{aligned} g^{-1} \circ f^{-1} &= g^{-1}(f^{-1}(x)) \\ &= g^{-1}(x - 4) \\ &= \frac{(x-4)+5}{2} \\ &= \frac{x+1}{2} \end{aligned}$$

57. Prove that if f is a one-to-one odd function, then f^{-1} is an odd function.

Solution:

Suppose f^{-1} is not an odd function. Then there exists a value $x = a$ such that

$$f^{-1}(-a) \neq -f^{-1}(a).$$

Since f is one-to-one, we have $f(f^{-1}(-a)) \neq f(-f^{-1}(a))$, but $f(f^{-1}(-a)) = -a$ since f and f^{-1} are inverses and

$$f(-f^{-1}(a)) = -f(f^{-1}(a)) = -a$$

since f is odd. Therefore,

$$f(f^{-1}(-a)) = f(-f^{-1}(a)),$$

which is a contradiction. Therefore, f^{-1} must be an odd function.

SECTION 2.8

Variation and Mathematical Models

You should know the following terms and formulas for variation.

- Direct Variation

  (a) $y = kx$
  (b) $y = kx^n$ (as nth power)

- Inverse Variation

  (a) $y = k/x$
  (b) $y = k/(x^n)$ (as nth power)

- Joint Variation

  (a) $z = kxy$
  (b) $z = kx^n y^m$ (as nth power of x and mth power of y)

- k is called the constant of proportionality.

Solutions to Selected Exercises

5. Find a mathematical model for the statement "z is proportional to the cube root of u".

Solution:

$$z = k\sqrt[3]{u}$$

9. Find a mathematical model for the statement "F varies directly as g and inversely as the square of r".

Solution:

$$F = \frac{kg}{r^2}$$

13. Find a mathematical model for **Newton's Law of Universal Gravitation:** The gravitational attraction F between two objects of masses m_1 and m_2 is proportional to the product of the masses and inversely proportional to the square of the distance r between the objects.

Solution:

$$F = \frac{km_1m_2}{r^2}$$

17. Find a mathematical model for the statement "A varies directly as the square of r". Determine the constant of proportionality given $A = 9\pi$ when $r = 3$.

Solution:

$$\begin{aligned} A &= kr^2 \\ 9\pi &= k(3)^2 \\ \pi &= k \\ A &= \pi r^2 \end{aligned}$$

21. Find a mathematical model for the statement "h is inversely proportional to the third power of t". Determine the constant of proportionality given $h = 3/16$ when $t = 4$.

Solution:

$$\begin{aligned} h &= \frac{k}{t^3} \\ \frac{3}{16} &= \frac{k}{(4)^3} \\ \frac{3}{16} &= \frac{k}{64} \\ k &= 12 \\ h &= \frac{12}{t^3} \end{aligned}$$

25. Find a mathematical model for the statement "F is jointly proportional to r and the third power of s". Determine the constant of proportionality given $F = 4158$ when $r = 11$ and $s = 3$.

Solution:

$$\begin{aligned} F &= krs^3 \\ 4158 &= k(11)(3)^3 \\ k &= 14 \\ F &= 14rs^3 \end{aligned}$$

29. Find a mathematical model for the statement "S varies directly as L and inversely as $L - S$". Determine the constant of proportionality given $S = 4$ when $L = 6$.

Solution:

$$S = \frac{kL}{L - S}$$

$$4 = \frac{k(6)}{6 - 4}$$

$$4 = 3k$$

$$k = \frac{4}{3}$$

$$S = \frac{4/3L}{L - S} = \frac{4L}{3(L - S)}$$

33. The coiled spring of a toy supports the weight of a child. The spring compresses a distance of 1.9 inches under the weight of a 25-pound child. The toy will not work properly if its spring is compressed more than 3 inches. What is the weight of the heaviest child who should be allowed to use the toy?

Solution:

From Example 1, we have

$$d = kF$$

$$1.9 = k(25) \implies k = 0.076$$

$$d = 0.076F$$

When the distance compressed is 3 inches, we have

$$3 = 0.076F$$

$$F \approx 39.4737$$

No child over 39 pounds should use the toy.

35. A stream with a velocity of 1/4 mile per hour can move coarse sand particles of about 0.02 inch diameter. What must the velocity be to carry particles with a diameter of 0.12 inch? Use the fact that the diameter of a particle moved by a stream varies approximately as the square of the velocity of the stream.

Solution:

$$d = kv^2$$

$$0.02 = k(1/4)^2$$

$$k = 0.32$$

$$d = 0.32v^2$$

$$0.12 = 0.32v^2$$

$$v^2 = 0.12/0.32 = 3/8$$

$$v = \sqrt{3}/(2\sqrt{2}) = \sqrt{6}/4 \approx 0.612 \text{ mi/hr}$$

39. The illumination from a light source varies inversely as the square of the distance from the light source. When the distance from a light source is doubled, how does the illumination change?

Solution:

$$l = \frac{k}{d^2}$$

When the distance is doubled:

$$l = \frac{k}{(2d)^2}$$

$$l = \frac{1}{4}\left(\frac{k}{d^2}\right)$$

The amount of illumination is 1/4 as bright.

41. The resistance of a wire carrying electrical current is directly proportional to its length and inversely proportional to its cross-sectional area. #28 copper wire (which has a diameter of 0.0126 inch) has a resistance of 66.17 ohms per thousand feet. A 14-foot piece of copper wire produces a resistance of 0.05 ohms. Find the diameter of the wire.

Solution:

$$r = \frac{kl}{A}, \qquad A = \pi r^2 = \frac{\pi d^2}{4}$$

$$r = \frac{4kl}{\pi d^2}$$

$$66.17 = \frac{4(1000)k}{\pi(0.0126/12)^2}$$

$$k = 5.7 \times 10^{-8}$$

$$r = \frac{4(5.7 \times 10^{-8})l}{\pi d^2}$$

$$r = 7.3 \times 10^{-8}\frac{l}{d^2} \qquad \text{From Exercise 40}$$

$$0.05 = (7.3 \times 10^{-8})\frac{14}{d^2}$$

$$d^2 = (7.3 \times 10^{-8})\frac{14}{0.05}$$

$$d^2 = 2.044 \times 10^{-5}$$

$$d = 0.00452 \text{ ft}$$

$$d = 0.05425 \text{ in}$$

REVIEW EXERCISES FOR CHAPTER 2

Solutions to Selected Exercises

3. For the points (2, 1) and (14, 6), find (a) the distance between the two points, (b) the coordinates of the midpoint of the line segment between the two points, (c) an equation of the line through the two points, and (d) an equation of the circle whose diameter is the line segment between the two points.

Solution:

(a)
$$\begin{aligned} d &= \sqrt{(14-2)^2 + (6-1)^2} \\ &= \sqrt{144+25} \\ &= \sqrt{169} \\ &= 13 \end{aligned}$$

(b)
$$\begin{aligned} m &= \left(\frac{2+14}{2}, \frac{1+6}{2}\right) \\ &= \left(8, \frac{7}{2}\right) \end{aligned}$$

(c)
$$\begin{aligned} y - 1 &= \frac{6-1}{14-2}(x-2) \\ &= \frac{5}{12}(x-2) \\ 12y - 12 &= 5x - 10 \\ 5x - 12y + 2 &= 0 \end{aligned}$$

(d) The length of the diameter is 13, so the length of the radius is $\frac{13}{2}$. The midpoint of the line segment is the center of the circle. Center: $(8, \frac{7}{2})$ Radius: $\frac{13}{2}$

$$\begin{aligned} (x-8)^2 + \left(y - \frac{7}{2}\right)^2 &= \left(\frac{13}{2}\right)^2 \\ x^2 - 16x + 64 + y^2 - 7y + \frac{49}{4} &= \frac{169}{4} \\ x^2 + y^2 - 16x - 7y + 34 &= 0 \end{aligned}$$

7. Find t so that the points $(-2, 5)$, $(0, t)$ and $(1, 1)$ are collinear.

Solution:

The line through $(-2, 5)$ and $(1, 1)$ is

$$\begin{aligned} y - 5 &= \frac{1-5}{1+2}(x+2) \\ y - 5 &= -\frac{4}{3}(x+2) \\ 3y - 15 &= -4x - 8 \\ 4x + 3y &= 7 \end{aligned}$$

For $(0, t)$ to be on this line also, it must satisfy the equation $4x + 3y = 7$.

$$4(0) + 3(t) = 7$$

Thus, $t = \frac{7}{3}$.

11. Show that the points $(1, 1)$, $(8, 2)$, $(9, 5)$, and $(2, 4)$ form the vertices of a parallelogram.

Solution:

$$d_1 = \sqrt{(2-1)^2 + (4-1)^2} = \sqrt{10}$$
$$d_2 = \sqrt{(9-2)^2 + (5-4)^2} = \sqrt{50} = 5\sqrt{2}$$
$$d_3 = \sqrt{(8-9)^2 + (2-5)^2} = \sqrt{10}$$
$$d_4 = \sqrt{(1-8)^2 + (1-2)^2} = \sqrt{50} = 5\sqrt{2}$$

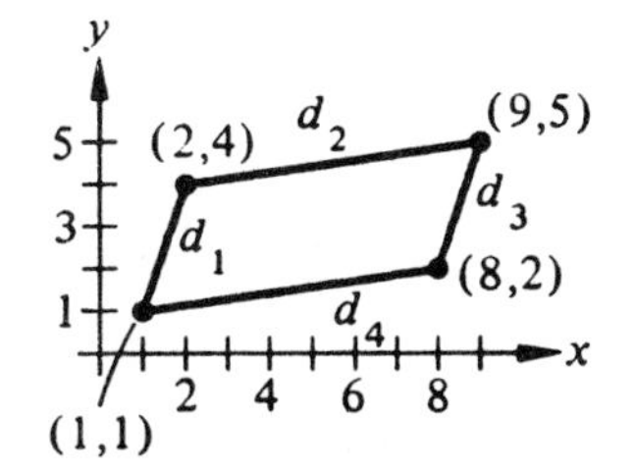

Since $d_1 = d_3$ and $d_2 = d_4$, these points are the vertices of a parallelogram.

15. Find the intercepts of the graph of $2y^2 = x^3$ and check for symmetry with respect to each of the coordinate axes and the origin.

Solution:
The only intercept is the origin, $(0, 0)$.
The graph is symmetric with respect to the x-axis since $2(-y)^2 = x^3$ results in the original equation. Replacing x with $-x$ or replacing both x and y with $-x$ and $-y$ does not yield equivalent equations. Thus, the graph is not symmetric with respect to either the y-axis or the origin.

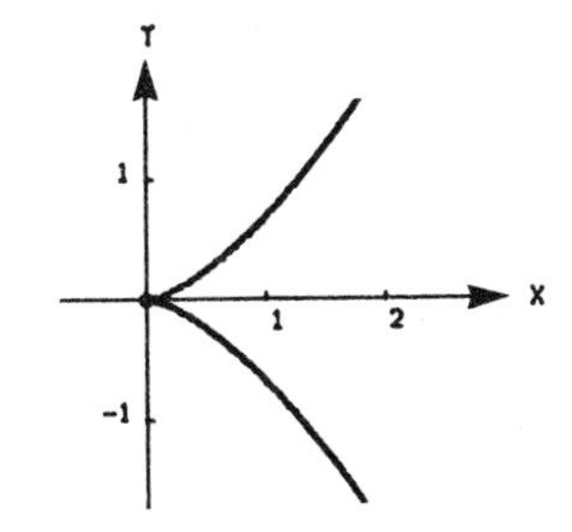

19. Find the intercepts of the graph of $y = x\sqrt{4 - x^2}$ and check for symmetry with respect to each of the coordinate axes and the origin.

Solution:
Let $y = 0$, then $0 = x\sqrt{4 - x^2}$ and $x = 0, \pm 2$.
x-intercepts: $(0, 0)$, $(2, 0)$, $(-2, 0)$
Let $x = 0$, then $y = 0\sqrt{4 - 0^2}$ and $y = 0$.
y-intercept: $(0, 0)$
The graph is symmetric with respect to the origin since

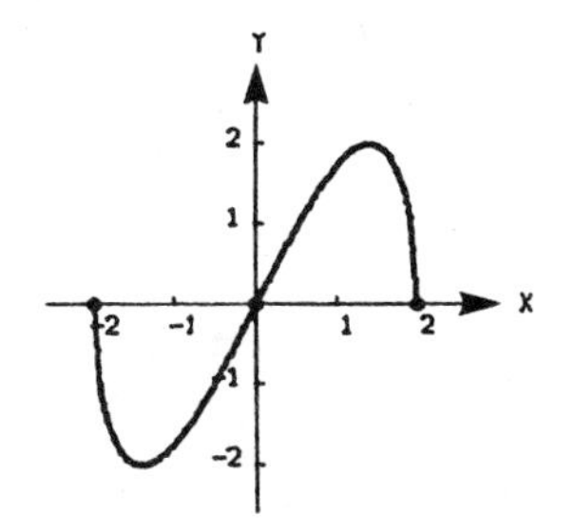

$$-y = -x\sqrt{4 - (-x)^2}$$
$$-y = -x\sqrt{4 - x^2}$$
$$y = x\sqrt{4 - x^2}$$

The graph is not symmetric with respect to either axis.

27. Determine the center and radius of the circle. Then, sketch the graph of $4x^2 + 4y^2 - 4x - 40y + 92 = 0$.

Solution:

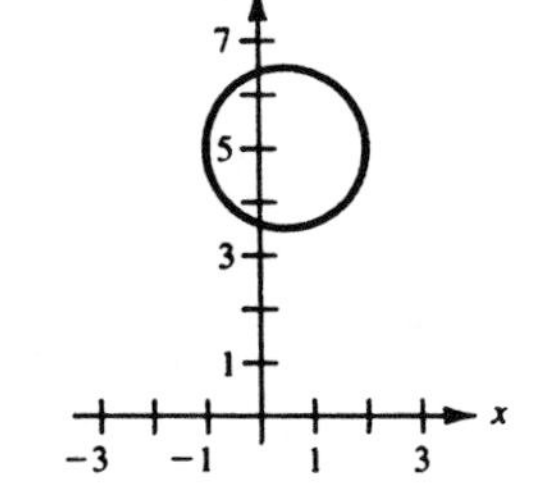

$$4x^2 + 4y^2 - 4x - 40y + 92 = 0$$
$$x^2 + y^2 - x - 10y + 23 = 0$$
$$\left(x^2 - x + \tfrac{1}{4}\right) + (y^2 - 10y + 25) = -23 + \tfrac{1}{4} + 25$$
$$\left(x - \tfrac{1}{2}\right)^2 + (y - 5)^2 = \tfrac{9}{4}$$

Center: $(\frac{1}{2}, 5)$ Radius: $\frac{3}{2}$

29. Sketch a graph of the equation $y - 2x - 3 = 0$.

Solution:
$y - 2x - 3 = 0$
x-intercept: $(-\frac{3}{2}, 0)$
y-intercept: $(0, 3)$

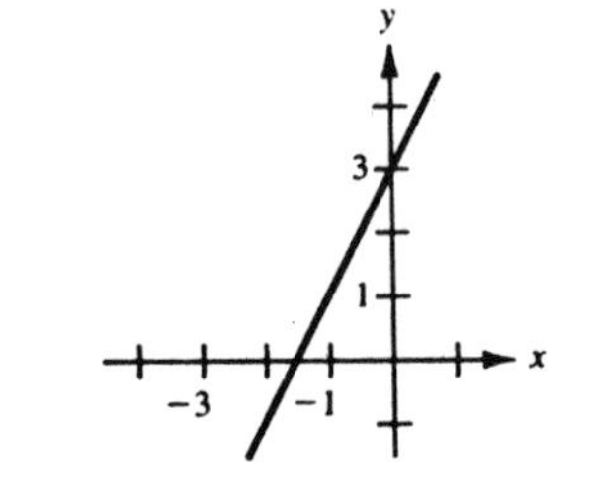

33. Sketch a graph of the equation $y = \sqrt{5 - x}$.

Solution:
$y = \sqrt{5 - x}$
x-intercept: $(5, 0)$
Domain: $(-\infty, 5]$
Range: $[0, \infty)$

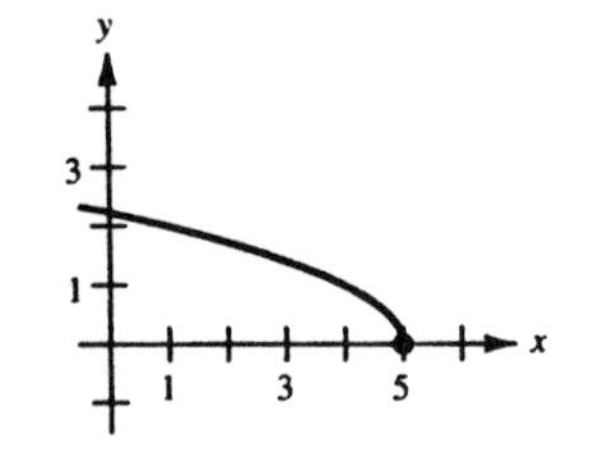

37. Sketch a graph of the equation $y = \sqrt{25 - x^2}$.

Solution:
$y = \sqrt{25 - x^2}$
x-intercepts: $(5, 0)$, $(-5, 0)$
y-intercept: $(0, 5)$
y-axis symmetry
Domain: $[-5, 5]$
Range: $[0, 5]$

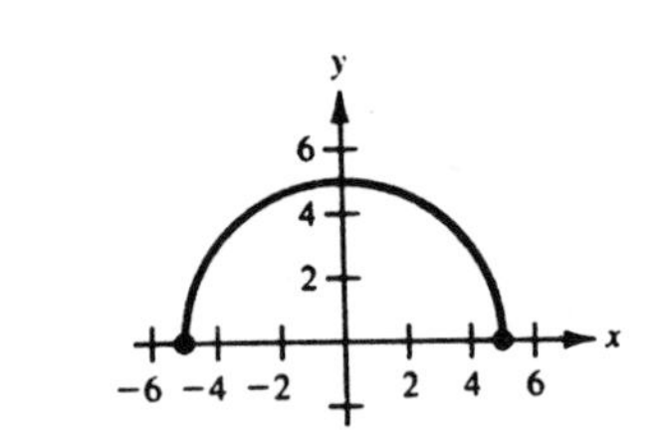

41. Sketch a graph of the equation $y = \frac{1}{4}(x+1)^3$.

Solution:

$y = \frac{1}{4}(x+1)^3$

x-intercept: $(-1,\ 0)$

y-intercept: $(0,\ \frac{1}{4})$

Domain: $(-\infty,\ \infty)$

Range: $(-\infty,\ \infty)$

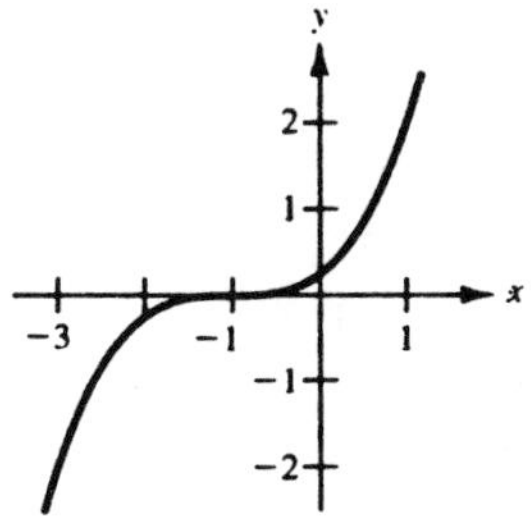

45. Evaluate the function $h(x) = 6 - 5x^2$ at the specified values of the independent variable and simplify your answers.

(a) $h(2)$

(b) $h(x+3)$

(c) $\dfrac{h(4) - h(2)}{4 - 2}$

(d) $\dfrac{h(x + \Delta x) - h(x)}{\Delta x}$

Solution:

(a) $h(2) = 6 - 5(2)^2 = -14$

(b)
$$\begin{aligned} h(x+3) &= 6 - 5(x+3)^2 \\ &= 6 - 5(x^2 + 6x + 9) \\ &= -5x^2 - 30x - 39 \end{aligned}$$

(c)
$$\begin{aligned} h(4) &= 6 - 5(4)^2 = -74 \\ h(2) &= -14 \quad \text{from part (a)} \\ \frac{h(4) - h(2)}{4 - 2} &= \frac{-74 - (-14)}{4 - 2} = \frac{-60}{2} = -30 \end{aligned}$$

(d)
$$\begin{aligned} h(x + \Delta x) &= 6 - 5(x + \Delta x)^2 \\ &= 6 - 5(x^2 + 2x\Delta x + (\Delta x)^2) \\ &= 6 - 5x^2 - 10x\Delta x - 5(\Delta x)^2 \\ h(x + \Delta x) - h(x) &= -10x\Delta x - 5(\Delta x)^2 \\ \frac{h(x + \Delta x) - h(x)}{\Delta x} &= \frac{-10x\Delta x - 5(\Delta x)^2}{\Delta x} = -10x - 5\Delta x \end{aligned}$$

49. Determine the domain of the function

$$g(s) = \frac{5}{3s - 9}.$$

Solution:

The domain of $g(s) = 5/(3s - 9)$ includes all real numbers except $s = 3$, since this value would yield a zero in the denominator.

55. For $f(x) = \sqrt{x+1}$ (a) find f^{-1}, (b) sketch the graphs of f and f^{-1} on the same coordinate plane, and (c) verify that $f^{-1}(f(x)) = x = f(f^{-1}(x))$.

Solution:

(a)

$$\begin{aligned} y &= \sqrt{x+1}, \quad y \geq 0 \\ y^2 &= x+1 \\ x &= y^2 - 1 \\ f^{-1}(x) &= x^2 - 1, \quad x \geq 0 \end{aligned}$$

(b)

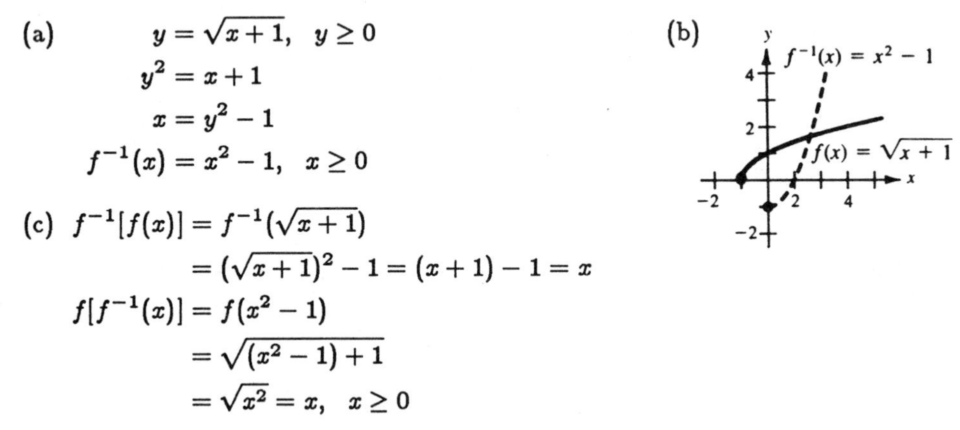

(c)

$$\begin{aligned} f^{-1}[f(x)] &= f^{-1}(\sqrt{x+1}) \\ &= (\sqrt{x+1})^2 - 1 = (x+1) - 1 = x \\ f[f^{-1}(x)] &= f(x^2 - 1) \\ &= \sqrt{(x^2-1)+1} \\ &= \sqrt{x^2} = x, \quad x \geq 0 \end{aligned}$$

59. Restrict the domain of the function $f(x) = 2(x-4)^2$ to an interval where the function is increasing and determine f^{-1} over that interval.

Solution:

$f(x) = 2(x-4)^2$ is increasing on the interval $[4, \infty)$. It is decreasing on the interval $(-\infty, 4)$.

$$\begin{aligned} f(x) &= 2(x-4)^2, \; x \geq 4 \\ y &= 2(x-4)^2, \; x \geq 4 \\ \sqrt{y} &= \sqrt{2}(x-4) \\ \sqrt{y/2} &= x - 4 \\ x &= \sqrt{y/2} + 4 \\ f^{-1}(x) &= \sqrt{x/2} + 4 \end{aligned}$$

63. Let $f(x) = 3 - 2x$, $g(x) = \sqrt{x}$, and $h(x) = 3x^2 + 2$. Find $(f-g)(4)$.

Solution:

$$\begin{aligned} (f-g)(4) &= f(4) - g(4) \\ &= [3 - 2(4)] - \sqrt{4} \\ &= -7 \end{aligned}$$

67. Let $f(x) = 3 - 2x$, $g(x) = \sqrt{x}$, and $h(x) = 3x^2 + 2$. Find $(h \circ g)(7)$.

Solution:

$$\begin{aligned} (h \circ g)(7) &= h(g(7)) \\ &= h(\sqrt{7}) \\ &= 3(\sqrt{7})^2 + 2 \\ &= 23 \end{aligned}$$

73. Find a mathematical model representing the statement "z varies directly as the square of x and inversely as y". Determine the constant of proportionality if $z = 16$ when $x = 5$ and $y = 2$.

Solution:

$$z = \frac{kx^2}{y}$$

$$16 = \frac{k(5)^2}{2}$$

$$32 = 25k$$

$$k = \frac{32}{25}, \quad \text{therefore, } z = \frac{32x^2}{25y}.$$

77. A wire 24 inches long is to be cut into four pieces to form a rectangle whose shortest side has a length of x. Express the area A of the rectangle as a function of x. Determine the domain of the function and sketch its graph over that domain.

Solution:

Let y be the longer side of the rectangle. Then we have $A = xy$. Since the perimeter is 24 inches, we have $2x + 2y = 24$ or $y = (24 - 2x)/2 = 12 - x$. The area equation now becomes: $A = xy = x(12 - x)$. To find the domain of A, we realize that area is a nonnegative quantity. Thus, $x(12 - x) \geq 0$. This gives us the interval $[0, 12]$. We also have the further restriction that x is the shortest side. This occurs on the interval $[0, 6]$.

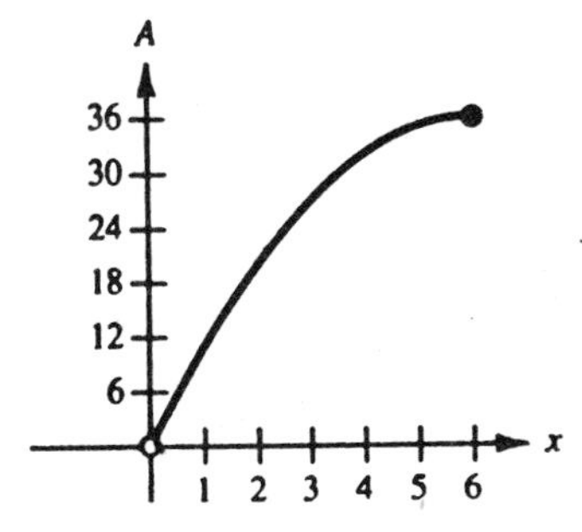

Practice Test for Chapter 2

1. Find the distance between (4, −1) and (0, 3).

2. Find the midpoint of the line segment joining (4, −1) and (0, 3).

3. Find x so that the distance from the origin to $(x, -2)$ is 6.

4. Given $y = \dfrac{x-2}{x+3}$, find the intercepts.

5. Given $xy^2 = 6$, list all symmetries.

6. Graph $y = x^3 - 4x$.

7. Find the center and radius of the circle $x^2 + y^2 - 6x + 2y + 6 = 0$.

8. Given $f(x) = x^2 - 2x + 1$, find $f(x-3)$.

9. Given $f(x) = 4x - 11$, find $\dfrac{f(x) - f(3)}{x-3}$.

10. Find the domain and range of $f(x) = \sqrt{36 - x^2}$.

11. Which equations determine y as a function of x?
(a) $6x - 5y + 4 = 0$
(b) $x^2 + y^2 = 9$
(c) $y^3 = x^2 + 6$

12. Sketch the graph of $f(x) = x^2 - 5$.

13. Sketch the graph of $f(x) = |x + 3|$.

14. Sketch the graph of $f(x) = \begin{cases} 2x+1 & \text{if } x \geq 0, \\ x^2 - x & \text{if } x < 0. \end{cases}$

15. Find the equation of the line through (2, 4) and (3, −1).

16. Find the equation of the line with slope $m = 4/3$ and y-intercept $b = -3$.

17. Find the equation of the line through (4, 1) and perpendicular to the line $2x + 3y = 0$.

18. If it costs a company $32 to produce 5 units of a product and $44 to produce 9 units, how much does it cost to produce 20 units? (Assume that the cost function is linear.)

19. Given $f(x) = x^2 - 2x + 16$ and $g(x) = 2x + 3$, find $f(g(x))$.

20. Given $f(x) = x^3 + 7$, find $f^{-1}(x)$.

21. Which of the following functions are one-to-one?
(a) $f(x) = |x - 6|$
(b) $f(x) = ax + b,\ a \neq 0$
(c) $f(x) = x^3 - 19$

22. Given $f(x) = \sqrt{\dfrac{3-x}{x}},\ 0 < x \leq 3$, find $f^{-1}(x)$.

23. Find the equation: y varies directly as x and $y = 30$ when $x = 5$.

24. Find the equation: y varies inversely as x and $y = 0.5$ when $x = 14$.

25. z varies directly as the square of x and inversely as y, and $z = 3$ when $x = 3$ and $y = -6$. Find the equation relating z to x and y.

CHAPTER 3

Polynomial Functions: Graphs and Zeros

SECTION 3.1

Quadratic Functions

You should know the following facts about parabolas.

- $f(x) = ax^2 + bx + c$, $a \neq 0$, is a quadratic function, and its graph is a parabola.
- If $a > 0$, the parabola opens upward. If $a < 0$, the parabola opens downward.
- The vertex is $(-b/2a,\ f(-b/2a))$.
- To find the x-intercepts (if any), solve

  $$ax^2 + bx + c = 0$$

- The standard form of the equation of a parabola is

  $$f(x) = a(x-h)^2 + k$$

  where $a \neq 0$.
  (a) The vertex is $(h,\ k)$.
  (b) The axis is the vertical line $x = h$.

Solutions to Selected Exercises

7. Find an equation for the given parabola.

Solution:
The vertex is (2, 0) and the parabola passes through the point (0, 4).

$$f(x) = a(x-2)^2 + 0$$
$$f(x) = a(x-2)^2$$
$$4 = a(0-2)^2 \implies a = 1$$
$$f(x) = (x-2)^2$$

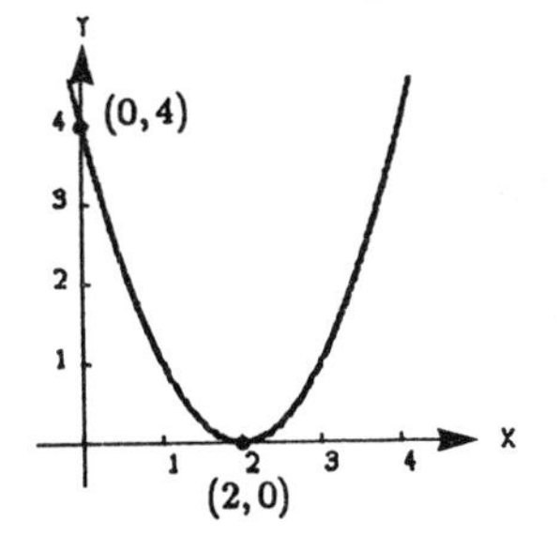

11. Find an equation for the given parabola.

Solution:

The vertex is $(-3,\ 3)$ and the parabola passes through $(-2,\ 1)$.

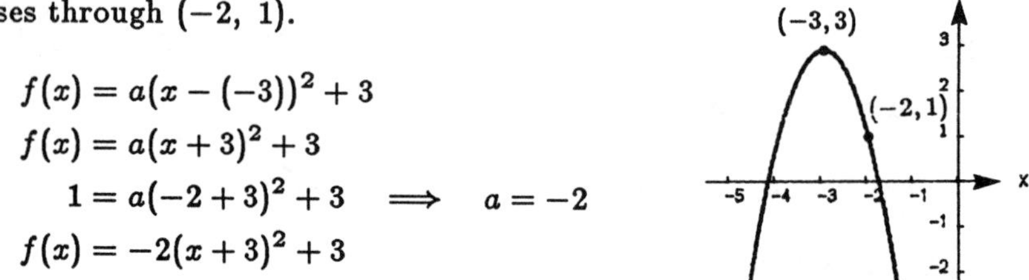

$$f(x) = a(x - (-3))^2 + 3$$
$$f(x) = a(x + 3)^2 + 3$$
$$1 = a(-2 + 3)^2 + 3 \quad \Longrightarrow \quad a = -2$$
$$f(x) = -2(x + 3)^2 + 3$$

17. Sketch the graph of $f(x) = (x + 5)^2 - 6$. Identify the vertex and x- and y-intercepts.

Solution:

Vertex: $(-5,\ -6)$

x-intercepts:

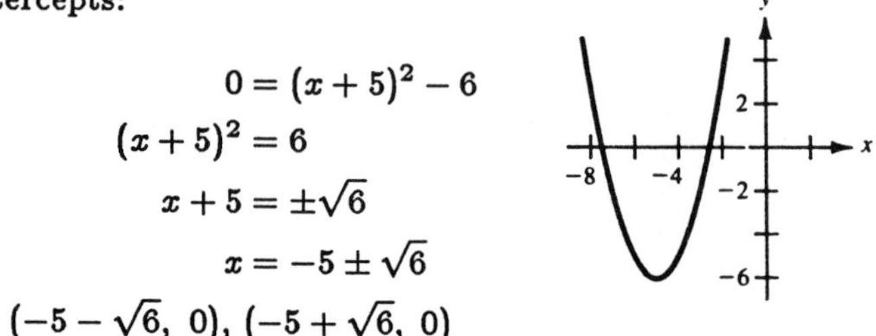

$$0 = (x + 5)^2 - 6$$
$$(x + 5)^2 = 6$$
$$x + 5 = \pm\sqrt{6}$$
$$x = -5 \pm \sqrt{6}$$
$$(-5 - \sqrt{6},\ 0),\ (-5 + \sqrt{6},\ 0)$$

y-intercept: $(0,\ 19)$

19. Sketch the graph of $h(x) = x^2 - 8x + 16$. Identify the vertex and x- and y-intercepts.

Solution:

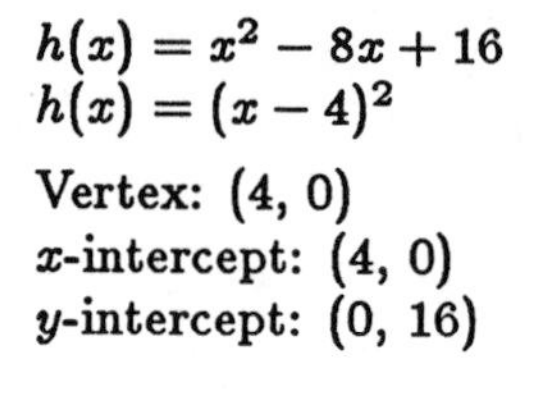

$h(x) = x^2 - 8x + 16$
$h(x) = (x - 4)^2$

Vertex: $(4,\ 0)$
x-intercept: $(4,\ 0)$
y-intercept: $(0,\ 16)$

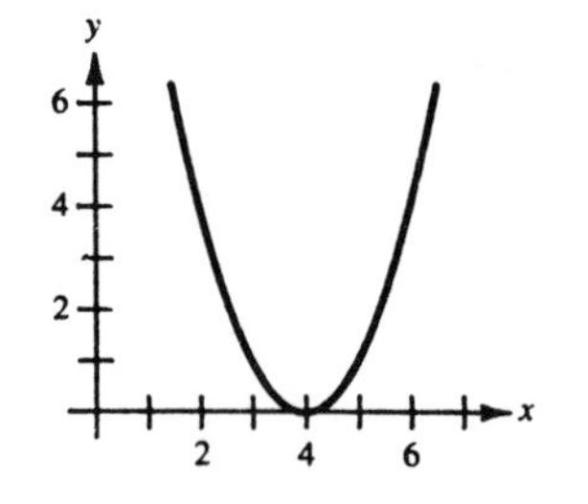

23. Sketch the graph of $f(x) = x^2 - x + \frac{5}{4}$. Identify the vertex and x- and y-intercepts.

Solution:

$f(x) = x^2 - x + \frac{5}{4}$

$f(x) = x^2 - x + \frac{1}{4} - \frac{1}{4} + \frac{5}{4}$

$f(x) = (x - \frac{1}{2})^2 + 1$

Vertex: $(\frac{1}{2}, 1)$

x-intercept: None since

$$0 = (x - \tfrac{1}{2})^2 + 1$$
$$-1 = (x - \tfrac{1}{2})^2$$
$$\pm\sqrt{-1} = x - \tfrac{1}{2} \quad \text{has no real solutions.}$$

y-intercept: $(0, \frac{5}{4})$

27. Sketch the graph of $h(x) = 4x^2 - 4x + 21$. Identify the vertex and x- and y-intercepts.

Solution:

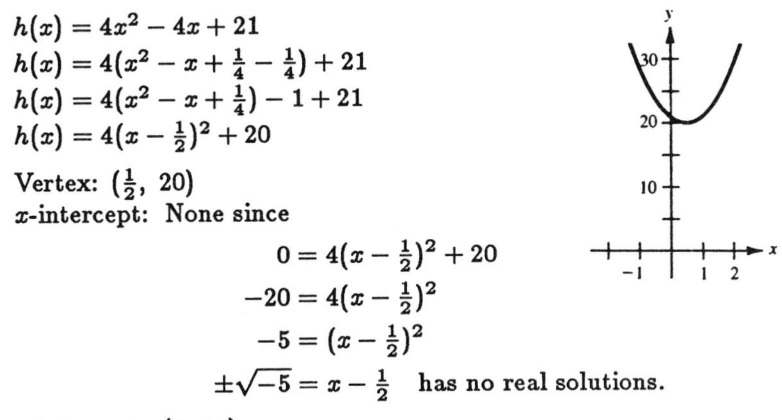

$h(x) = 4x^2 - 4x + 21$

$h(x) = 4(x^2 - x + \frac{1}{4} - \frac{1}{4}) + 21$

$h(x) = 4(x^2 - x + \frac{1}{4}) - 1 + 21$

$h(x) = 4(x - \frac{1}{2})^2 + 20$

Vertex: $(\frac{1}{2}, 20)$

x-intercept: None since

$$0 = 4(x - \tfrac{1}{2})^2 + 20$$
$$-20 = 4(x - \tfrac{1}{2})^2$$
$$-5 = (x - \tfrac{1}{2})^2$$
$$\pm\sqrt{-5} = x - \tfrac{1}{2} \quad \text{has no real solutions.}$$

y-intercept: $(0, 21)$

33. Find the quadratic function with a vertex of $(5, 12)$ and whose graph passes through the point $(7, 15)$.

Solution:

$(5, 12)$ is the vertex.

$$f(x) = a(x - 5)^2 + 12$$

Since the graph passes through the point $(7, 15)$, we have

$$15 = a(7 - 5)^2 + 12$$
$$3 = 4a \implies a = \tfrac{3}{4}$$
$$f(x) = \tfrac{3}{4}(x - 5)^2 + 12.$$

37. Find two quadratic functions whose graphs have the x-intercepts $(0, 0)$ and $(10, 0)$. (One function has a graph that opens upward and the other has a graph that opens downward.)

Solution:

$$\begin{aligned} f(x) &= (x-0)(x-10) && \text{opens upward} \\ &= x^2 - 10x \\ g(x) &= -(x-0)(x-10) && \text{opens downward} \\ &= -x^2 + 10x \end{aligned}$$

43. Find two positive real numbers satisfying the requirements "the sum of the first and twice the second is 24 and the product is a maximum".

Solution:

Let $x =$ the first number and $y =$ the second number. Then

$$x + 2y = 24 \quad \Longrightarrow \quad y = \frac{24 - x}{2}.$$

The product is $P(x) = xy = x\left(\frac{24-x}{2}\right)$.

$$\begin{aligned} P(x) &= \frac{1}{2}(-x^2 + 24x) \\ &= -\frac{1}{2}(x^2 - 24x + 144 - 144) \\ &= -\frac{1}{2}[(x-12)^2 - 144] \\ &= -\frac{1}{2}(x-12)^2 + 72 \end{aligned}$$

The maximum value of the product occurs at the vertex of $P(x)$ and is 72. This happens when $x = 12$ and $y = (24 - 12)/2 = 6$. Thus, the numbers are 12 and 6.

47. A rancher has 200 feet of fencing to enclose two adjacent rectangular corrals, as shown in the figure. What dimensions will produce a maximum enclosed area?

Solution:

Since the rancher has 200 feet of fencing, we have the equation $4x+3y = 200$ or $y = (200-4x)/3$. The area is

$$A = 2xy = 2x\left(\frac{200 - 4x}{3}\right).$$

$$\begin{aligned} A &= \frac{2}{3}(-4x^2 + 200x) \\ &= -\frac{8}{3}(x^2 - 50x) \\ &= -\frac{8}{3}(x^2 - 50x + 625 - 625) \\ &= -\frac{8}{3}[(x - 25)^2 - 625] \\ &= -\frac{8}{3}(x - 25)^2 + \frac{5000}{3} \end{aligned}$$

The maximum area occurs at the vertex and is 5000/3 square feet. This happens when $x = 25$ feet and $y = (200 - 4(25))/3 = 100/3$ feet. The dimensions are $2x = 50$ feet by $33\frac{1}{3}$ feet.

51. Let x be the amount (in hundreds of dollars) a company spends on advertising, and let P be the profit, where $P = 230 + 20x - 0.5x^2$. What expenditure for advertising gives a maximum profit?

Solution:

$$\begin{aligned} P &= 230 + 20x - 0.5x^2 \\ &= -0.5(x^2 - 40x - 460) \\ &= -0.5(x^2 - 40x + 400 - 400 - 460) \\ &= -0.5[(x - 20)^2 - 860] \\ &= -0.5(x - 20)^2 + 430 \end{aligned}$$

The profit is maximized when $x = 20$, or when \$2000 is spent on advertising.

55. Assume that the function $f(x) = ax^2 + bx + c \quad (a \neq 0)$ has two real zeros. Show that the x-coordinate of the vertex of the graph is the average of the zeros of f. [*Hint:* Use the Quadratic Formula.]

Solution:

If $f(x) = ax^2 + bx + c$ has two real zeros, then by the Quadratic Formula they are

$$x = \frac{-b \pm \sqrt{b^2 - 4ac}}{2a}.$$

The average of the zeros of f is

$$\frac{\dfrac{-b - \sqrt{b^2 - 4ac}}{2a} + \dfrac{-b + \sqrt{b^2 - 4ac}}{2a}}{2} = \frac{\dfrac{-2b}{2a}}{2} = -\frac{b}{2a}.$$

This is the x-coordinate of the vertex of the graph.

SECTION 3.2

Polynomial Functions of Higher Degree

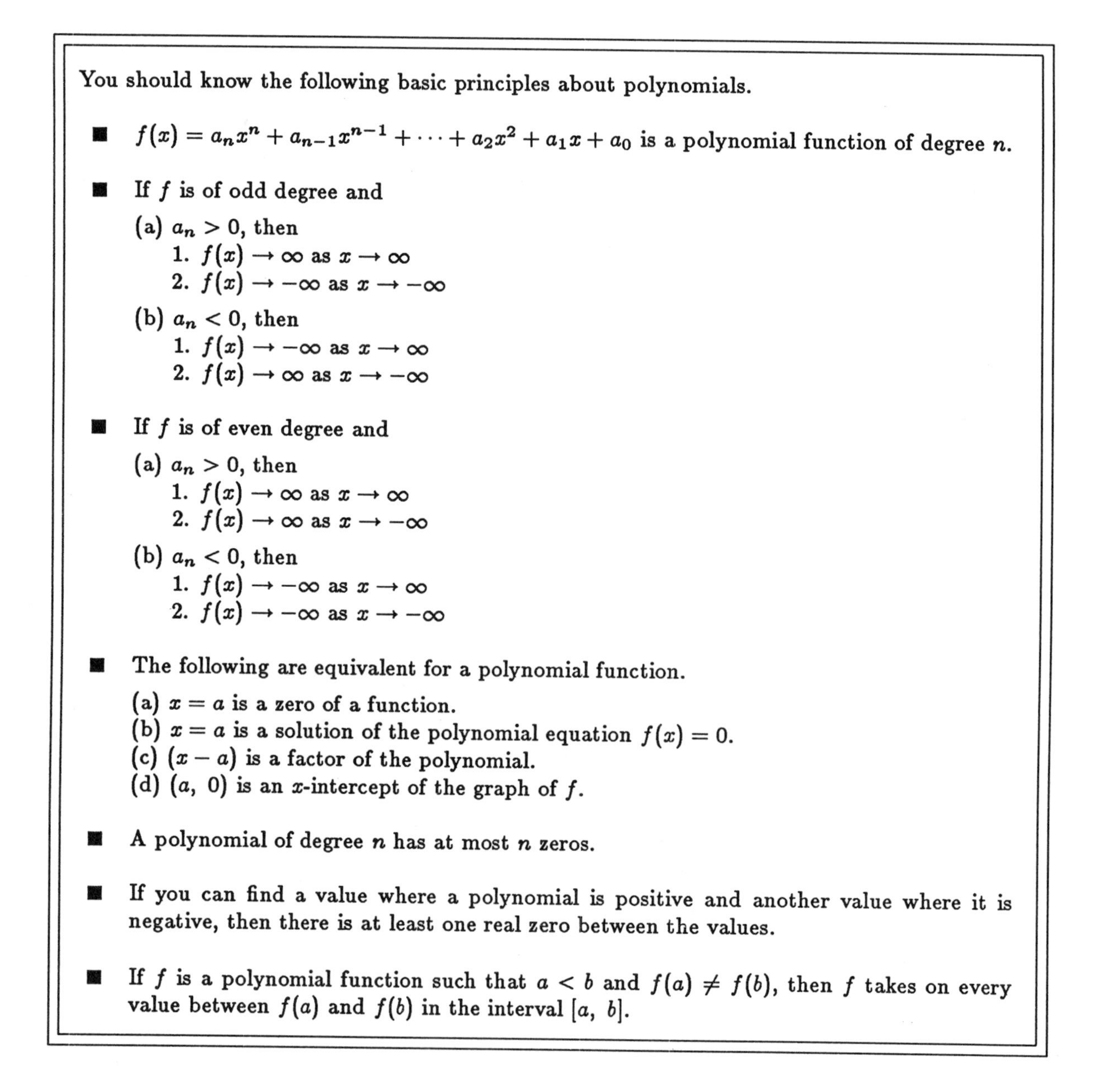

You should know the following basic principles about polynomials.

- $f(x) = a_n x^n + a_{n-1}x^{n-1} + \cdots + a_2x^2 + a_1x + a_0$ is a polynomial function of degree n.
- If f is of odd degree and
  (a) $a_n > 0$, then
     1. $f(x) \to \infty$ as $x \to \infty$
     2. $f(x) \to -\infty$ as $x \to -\infty$

  (b) $a_n < 0$, then
     1. $f(x) \to -\infty$ as $x \to \infty$
     2. $f(x) \to \infty$ as $x \to -\infty$
- If f is of even degree and
  (a) $a_n > 0$, then
     1. $f(x) \to \infty$ as $x \to \infty$
     2. $f(x) \to \infty$ as $x \to -\infty$

  (b) $a_n < 0$, then
     1. $f(x) \to -\infty$ as $x \to \infty$
     2. $f(x) \to -\infty$ as $x \to -\infty$
- The following are equivalent for a polynomial function.
  (a) $x = a$ is a zero of a function.
  (b) $x = a$ is a solution of the polynomial equation $f(x) = 0$.
  (c) $(x - a)$ is a factor of the polynomial.
  (d) $(a, 0)$ is an x-intercept of the graph of f.
- A polynomial of degree n has at most n zeros.
- If you can find a value where a polynomial is positive and another value where it is negative, then there is at least one real zero between the values.
- If f is a polynomial function such that $a < b$ and $f(a) \neq f(b)$, then f takes on every value between $f(a)$ and $f(b)$ in the interval $[a, b]$.

Solutions to Selected Exercises

7. Match the polynomial function $f(x) = 3x^4 + 4x^3$ with the correct graph.

Solution:

$$f(x) = 3x^4 + 4x^3 = x^3(3x + 4)$$

Zeros: $0, \ -\frac{4}{3}$
$f(x) \to \infty$ as $x \to \infty$
$f(x) \to \infty$ as $x \to -\infty$
Matches (d)

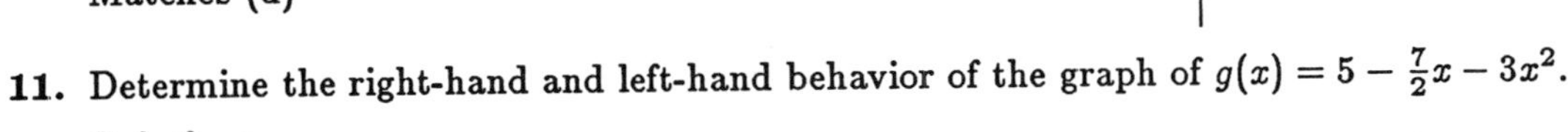

11. Determine the right-hand and left-hand behavior of the graph of $g(x) = 5 - \frac{7}{2}x - 3x^2$.

Solution:

$$g(x) = 5 - \tfrac{7}{2}x - 3x^2$$

Even degree with leading coefficient of -3
Left: $g(x) \to -\infty$ as $x \to -\infty$, so the graph moves down to the left.
Right: $g(x) \to -\infty$ as $x \to +\infty$, so the graph moves down to the right.

15. Determine the right-hand and left-hand behavior of the graph of $f(x) = 6 - 2x + 4x^2 - 5x^3$.

Solution:

$$f(x) = 6 - 2x + 4x^2 - 5x^3$$

Odd degree with a negative leading coefficient of -5
Left: $f(x) \to \infty$ as $x \to -\infty$, so the graph moves up to the left.
Right: $f(x) \to -\infty$ as $x \to \infty$, so the graph moves down to the right.

21. Find all the real zeros of $h(t) = t^2 - 6t + 9$.

Solution:

$$\begin{aligned} h(t) &= t^2 - 6t + 9 \\ 0 &= t^2 - 6t + 9 \\ 0 &= (t-3)^2 \\ t &= 3 \end{aligned}$$

25. Find all the real zeros of $f(x) = 3x^2 - 12x + 3$.

Solution:

$$\begin{aligned} f(x) &= 3x^2 - 12x + 3 \\ 0 &= 3(x^2 - 4x + 1) \\ x &= \frac{4 \pm \sqrt{12}}{2} \quad \text{by the Quadratic Formula} \\ x &= 2 \pm \sqrt{3} \end{aligned}$$

29. Find all the real zeros of $g(t) = \frac{1}{2}t^4 - \frac{1}{2}$.

Solution:

$$\begin{aligned} g(t) &= \tfrac{1}{2}t^4 - \tfrac{1}{2} \\ 0 &= \tfrac{1}{2}(t^4 - 1) \\ 0 &= \tfrac{1}{2}(t^2 + 1)(t^2 - 1) \\ 0 &= \tfrac{1}{2}(t^2 + 1)(t + 1)(t - 1) \\ t &= \pm 1 \end{aligned}$$

33. Find all the real zeros of $f(x) = 5x^4 + 15x^2 + 10$.

Solution:

$$\begin{aligned} f(x) &= 5x^4 + 15x^2 + 10 \\ 0 &= 5(x^4 + 3x^2 + 2) \\ 0 &= 5(x^2 + 1)(x^2 + 2) \end{aligned}$$

No real zeros

35. Find a polynomial function that has the zeros 0 and 10.

Solution:

$$\begin{aligned} f(x) &= (x - 0)(x - 10) \\ f(x) &= x^2 - 10x \end{aligned}$$

41. Find a polynomial function that has the zeros 4, -3, 3 and 0.

Solution:

$$\begin{aligned} f(x) &= (x - 4)(x + 3)(x - 3)(x - 0) \\ &= (x - 4)(x^2 - 9)x \\ &= x^4 - 4x^3 - 9x^2 + 36x \end{aligned}$$

47. Sketch the graph of $f(x) = -\frac{3}{2}$.

Solution:

$f(x) = -\frac{3}{2}$ is a horizontal line.

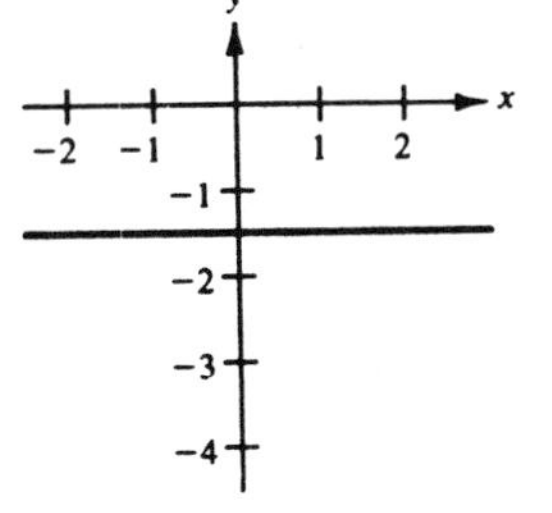

51. Sketch the graph of $f(x) = x^3 - 3x^2$.

Solution:

$$f(x) = x^3 - 3x^2 = x^2(x-3)$$

Zeros: 0 and 3
Right: Moves up
Left: Moves down

x	0	1	2	3	−1
$f(x)$	0	−2	−4	0	−4

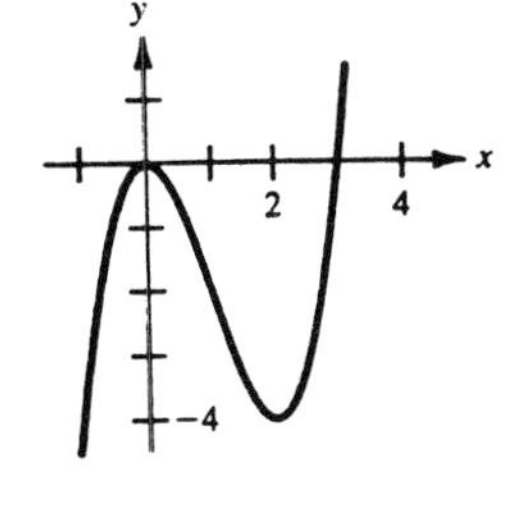

55. Sketch the graph of $g(t) = -\frac{1}{4}(t-2)^2(t+2)^2$.

Solution:

$$g(t) = -\tfrac{1}{4}(t-2)^2(t+2)^2$$

Zeros: 2 and −2
Right: Moves down
Left: Moves down

t	−3	−2	−1	0	1	2	3
$g(t)$	$-\frac{25}{4}$	0	$-\frac{9}{4}$	−4	$-\frac{9}{4}$	0	$-\frac{25}{4}$

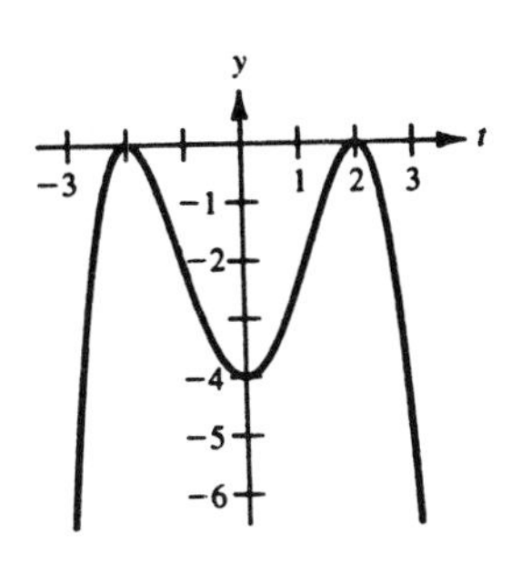

61. Follow the procedure given in Example 8 to estimate the zero of $f(x) = x^4 - 10x^2 - 11$ in the interval [3, 4]. (Give your approximation to the nearest tenth.)

Solution:

$$f(x) = x^4 - 10x^2 - 11, \quad [3, 4]$$

x	3	3.1	3.2	3.3	3.4	3.5	3.6	3.7	3.8	3.9	4
$f(x)$	−20	−14.748	−8.542	−1.308	7.034	16.563	27.362	39.516	53.114	68.244	85

The zero lies between 3.3 and 3.4. It is closer to 3.3.

SECTION 3.3

Polynomial Division and Synthetic Division

You should know the following basic techniques and principles of polynomial division.

- The Division Algorithm (Long Division of Polynomials)
- Synthetic Division
- $f(k)$ is equal to the remainder of $f(x)$ divided by $(x - k)$.
- $f(k) = 0$ if and only if $(x - k)$ is a factor of $f(x)$.
- Horner's Method

Solutions to Selected Exercises

5. Divide $x^4 + 5x^3 + 6x^2 - x - 2$ by $x + 2$ using long division.

Solution:

$$\begin{array}{r}
x^3 + 3x^2 \quad - 1 \\
x + 2 \overline{\big) \; x^4 + 5x^3 + 6x^2 - x - 2} \\
\underline{-(x^4 + 2x^3)} \qquad\qquad\quad \\
3x^3 + 6x^2 \qquad\quad \\
\underline{-(3x^3 + 6x^2)} \qquad\quad \\
-x - 2 \\
\underline{-(-x - 2)} \\
0
\end{array}$$

Thus, $\dfrac{x^4 + 5x^3 + 6x^2 - x - 2}{x + 2} = x^3 + 3x^2 - 1$.

7. Divide $7x + 3$ by $x + 2$ using long division.

Solution:

$$\begin{array}{r}
7 \quad\; \\
x + 2 \overline{\big) \; 7x + 3} \\
\underline{-(7x + 14)} \\
-11
\end{array}$$

Thus, $\dfrac{7x + 3}{x + 2} = 7 - \dfrac{11}{x + 2}$.

11. Divide $x^4 + 3x^2 + 1$ by $x^2 - 2x + 3$ using long division.

Solution:

$$
\begin{array}{r|l}
 & \quad x^2 + 2x + 4 \\
\hline
x^2 - 2x + 3 & \quad x^4 + 0x^3 + 3x^2 + 0x + 1 \\
 & -(x^4 - 2x^3 + 3x^2) \\
 & \qquad 2x^3 + 0x^2 + 0x \\
 & \qquad -(2x^3 - 4x^2 + 6x) \\
 & \qquad\qquad 4x^2 - 6x + 1 \\
 & \qquad\qquad -(4x^2 - 8x + 12) \\
 & \qquad\qquad\qquad 2x - 11
\end{array}
$$

Thus, $\dfrac{x^4 + 3x^2 + 1}{x^2 - 2x + 3} = x^2 + 2x + 4 + \dfrac{2x - 11}{x^2 - 2x + 3}$.

15. Divide $3x^3 - 17x^2 + 15x - 25$ by $x - 5$ using synthetic division.

Solution:

$$
\begin{array}{r|rrrr}
5 & 3 & -17 & 15 & -25 \\
 & & 15 & -10 & 25 \\
\hline
 & 3 & -2 & 5 & 0
\end{array}
$$

Thus, $\dfrac{3x^3 - 17x^2 + 15x - 25}{x - 5} = 3x^2 - 2x + 5$.

19. Divide $-x^3 + 75x - 250$ by $x + 10$ using synthetic division.

Solution:

$$
\begin{array}{r|rrrr}
-10 & -1 & 0 & 75 & -250 \\
 & & 10 & -100 & 250 \\
\hline
 & -1 & 10 & -25 & 0
\end{array}
$$

Thus, $\dfrac{-x^3 + 75x - 250}{x + 10} = -x^2 + 10x - 25$.

23. Divide $10x^4 - 50x^3 - 800$ by $x - 6$ using synthetic division.

Solution:

$$
\begin{array}{r|rrrrr}
6 & 10 & -50 & 0 & 0 & -800 \\
 & & 60 & 60 & 360 & 2160 \\
\hline
 & 10 & 10 & 60 & 360 & 1360
\end{array}
$$

Thus, $\dfrac{10x^4 - 50x^3 - 800}{x - 6} = 10x^3 + 10x^2 + 60x + 360 + \dfrac{1360}{x - 6}$.

27. Divide $-3x^4$ by $x-2$ using synthetic division.

Solution:

$$\begin{array}{r|rrrrr} 2 & -3 & 0 & 0 & 0 & 0 \\ & & -6 & -12 & -24 & -48 \\ \hline & -3 & -6 & -12 & -24 & -48 \end{array}$$

Thus, $\dfrac{-3x^4}{x-2} = -3x^3 - 6x^2 - 12x - 24 - \dfrac{48}{x-2}$.

31. Divide $4x^3 + 16x^2 - 23x - 15$ by $x + \frac{1}{2}$ using synthetic division.

Solution:

$$\begin{array}{r|rrrr} -\frac{1}{2} & 4 & 16 & -23 & -15 \\ & & -2 & -7 & 15 \\ \hline & 4 & 14 & -30 & 0 \end{array}$$

Thus, $\dfrac{4x^3 + 16x^2 - 23x - 15}{x + \frac{1}{2}} = 4x^2 + 14x - 30$.

35. Use synthetic division to show that $x = \frac{1}{2}$ is a solution of $2x^3 - 15x^2 + 27x - 10 = 0$, and use the result to factor the polynomial completely.

Solution:

$$\begin{array}{r|rrrr} \frac{1}{2} & 2 & -15 & 27 & -10 \\ & & 1 & -7 & 10 \\ \hline & 2 & -14 & 20 & 0 \end{array}$$

$$\begin{aligned} 2x^3 - 15x^2 + 27x - 10 &= \left(x - \frac{1}{2}\right)(2x^2 - 14x + 20) \\ &= \left(x - \frac{1}{2}\right)2(x^2 - 7x + 10) \\ &= (2x-1)(x-2)(x-5) \end{aligned}$$

37. Use synthetic division to show that $x = 1+\sqrt{3}$ is a solution of $x^3 - 3x^2 + 2 = 0$, and use the result to factor the polynomial completely.

Solution:

$$\begin{array}{r|rrrr} 1+\sqrt{3} & 1 & -3 & 0 & 2 \\ & & 1+\sqrt{3} & 1-\sqrt{3} & -2 \\ \hline & 1 & -2+\sqrt{3} & 1-\sqrt{3} & 0 \end{array}$$

$$\begin{aligned} x^3 - 3x^2 + 2 &= \left[x - (1+\sqrt{3})\right]\left[x^2 + (-2+\sqrt{3})x + 1 - \sqrt{3}\right] \\ &= \left[x - (1+\sqrt{3})\right]\left[x - (1-\sqrt{3})\right](x-1) \end{aligned}$$

41. Express the function $f(x) = x^3 - x^2 - 14x + 11$ in the form $f(x) = (x-k)q(x) + r$ for $k = 4$, and demonstrate that $f(k) = r$.

Solution:

$$\begin{array}{r|rrrr} 4 & 1 & -1 & -14 & 11 \\ & & 4 & 12 & -8 \\ \hline & 1 & 3 & -2 & 3 \end{array}$$

$$f(x) = (x-4)(x^2 + 3x - 2) + 3$$

$$r = 3$$

$$\begin{aligned} f(4) &= 4^3 - 4^2 - 14(4) + 11 \\ &= 64 - 16 - 56 + 11 \\ &= 3 \end{aligned}$$

45. Use synthetic division to find the required function values of $f(x) = 4x^3 - 13x + 10$.

(a) $f(1)$ (b) $f(-2)$ (c) $f(1/2)$ (d) $f(8)$

Solution:

(a)
$$\begin{array}{r|rrrr} 1 & 4 & 0 & -13 & 10 \\ & & 4 & 4 & -9 \\ \hline & 4 & 4 & -9 & 1 \end{array}$$

Thus, $f(1) = 1$.

(b)
$$\begin{array}{r|rrrr} -2 & 4 & 0 & -13 & 10 \\ & & -8 & 16 & -6 \\ \hline & 4 & -8 & 3 & 4 \end{array}$$

Thus, $f(-2) = 4$.

(c)
$$\begin{array}{r|rrrr} \frac{1}{2} & 4 & 0 & -13 & 10 \\ & & 2 & 1 & -6 \\ \hline & 4 & 2 & -12 & 4 \end{array}$$

Thus, $f(1/2) = 4$.

(d)
$$\begin{array}{r|rrrr} 8 & 4 & 0 & -13 & 10 \\ & & 32 & 256 & 1944 \\ \hline & 4 & 32 & 243 & 1954 \end{array}$$

Thus, $f(8) = 1954$.

49. Use synthetic division to find the required function values of $f(x) = x^3 - 2x^2 - 11x + 52$.

(a) $f(5)$ (b) $f(-4)$ (c) $f(1.2)$ (d) $f(2)$

Solution:

(a)

5	1	−2	−11	52
		5	15	20
	1	3	4	72

Thus, $f(5) = 72$.

(b)

−4	1	−2	−11	52
		−4	24	−52
	1	−6	13	0

Thus, $f(-4) = 0$.

(c)

1.2	1	−2	−11	52
		1.2	−0.96	−14.352
	1	−0.8	−11.96	37.648

Thus, $f(1.2) = 37.648$.

(d)

2	1	−2	−11	52
		2	0	−22
	1	0	−11	30

Thus, $f(2) = 30$.

53. Use Horner's Method to find the required function values of $f(x) = -5x^4 + 8.5x^3 + 10x - 3$.

(a) $f(1.08)$ (b) $f(-5.4)$

Solution:

$$\begin{aligned} f(x) &= -5x^4 + 8.5x^3 + 10x - 3 \\ &= (-5x^3 + 8.5x^2 + 10)x - 3 \\ &= ([-5x^2 + 8.5x]x + 10)x - 3 \\ &= ([-5x + 8.5]x^2 + 10)x - 3 \end{aligned}$$

(a) $f(1.08) \approx 11.705$

(b) $f(-5.4) \approx -5646.972$

SECTION 3.4

Real Zeros of Polynomial Functions

- You should know Descartes's Rule of Signs.
  (a) The number of positive real zeros of f is either equal to the number of variations of sign of f or is less than that number by an even integer.
  (b) The number of negative real zeros of f is either equal to the number of variations in sign of $f(-x)$ or is less than that number by an even integer.
  (c) When there is only one variation in sign, there is exactly one positive (or negative) real zero.

- You should know the Rational Zero Test.

- You should know shortcuts for the Rational Zero Test.
  (a) Use a programmable calculator.
  (b) Sketch a graph.
  (c) After finding a root, use synthetic division to reduce the degree of the polynomial.

- You should be able to observe the last row obtained from synthetic division in order to determine upper or lower bounds.
  (a) If the test value is positive and all of the entries in the last row are positive, then the test value is an upper bound.
  (b) If the test value is negative and the entries in the last row alternate from positive to negative, then the test value is a lower bound.

Solutions to Selected Exercises

5. Use Descartes's Rule of Signs to determine the possible number of positive and negative zeros of $g(x) = 2x^3 - 3x^2 - 3$.

Solution:

$$g(x) = 2x^3 - 3x^2 - 3$$
$$g(-x) = -2x^3 - 3x^2 - 3$$

Since $g(x)$ has one variation in sign, g has exactly one positive real zero. Since $g(-x)$ has no variations in sign, there are no negative real zeros.

9. Use Descartes's Rule of Signs to determine the possible number of positive and negative real zeros of $h(x) = 4x^2 - 8x + 3$.

Solution:

$$h(x) = 4x^2 - 8x + 3$$
$$h(-x) = 4x^2 + 8x + 3$$

Since $h(x)$ has two variations in sign, h has either two or zero positive real zeros. Since $h(-x)$ has no variations in sign, there are no negative real zeros.

13. Use the Rational Zero Test to list all the possible rational zeros of $f(x) = -4x^3 + 15x^2 - 8x - 3$ and verify that the zeros of f shown on the graph are contained in the list.

Solution:

Since the leading coefficient is -4 and the constant term is -3, the possible rational zeros of f are

$$\frac{\text{factors of } -3}{\text{factors of } -4} = \frac{\pm 1, \pm 3}{\pm 1, \pm 2, \pm 4} = \pm 1, \pm 3, \pm\frac{1}{2}, \pm\frac{3}{2}, \pm\frac{1}{4}, \pm\frac{3}{4}$$

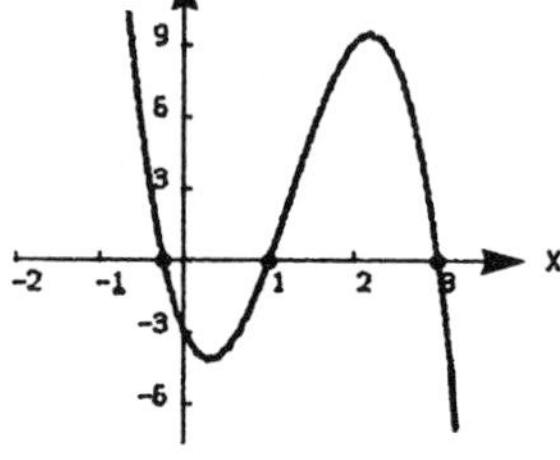

The zeros shown on the graph are $-\frac{1}{4}$, 1 and 3 and are contained in the list.

17. Use synthetic division to determine if the given x-value is an upper bound or lower bound of the zeros of $f(x) = x^4 - 4x^3 + 15$.

(a) $x = 4$ (b) $x = -1$ (c) $x = 3$

Solution:

(a)
$$\begin{array}{r|rrrrr} 4 & 1 & -4 & 0 & 0 & 15 \\ & & 4 & 0 & 0 & 0 \\ \hline & 1 & 0 & 0 & 0 & 15 \end{array}$$

Since the test value is positive and all the entries in the last row are positive, $x = 4$ is an upper bound.

(b)
$$\begin{array}{r|rrrrr} -1 & 1 & -4 & 0 & 0 & 15 \\ & & -1 & 5 & -5 & 5 \\ \hline & 1 & -5 & 5 & -5 & 20 \end{array}$$

Since the test value is negative and the entries in the last row alternate in sign, $x = -1$ is a lower bound.

(c) $$\begin{array}{r|rrrrr} 3 & 1 & -4 & 0 & 0 & 15 \\ & & 3 & -3 & -9 & -27 \\ \hline & 1 & -1 & -3 & -9 & -12 \end{array}$$

$x = 3$ is neither an upper nor a lower bound.

21. Find the real zeros of $f(x) = x^3 - 6x^2 + 11x - 6$.

Solution:

Possible rational zeros: ±1, ±2, ±3, ±6

$$\begin{array}{r|rrrr} 1 & 1 & -6 & 11 & -6 \\ & & 1 & -5 & 6 \\ \hline & 1 & -5 & 6 & 0 \end{array}$$

$$x^3 - 6x^2 + 11x - 6 = (x-1)(x^2 - 5x + 6) = (x-1)(x-2)(x-3)$$

The zeros are 1, 2, and 3.

25. Find the real zeros of $h(t) = t^3 + 12t^2 + 21t + 10$.

Solution:

Possible rational zeros: ±1, ±2, ±5, ±10

$$\begin{array}{r|rrrr} -1 & 1 & 12 & 21 & 10 \\ & & -1 & -11 & -10 \\ \hline & 1 & 11 & 10 & 0 \end{array}$$

$$t^3 + 12t^2 + 21t + 10 = (t+1)(t^2 + 11t + 10) = (t+1)(t+1)(t+10)$$

Thus, the zeros are -1 and -10.

27. Find the real zeros of $f(x) = x^3 - 4x^2 + 5x - 2$.

Solution:

Possible rational zeros: ±1, ±2

$$\begin{array}{r|rrrr} 1 & 1 & -4 & 5 & -2 \\ & & 1 & -3 & 2 \\ \hline & 1 & -3 & 2 & 0 \end{array}$$

$$x^3 - 4x^2 + 5x - 2 = (x-1)(x^2 - 3x + 2) = (x-1)(x-1)(x-2)$$

Thus, the zeros are 1 and 2.

31. Find the real zeros of $f(x) = 4x^3 - 3x - 1$.

Solution:

Possible rational zeros: $\pm 1,\ \pm\frac{1}{2},\ \pm\frac{1}{4}$

$$\begin{array}{r|rrrr} 1 & 4 & 0 & -3 & -1 \\ & & 4 & 4 & 1 \\ \hline & 4 & 4 & 1 & 0 \end{array}$$

$$4x^3 - 3x - 1 = (x-1)(4x^2 + 4x + 1) = (x-1)(2x+1)^2$$

Thus, the zeros are 1 and $-\frac{1}{2}$.

33. Find the real zeros of $f(y) = 4y^3 + 3y^2 + 8y + 6$.

Solution:

Possible rational zeros: $\pm 1,\ \pm 2,\ \pm 3,\ \pm 6,\ \pm\frac{1}{2},\ \pm\frac{3}{2},\ \pm\frac{1}{4},\ \pm\frac{3}{4}$

$$\begin{array}{r|rrrr} -\frac{3}{4} & 4 & 3 & 8 & 6 \\ & & -3 & 0 & -6 \\ \hline & 4 & 0 & 8 & 0 \end{array}$$

$$4y^3 + 3y^2 + 8y + 6 = (y + \tfrac{3}{4})(4y^2 + 8) = (y + \tfrac{3}{4})4(y^2 + 2) = (4y+3)(y^2+2)$$

Thus, the only zero is $-\frac{3}{4}$.

35. Find the real zeros of $f(x) = x^4 - 3x^2 + 2$.

Solution:

$$\begin{aligned} f(x) &= x^4 - 3x^2 + 2 \\ &= (x^2 - 1)(x^2 - 2) \\ &= (x+1)(x-1)(x+\sqrt{2})(x-\sqrt{2}) \end{aligned}$$

Thus, the zeros are ± 1 and $\pm\sqrt{2}$.

39. Find all the real solutions of $x^4 - 13x^2 - 12x = 0$.

Solution:

$$f(x) = x^4 - 13x^2 - 12x = x(x^3 - 13x - 12)$$

0 is a zero.

Possible rational zeros: $\pm 1,\ \pm 2,\ \pm 3,\ \pm 4,\ \pm 6,\ \pm 12$

$$\begin{array}{r|rrrr} -1 & 1 & 0 & -13 & -12 \\ & & -1 & 1 & 12 \\ \hline & 1 & -1 & -12 & 0 \end{array}$$

$$x^4 - 13x^2 - 12x = x(x^3 - 13x - 12) = x(x+1)(x^2 - x - 12) = x(x+1)(x+3)(x-4)$$

Thus, the zeros are 0, -1, -3, and 4.

43. Find all the real solutions of $x^5 - 7x^4 + 10x^3 + 14x^2 - 24x = 0$.

Solution:

$$f(x) = x^5 - 7x^4 + 10x^3 + 14x^2 - 24x = x(x^4 - 7x^3 + 10x^2 + 14x - 24)$$

0 is a zero.

Possible rational zeros: ± 1, ± 2, ± 3, ± 4, ± 6, ± 8, ± 12, ± 24

$$\begin{array}{r|rrrrr} 4 & 1 & -7 & 10 & 14 & -24 \\ & & 4 & -12 & -8 & 24 \\ \hline 3 & 1 & -3 & -2 & 6 & 0 \\ & & 3 & 0 & 0 & \\ \hline & 1 & 0 & -2 & 0 & \end{array}$$

$$\begin{aligned} x^5 - 7x^4 + 10x^3 + 14x^2 - 24x &= x(x^4 - 7x^3 + 10x^2 + 14x - 24) \\ &= x(x-4)(x-3)(x^2-2) \\ &= x(x-4)(x-3)(x+\sqrt{2})(x-\sqrt{2}) \end{aligned}$$

Thus, the zeros are 0, 4, 3, and $\pm\sqrt{2}$.

47. For $f(x) = 4x^3 + 7x^2 - 11x - 18$

(a) list all the possible rational zeros of f,
(b) sketch the graph of f so that some of the possible zeros in part (a) can be disregarded, and
(c) then determine all the real zeros of f.

Solution:

(a) Possible rational roots: ± 1, ± 2, ± 3, ± 6, ± 9, ± 18, $\pm\frac{1}{2}$, $\pm\frac{3}{2}$, $\pm\frac{9}{2}$, $\pm\frac{1}{4}$, $\pm\frac{3}{4}$, $\pm\frac{9}{4}$

(b)

x	0	1	-1	$\frac{1}{2}$	-2	-3	$-\frac{3}{2}$
$f(x)$	-18	-18	-4	-21.25	0	-30	0.75

By testing values using synthetic division, we find that 2 is an upper bound and -3 is a lower bound. This eliminates 3, ± 6, ± 9, ± 18, $\pm\frac{9}{2}$, and $\frac{9}{4}$ as possible zeros.

(c) -2 is a zero.

$4x^3 + 7x^2 - 11x - 18 = (x+2)(4x^2 - x - 9) = 0$

$4x^2 - x - 9$ does not factor, so by the Quadratic Formula

$$x = \frac{1 \pm \sqrt{145}}{8} \text{ are also zeros.}$$

51. Find all the rational zeros of $f(x) = x^3 - \frac{1}{4}x^2 - x + \frac{1}{4}$.

Solution:

$$f(x) = x^3 - \tfrac{1}{4}x^2 - x + \tfrac{1}{4} = \tfrac{1}{4}(4x^3 - x^2 - 4x + 1)$$

Possible rational zeros: ± 1, $\pm\frac{1}{4}$, $\pm\frac{1}{2}$

By testing these values, we see that $x = \pm 1$ and $x = \frac{1}{4}$ work.

SECTION 3.5

Complex Numbers

- You should know how to work with complex numbers.
- Operations on Complex Numbers

  (a) Addition: $(a+bi)+(c+di)=(a+c)+(b+d)i$
  (b) Subtraction: $(a+bi)-(c+di)=(a-c)+(b-d)i$
  (c) Multiplication: $(a+bi)(c+di)=(ac-bd)+(ad+bc)i$
  (d) Division: $\frac{a+bi}{c+di}=\frac{a+bi}{c+di}\cdot\frac{c-di}{c-di}=\frac{ac+bd}{c^2+d^2}+\frac{bc-ad}{c^2+d^2}i$

- The complex conjugate of $a+bi$ is $a-bi$:

  $$(a+bi)(a-bi)=a^2+b^2$$

- The additive inverse of $a+bi$ is $-a-bi$.
- The multiplicative inverse of $a+bi$ is

  $$\frac{a-bi}{a^2+b^2}.$$

- $\sqrt{-a}=\sqrt{a}\,i$ for $a>0$.

Solutions to Selected Exercises

1. Write out the first 16 positive powers of i and express each as i, $-i$, 1, or -1.

Solution:

$i=i$	$i^5=i$	$i^9=i$	$i^{13}=i$	$i^{17}=i$
$i^2=-1$	$i^6=-1$	$i^{10}=-1$	$i^{14}=-1$	$i^{18}=-1$
$i^3=-i$	$i^7=-i$	$i^{11}=-i$	$i^{15}=-i$	$i^{19}=-i$
$i^4=1$	$i^8=1$	$i^{12}=1$	$i^{16}=1$	$i^{20}=1$

5. Find real numbers a and b so that the equation $(a-1)+(b+3)i=5+8i$ is true.

Solution:
$(a-1)+(b+3)i=5+8i$

$$a-1=5 \quad \Rightarrow \quad a=6$$
$$b+3=8 \quad \Rightarrow \quad b=5$$

9. Write $2-\sqrt{-27}$ in standard form and find its complex conjugate.

Solution:

$$2-\sqrt{-27}=2-\sqrt{27}\,i=2-3\sqrt{3}\,i$$

Complex conjugate: $2+3\sqrt{3}\,i$

13. Write $-6i+i^2$ in standard form and find its complex conjugate.

Solution:

$$-6i+i^2=-6i+(-1)=-1-6i$$

Complex conjugate: $-1+6i$

21. Perform the indicated operation and write the result in standard form.

$$(8-i)-(4-i)$$

Solution:

$$(8-i)-(4-i)=8-i-4+i=4$$

23. Perform the indicated operation and write the result in standard form.

$$(-2+\sqrt{-8})+(5-\sqrt{-50})$$

Solution:

$$(-2+\sqrt{-8})+(5-\sqrt{-50})=-2+2\sqrt{2}\,i+5-5\sqrt{2}\,i=3-3\sqrt{2}\,i$$

27. Perform the indicated operation and write the result in standard form.

$$\sqrt{-6}\sqrt{-2}$$

Solution:

$$\sqrt{-6}\sqrt{-2}=(\sqrt{6}\,i)(\sqrt{2}\,i)=\sqrt{12}\,i^2=2\sqrt{3}(-1)=-2\sqrt{3}$$

31. Perform the indicated operation and write the result in standard form.

$$(1+i)(3-2i)$$

Solution:

$$(1+i)(3-2i) = 3-2i+3i-2i^2 = 3+i+2 = 5+i$$

35. Perform the indicated operation and write the result in standard form.

$$6i(5-2i)$$

Solution:

$$6i(5-2i) = 30i-12i^2 = 12+30i$$

39. Perform the indicated operation and write the result in standard form.

$$(\sqrt{14}+\sqrt{10}\,i)(\sqrt{14}-\sqrt{10}\,i)$$

Solution:

$$(\sqrt{14}+\sqrt{10}\,i)(\sqrt{14}-\sqrt{10}\,i) = 14-10i^2 = 14+10 = 24$$

45. Perform the indicated operation and write the result in standard form.

$$\frac{2+i}{2-i}$$

Solution:

$$\frac{2+i}{2-i} = \frac{2+i}{2-i}\cdot\frac{2+i}{2+i} = \frac{4+4i+i^2}{4+1} = \frac{3+4i}{5} = \frac{3}{5}+\frac{4}{5}i$$

53. Perform the indicated operation and write the result in standard form.

$$\frac{(21-7i)(4+3i)}{2-5i}$$

Solution:

$$\begin{aligned}\frac{(21-7i)(4+3i)}{2-5i} &= \frac{(84+63i-28i-21i^2)}{2-5i}\cdot\frac{2+5i}{2+5i}\\ &= \frac{(105+35i)(2+5i)}{4+25}\\ &= \frac{210+525i+70i+175i^2}{29}\\ &= \frac{35+595i}{29} = \frac{35}{29}+\frac{595}{29}i\end{aligned}$$

59. Use the quadratic formula to solve $4x^2 + 16x + 17 = 0$.

Solution:

$4x^2 + 16x + 17 = 0;\ a = 4,\ b = 16,\ c = 17$

$$\begin{aligned} x &= \frac{-16 \pm \sqrt{(16)^2 - 4(4)(17)}}{2(4)} \\ &= \frac{-16 \pm \sqrt{-16}}{8} = \frac{-16 \pm 4i}{8} \\ &= -2 \pm \frac{1}{2}i \end{aligned}$$

63. Use the quadratic formula to solve $16t^2 - 4t + 3 = 0$.

Solution:

$16t^2 - 4t + 3 = 0;\ a = 16,\ b = -4,\ c = 3$

$$\begin{aligned} x &= \frac{-(-4) \pm \sqrt{(-4)^2 - 4(16)(3)}}{2(16)} \\ &= \frac{4 \pm \sqrt{-176}}{32} = \frac{4 \pm 4\sqrt{11}i}{32} \\ &= \frac{1}{8} \pm \frac{\sqrt{11}}{8}i \end{aligned}$$

65. Prove that the sum of a complex number and its conjugate is a real number.

Solution:

$$\begin{aligned} (a + bi) + (a - bi) &= (a + a) + (b - b)i \\ &= 2a + 0i = 2a \qquad \text{which is a real number.} \end{aligned}$$

69. Prove that the conjugate of the sum of two complex numbers is the sum of their conjugates.

Solution:

$(a + bi) + (c + di) = (a + c) + (b + d)i$

The complex conjugate of the sum is $(a + c) - (b + d)i$, and the sum of the conjugates is

$$\begin{aligned} (a - bi) + (c - di) &= (a + c) + (-b - d)i \\ &= (a + c) - (b + d)i \end{aligned}$$

Thus, the conjugate of the sum is the sum of the conjugates.

SECTION 3.6

Complex Zeros and the Fundamental Theorem of Algebra

- You should know that if f is a polynomial of degree $n > 0$, then f has exactly n zeros (roots) in the complex number system.
- You should know that if $a + bi$ is a complex zero of a polynomial f, with real coefficients, then $a - bi$ is also a complex zero of f.
- You should know the difference between a factor that is irreducible over the rationals (such as $x^2 - 7$) and a factor that is irreducible over the reals (such as $x^2 + 9$).

Solutions to Selected Exercises

3. Find all the zeros of $h(x) = x^2 - 4x + 1$ and write the polynomial as a product of linear factors.

Solution:
h has no rational zeros.
By the Quadratic Formula, the zeros are $x = \dfrac{4 \pm \sqrt{16-4}}{2} = 2 \pm \sqrt{3}$.

$$h(x) = [x - (2 + \sqrt{3})][x - (2 - \sqrt{3})] = (x - 2 - \sqrt{3})(x - 2 + \sqrt{3})$$

7. Find all the zeros of $f(z) = z^2 - 2z + 2$ and write the polynomial as a product of linear factors.

Solution:
f has no rational zeros.
By the Quadratic Formula, the zeros are $z = \dfrac{2 \pm \sqrt{4-8}}{2} = 1 \pm i$.

$$f(z) = [z - (1 + i)][z - (1 - i)] = (z - 1 - i)(z - 1 + i)$$

9. Find all the zeros of $g(x) = x^3 - 6x^2 + 13x - 10$ and write the polynomial as a product of linear factors.

Solution:
Possible rational zeros: $\pm 1, \ \pm 2, \ \pm 5, \ \pm 10$

$$\begin{array}{r|rrrr} 2 & 1 & -6 & 13 & -10 \\ & & 2 & -8 & 10 \\ \hline & 1 & -4 & 5 & 10 \end{array}$$

$$g(x) = (x - 2)(x^2 - 4x + 5)$$

$x = 2$ is a zero, and by the Quadratic Formula $x = \dfrac{4 \pm \sqrt{16 - 20}}{2} = 2 \pm i$ are also zeros.

$$g(x) = (x - 2)[x - (2 + i)][x - (2 - i)] = (x - 2)(x - 2 - i)(x - 2 + i)$$

15. Find all the zeros of $f(x) = 16x^3 - 20x^2 - 4x + 15$ and write the polynomial as a product of linear factors.

Solution:
Possible rational zeros: $\pm 1, \pm 3, \pm 5, \pm 15, \pm\frac{1}{2}, \pm\frac{3}{2}, \pm\frac{5}{2}, \pm\frac{15}{2}, \pm\frac{1}{4}, \pm\frac{3}{4}, \pm\frac{5}{4}, \pm\frac{15}{4}, \pm\frac{1}{8}, \pm\frac{3}{8}, \pm\frac{5}{8}, \pm\frac{15}{8}, \pm\frac{1}{16}, \pm\frac{3}{16}, \pm\frac{5}{16}, \pm\frac{15}{16}$

$$\begin{array}{r|rrrr} -\frac{3}{4} & 16 & -20 & -4 & 15 \\ & & -12 & 24 & -15 \\ \hline & 16 & -32 & 20 & 0 \end{array}$$

$$f(x) = \left(x + \frac{3}{4}\right)(16x^2 - 32x + 20) = 4\left(x + \frac{3}{4}\right)(4x^2 - 8x + 5) = (4x + 3)(4x^2 - 8x + 5)$$

$x = -\frac{3}{4}$ is a zero, and by the Quadratic Formula $x = \dfrac{8 \pm \sqrt{64 - 80}}{8} = 1 \pm \frac{1}{2}i$ are also zeros.

$$f(x) = 16\left(x + \frac{3}{4}\right)\left[x - \left(1 + \frac{1}{2}i\right)\right]\left[x - \left(1 - \frac{1}{2}i\right)\right] = (4x + 3)(2x - 2 - i)(2x - 2 + i)$$

21. Find all the zeros of $g(x) = x^4 - 4x^3 + 8x^2 - 16x + 16$ and write the polynomial as a product of linear factors.

Solution:
Possible rational zeros: $\pm 1, \pm 2, \pm 4, \pm 8, \pm 16$

$$\begin{array}{r|rrrrr} 2 & 1 & -4 & 8 & -16 & 16 \\ & & 2 & -4 & 8 & -16 \\ \hline 2 & 1 & -2 & 4 & -8 & 0 \\ & & 2 & 0 & 8 & \\ \hline & 1 & 0 & 4 & 0 & \end{array}$$

$$g(x) = (x - 2)(x - 2)(x^2 + 4) = (x - 2)^2(x + 2i)(x - 2i)$$

The zeros of g are 2 and $\pm 2i$.

27. Find a polynomial with integer coefficients that has the zeros 1, $5i$, and $-5i$.

Solution:

$$\begin{aligned} f(x) &= (x - 1)(x - 5i)(x + 5i) \\ &= (x - 1)(x^2 + 25) \\ &= x^3 - x^2 + 25x - 25 \end{aligned}$$

31. Find a polynomial with integer coefficients that has the zeros i, $-i$, $6i$, and $-6i$.

Solution:

$$\begin{aligned} f(x) &= (x-i)(x+i)(x-6i)(x+6i) \\ &= (x^2+1)(x^2+36) \\ &= x^4+37x^2+36 \end{aligned}$$

35. Find a polynomial with integer coefficients that has the zeros $\frac{3}{4}$, -2, and $-\frac{1}{2}+i$.

Solution:

Since $-\frac{1}{2}+i$ is a zero, so is $-\frac{1}{2}-i$.

$$\begin{aligned} f(x) &= 16(x-\tfrac{3}{4})(x+2)\left[x-(-\tfrac{1}{2}+i)\right]\left[x-(-\tfrac{1}{2}-i)\right] \\ &= 4(4x-3)(x+2)\left[x^2+x+(\tfrac{1}{4}+1)\right] \\ &= (4x^2+5x-6)(4x^2+4x+5) \\ &= 16x^4+36x^3+16x^2+x-30 \end{aligned}$$

39. Write $f(x)=x^4-4x^3+5x^2-2x-6$

(a) as the product of factors that are irreducible over the rationals,

(b) as the product of linear and quadratic factors that are irreducible over the reals, and

(c) in completely factored form. [*Hint:* One factor is x^2-2x-2.]

Solution:

$$\begin{array}{r|l} & x^2 - 2x + 3 \\ \hline x^2-2x-2 & x^4 - 4x^3 + 5x^2 - 2x - 6 \\ & -(x^4 - 2x^3 - 2x^2) \\ & -2x^3 + 7x^2 - 2x \\ & -(-2x^3 + 4x^2 + 4x) \\ & 3x^2 - 6x - 6 \\ & -(3x^2 - 6x - 6) \\ & 0 \end{array}$$

$$f(x)=(x^2-2x+3)(x^2-2x-2)$$

(a) $f(x)=(x^2-2x+3)(x^2-2x-2)$

(b) $f(x)=(x^2-2x+3)(x-1+\sqrt{3})(x-1-\sqrt{3})$

(c) $f(x)=(x-1+\sqrt{2}\,i)(x-1-\sqrt{2}\,i)(x-1+\sqrt{3})(x-1-\sqrt{3})$

Note: Use the Quadratic Formula for (b) and (c).

43. Use the zero, $r = 2i$, to find all the zeros of $f(x) = 2x^4 - x^3 + 7x^2 - 4x - 4$.

Solution:

Since $2i$ is a zero of f, so is $-2i$.

$$\begin{array}{r|rrrrr} 2i & 2 & -1 & 7 & -4 & -4 \\ & & 0+4i & -8-2i & 4-2i & 4 \\ \hline -2i & 2 & -1+4i & -1-2i & -2i & 0 \\ & & 0-4i & 0+2i & 2i & \\ \hline & 2 & -1 & -1 & 0 & \end{array}$$

$$f(x) = (x - 2i)(x + 2i)(2x^2 - x - 1) = (x - 2i)(x + 2i)(2x + 1)(x - 1)$$

The zeros of f are $\pm 2i$, $-\frac{1}{2}$, and 1.

47. Use the zero, $r = -3 + \sqrt{2}i$, to find all the zeros of $f(x) = x^4 + 3x^3 - 5x^2 - 21x + 22$.

Solution:

Since $-3 + \sqrt{2}i$ is a zero of f, so is $-3 - \sqrt{2}i$.

$$\begin{array}{r|rrrrr} -3+\sqrt{2}i & 1 & 3 & -5 & -21 & 22 \\ & & -3+\sqrt{2}i & -2-3\sqrt{2}i & 27+2\sqrt{2}i & 22 \\ \hline -3-\sqrt{2}i & 1 & \sqrt{2}i & -7-3\sqrt{2}i & 6+2\sqrt{2}i & 0 \\ & & -3-\sqrt{2}i & 9+3\sqrt{2}i & -6-2\sqrt{2}i & \\ \hline & 1 & -3 & 2 & 0 & \end{array}$$

$$\begin{aligned} f(x) &= [x - (-3 + \sqrt{2}i)][x - (-3 - \sqrt{2}i)](x^2 - 3x + 2) \\ &= [x - (-3 + \sqrt{2}i)][x - (-3 - \sqrt{2}i)](x - 1)(x - 2) \end{aligned}$$

The zeros of f are $-3 + \sqrt{2}i$, $-3 - \sqrt{2}i$, 1, and 2.

51. Find a quadratic function f (with integer coefficients) that has $\pm\sqrt{b}i$ as zeros. Assume that b is a positive integer.

Solution:

$$f(x) = (x - \sqrt{b}i)(x + \sqrt{b}i) = x^2 + b$$

SECTION 3.7

Rational Functions and Their Graphs

- You should know the following basic facts about rational functions.
  (a) A function of the form $f(x) = P(x)/Q(x)$, $Q(x) \neq 0$, where $P(x)$ and $Q(x)$ are polynomials, is called a rational function.
  (b) The domain of a rational function is the set of all real numbers except those which make the denominator zero.
  (c) If $f(x) = P(x)/Q(x)$ is in reduced form, and a is a value such that $Q(a) = 0$, then the line $x = a$ is a vertical asymptote of the graph of f.
  (d) The line $y = b$ is a horizontal asymptote of the graph of f if $f(x) \to b$ as $x \to \infty$ or $x \to -\infty$.
  (e) If $f(x) = P(x)/Q(x) = mx + b + R(x)/Q(x)$, then the line $y = mx + b$ is a slant asymptote of the graph of f.
- Be able to graph rational functions.

Solutions to Selected Exercises

3. Match $f(x) = (x+1)/x$ with its graph.

Solution:

$$f(x) = \frac{x+1}{x} = 1 + \frac{1}{x}$$

Vertical asymptote: $x = 0$
Horizontal asymptote: $y = 1$
x-intercept: $(-1, 0)$
Matches graph (a)

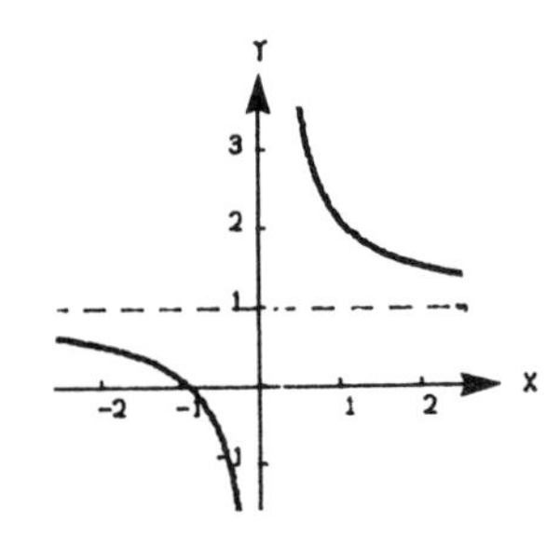

7. Match $f(x) = (x^2+1)/x$ with its graph.

Solution:

$$f(x) = \frac{x^2+1}{x} = x + \frac{1}{x}$$

Vertical asymptote: $x = 0$
Slant asymptote: $y = x$
No intercepts
Matches graph (h)

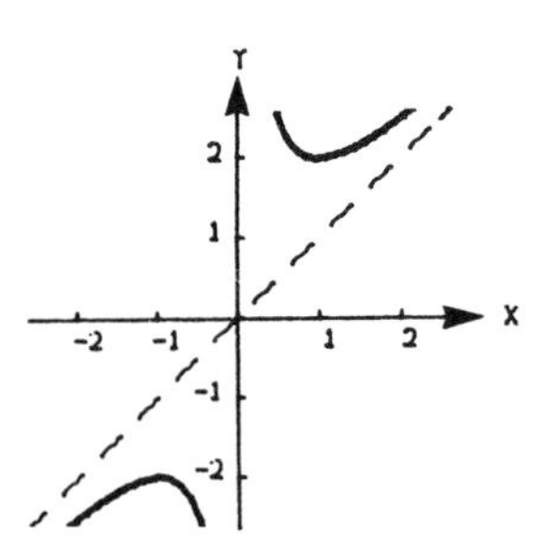

11. Find the domain of the following function and identify any horizontal, vertical, or slant asymptotes.

$$f(x) = \frac{2+x}{2-x}$$

Solution:

$$f(x) = \frac{2+x}{2-x} = -1 + \frac{4}{2-x}$$

Domain: all real numbers except 2
Vertical asymptote: $x = 2$
Horizontal asymptote: $y = -1$

15. Find the domain of the following function and identify any horizontal, vertical, or slant asymptotes.

$$f(x) = \frac{3x^2+1}{x^2+9}$$

Solution:

$$f(x) = \frac{3x^2+1}{x^2+9} = 3 - \frac{26}{x^2+9}$$

Domain: all real numbers
Horizontal asymptotes: $y = 3$

23. Sketch the graph of the following rational function. As sketching aids, check for intercepts, symmetry, vertical asymptotes, and horizontal asymptotes.

$$h(x) = \frac{-1}{x+2}$$

Solution:
Vertical asymptote: $x = -2$
Horizontal asymptote: $y = 0$
y-intercept: $(0, -\frac{1}{2})$

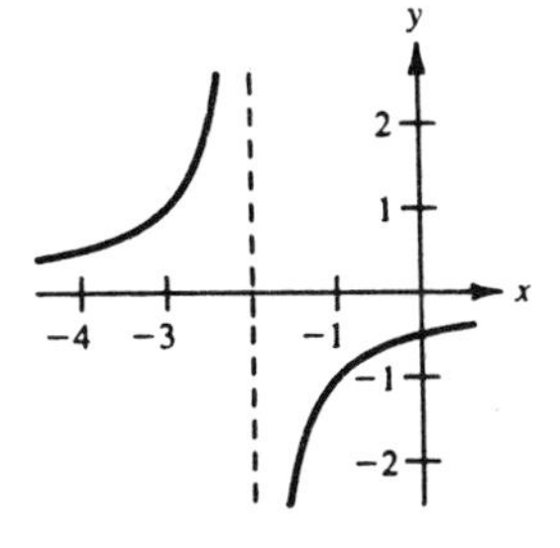

25. Sketch the graph of the following rational function. As sketching aids, check for intercepts, symmetry, vertical asymptotes, and horizontal asymptotes.

$$f(x) = \frac{x+1}{x+2}$$

Solution:

$$f(x) = \frac{x+1}{x+2} = 1 - \frac{1}{x+2}$$

Vertical asymptote: $x = -2$
Horizontal asymptote: $y = 1$
x-intercept: $(-1,\ 0)$
y-intercept: $(0,\ \frac{1}{2})$

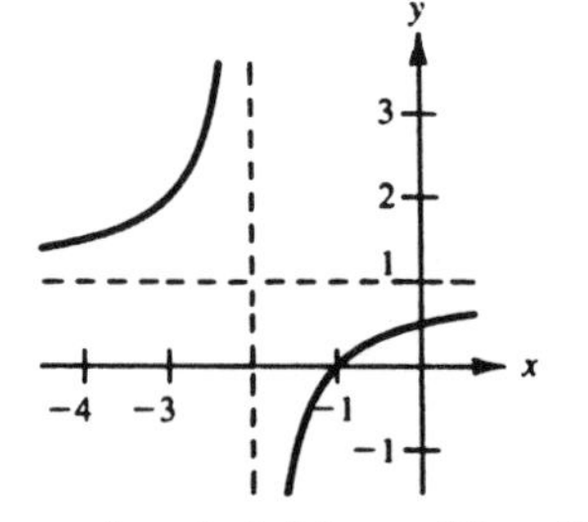

29. Sketch the graph of the following rational function. As sketching aids, check for intercepts, symmetry, vertical asymptotes, and horizontal asymptotes.

$$f(t) = \frac{3t+1}{t}$$

Solution:

$$f(t) = \frac{3t+1}{t} = 3 + \frac{1}{t}$$

Vertical asymptote: $x = 0$
Horizontal asymptote: $y = 3$
x-intercept: $(-\frac{1}{3},\ 0)$

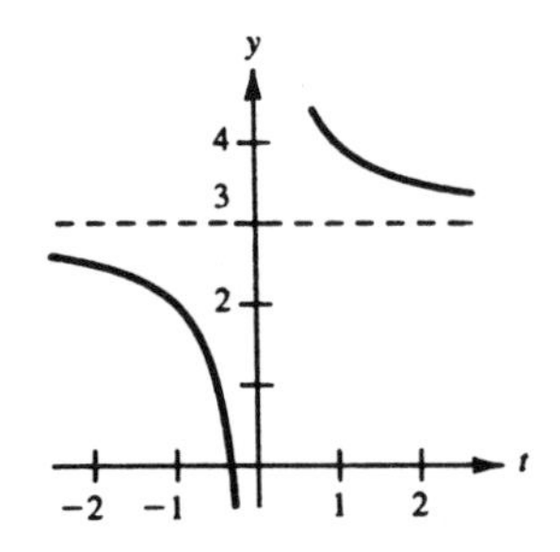

33. Sketch the graph of the following rational function. As sketching aids, check for intercepts, symmetry, vertical asymptotes, and horizontal asymptotes.

$$C(x) = \frac{5+2x}{1+x}$$

Solution:

$$C(x) = \frac{5+2x}{1+x} = 2 + \frac{3}{1+x}$$

Vertical asymptote: $x = -1$
Horizontal asymptote: $y = 2$
x-intercept: $(-\frac{5}{2},\ 0)$
y-intercept: $(0,\ 5)$

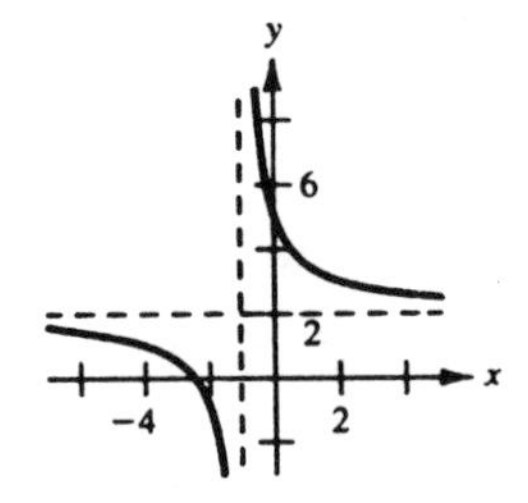

37. Sketch the graph of the following rational function. As sketching aids, check for intercepts, symmetry, vertical asymptotes, and horizontal asymptotes.

$$h(x) = \frac{x^2}{x^2 - 9}$$

Solution:

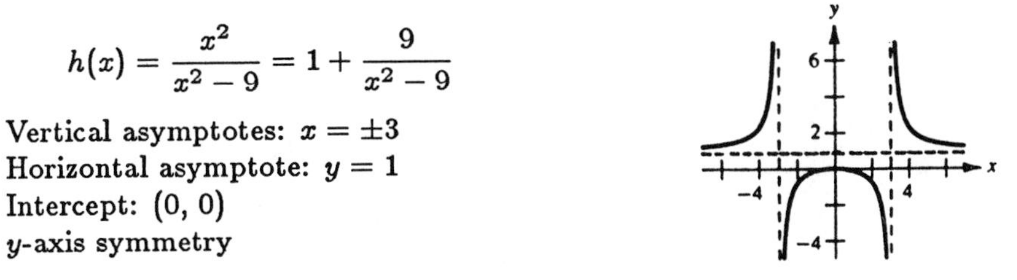

$$h(x) = \frac{x^2}{x^2 - 9} = 1 + \frac{9}{x^2 - 9}$$

Vertical asymptotes: $x = \pm 3$
Horizontal asymptote: $y = 1$
Intercept: $(0, 0)$
y-axis symmetry

41. Sketch the graph of the following rational function. As sketching aids, check for intercepts, symmetry, vertical asymptotes, and horizontal asymptotes.

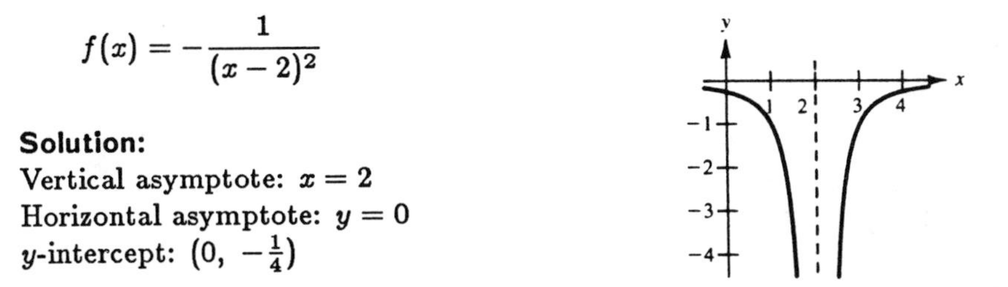

$$f(x) = -\frac{1}{(x - 2)^2}$$

Solution:
Vertical asymptote: $x = 2$
Horizontal asymptote: $y = 0$
y-intercept: $(0, -\frac{1}{4})$

45. Sketch the graph of the following rational function. As sketching aids, check for intercepts, symmetry, vertical asymptotes, and slant asymptotes.

$$f(x) = \frac{2x^2 + 1}{x}$$

Solution:

$$f(x) = \frac{2x^2 + 1}{x} = 2x + \frac{1}{x}$$

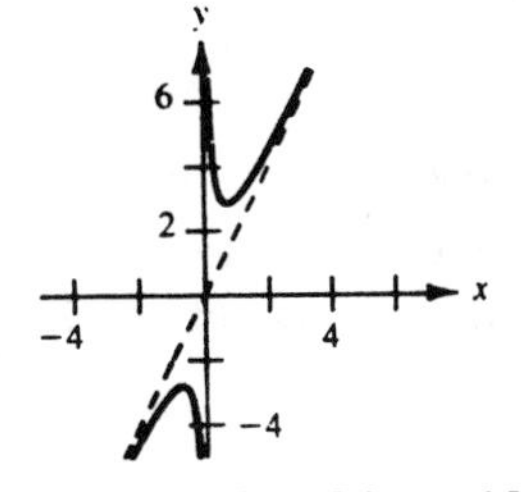

Vertical asymptote: $x = 0$
Slant asymptote: $y = 2x$
Origin symmetry

49. Sketch the graph of the following rational function. As sketching aids, check for intercepts, symmetry, vertical asymptotes, and slant asymptotes.

$$f(x) = \frac{x^3}{x^2 - 1}$$

Solution:

$$f(x) = \frac{x^3}{x^2-1} = x + \frac{x}{x^2-1}$$

Vertical asymptotes: $x = \pm 1$
Slant asymptote: $y = x$
Intercept: $(0, 0)$
Origin symmetry

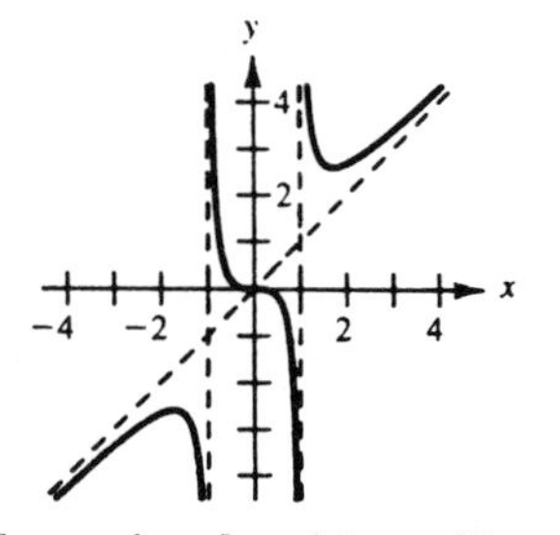

51. Sketch the graph of the following rational function. As sketching aids, check for intercepts, symmetry, vertical asymptotes, and slant asymptotes.

$$f(x) = \frac{x^2 - x + 1}{x - 1}$$

Solution:

$$f(x) = \frac{x^2 - x + 1}{x - 1} = x + \frac{1}{x-1}$$

Vertical asymptote: $x = 1$
Slant asymptote: $y = x$
y-intercept: $(0, -1)$

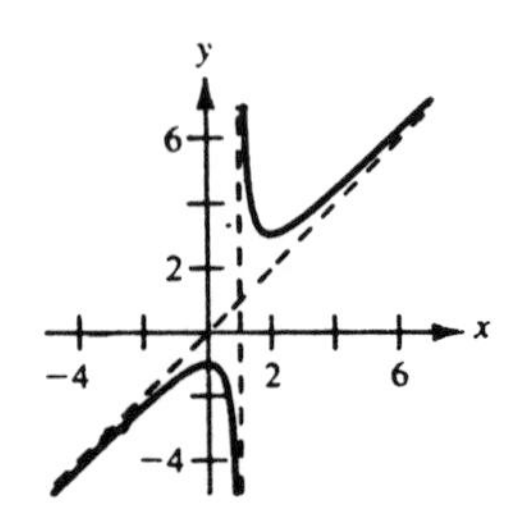

55. The game commission introduces 50 deer into newly acquired state game lands. The population of the herd is given by

$$N = \frac{10(5+3t)}{1+0.04t}, \qquad 0 \le t$$

where t is time in years.

(a) Find the population when t is 5, 10, and 25.
(b) What is the limiting size of the herd as time increases?

Solution:

$$N = \frac{10(5+3t)}{1+0.04t} = 750 - \frac{700}{1+0.04t}$$

(a) $N(5) \approx 167$ deer
$N(10) = 250$ deer
$N(25) = 400$ deer

(b) As $t \to \infty$, $N \to 750$ deer.

59. A right triangle is formed in the first quadrant by the x-axis, the y-axis, and a line segment through the point (2, 3), as shown in the figure.

(a) Show that an equation of the line segment is

$$y = \frac{3(x-a)}{2-a}, \qquad 0 \le x \le a.$$

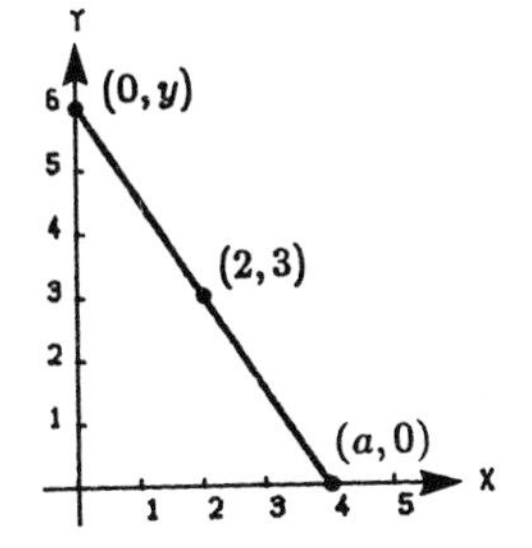

(b) Show that the area of the triangle is

$$A = \frac{-3a^2}{2(2-a)}.$$

(c) Sketch the graph of the area function of part (b), and from the graph estimate the value of a that yields a minimum area.

Solution:

(a) The line passes through the points $(a, 0)$ and $(2, 3)$.

$$m = \frac{3-0}{2-a} = \frac{3}{2-a}$$

$$y - 0 = \frac{3}{2-a}(x-a) \qquad \text{By the point slope equation}$$

$$y = \frac{3(x-a)}{2-a}$$

(b) The area of a triangle is $A = \frac{1}{2}bh$.

$$b = a$$

$$h = y \text{ when } x = 0, \text{ so } h = \frac{3(0-a)}{2-a} = \frac{-3a}{2-a}$$

$$A = \left(\frac{1}{2}a\right)\left(\frac{-3a}{2-a}\right)$$

$$= \frac{-3a^2}{2(2-a)}$$

(c) $A = \dfrac{-3a^2}{2(2-a)} = \dfrac{3}{2}a + 3 + \dfrac{6}{a-2}, \quad a > 2$

A is minimum when $a \approx 4$.

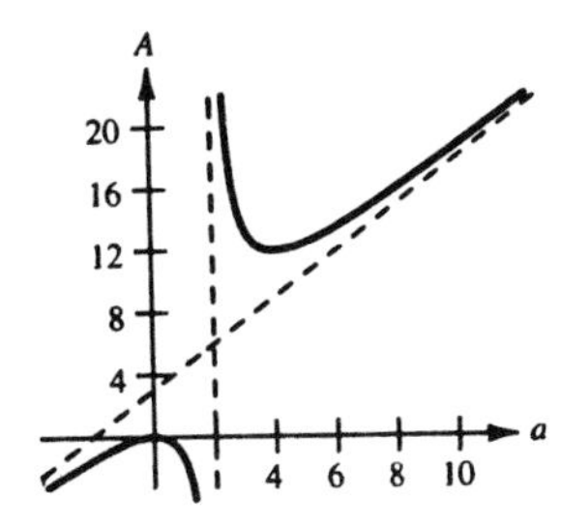

SECTION 3.8

Partial Fractions

- You should know how to decompose a rational function $\frac{N(x)}{D(x)}$ into partial fractions.

  (a) If the fraction is improper, divide to obtain

  $$\frac{N(x)}{D(x)} = p(x) + \frac{N_1(x)}{D(x)}$$

  where $p(x)$ is a polynomial.

  (b) Factor the denominator completely into linear and irreducible (over the reals) quadratic factors.

  (c) For each factor of the form $(px+q)^m$, the partial fraction decomposition includes the terms

  $$\frac{A_1}{(px+q)} + \frac{A_2}{(px+q)^2} + \cdots + \frac{A_m}{(px+q)^m}.$$

  (d) For each factor of the form $(ax^2+bx+c)^n$, the partial fraction decomposition includes the terms

  $$\frac{B_1x+C_1}{ax^2+bx+c} + \frac{B_2x+C_2}{(ax^2+bx+c)^2} + \cdots + \frac{B_nx+C_n}{(ax^2+bx+c)^n}.$$

- You should know how to determine the values of the constants in the numerators.

  (a) Set $\frac{N_1(x)}{D(x)}$ = partial fraction decomposition.

  (b) Multiply both sides by $D(x)$. This is called the basic equation.

  (c) For distinct linear factors, substitute the roots of the distinct linear factors into the basic equation.

  (d) For repeated linear factors, use the coefficients found in part (c) to rewrite the basic equation. Then use other values of x to solve for the remaining coefficients.

  (e) For quadratic factors, expand the basic equation, collect like terms, and then equate the coefficients of like powers.

Solutions to Selected Exercises

3. Write the partial fraction decomposition for the rational expression

$$\frac{1}{x^2 + x}.$$

Solution:
Since $x^2 + x = x(x + 1)$,

$$\frac{1}{x^2 + x} = \frac{A}{x} + \frac{B}{x + 1}$$
$$1 = A(x + 1) + Bx.$$

Let $x = 0:$ $\quad 1 = A$

Let $x = -1:$ $\quad 1 = -B \Rightarrow B = -1$

Thus, $\dfrac{1}{x^2 + x} = \dfrac{1}{x} - \dfrac{1}{x + 1}.$

7. Write the partial fraction decomposition for the rational expression

$$\frac{3}{x^2 + x - 2}.$$

Solution:
Since $x^2 + x - 2 = (x - 1)(x + 2)$,

$$\frac{3}{x^2 + x - 2} = \frac{A}{x - 1} + \frac{B}{x + 2}$$
$$3 = A(x + 2) + B(x - 1).$$

Let $x = 1:$ $\quad 3 = 3A$

$\quad 1 = A$

Let $x = -2:$ $\quad 3 = -3B$

$\quad -1 = B$

Thus, $\dfrac{3}{x^2 + x - 2} = \dfrac{1}{x - 1} - \dfrac{1}{x + 2}.$

11. Write the partial fraction decomposition for the rational expression

$$\frac{x^2 + 12x + 12}{x^3 - 4x}.$$

Solution:
Since $x^3 - 4x = x(x+2)(x-2)$,

$$\frac{x^2+12x+12}{x^3-4x} = \frac{A}{x} + \frac{B}{x+2} + \frac{C}{x-2}$$
$$x^2+12x+12 = A(x+2)(x-2) + Bx(x-2) + Cx(x+2).$$

Let $x = 0$: $\quad 12 = -4A$

$-3 = A$

Let $x = -2$: $\quad -8 = 8B$

$-1 = B$

Let $x = 2$: $\quad 40 = 8C$

$5 = C$

Thus, $\dfrac{x^2+12x+12}{x^3-4x} = -\dfrac{3}{x} - \dfrac{1}{x+2} + \dfrac{5}{x-2}$.

13. Write the partial fraction decomposition for the rational expression

$$\frac{4x^2+2x-1}{x^2(x+1)}.$$

Solution:

$$\frac{4x^2+2x-1}{x^2(x+1)} = \frac{A}{x} + \frac{B}{x^2} + \frac{C}{x+1}$$
$$4x^2+2x-1 = Ax(x+1) + B(x+1) + Cx^2$$

Let $x = 0$: $\quad -1 = B$

Let $x = -1$: $\quad 1 = C$

Let $x = 1$: $\quad 5 = 2A + 2B + C$

$5 = 2A - 2 + 1$

$6 = 2A$

$3 = A$

Thus, $\dfrac{4x^2+2x-1}{x^2(x+1)} = \dfrac{3}{x} - \dfrac{1}{x^2} + \dfrac{1}{x+1}$.

Note: $x^2 = (x-0)^2$ and is a linear factor squared. It is not an irreducible quadratic factor.
Do not write $\dfrac{Bx+C}{x^2}$ in the partial fraction decomposition.

19. Write the partial fraction decomposition for the rational expression

$$\frac{x^2-1}{x(x^2+1)}.$$

Solution:

$$\frac{x^2-1}{x(x^2+1)} = \frac{A}{x} + \frac{Bx+C}{x^2+1}$$
$$x^2-1 = A(x^2+1) + (Bx+C)x$$

Let $x = 0$: $\quad -1 = A$

$$x^2-1 = Ax^2 + A + Bx^2 + Cx = -x^2 - 1 + Bx^2 + Cx = x^2(B-1) + Cx - 1$$

Equating coefficients of like powers,

$$1 = B - 1$$
$$2 = B \quad \text{and} \quad 0 = C$$

Thus, $\dfrac{x^2-1}{x(x^2+1)} = \dfrac{-1}{x} + \dfrac{2x}{x^2+1}$.

23. Write the partial fraction decomposition for the rational expression

$$\frac{x}{16x^4-1}.$$

Solution:
Since $16x^4 - 1 = (4x^2+1)(2x+1)(2x-1)$,

$$\frac{x}{16x^4-1} = \frac{A}{2x+1} + \frac{B}{2x-1} + \frac{Cx+D}{4x^2+1}$$
$$x = A(2x-1)(4x^2+1) + B(2x+1)(4x^2+1) + (Cx+D)(2x+1)(2x-1).$$

Let $x = -\frac{1}{2}$: $\quad -\frac{1}{2} = -4A$

$\frac{1}{8} = A$

Let $x = \frac{1}{2}$: $\quad \frac{1}{2} = 4B$

$\frac{1}{8} = B$

Let $x = 0$: $\quad 0 = -A + B - D$

$0 = -\frac{1}{8} + \frac{1}{8} - D$

$0 = D$

$$\begin{aligned}
\text{Let } x = 1: \quad 1 &= 5A + 15B + 3C + 3D \\
1 &= \tfrac{5}{8} + \tfrac{15}{8} + 3C + 0 \\
1 &= \tfrac{20}{8} + 3C \\
-\tfrac{3}{2} &= 3C \\
-\tfrac{1}{2} &= C
\end{aligned}$$

Thus, $\dfrac{x}{16x^4 - 1} = \dfrac{1/8}{2x+1} + \dfrac{1/8}{2x-1} - \dfrac{x/2}{4x^2+1} = \dfrac{1}{8}\left[\dfrac{1}{2x+1} + \dfrac{1}{2x-1} - \dfrac{4x}{4x^2+1}\right].$

27. Write the partial fraction decomposition for the rational expression

$$\frac{x^2+5}{(x+1)(x^2-2x+3)}.$$

Solution:

$$\frac{x^2+5}{(x+1)(x^2-2x+3)} = \frac{A}{x+1} + \frac{Bx+C}{x^2-2x+3}$$

$$x^2 + 5 = A(x^2 - 2x + 3) + (Bx + C)(x + 1)$$

$$\begin{aligned}
\text{Let } x = -1: \quad 6 &= 6A \\
1 &= A
\end{aligned}$$

$$\begin{aligned}
x^2 + 5 &= x^2 - 2x + 3 + Bx^2 + Bx + Cx + C \\
&= x^2(1 + B) + x(-2 + B + C) + (3 + C)
\end{aligned}$$

Equating coefficients of like powers,

$$1 = 1 + B, \quad 0 = -2 + B + C, \quad \text{and} \quad 5 = 3 + C$$

$$0 = B \qquad\qquad\qquad\qquad 2 = C$$

Thus, $\dfrac{x^2+5}{(x+1)(x^2-2x+3)} = \dfrac{1}{x+1} + \dfrac{2}{x^2-2x+3}.$

31. Write the partial fraction decomposition for the rational expression

$$\frac{x^4}{(x-1)^3}.$$

Solution:

$$\frac{x^4}{(x-1)^3} = \frac{x^4}{x^3 - 3x^2 + 3x - 1} \quad \text{By division}$$

$$= x + 3 + \frac{6x^2 - 8x + 3}{(x-1)^3}$$

$$\frac{6x^2 - 8x + 3}{(x-1)^3} = \frac{A}{x-1} + \frac{B}{(x-1)^2} + \frac{C}{(x-1)^3}$$

$$6x^2 - 8x + 3 = A(x-1)^2 + B(x-1) + C$$

Let $x = 1$: $\quad 1 = C$

Let $x = 0$: $\quad 3 = A - B + C$

$3 = A - B + 1$

$2 = A - B$

Let $x = -1$: $\quad 17 = 4A - 2B + C$

$17 = 4A - 2B + 1$

$16 = 4A - 2B$

$8 = 2A - B$

$-2 = -A + B$

$6 = A$

$2 = A - B$

$2 = 6 - B$

$4 = B$

Thus, $\dfrac{x^4}{(x-1)^3} = x + 3 + \dfrac{6}{x-1} + \dfrac{4}{(x-1)^2} + \dfrac{1}{(x-1)^3}$.

35. Write the partial fraction decomposition for the rational expression

$$\frac{1}{y(L-y)}, \quad L \text{ is a constant.}$$

Solution:

$$\frac{1}{y(L-y)} = \frac{A}{y} + \frac{B}{L-y}$$

$$1 = A(L-y) + By$$

Let $y = 0$: $\quad 1 = LA$

$1/L = A$

Let $y = L$: $\quad 1 = LB$

$1/L = B$

Thus, $\dfrac{1}{y(L-y)} = \dfrac{1/L}{y} + \dfrac{1/L}{L-y} = \dfrac{1}{L}\left(\dfrac{1}{y} + \dfrac{1}{L-y}\right)$.

REVIEW EXERCISES FOR CHAPTER 3

Solutions to Selected Exercises

3. Sketch the graph of $g(x) = x^4 - x^3 - 2x^2$.

Solution:

$$\begin{aligned} g(x) &= x^4 - x^3 - 2x^2 \\ &= x^2(x^2 - x - 2) \\ &= x^2(x + 1)(x - 2) \end{aligned}$$

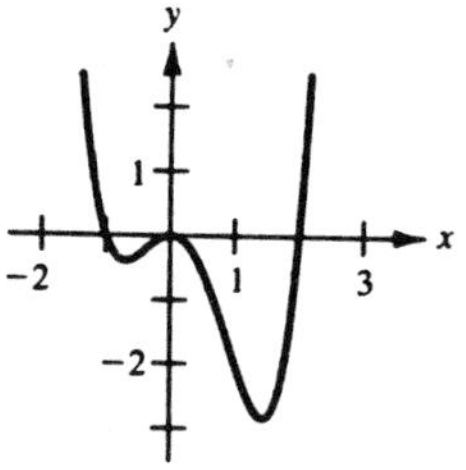

The zeros of g are 0, -1, and 2, so the graph of g crosses the x-axis at these points. Since the degree of g is even and the leading coefficient is positive, the graph moves up to the right and left.

x	0	-1	2	1	$-\frac{1}{2}$	$\frac{1}{2}$
$g(x)$	0	0	0	-2	$-\frac{5}{16}$	$-\frac{9}{16}$

9. Find the maximum or minimum value of $g(x) = x^2 - 2x$.

Solution:

$$\begin{aligned} g(x) &= x^2 - 2x \qquad \text{has a minimum since } a > 0 \\ &= x^2 - 2x + 1 - 1 \\ &= (x - 1)^2 - 1 \end{aligned}$$

Minimum value is $g(1) = -1$.

13. Find the maximum or minimum value of $f(t) = -2t^2 + 4t + 1$.

Solution:

$$\begin{aligned} f(t) &= -2t^2 + 4t + 1 \qquad \text{has a maximum since } a < 0 \\ &= -2(t^2 - 2t - \tfrac{1}{2}) \\ &= -2(t^2 - 2t + 1 - 1 - \tfrac{1}{2}) \\ &= -2[(t - 1)^2 - \tfrac{3}{2}] \\ &= -2(t - 1)^2 + 3 \end{aligned}$$

Maximum value is $f(1) = 3$.

19. Perform the indicated division.

$$\frac{x^4 + x^3 - x^2 + 2x}{x^2 + 2x}$$

Solution:

$$\begin{array}{r|rrrr}
 & x^2 - & x & +1 & \\
\hline
x^2 + 2x & x^4 + & x^3 & - \; x^2 & + 2x \\
 & -(x^4 + & 2x^3) & & \\
\hline
 & & -x^3 & - \; x^2 & \\
 & & -(-x^3 & - \; 2x^2) & \\
\hline
 & & & x^2 & + 2x \\
 & & & -(x^2 & + 2x) \\
\hline
 & & & & 0
\end{array}$$

Thus, $\dfrac{x^4 + x^3 - x^2 + 2x}{x^2 + 2x} = x^2 - x + 1.$

25. Use synthetic division to perform the indicated division.

$$\frac{6x^4 - 4x^3 - 27x^2 + 18x}{x - (2/3)}$$

Solution:

$$\begin{array}{c|rrrrr}
\frac{2}{3} & 6 & -4 & -27 & 18 & 0 \\
 & & 4 & 0 & -18 & 0 \\
\hline
 & 6 & 0 & -27 & 0 & 0
\end{array}$$

Thus, $\dfrac{6x^4 - 4x^3 - 27x^2 + 18x}{x - (2/3)} = 6x^3 - 27x.$

29. Use synthetic division to determine whether the given values of x are zeros of

$$f(x) = 2x^3 + 7x^2 - 18x - 30.$$

(a) $x = 1$ (b) $x = \frac{5}{2}$ (c) $x = -3 + \sqrt{3}$ (d) $x = 0$

Solution:

(a)
$$\begin{array}{r|rrrr} 1 & 2 & 7 & -18 & -30 \\ & & 2 & 9 & -9 \\ \hline & 2 & 9 & -9 & -39 \end{array}$$

$x = 1$ is *not* a zero of f.

(b)
$$\begin{array}{r|rrrr} \frac{5}{2} & 2 & 7 & -18 & -30 \\ & & 5 & 30 & 30 \\ \hline & 2 & 12 & 12 & 0 \end{array}$$

$x = \frac{5}{2}$ *is* a zero of f.

(c)
$$\begin{array}{r|rrrr} -3+\sqrt{3} & 2 & 7 & -18 & -30 \\ & & -6+2\sqrt{3} & 3-5\sqrt{3} & 30 \\ \hline & 2 & 1+2\sqrt{3} & -15-5\sqrt{3} & 0 \end{array}$$

$x = -3 + \sqrt{3}$ *is* a zero of f.

(d)
$$\begin{array}{r|rrrr} 0 & 2 & 7 & -18 & -30 \\ & & 0 & 0 & 0 \\ \hline & 2 & 7 & -18 & -30 \end{array}$$

$x = 0$ is *not* a zero of f.

33. Perform the indicated operations and write the result in standard form.

$$\left(\frac{\sqrt{2}}{2} - \frac{\sqrt{2}}{2}i\right) - \left(\frac{\sqrt{2}}{2} + \frac{\sqrt{2}}{2}i\right)$$

Solution:

$$\left(\frac{\sqrt{2}}{2} - \frac{\sqrt{2}}{2}i\right) - \left(\frac{\sqrt{2}}{2} + \frac{\sqrt{2}}{2}i\right) = \left(\frac{\sqrt{2}}{2} - \frac{\sqrt{2}}{2}\right) + \left(-\frac{\sqrt{2}}{2} - \frac{\sqrt{2}}{2}\right)i = 0 - \sqrt{2}i = -\sqrt{2}i$$

37. Perform the indicated operations and write the result in standard form.

$$(10 - 8i)(2 - 3i)$$

Solution:

$$(10 - 8i)(2 - 3i) = 20 - 30i - 16i + 24i^2 = -4 - 46i$$

41. Perform the indicated operations and write the result in standard form.

$$\frac{4}{-3i}$$

Solution:

$$\frac{4}{-3i} = \frac{4}{-3i} \cdot \frac{3i}{3i} = \frac{12i}{9} = \frac{4i}{3} = 0 + \frac{4}{3}i$$

45. Use synthetic division to find the specified value of $f(x) = x^4 + 10x^3 - 24x^2 + 20x + 44$.

(a) $f(-3)$ (b) $f(\sqrt{2}i)$

Solution:

(a)

-3	1	10	-24	20	44
		-3	-21	135	-465
	1	7	-45	155	-421

Thus, $f(-3) = -421$.

(b)

$\sqrt{2}i$	1	10	-24	20	44
		$0 + \sqrt{2}i$	$-2 + 10\sqrt{2}i$	$-20 - 26\sqrt{2}i$	52
	1	$10 + \sqrt{2}i$	$-26 + 10\sqrt{2}i$	$-26\sqrt{2}i$	96

Thus, $f(\sqrt{2}i) = 96$.

47. Find a fourth degree polynomial with the zeros -1, -1, $\frac{1}{3}$, and $-\frac{1}{2}$.

Solution:

$$
\begin{aligned}
f(x) &= 6(x+1)^2\left(x - \tfrac{1}{3}\right)\left(x + \tfrac{1}{2}\right) && \text{Multiply by 6 to clear the fractions.} \\
&= (x+1)^2 3\left(x - \tfrac{1}{3}\right) 2\left(x + \tfrac{1}{2}\right) \\
&= (x^2 + 2x + 1)(3x - 1)(2x + 1) \\
&= (x^2 + 2x + 1)(6x^2 + x - 1) \\
&= 6x^4 + 13x^3 + 7x^2 - x - 1
\end{aligned}
$$

51. Find all the zeros of $f(x) = 6x^3 - 5x^2 + 24x - 20$.

Solution:

Possible rational zeros: ± 1, ± 2, ± 4, ± 5, ± 10, ± 20, $\pm\frac{1}{2}$, $\pm\frac{5}{2}$, $\pm\frac{1}{3}$, $\pm\frac{2}{3}$, $\pm\frac{4}{3}$, $\pm\frac{5}{3}$, $\pm\frac{10}{3}$, $\pm\frac{20}{3}$, $\pm\frac{1}{6}$, $\pm\frac{5}{6}$

$\frac{5}{6}$	6	-5	24	-20
		5	0	20
	6	0	24	0

$$f(x) = \left(x - \tfrac{5}{6}\right)(6x^2 + 24) = 6\left(x - \tfrac{5}{6}\right)(x^2 + 4) = 6\left(x - \tfrac{5}{6}\right)(x + 2i)(x - 2i)$$

The zeros of f are $\frac{5}{6}$ and $\pm 2i$.

This problem can also be solved by factoring by grouping.

$$
\begin{aligned}
f(x) &= 6x^3 - 5x^2 + 24x - 20 = x^2(6x - 5) + 4(6x - 5) \\
&= (6x - 5)(x^2 + 4) = (6x - 5)(x + 2i)(x - 2i)
\end{aligned}
$$

The zeros of f are $\frac{5}{6}$ and $\pm 2i$.

59. Analyze the following equation and sketch its graph.

$$y = \frac{x^2}{x^2+1}$$

Solution:

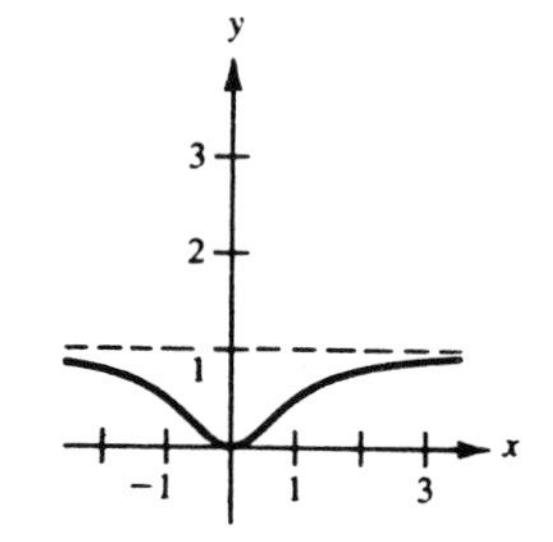

$$y = \frac{x^2}{x^2+1} = 1 - \frac{1}{x^2+1}$$

Intercept: (0, 0)
y-axis symmetry
Horizontal asymptote: $y = 1$

63. Analyze the following equation and sketch its graph.

$$y = \frac{2x^2}{x^2-4}$$

Solution:

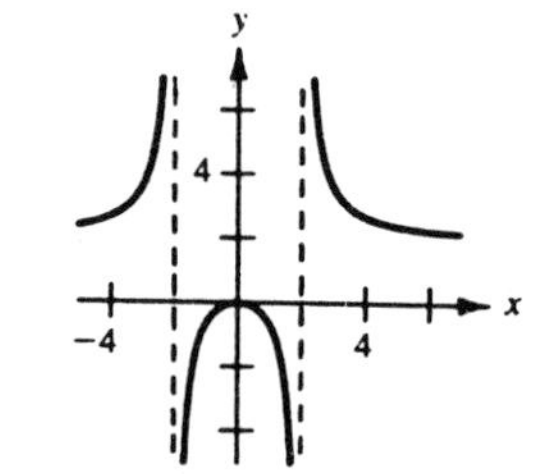

$$y = \frac{2x^2}{x^2-4} = 2 + \frac{8}{x^2-4}$$

Intercept: (0, 0)
y-axis symmetry
Vertical asymptotes: $x = 2$ and $x = -2$
Horizontal asymptote: $y = 2$

65. Write the partial fraction decomposition for $\dfrac{4-x}{x^2+6x+8}$.

Solution:

$$\frac{4-x}{x^2+6x+8} = \frac{4-x}{(x+2)(x+4)}$$

$$\frac{4-x}{(x+2)(x-4)} = \frac{A}{x+2} + \frac{B}{x+4}$$

$$4 - x = A(x+4) + B(x+2)$$

Let $x = -2$: $\quad 6 = 2A \Rightarrow A = 3$
Let $x = -4$: $\quad 8 = -2B \Rightarrow B = -4$

$$\frac{4-x}{x^2+6x+8} = \frac{3}{x+2} - \frac{4}{x+4}$$

69. Write the partial fraction decomposition for $\dfrac{x^2+2x}{x^3-x^2+x-1}$.

Solution:

$$\frac{x^2+2x}{x^3-x^2+x-1} = \frac{x^2+2x}{(x-1)(x^2+1)}$$

$$\frac{x^2+2x}{(x-1)(x^2+1)} = \frac{A}{x-1} + \frac{Bx+C}{x^2+1}$$

$$x^2+2x = A(x^2+1)+(Bx+C)(x-1)$$

Let $x=1$: $\quad 3 = 2A,\ A = 3/2$

Let $x=0$: $\quad 0 = A - C,\ C = 3/2$

Let $x=2$: $\quad 8 = 5A + 2B + C$

$$8 = (15/2) + 2B + (3/2),\ B = -1/2$$

$$\frac{3/2}{x-1} + \frac{-(1/2)x+3/2}{x^2+1} = \frac{1}{2}\left(\frac{3}{x-1} - \frac{x-3}{x^2+1}\right)$$

73. Find the number of units x that produce a maximum revenue R for $R = 900x - 0.1x^2$.

Solution:

$$R = 900x - 0.1x^2 = -0.1(x^2 - 9000x + 4500^2) + 2,025,000$$

$$= 2,025,000 - 0.1(x-4500)^2$$

$$x = 4500 \text{ units}$$

77. A rectangle is bounded by the x-axis, the y-axis, and the graph of $x+2y-6=0$, as shown in the figure. What length and width should the rectangle have so that its area is maximum?

Solution:

$$x + 2y - 6 = 0 \quad \Longrightarrow \quad y = \frac{6-x}{2}$$

The area of the rectangle is

$$A = xy = x\left(\frac{6-x}{2}\right) = -\frac{1}{2}x^2 + 3x$$

$$= -\frac{1}{2}(x^2 - 6x + 9 - 9)$$

$$= -\frac{1}{2}[(x-3)^2 - 9]$$

$$= -\frac{1}{2}(x-3)^2 + \frac{9}{2}$$

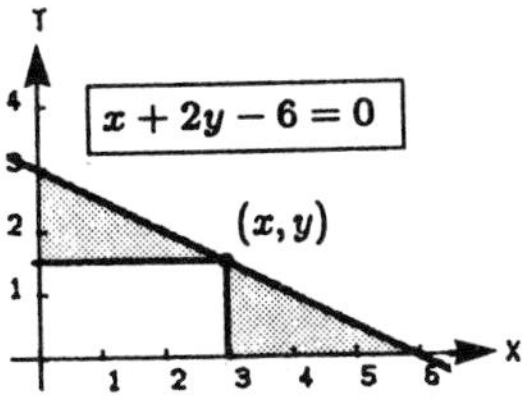

The maximum value of the area is 9/2 square units and this occurs when $x = 3$ and $y = (6-3)/2 = 3/2$.

79. The cost in millions of dollars for the government to seize $p\%$ of a certain illegal drug as it enters the country is given by

$$C = \frac{528p}{100-p}, \quad 0 \leq p < 100.$$

(a) Find the cost of seizing 25%.
(b) Find the cost of seizing 50%.
(c) Find the cost of seizing 75%.
(d) What does the cost approach as p approaches 100?

Solution:

(a) When $p = 25$, $C = \dfrac{528(25)}{100-25} = \$176$ million.

(b) When $p = 50$, $C = \dfrac{528(50)}{100-50} = \$528$ million.

(c) When $p = 75$, $C = \dfrac{528(75)}{100-75} = \$1584$ million.

(d) As $p \to 100$, $C \to \infty$.

Practice Test for Chapter 3

1. Sketch the graph of $f(x) = x^2 - 9$. Identify the vertex and intercepts.

2. Sketch the graph of $g(x) = 3x^2 - 5x - 28$. Identify the vertex and intercepts.

3. Find the quadratic function that has a vertex at $(-2, 1)$ and passes through the point $(3, 4)$.

4. Find all the real zeros of $f(x) = x^4 - 5x^2 + 4$.

5. Find a polynomial that has the zeros -1, $\sqrt{3}$, and $-\sqrt{3}$.

6. Sketch the graph of $f(x) = x^3 - 9x$.

7. Use long division to divide $6x^3 - 5x + 1$ by $x^2 - 2$.

8. Use synthetic division to divide $x^4 + 3x^3 - 9x^2 + 1$ by $x + 3$.

9. Use synthetic division to find $f(-2)$ for $f(x) = x^6 - 3x^3 + x - 5$.

10. Find all real zeros of $p(x) = x^3 - 2x^2 - 5x + 6$.

11. Find all real zeros of $x^3 + x^2 - 5x - 6 = 0$.

12. List all the possible rational zeros of $p(x) = 4x^3 - 7x^2 + x - 20$.

13. Write $5i^3 - (\sqrt{-9})^2$ in standard form.

14. Multiply $(6 + 5i)(-2 + 9i)$ and write the result in standard form.

15. Divide $\dfrac{3 + 2i}{4 - 7i}$ and write the result in standard form.

16. Find all the zeros of $f(x) = x^2 - 3x + 5$.

17. Find all the zeros of $p(x) = x^5 + x^3 - 8x^2 - 8$.

18. Find a polynomial with integer coefficients that has the zeros 0, 4, $2 + i$.

19. Find any horizontal or vertical asymptotes of $h(x) = \dfrac{x^2 + 1}{x^2 - 16}$.

20. Graph $f(x) = \dfrac{x + 3}{x - 2}$.

21. Graph $g(x) = \dfrac{x}{x^2 - 1}$.

22. Find any slant asymptotes of $h(x) = \dfrac{x^2 + 6x - 7}{x + 3}$.

23. Write the partial fraction decomposition for $\dfrac{8x - 2}{x^2 - 2x - 8}$.

24. Write the partial fraction decomposition for $\dfrac{3x^2 + 15}{x^3 + 3x}$.

25. Write the partial fraction decomposition for $\dfrac{2x^3 + 8x^2 + 9x - 1}{x^2 + 4x + 4}$.

CHAPTER 4

Exponential and Logarithmic Functions

SECTION 4.1

Exponential Functions

- You should know that a function of the form $y = a^x$, where $a > 0$, $a \neq 1$, is called an exponential function with base a.
- You should be able to graph exponential functions.
- You should know some properties of exponential functions where $a > 0$ and $a \neq 1$.
  (a) If $a^x = a^y$, then $x = y$.
  (b) If $a^x = b^x$ and $x \neq 0$, then $a = b$.
- You should know formulas for compound interest.
  (a) For n compoundings per year: $A = P\left(1 + \frac{r}{n}\right)^{nt}$.
  (b) For continuous compoundings: $A = Pe^{rt}$.

Solutions to Selected Exercises

3. Use a calculator to evaluate $1000(1.06)^{-5}$. Round your answer to three decimal places.

Solution:

$$1000(1.06)^{-5} \approx 747.258$$

7. Use a calculator to evaluate $8^{2\pi}$. Round your answer to three decimal places.

Solution:

$$8^{2\pi} \approx 472,369.379$$

11. Use a calculator to evaluate e^2. Round your answer to three decimal places.

Solution:

$$e^2 \approx 7.389$$

15. Match $f(x) = 3^x$ with its graph.

Solution:
$f(x) = 3^x$
y-intercept: $(0, 1)$
3^x increases as x increases
Matches graph (g)

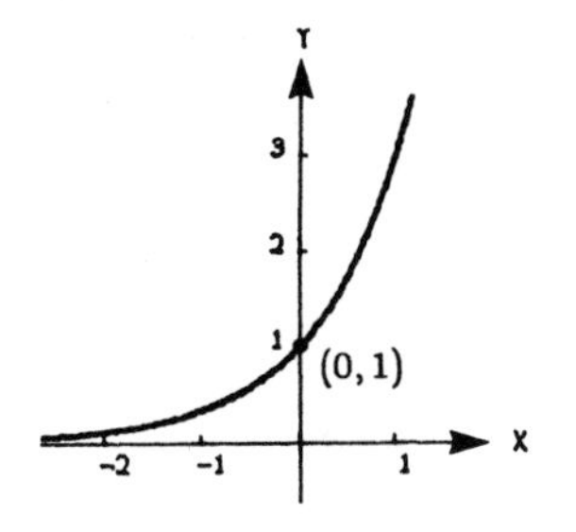

19. Match $f(x) = 3^x - 4$ with its graph.

Solution:
$f(x) = 3^x - 4$
y-intercept: $(0, -3)$
$3^x - 4$ increases as x increases
Matches graph (d)

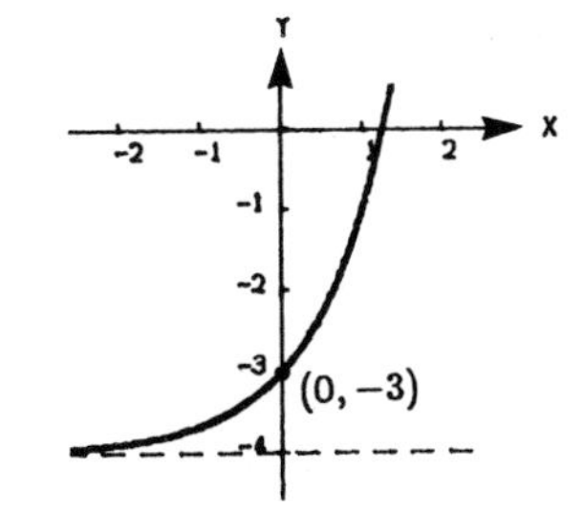

23. Sketch the graph of $g(x) = 5^x$.

Solution:
$g(x) = 5^x$

x	-2	-1	0	1	2
$g(x)$	$\frac{1}{25}$	$\frac{1}{5}$	1	5	25

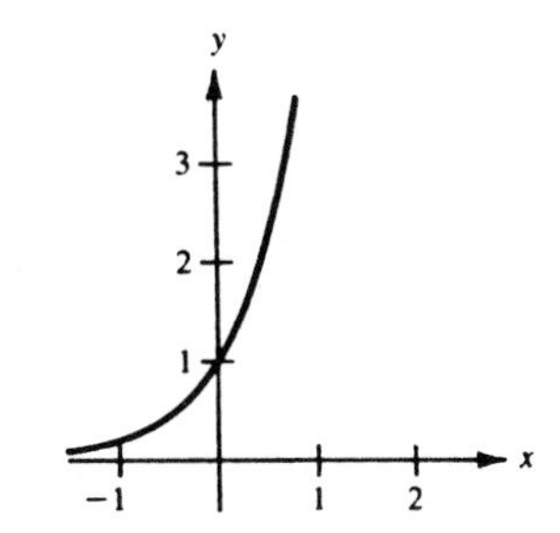

27. Sketch the graph of $h(x) = 5^{x-2}$.

Solution:
$h(x) = 5^{x-2}$

x	-1	0	1	2	3
$h(x)$	$\frac{1}{125}$	$\frac{1}{25}$	$\frac{1}{5}$	1	5

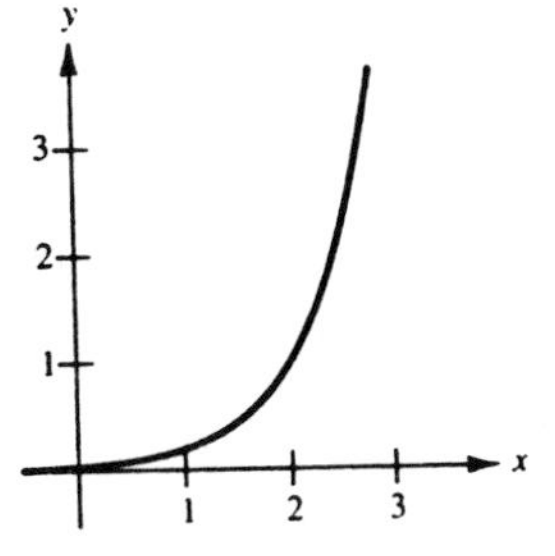

31. Sketch the graph of $y = 2^{-x^2}$.

Solution:

$y = 2^{-x^2} = \left(\frac{1}{2}\right)^{x^2}$

x	0	1	-1	2	-2
y	1	$\frac{1}{2}$	$\frac{1}{2}$	$\frac{1}{16}$	$\frac{1}{16}$

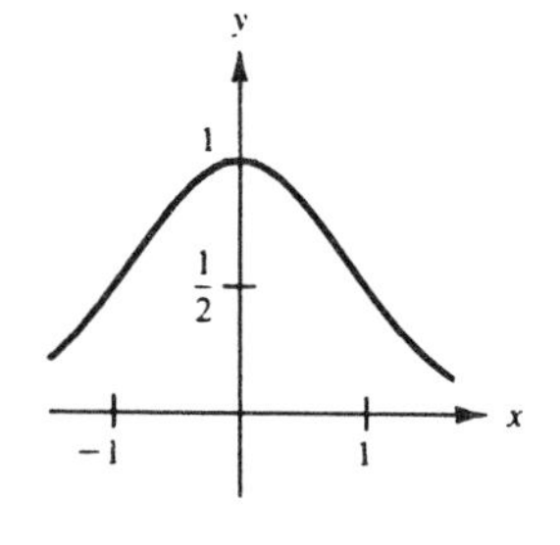

33. Sketch the graph of $y = 3^{|x|}$.

Solution:

$y = 3^{|x|}$

x	0	1	2	-1	-2
y	1	3	9	3	9

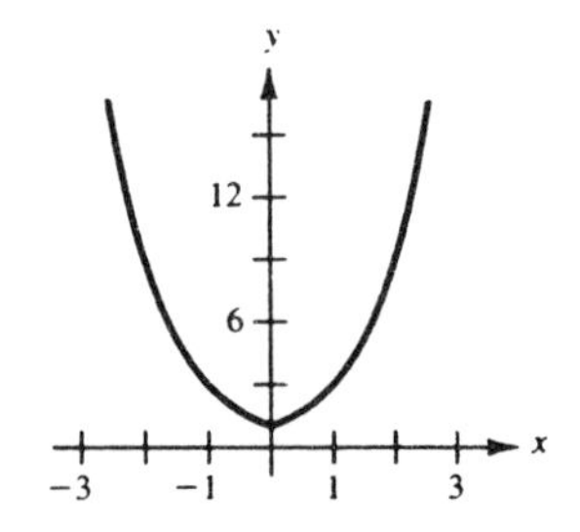

37. Sketch the graph of $f(x) = e^{2x}$.

Solution:

$f(x) = e^{2x}$

x	0	1	2	-1	-2
$f(x)$	1	7.39	54.60	0.135	0.02

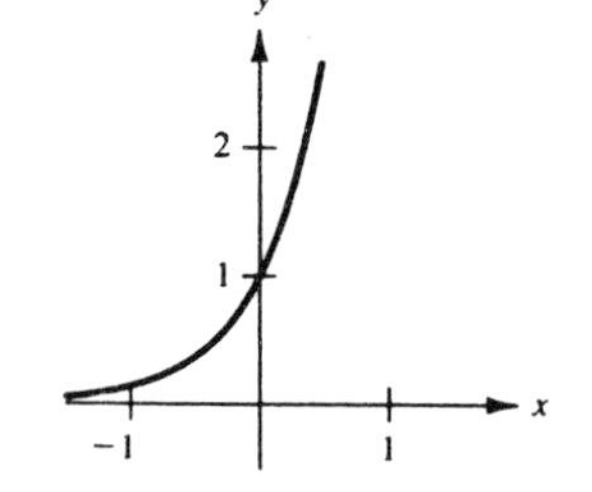

43. Complete the following table to determine the balance A for \$2500 invested at 12% for 20 years and compounded n times per year.

n	1	2	4	12	365	Continuous compounding
A						

Solution:

$A = P\left(1 + \frac{r}{n}\right)^{nt}$

$P = 2500, \ r = 0.12, \ t = 20$

When $n = 1, \ A = 2500\left(1 + \frac{0.12}{1}\right)^{(1)(20)} \approx \$24,115.73$

When $n = 2, \ A = 2500\left(1 + \frac{0.12}{2}\right)^{(2)(20)} \approx \$25,714.29$

When $n = 4, \ A = 2500\left(1 + \frac{0.12}{4}\right)^{(4)(20)} \approx \$26,602.23$

When $n = 12, \ A = 2500\left(1 + \frac{0.12}{12}\right)^{(12)(20)} \approx \$27,231.38$

When $n = 365, \ A = 2500\left(1 + \frac{0.12}{365}\right)^{(365)(20)} \approx \$27,547.07$

For continuous compounding, $A = Pe^{rt}, \ \ A = 2500e^{(0.12)(20)} \approx \$27,557.94$

n	1	2	4	12	365	Continuous compounding
A	\$24,115.73	\$25,714.29	\$26,602.23	\$27,231.38	\$27,547.07	\$27,557.94

49. The demand equation for a certain product is given by $p = 500 - 0.5e^{0.004x}$. Find the price p for a demand of (a) $x = 1000$ units and (b) $x = 1500$ units.

Solution:

(a) $x = 1000$

$$p = 500 - 0.5e^4 \approx \$472.70$$

(b) $x = 1500$

$$p = 500 - 0.5e^6 \approx \$298.29$$

53. Given the exponential function $f(x) = a^x$, show that (a) $f(u + v) = f(u) \cdot f(v)$ and (b) $f(2x) = [f(x)]^2$.

Solution:

(a) $f(u+v) = a^{u+v} = a^u \cdot a^v = f(u) \cdot f(v)$

(b) $f(2x) = a^{2x} = (a^x)^2 = [f(x)]^2$

SECTION 4.2

Logarithmic Functions

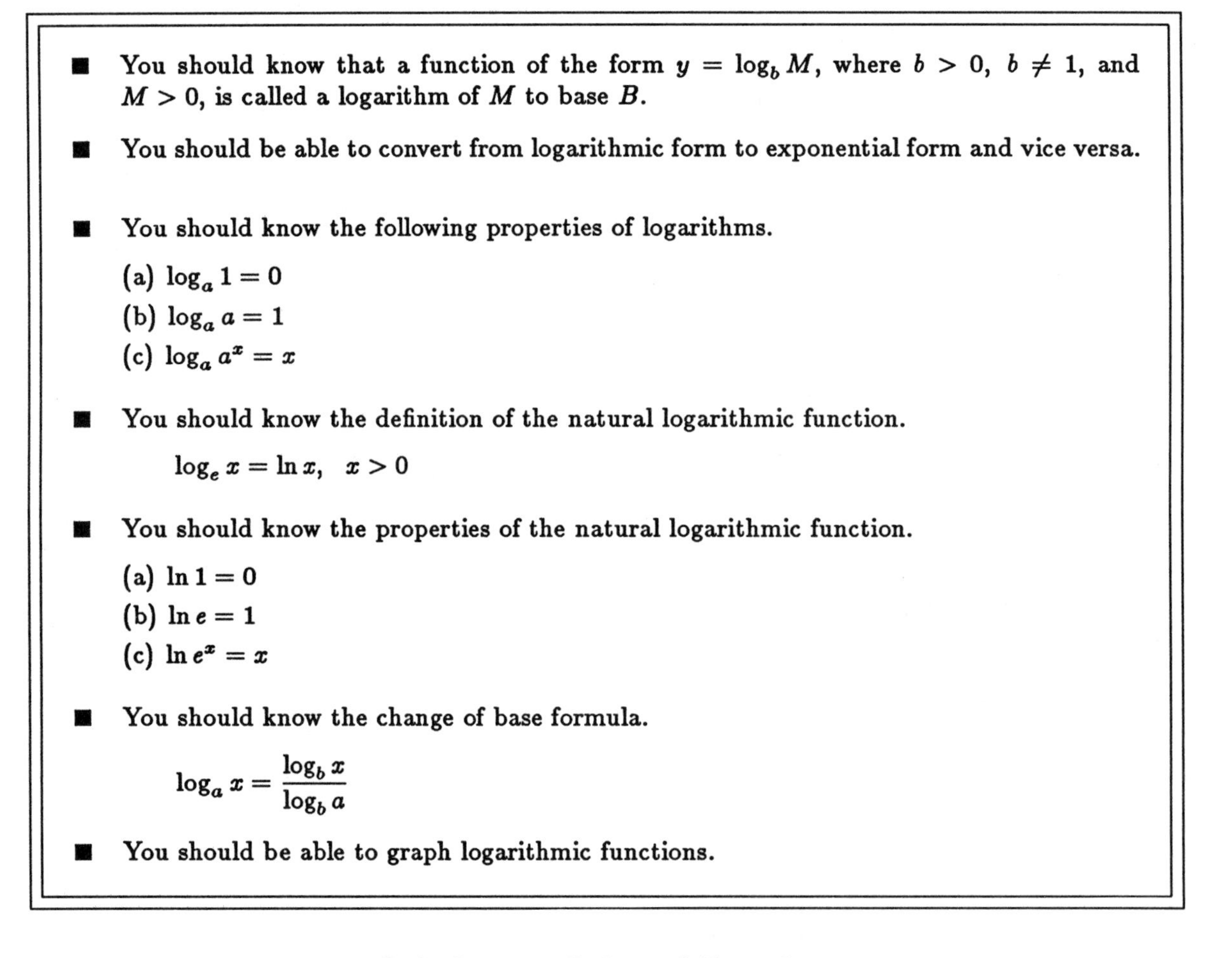

- You should know that a function of the form $y = \log_b M$, where $b > 0$, $b \neq 1$, and $M > 0$, is called a logarithm of M to base B.
- You should be able to convert from logarithmic form to exponential form and vice versa.
- You should know the following properties of logarithms.
  (a) $\log_a 1 = 0$
  (b) $\log_a a = 1$
  (c) $\log_a a^x = x$
- You should know the definition of the natural logarithmic function.

  $$\log_e x = \ln x, \quad x > 0$$

- You should know the properties of the natural logarithmic function.
  (a) $\ln 1 = 0$
  (b) $\ln e = 1$
  (c) $\ln e^x = x$
- You should know the change of base formula.

  $$\log_a x = \frac{\log_b x}{\log_b a}$$

- You should be able to graph logarithmic functions.

Solutions to Selected Exercises

5. Evaluate $\log_{16} 4$ without using a calculator.

Solution:

$$\log_{16} 4 = \log_{16} \sqrt{16} = \log_{16} 16^{1/2} = \tfrac{1}{2}$$

9. Evaluate $\log_{10} 0.01$ without using a calculator.

Solution:

$$\log_{10} 0.01 = \log_{10} \tfrac{1}{100} = \log_{10} 10^{-2} = -2$$

13. Evaluate $\ln e^{-2}$ without using a calculator.

Solution:

$$\ln e^{-2} = -2$$

17. Use the definition of a logarithm to write $5^3 = 125$ in logarithmic form.

Solution:

$$5^3 = 125$$
$$\log_5 125 = 3$$

23. Use the definition of a logarithm to write $e^3 = 20.0855\ldots$ in logarithmic form.

Solution:

$$e^3 = 20.0855\ldots$$
$$\log_e 20.0855\ldots = 3$$
$$\ln 20.0855\ldots \approx 3$$

27. Use a calculator to evaluate $\log_{10} 345$. Round your answer to three decimal places.

Solution:

$$\log_{10} 345 = 2.537819095\ldots$$
$$\approx 2.538$$

31. Use a calculator to evaluate $\ln(1 + \sqrt{3})$. Round your answer to three decimal places.

Solution:

$$\ln(1 + \sqrt{3}) = 1.005052539\ldots$$
$$\approx 1.005$$

35. Demonstrate that $f(x) = e^x$ and $g(x) = \ln x$ are inverses of each other by sketching their graphs on the same coordinate plane.

Solution:

x	-2	-1	0	1	2	3
$f(x)$	0.135	0.368	1	2.718	7.389	20.086
$g(x)$	—	—	—	0	0.693	1.097

The graph of g is obtained by reflecting the graph of f about the line $y = x$.

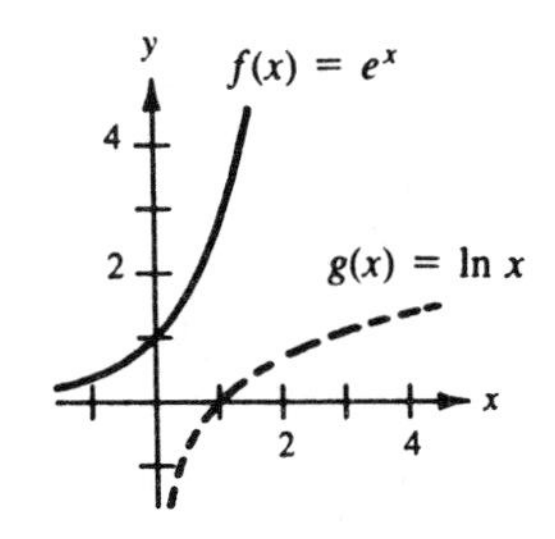

39. Use the graph of $y = \ln x$ to match $f(x) = -\ln(x + 2)$ to its graph.

Solution:
$f(x) = -\ln(x + 2)$
Vertical asymptote: $x = -2$
x-intercept: $(-1,\ 0)$
Matches graph (a)

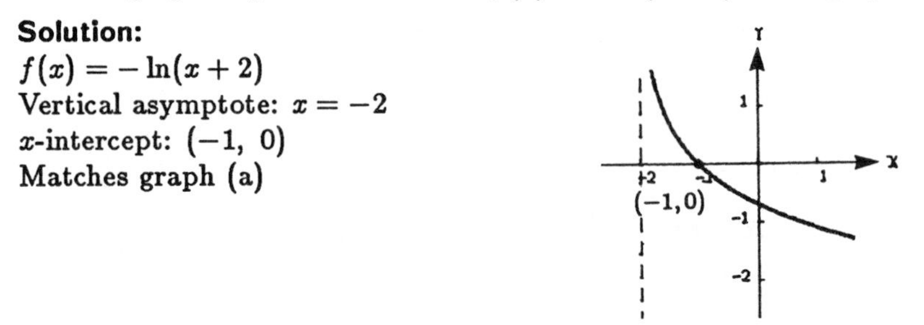

45. Find the domain, vertical asymptote, and x-intercept of $h(x) = \log_4(x - 3)$, and sketch its graph.

Solution:

Domain: $x - 3 > 0 \Rightarrow x > 3$
The domain is $(3,\ \infty)$.

Vertical asymptote: $x - 3 = 0 \Rightarrow x = 3$
The vertical asymptote is the line $x = 3$.

x-intercept: $\log_4(x - 3) = 0$
$x - 3 = 1 \Rightarrow x = 4$
The x-intercept is $(4,\ 0)$.

x	3.5	4	5	7
$h(x)$	-0.5	0	0.5	1

51. Use the change of base formula to write $\log_3 5$ as a multiple of a common logarithm.

Solution:

$$\log_3 5 = \frac{\log_{10} 5}{\log_{10} 3}$$

55. Use the change of base formula to write $\log_3 5$ as a multiple of a natural logarithm.

Solution:

$$\log_3 5 = \frac{\ln 5}{\ln 3}$$

59. Evaluate $\log_3 7$ using the change of base formula. Do the problem twice; once with common logarithms and once with natural logarithms. Round your answer to three decimal places.

Solution:

$$\log_3 7 = \frac{\log_{10} 7}{\log_{10} 3} = 1.771243749\ldots$$
$$\approx 1.771$$
$$\log_3 7 = \frac{\ln 7}{\ln 3} = 1.771243749\ldots$$
$$\approx 1.771$$

67. Students in a mathematics class were given an exam and then retested monthly with an equivalent exam. The average score for the class was given by the human memory model

$$f(t) = 80 - 17\log_{10}(t+1), \quad 0 \le t \le 12$$

where t is the time in months.

(a) What was the average score on the original exam ($t = 0$)?
(b) What was the average score after four months?
(c) What was the average score after ten months?

Solution:
(a) $f(0) = 80 - 17\log_{10} 1 = 80.0$
(b) $f(4) = 80 - 17\log_{10} 5 \approx 68.1$
(c) $f(10) = 80 - 17\log_{10} 11 \approx 62.3$

69. The population of a town will double in

$$t = \frac{10 \ln 2}{\ln 67 - \ln 50}$$

years. Find t.

Solution:

$$t = \frac{10 \ln 2}{\ln 67 - \ln 50}$$
$$t \approx \frac{6.931471806}{4.204692619 - 3.912023005}$$
$$t \approx 23.68 \text{ years}$$

73. (a) Use a calculator to complete the following table for the function

$$f(x) = \frac{\ln x}{x}.$$

x	1	5	10	10^2	10^4	10^6
$f(x)$						

(b) Use the table in part (a) to determine what $f(x)$ approaches as x increases without bound.

Solution:

(a)

x	1	5	10	10^2	10^4	10^6
$f(x)$	0	0.322	0.230	0.046	0.00092	0.0000138

(b) As $x \to \infty$, $f(x) \to 0$.

SECTION 4.3

Properties of Logarithms

- You should know the following properties of logarithms.
  (a) $\log_a(uv) = \log_a u + \log_a v$
  (b) $\log_a(u/v) = \log_a u - \log_a v$
  (c) $\log_a u^n = n \log_a u$
- You should be able to rewrite logarithmic expressions.

Solutions to Selected Exercises

5. Use the properties of logarithms to write $\log_8 x^4$ as a sum, difference, or multiple of logarithms.

Solution:

$$\log_8 x^4 = 4 \log_8 x$$

9. Use the properties of logarithms to write $\log_2 xyz$ as a sum, difference, or multiple of logarithms.

Solution:

$$\log_2 xyz = \log_2[x(yz)] = \log_2 x + \log_2 yz = \log_2 x + \log_2 y + \log_2 z$$

15. Use the properties of logarithms to write the following expression as a sum, difference, or multiple of logarithms.

$$\log_b \frac{x^2}{y^2 z^3}$$

Solution:

$$\log_b \frac{x^2}{y^2 z^3} = \log_b x^2 - \log_b y^2 z^3 = \log_b x^2 - [\log_b y^2 + \log_b z^3] = 2 \log_b x - 2 \log_b y - 3 \log_b z$$

19. Use the properties of logarithms to write the following expression as a sum, difference, or multiple of logarithms.

$$\log_9 \frac{x^4 \sqrt{y}}{z^5}$$

Solution:

$$\log_9 \frac{x^4 \sqrt{y}}{z^5} = \log_9 x^4 \sqrt{y} - \log_9 z^5 = \log_9 x^4 + \log_9 \sqrt{y} - \log_9 z^5 = 4 \log_9 x + \frac{1}{2} \log_9 y - 5 \log_9 z$$

23. Write $\log_4 z - \log_4 y$ as the logarithm of a single quantity.

Solution:

$$\log_4 z - \log_4 y = \log_4 \frac{z}{y}$$

27. Write $\ln x - 3\ln(x+1)$ as the logarithm of a single quantity.

Solution:

$$\ln x - 3\ln(x+1) = \ln x - \ln(x+1)^3 = \ln \frac{x}{(x+1)^3}$$

33. Write $\ln x - 2[\ln(x+2) + \ln(x-2)]$ as the logarithm of a single quantity.

Solution:

$$\begin{aligned}\ln x - 2[\ln(x+2) + \ln(x-2)] &= \ln x - 2\ln(x+2)(x-2)\\ &= \ln x - 2\ln(x^2-4)\\ &= \ln x - \ln(x^2-4)^2\\ &= \ln \frac{x}{(x^2-4)^2}\end{aligned}$$

37. Write $\frac{1}{3}[\ln y + 2\ln(y+4)] - \ln(y-1)$ as the logarithm of a single quantity.

Solution:

$$\begin{aligned}\frac{1}{3}[\ln y + 2\ln(y+4)] - \ln(y-1) &= \frac{1}{3}[\ln y + \ln(y+4)^2] - \ln(y-1)\\ &= \frac{1}{3}\ln[y(y+4)^2] - \ln(y-1)\\ &= \ln \sqrt[3]{y(y+4)^2} - \ln(y-1)\\ &= \ln \frac{\sqrt[3]{y(y+4)^2}}{y-1}\end{aligned}$$

41. Approximate $\log_b 6$ using the properties of logarithms, given $\log_b 2 \approx 0.3562$ and $\log_b 3 \approx 0.5646$.

Solution:

$$\log_b 6 = \log_b(2 \cdot 3) = \log_b 2 + \log_b 3 \approx 0.3562 + 0.5646 = 0.9208$$

47. Approximate $\log_b \sqrt{2}$ using the properties of logarithms, given $\log_b 2 \approx 0.3562$.

Solution:

$$\log_b \sqrt{2} = \log_b(2^{1/2}) = \tfrac{1}{2}\log_b 2 \approx \tfrac{1}{2}(0.3562) = 0.1781$$

51. Approximate $\log_b \frac{1}{4}$ using the properties of logarithms, given $\log_b 2 \approx 0.3562$.

Solution:

$$\log_b \tfrac{1}{4} = \log_b 1 - \log_b 4 = 0 - \log_b 2^2 = -2\log_b 2 \approx -2(0.3562) = -0.7124$$

55. Approximate the following using the properties of logarithms, given $\log_b 2 \approx 0.3562$ and $\log_b 3 \approx 0.5646$.

$$\log_b\left[\frac{(4.5)^3}{\sqrt{3}}\right]$$

Solution:

$$\begin{aligned}\log_b\left[\frac{(4.5)^3}{\sqrt{3}}\right] &= 3\log_b 4.5 - \frac{1}{2}\log_b 3\\ &= 3\log_b \frac{9}{2} - \frac{1}{2}\log_b 3\\ &= 3[\log_b 9 - \log_b 2] - \frac{1}{2}\log_b 3\\ &= 3[2\log_b 3 - \log_b 2] - \frac{1}{2}\log_b 3\\ &= 3[2(0.5646) - 0.3562] - \frac{1}{2}(0.5646)\\ &= 2.0367\end{aligned}$$

59. Find the exact value of $\log_4 16^{1.2}$.

Solution:

$$\log_4 16^{1.2} = 1.2\log_4 16 = 1.2\log_4 4^2 = (1.2)(2)\log_4 4 = (2.4)(1) = 2.4$$

63. Use the properties of logarithms to simplify $\log_4 8$.

Solution:

$$\log_4 8 = \log_4 2^3 = 3\log_4 2 = 3\log_4 \sqrt{4} = 3\log_4 4^{1/2} = 3\left(\tfrac{1}{2}\right)\log_4 4 = \tfrac{3}{2}$$

67. Use the properties of logarithms to simplify $\log_5 \frac{1}{250}$.

Solution:

$$\begin{aligned}\log_5 \tfrac{1}{250} &= \log_5 1 - \log_5 250 = 0 - \log_5(125 \cdot 2)\\ &= -\log_5(5^3 \cdot 2) = -[\log_5 5^3 + \log_5 2]\\ &= -[3\log_5 5 + \log_5 2] = -3 - \log_5 2\end{aligned}$$

71. Prove that $\log_b \dfrac{u}{v} = \log_b u - \log_b v$.

Solution:

Let $x = \log_b u$ and $y = \log_b v$, then $b^x = u$ and $b^y = v$.

$$\frac{u}{v} = \frac{b^x}{b^y} = b^{x-y}$$

$$\log_b\left(\frac{u}{v}\right) = \log_b(b^{x-y}) = x - y = \log_b u - \log_b v$$

SECTION 4.4

Solving Exponential and Logarithmic Equations

- You should be able to solve exponential and logarithmic equations.
- To solve an exponential equation, take the logarithm of both sides.
- To solve a logarithmic equation, rewrite it in exponential form.

Solutions to Selected Exercises

5. Solve $\left(\frac{3}{4}\right)^x = \frac{27}{64}$ for x.

Solution:

$$\left(\tfrac{3}{4}\right)^x = \tfrac{27}{64}$$
$$\left(\tfrac{3}{4}\right)^x = \left(\tfrac{3}{4}\right)^3$$
$$x = 3$$

9. Solve $\log_{10} x = -1$ for x.

Solution:

$$\log_{10} x = -1$$
$$x = 10^{-1} = \tfrac{1}{10}$$

13. Apply the inverse properties of $\ln x$ and e^x to simplify $e^{\ln(5x+2)}$.

Solution:

$$e^{\ln(5x+2)} = 5x + 2$$

17. Solve $10^x = 42$.

Solution:

$$10^x = 42$$
$$x = \log_{10} 42 \approx 1.6232$$

21. Solve $3(10^{x-1}) = 2$.

Solution:

$$\begin{aligned} 3(10^{x-1}) &= 2 \\ 10^{x-1} &= \tfrac{2}{3} \\ x - 1 &= \log_{10} \tfrac{2}{3} \\ x &= 1 + \log_{10} \tfrac{2}{3} \approx 0.8239 \end{aligned}$$

25. Solve $500e^{-x} = 300$.

Solution:

$$\begin{aligned} 500e^{-x} &= 300 \\ e^{-x} &= \tfrac{3}{5} \\ -x &= \ln \tfrac{3}{5} \\ x &= -\ln \tfrac{3}{5} = \ln \tfrac{5}{3} \approx 0.5108 \end{aligned}$$

29. Solve $25e^{2x+1} = 962$.

Solution:

$$\begin{aligned} 25e^{2x+1} &= 962 \\ e^{2x+1} &= \tfrac{962}{25} \\ 2x + 1 &= \ln \tfrac{962}{25} \\ 2x &= -1 + \ln \tfrac{962}{25} \\ x &= \tfrac{1}{2}\left[-1 + \ln \tfrac{962}{25}\right] = -\tfrac{1}{2} + \tfrac{1}{2} \ln \tfrac{962}{25} \approx 1.3251 \end{aligned}$$

35. Solve $\left(1 + \frac{0.10}{12}\right)^{12t} = 2$.

Solution:

$$\begin{aligned} \left(1 + \frac{0.10}{12}\right)^{12t} &= 2 \\ \ln\left(1 + \frac{0.10}{12}\right)^{12t} &= \ln 2 \\ 12t \ln\left(1 + \frac{0.10}{12}\right) &= \ln 2 \\ t &= \frac{\ln 2}{12 \ln\left(1 + \frac{0.10}{12}\right)} \approx 6.9603 \end{aligned}$$

39. Solve $\left(\frac{1}{1.0775}\right)^N = 0.2247$.

Solution:

$$\left(\frac{1}{1.0775}\right)^N = 0.2247$$

$$N = \frac{\ln(0.2247)}{\ln\left(\frac{1}{1.0775}\right)} = \frac{\ln(0.2247)}{\ln 1 - \ln(1.0775)} = \frac{\ln(0.2247)}{-\ln(1.0775)} \approx 20.0016$$

43. Solve $3(1 + e^{2x}) = 4$.

Solution:

$$\begin{aligned} 3(1 + e^{2x}) &= 4 \\ 1 + e^{2x} &= \tfrac{4}{3} \\ e^{2x} &= \tfrac{1}{3} \\ 2x &= \ln \tfrac{1}{3} \\ x &= \tfrac{1}{2} \ln \tfrac{1}{3} \approx -0.5493 \end{aligned}$$

47. Solve $\dfrac{e^x + e^{-x}}{e^x - e^{-x}} = 2$.

Solution:

$$\begin{aligned} \frac{e^x + e^{-x}}{e^x - e^{-x}} &= 2 \\ \frac{e^x(e^x + e^{-x})}{e^x(e^x - e^{-x})} &= 2 \\ \frac{e^{2x} + 1}{e^{2x} - 1} &= 2 \\ e^{2x} + 1 &= 2(e^{2x} - 1) \\ 3 &= e^{2x} \\ 2x &= \ln 3 \\ x &= \frac{1}{2} \ln 3 \approx 0.549 \end{aligned}$$

51. Solve $2 \ln x = 7$.

Solution:

$$\begin{aligned} 2 \ln x &= 7 \\ \ln x &= \tfrac{7}{2} \\ x &= e^{7/2} \approx 33.1154 \end{aligned}$$

55. Solve $\log_{10}(z - 3) = 2$.

Solution:

$$\begin{aligned} \log_{10}(z - 3) &= 2 \\ z - 3 &= 10^2 \\ z &= 10^2 + 3 = 103 \end{aligned}$$

59. Solve $\log_{10}(x+4) - \log_{10} x = \log_{10}(x+2)$.

Solution:

$$\begin{aligned}
\log_{10}(x+4) - \log_{10} x &= \log_{10}(x+2) \\
\log_{10}\left(\frac{x+4}{4}\right) &= \log_{10}(x+2) \\
\frac{x+4}{x} &= x+2 \\
x+4 &= x^2+2x \\
0 &= x^2+x-4 \\
x &= \frac{-1 \pm \sqrt{17}}{2} = -\frac{1}{2} \pm \frac{\sqrt{17}}{2} \qquad \text{Quadratic Formula}
\end{aligned}$$

63. Solve $\ln x^2 = (\ln x)^2$.

Solution:

$$\begin{aligned}
\ln x^2 &= (\ln x)^2 \\
2\ln x &= (\ln x)^2 \\
0 &= (\ln x)^2 - 2\ln x \\
0 &= \ln x(\ln x - 2)
\end{aligned}$$

$$\begin{array}{rcr}
\ln x = 0 & \text{or} & \ln x - 2 = 0 \\
x = e^0 & \text{or} & \ln x = 2 \\
x = 1 & \text{or} & x = e^2
\end{array}$$

67. The demand equation for a certain product is given by $p = 500 - 0.5(e^{0.004x})$. Find the demand x for a price of (a) $p = \$350$ and (b) $p = \$300$.

Solution:

(a)

$$\begin{aligned}
350 &= 500 - 0.5(e^{0.004x}) \\
-150 &= -0.5(e^{0.004x}) \\
300 &= e^{0.004x} \\
0.004x &= \ln 300 \\
x &= \frac{\ln 300}{0.004} \approx 1426 \text{ units}
\end{aligned}$$

(b)

$$\begin{aligned}
300 &= 500 - 0.5(e^{0.004x}) \\
-200 &= -0.5(e^{0.004x}) \\
400 &= e^{0.004x} \\
0.004x &= \ln 400 \\
x &= \frac{\ln 400}{0.004} \approx 1498 \text{ units}
\end{aligned}$$

SECTION 4.5

Applications of Exponential and Logarithmic Functions

- You should be able to solve compound interest problems.
  (a) Compound interest formulas:
    1. $A = P\left(1 + \frac{r}{n}\right)^{nt}$
    2. $A = Pe^{rt}$

  (b) Doubling time:
    1. $t = \frac{\ln 2}{n \ln[1 + (r/n)]}$, n compoundings per year
    2. $t = \frac{\ln 2}{r}$, continuous compounding

  (c) Effective yield:
    1. Effective yield $= \left(1 + \frac{r}{n}\right)^{n} - 1$, n compoundings per year
    2. Effective yield $= e^{r} - 1$, continuous compounding

- You should be able to solve growth and decay problems.

  $$Q(t) = Ce^{kt}$$

  (a) If $k > 0$, the population grows.
  (b) If $k < 0$, the population decays.
  (c) Ratio of Carbon 12 to Carbon 14 is $R(t) = \frac{1}{10^{12}} 2^{-t/5700}$

- You should be able to solve logistics model problems.

  $$Q(t) = \frac{M}{1 + \left(\frac{M}{Q_0} - 1\right)e^{-kt}}$$

- You should be able to solve intensity model problems.

  $$S = k \log_{10} \frac{I}{I_0}$$

Solutions to Selected Exercises

5. \$500 is deposited into an account with continuously compounded interest. If the balance is \$1292.85 after ten years, find the annual percentage rate, the effective yield, and the time to double.

Solution:
$P = 500, \; A = 1292.85, \; t = 10$
$A = Pe^{rt}$

$$1292.85 = 500e^{10r}$$
$$\frac{1292.85}{500} = e^{10r}$$
$$10r = \ln\left(\frac{1292.85}{500}\right)$$
$$r = \frac{1}{10}\ln\left(\frac{1292.85}{100}\right) \approx 0.095 = 9.5\%$$

Effective yield $= e^{0.095} - 1 \approx 0.09966 \approx 9.97\%$

Time to double: $1000 = 500e^{0.095t}$
$$2 = e^{0.095t}$$
$$0.095t = \ln 2$$
$$t = \frac{\ln 2}{0.095} \approx 7.30 \text{ years}$$

9. \$5000 is deposited into an account with continuously compounded interest. If the effective yield is 8.33%, find the annual percentage rate, the time to double, and the amount after 10 years.

Solution:
$P = 5000$
Effective yield $= 8.33\%$
$0.0833 = e^r - 1$

$$r = \ln 1.0833 \approx 0.0800 = 8\%$$

Time to double: $10,000 = 5000e^{0.08t}$
$$t = \frac{\ln 2}{0.08} \approx 8.66 \text{ years}$$

After 10 years: $A = 5000e^{0.08(10)} \approx \$11,127.70$

13. Determine the time necessary for \$1000 to double if it is invested at 11% compounded (a) annually, (b) monthly, (c) daily, and (d) continuously.

Solution:

$P = 1000, \ r = 11\%$

(a) $n = 1$

$$t = \frac{\ln 2}{\ln(1 + 0.11)} \approx 6.642 \text{ years}$$

(b) $n = 12$

$$t = \frac{\ln 2}{12 \ln\left(1 + \frac{0.11}{12}\right)} \approx 6.330 \text{ years}$$

(c) $n = 365$

$$t = \frac{\ln 2}{365 \ln\left(1 + \frac{0.11}{365}\right)} \approx 6.302 \text{ years}$$

(d) Continuously

$$t = \frac{\ln 2}{0.11} \approx 6.301 \text{ years}$$

17. \$50 is deposited monthly into a savings account at an annual rate of 7% compounded monthly. Find the balance, A, after 20 years given that

$$A = \frac{P(e^{rt} - 1)}{e^{r/12} - 1}.$$

Solution:

$p = 50, \ r = 7\%, \ t = 20$

$$A = \frac{50(e^{0.07(20)} - 1)}{e^{0.07/12} - 1} \approx \$26,111.12$$

21. The population P of a city is given by $P = 105,300e^{0.015t}$, where t is the time in years with $t = 0$ corresponding to 1985. According to this model, in what year will the city have a population of 150,000?

Solution:

$$150,000 = 105,300e^{0.015t}$$

$$0.015t = \ln\left(\frac{150,000}{105,300}\right)$$

$$t = \frac{1}{0.015} \ln\left(\frac{150,000}{105,300}\right) \approx 23.588 \text{ years}$$

$$1985 + 24 = 2009$$

The city will have a population of 150,000 in the year 2009.

25. The half-life of the isotope Ra^{226} is 1,620 years. If the initial quantity is 10 grams, how much will remain after 1000 years, and after 10,000 years?

Solution:

$$Q(t) = Ce^{kt}$$
$$Q = 10 \qquad \text{when } t = 0 \Rightarrow 10 = Ce^{\circ} \Rightarrow 10 = C$$
$$Q(t) = 10e^{kt}$$
$$Q = 5 \qquad \text{when } t = 1620$$
$$5 = 10e^{1620k}$$
$$k = \frac{1}{1620}\ln\left(\frac{1}{2}\right)$$
$$Q(t) = 10e^{[\ln(1/2)/1620]t}$$

When $t = 1000$, $Q(t) = 10e^{[\ln(1/2)/1620](1000)} \approx 6.52$ grams.
When $t = 10,000$, $Q(t) = 10e^{[\ln(1/2)/1620](10000)} \approx 0.14$ gram.

29. The half-life of the isotope Pu^{230} is 24,360 years. If 2.1 grams remain after 1000 years, what is the initial quantity and how much will remain after 10,000 years?

Solution:

$$y = Ce^{[\ln(1/2)/24360]t}$$
$$2.1 = Ce^{[\ln(1/2)/24360](1000)}$$
$$C \approx 2.16$$

The initial quantity is 2.16 grams.
When $t = 10,000$, $y = 2.16e^{[\ln(1/2)/24360](10000)} \approx 1.62$ grams.

33. Find the constant k such that the exponential function $y = Ce^{kt}$ passes through the points (0, 1) and (4, 10).

Solution:

$$y = Ce^{kt}$$
$$1 = Ce^{k(0)}, \quad (0,\ 1)$$
$$1 = C$$
$$y = e^{kt}$$
$$10 = e^{4k}, \quad (4,\ 10)$$
$$4k = \ln 10$$
$$k = \frac{\ln 10}{4} \approx 0.5756$$

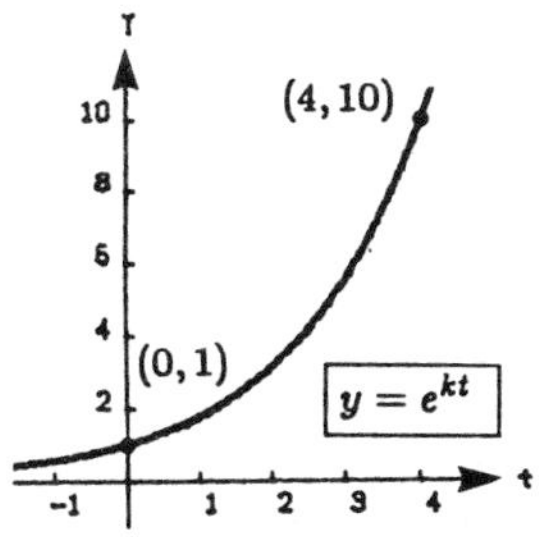

37. The sales S (in thousands of units) of a new product after it has been on the market t years are given by

$$S(t) = 100(1 - e^{kt}).$$

(a) Find S as a function of t if 15,000 units have been sold after one year.
(b) How many units will be sold after five years?

Solution:

$S(t) = 100(1 - e^{kt})$, $S = 15$ when $t = 1$

(a)
$$15 = 100(1 - e^{k})$$
$$0.15 = 1 - e^{k}$$
$$e^{k} = 0.85$$
$$k = \ln 0.85$$
$$S(t) = 100[1 - e^{(\ln 0.85)t}] = 100(1 - e^{-0.1625t})$$

(b) $S(5) = 100[1 - e^{-(0.1625)(5)}] \approx 55.625$ thousands of units $= 55,625$ units

41. The intensity level β, in decibels, of a sound wave is defined by

$$\beta(I) = 10 \log_{10} \frac{I}{I_0}$$

where I_0 is an intensity of 10^{-16} watts per square centimeter, corresponding roughly to the faintest sound that can be heard. Determine $\beta(I)$ for the following conditions.

(a) $I = 10^{-14}$ watts per centimeter (whisper)
(b) $I = 10^{-9}$ watts per centimeter (busy street corner)
(c) $I = 10^{-6.5}$ watts per centimeter (air hammer)
(d) $I = 10^{-4}$ watts per centimeter (threshold of pain)

Solution:

$\beta(I) = 10 \log_{10} \dfrac{I}{I_0}$ where $I_0 = 10^{-16}$ watt/cm^2.

(a) $\beta(10^{-14}) = 10 \log_{10} \dfrac{10^{-14}}{10^{-16}} = 10 \log_{10} 10^{2} = 20$

(b) $\beta(10^{-9}) = 10 \log_{10} \dfrac{10^{-9}}{10^{-16}} = 10 \log_{10} 10^{7} = 70$

(c) $\beta(10^{-6.5}) = 10 \log_{10} \dfrac{10^{-6.5}}{10^{-16}} = 10 \log_{10} 10^{9.5} = 95$

(d) $\beta(10^{-4}) = 10 \log_{10} \dfrac{10^{-4}}{10^{-16}} = 10 \log_{10} 10^{12} = 120$

45. Use the acidity model $\text{pH} = -\log_{10}[\text{H}^+]$, where acidity (pH) is a measure of the hydrogen ion concentration $[\text{H}^+]$ (measured in moles of hydrogen per liter) of a solution. Find the pH if $[\text{H}^+] = 2.3 \times 10^{-5}$.

Solution:

$$\text{pH} = -\log_{10}[\text{H}^+] = -\log_{10}[2.3 \times 10^{-5}] \approx 4.64$$

49. Use **Newton's Law of Cooling** which states that the rate of change in the temperature of an object is proportional to the difference between its temperature and the temperature of its environment. If $T(t)$ is the temperature of the object at time t in minutes, T_0 is the initial temperature, and T_e is the constant temperature of the environment, then

$$T(t) = T_e + (T_0 - T_e)e^{-kt}.$$

An object in a room at 70° F cools from 350° F to 150° F in 45 minutes.

(a) Find the temperature of the object as a function of time.
(b) Find the temperature after it has cooled for one hour.
(c) Find the time necessary for the object to cool to 80° F.

Solution:

(a) $T_e = 70, \quad T_0 = 350, \quad T = 150$ when $t = 45$

$$150 = 70 + (350 - 70)e^{-45k}$$

$$80 = 280e^{-45k}$$

$$\frac{2}{7} = e^{-45k}$$

$$k = \frac{\ln(2/7)}{-45}$$

$$T(t) = 70 + 280e^{-[\ln(2/7)/-45]t} = 70 + 280e^{-0.02784t}$$

(b) $T(60) = 70 + 280e^{[\ln(2/7)/45](60)} \approx 122.7°$

(c)

$$80 = 70 + 280e^{[\ln(2/7)/45]t}$$

$$\frac{1}{28} = e^{[\ln(2/7)/45]t}$$

$$\frac{\ln(2/7)}{45}t = \ln\left(\frac{1}{28}\right)$$

$$t = \frac{45\ln(1/28)}{\ln(2/7)} \approx 119.7 \text{ minutes}$$

REVIEW EXERCISES FOR CHAPTER 4

Solutions to Selected Exercises

3. Sketch the graph of $g(x) = 6^{-x}$.

Solution:

$$g(x) = 6^{-x} = \left(\tfrac{1}{6}\right)^x$$

x	0	1	-1
$g(x)$	1	$\frac{1}{6}$	6

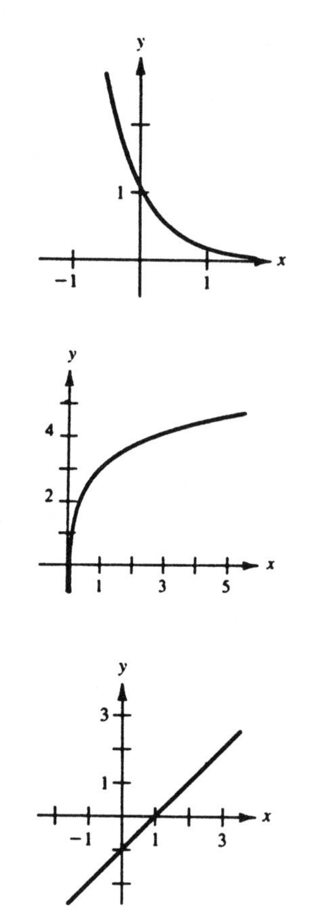

11. Sketch the graph of $f(x) = \ln x + 3$.

Solution:

$$f(x) = \ln x + 3$$

x	1	2	3	$\frac{1}{2}$	$\frac{1}{4}$
$f(x)$	3	3.69	4.10	2.31	1.61

15. Sketch the graph of $h(x) = \ln(e^{x-1})$.

Solution:

$$\begin{aligned} h(x) &= \ln(e^{x-1}) \\ &= (x-1)\ln e \\ &= x - 1 \end{aligned}$$

19. Use the properties of logarithms to write the following expression as a sum, difference, or multiple of logarithms.

$$\log_{10} \frac{5\sqrt{y}}{x^2}$$

Solution:

$$\begin{aligned} \log_{10} \frac{5\sqrt{y}}{x^2} &= \log_{10} 5\sqrt{y} - \log_{10} x^2 \\ &= \log_{10} 5 + \log_{10} \sqrt{y} - \log_{10} x^2 \\ &= \log_{10} 5 + \frac{1}{2}\log_{10} y - 2\log_{10} x \end{aligned}$$

23. Use the properties of logarithms to write the following expression as a sum, difference, or multiple of logarithms.

$$\ln[(x^2+1)(x-1)]$$

Solution:

$$\ln[(x^2+1)(x-1)] = \ln(x^2+1) + \ln(x-1)$$

27. Write $\frac{1}{2}\ln|2x-1| - 2\ln|x+1|$ as the logarithm of a single quantity.

Solution:

$$\frac{1}{2}\ln|2x-1| - 2\ln|x+1| = \ln\sqrt{|2x-1|} - \ln(x+1)^2 = \ln\frac{\sqrt{|2x-1|}}{(x+1)^2}$$

31. Write $\ln 3 + \frac{1}{3}\ln(4-x^2) - \ln x$ as the logarithm of a single quantity.

Solution:

$$\ln 3 + \frac{1}{3}\ln(4-x^2) - \ln x = \ln 3 + \ln\sqrt[3]{4-x^2} - \ln x = \ln(3\sqrt[3]{4-x^2}) - \ln x = \ln\frac{3\sqrt[3]{4-x^2}}{x}$$

35. Determine whether the equation $\ln(x+y) = \ln x + \ln y$ is true or false.

Solution:
False, since $\ln x + \ln y = \ln(xy)$.

39. Determine whether the following equation is true or false.

$$\frac{e^{2x}-1}{e^x-1} = e^x + 1$$

Solution:
True, since

$$\frac{e^{2x}-1}{e^x-1} = \frac{(e^x+1)(e^x-1)}{e^x-1} = e^x + 1.$$

41. A solution of a certain drug contained 500 units per milliliter when prepared. It was analyzed after 40 days and found to contain 300 units per milliliter. Assuming that the rate of decomposition is proportional to the amount present, the equation giving the amount A after t days is

$$A = 500e^{-0.013t}.$$

Use this model to find A when $t = 60$.

Solution:

$$A = 500e^{-0.013(60)} \approx 229.2 \text{ units per milliliter}$$

45. A certain automobile gets 28 miles per gallon of gasoline for speeds up to 50 miles per hour. Over 50 miles per hour, the number of miles per gallon drops at the rate of 12% for each 10 miles per hour. If s is the speed and y is the number of miles per gallon, then

$$y = 28e^{0.6-0.012s}, \quad s \geq 50.$$

Use this function to complete the following table.

Speed	50	55	60	65	70
Miles per gallon					

Solution:

When $s = 50$, $y = 28e^{0.6-0.012(50)} = 28$ miles per gallon

When $s = 55$, $y = 28e^{0.6-0.012(55)} \approx 26.369$ miles per gallon

When $s = 60$, $y = 28e^{0.6-0.012(60)} \approx 24.834$ miles per gallon

When $s = 65$, $y = 28e^{0.6-0.012(65)} \approx 23.388$ miles per gallon

When $s = 70$, $y = 28e^{0.6-0.012(70)} \approx 22.026$ miles per gallon

Speed	50	55	60	65	70
Miles per gallon	28	26.4	24.8	23.4	22.0

49. Find the exponential function $y = Ce^{kt}$ that passes through the points $(0, 4)$ and $(5, \frac{1}{2})$.

Solution:

$$4 = Ce^{k(0)}, \quad (0, 4)$$
$$4 = C(1) \text{ so } y = 4e^{kt}$$
$$\frac{1}{2} = 4e^{5k}, \quad (5, \tfrac{1}{2})$$
$$\frac{1}{8} = e^{5k}$$
$$5k = \ln \frac{1}{8}$$
$$k \approx -0.4159$$

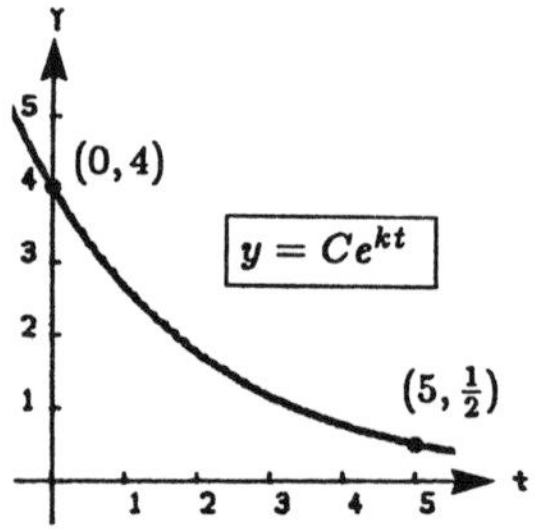

Thus, $y = 4e^{-0.4159t}$.

51. The demand equation for a certain product is given by

$$p = 500 - 0.5e^{0.004x}.$$

Find the demand x for a price of (a) $p = \$450$ and (b) $p = \$400$.

Solution:

(a)
$$p = 450$$
$$450 = 500 - 0.5e^{0.004x}$$
$$0.5e^{0.004x} = 50$$
$$e^{0.004x} = 100$$
$$0.004x = \ln 100$$
$$x \approx 1151 \text{ units}$$

(b)
$$p = 400$$
$$400 = 500 - 0.5e^{0.004x}$$
$$0.5e^{0.004x} = 100$$
$$e^{0.004x} = 200$$
$$0.004x = \ln 200$$
$$x \approx 1325 \text{ units}$$

57. In calculus it can be shown that

$$e^x \approx 1 + x + \frac{x^2}{2} + \frac{x^3}{6} + \frac{x^4}{24}.$$

Use this equation to approximate the following and compare the results to those obtained with a calculator.

(a) e (b) $e^{1/2}$ (c) $e^{-1/2}$

Solution:

(a) $e = e^1 \approx 1 + 1 + \frac{1}{2} + \frac{1}{6} + \frac{1}{24} \approx 2.7083$

By calculator: $e \approx 2.7183$

(b) $e^{1/2} \approx 1 + \frac{1}{2} + \frac{1/4}{2} + \frac{1/8}{6} + \frac{1/16}{24} \approx 1.6484$

By calculator: $e^{1/2} \approx 1.6487$

(c) $e^{-1/2} \approx 1 - \frac{1}{2} + \frac{1/4}{2} - \frac{1/8}{6} + \frac{1/16}{24} \approx 0.6068$

By calculator: $e^{-1/2} \approx 0.6065$

Practice Test for Chapter 4

1. Solve for x: $x^{3/5} = 8$.

2. Solve for x: $3^{x-1} = \frac{1}{81}$.

3. Graph $f(x) = 2^{-x}$.

4. Graph $g(x) = e^x + 1$.

5. If \$5000 is invested at 9% interest, find the amount after three years if the interest is compounded

(a) monthly (b) quarterly (c) continuously.

6. Write the equation in logarithmic form: $7^{-2} = \frac{1}{49}$.

7. Solve for $x : x - 4 = \log_2 \frac{1}{64}$.

8. Given $\log_b 2 = 0.3562$ and $\log_b 5 = 0.8271$, evaluate $\log_b \sqrt[4]{8/25}$.

9. Write $5 \ln x - \frac{1}{2} \ln y + 6 \ln z$ as a single logarithm.

10. Using your calculator and the change of base formula, evaluate $\log_9 28$.

11. Use your calculator to solve for $N : \log_{10} N = 0.6646$.

12. Graph $y = \log_4 x$.

13. Determine the domain of $f(x) = \log_3(x^2 - 9)$.

14. Graph $y = \ln(x - 2)$.

15. True or False: $\dfrac{\ln x}{\ln y} = \ln(x - y)$.

16. Solve for $x : 5^x = 41$.

17. Solve for $x : x - x^2 = \log_5 \frac{1}{25}$.

18. Solve for $x : \log_2 x + \log_2(x - 3) = 2$.

19. Solve for $x : \dfrac{e^x + e^{-x}}{3} = 4$.

20. \$6000 is deposited into a fund at an annual percentage rate of 13%. Find the time required for the investment to double if the interest is compounded continuously.

CHAPTER 5

Trigonometry

SECTION 5.1

Radian and Degree Measure

- If two angles have the same initial and terminal sides, they are coterminal angles.
- The radian measure of a central angle θ is found by taking the arc length s and dividing it by the radius r.

$$\theta = \frac{s}{r}$$

- You should know the following about angles:
  (a) θ is acute if $0 < \theta < \pi/2$.
  (b) θ is a right angle if $\theta = \pi/2$.
  (c) θ is obtuse if $\pi/2 < \theta < \pi$.
  (d) α and β are complementary if $\alpha + \beta = \pi/2$.
  (e) α and β are supplementary if $\alpha + \beta = \pi$.
- To convert degrees to radians, multiply by $\pi/180$.
- To convert radians to degrees, multiply by $180/\pi$.
- You should be able to convert angles to degrees, minutes, and seconds.
  (a) One minute: $1' = \frac{1}{60}(1°)$
  (b) One second: $1'' = \frac{1}{60}(1') = \frac{1}{3600}(1°)$
- Speed $= \dfrac{\text{distance}}{\text{time}} = \dfrac{s}{t}$
- Angular speed $= \dfrac{\theta}{t}$

Solutions to Selected Exercises

3. Determine the quadrant in which the following angles lie. (The angle measure is given in radians.)

(a) $-\frac{\pi}{12}$ (b) $-\frac{11\pi}{9}$

Solution:

(a) $-\frac{\pi}{12}$ lies in Quadrant IV. (b) $-\frac{11\pi}{9}$ lies in Quadrant II.

9. Determine the quadrant in which the following angles lie.

(a) $-132°50'$ (b) $-336°$

Solution:

(a) $-132°50'$ lies in Quadrant III. (b) $-336°$ lies in Quadrant I.

13. Sketch the given angle in standard position.

(a) $-\frac{7\pi}{4}$ (b) $-\frac{5\pi}{2}$

Solution:

(a) (b)

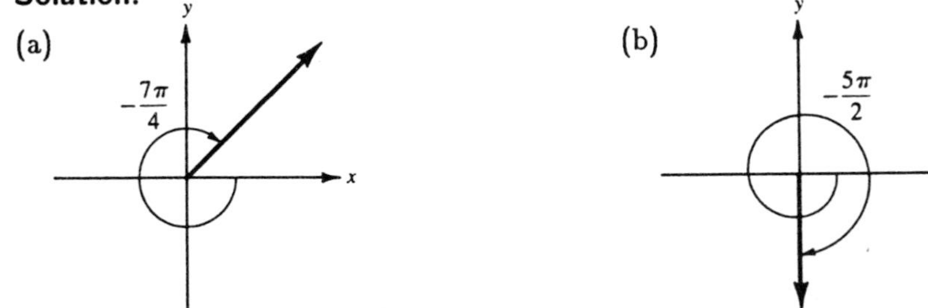

17. Determine two coterminal angles (one positive and one negative) for (a) $\theta = \pi/9$ and (b) $\theta = 4\pi/3$. Give the answers in radians.

Solution:

(a) Coterminal angles for $\frac{\pi}{9}$:

$$2\pi + \frac{\pi}{9} = \frac{19\pi}{9}$$

$$\frac{\pi}{9} - 2\pi = -\frac{17\pi}{9}$$

(b) Coterminal angles for $\frac{4\pi}{3}$:

$$2\pi + \frac{4\pi}{3} = \frac{10\pi}{3}$$

$$\frac{4\pi}{3} - 2\pi = -\frac{2\pi}{3}$$

21. Determine two coterminal angles (one positive and one negative) for (a) $\theta = 36°$ and (b) $\theta = -45°$. Give the answers in degrees.

Solution:

(a) Coterminal angles for $36°$:
$360° + 36° = 396°$
$36° - 360° = -324°$

(b) Coterminal angles for $-45°$:
$360° + (-45°) = 315°$
$-45° - 360° = -405°$

27. Express (a) $-20°$ and (b) $-240°$ in radian measure as a multiple of π. (Do not use a calculator.)

Solution:

(a) $-20° = -20\left(\frac{\pi}{180}\right) = -\frac{\pi}{9}$ (b) $-240° = -240\left(\frac{\pi}{180}\right) = -\frac{4\pi}{3}$

31. Express (a) $7\pi/3$ and (b) $-11\pi/30$ in degree measure. (Do not use a calculator.)

Solution:

(a) $\frac{7\pi}{3} = \frac{7\pi}{3}\left(\frac{180}{\pi}\right) = 420°$ (b) $-\frac{11\pi}{30} = -\frac{11\pi}{30}\left(\frac{180}{\pi}\right) = -66°$

35. Convert $-216.35°$ from degrees to radian measure. List your answer to three decimal places.

Solution:

$$-216.35° = -216.35\left(\frac{\pi}{180}\right) \approx -3.776 \text{ radians}$$

39. Convert $-0.83°$ from degrees to radian measure. List your answer to three decimal places.

Solution:

$$-0.83° = -0.83\left(\frac{\pi}{180}\right) \approx -0.014 \text{ radians}$$

41. Convert $\pi/7$ from radian to degree measure. List your answer to three decimal places.

Solution:

$$\frac{\pi}{7} = \frac{\pi}{7}\left(\frac{180}{\pi}\right) \approx 25.714°$$

45. Convert -4.2π from radian to degree measure. List your answer to three decimal places.

Solution:

$$-4.2\pi = -4.2\pi\left(\frac{180}{\pi}\right) = -756°$$

51. Convert (a) $85°\,18'\,30''$ and (b) $330°\,25''$ to decimal form.

Solution:

(a) $85°\,18'\,30'' = 85 + \frac{18}{60} + \frac{30}{3600} \approx 85.308°$ (b) $330°\,25'' = 330 + \frac{25}{3600} \approx 330.007°$

53. Convert (a) $240.6°$ and (b) $-145.8°$ to $D°\,M'\,S''$ form.

Solution:

(a) $240.6° = 240 + 0.6(60) = 240°\,36'$ (b) $-145.8° = -[145 + 0.8(60)] = -145°\,48'$

57. Find the radian measure of the central angle using radius $r = 10$ inches and arc length $s = 4$ inches.

Solution:

$$s = r\theta \Rightarrow \theta = \frac{s}{r}$$

$$\theta = \frac{4 \text{ inches}}{10 \text{ inches}} = \frac{2}{5} \text{ radians}$$

63. Find the length of the arc on the circle of radius $r = 6$ meters subtended by the central angle $\theta = 2$ radians.

Solution:

$$s = r\theta$$

$$s = (6)(2) = 12 \text{ m}$$

69. Assuming that the earth is a sphere of radius 4000 miles, what is the difference in latitude of two cities, one of which is 325 miles due north of the other?

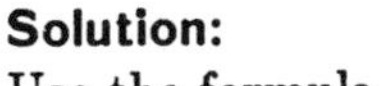

Solution:
Use the formula $\theta = s/r$.

$$\theta = \frac{325}{4000} \cdot \frac{180}{\pi} = 4.655°$$

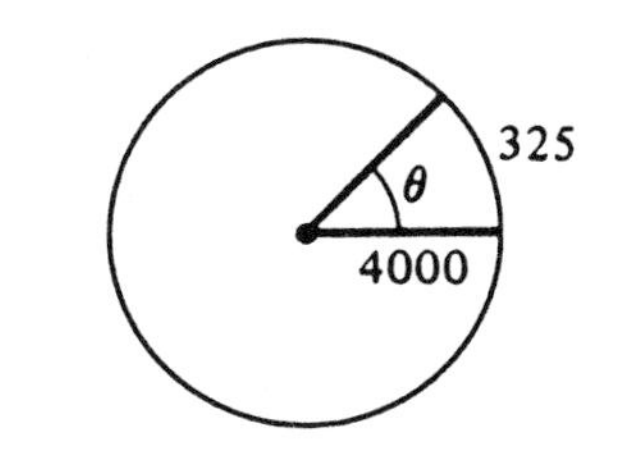

73. A car is moving at the rate of 50 miles per hour, and the diameter of each of its wheels is 2.5 feet.

(a) Find the number of revolutions per minute that the wheels are rotating.
(b) Find the *angular speed* of the wheels.

Solution:
(a) 50 miles per hour $= 50(5280)/60 = 4400$ feet per minute
The circumference of the tire is $C = 2.5\pi$ feet.
The number of revolutions per minute is $r = 4400/2.5\pi \approx 560.2$

(b) The angular speed is θ/t.

$$\theta = \frac{4400}{2.5\pi}(2\pi) = 3520 \text{ radians}$$

$$\text{Angular speed} = \frac{3520 \text{ radians}}{1 \text{ minute}} = 3520 \text{ rad/min}$$

SECTION 5.2

Trigonometric Functions and the Unit Circle

- You should know the definition of the trigonometric functions of t.

  (a) $\sin t = y$
  (b) $\cos t = x$
  (c) $\tan t = y/x, \quad x \neq 0$
  (d) $\cot t = x/y, \quad y \neq 0$
  (e) $\sec t = 1/x, \quad x \neq 0$
  (f) $\csc t = 1/y, \quad y \neq 0$

- The cosine and secant functions are even.

  (a) $\cos(-t) = \cos t$
  (b) $\sec(-t) = \sec t$

- The other four trigonometric functions are odd.

  (a) $\sin(-t) = -\sin t$
  (b) $\tan(-t) = -\tan t$
  (c) $\cot(-t) = -\cot t$
  (d) $\csc(-t) = -\csc t$

- You should be able to evaluate trigonometric functions with a calculator.

Solutions to Selected Exercises

3. Find the point (x, y) on the unit circle that corresponds to $t = 5\pi/6$.

Solution:
Using Figure 5.19 in the text, move counterclockwise to obtain the second quadrant point

$$(x, y) = \left(-\frac{\sqrt{3}}{2}, \frac{1}{2}\right).$$

5. Find the point (x, y) on the unit circle that corresponds to $\theta = 4\pi/3$.

Solution:
Using Figure 5.19 again and the fact that $4\pi/3 = 8\pi/6$, we obtain the point

$$(x, y) = \left(-\frac{1}{2}, -\frac{\sqrt{3}}{2}\right).$$

11. Evaluate the given trigonometric function.

(a) $\sin\left(-\frac{\pi}{6}\right)$ (b) $\cos\left(-\frac{\pi}{6}\right)$ (c) $\tan\left(-\frac{\pi}{6}\right)$

Solution:

$t = \frac{\pi}{6}$ corresponds to the point $\left(\frac{\sqrt{3}}{2}, \frac{1}{2}\right)$.

(a) $\sin\left(-\frac{\pi}{6}\right) = -\sin\frac{\pi}{6} = -y = -\frac{1}{2}$

(b) $\cos\left(-\frac{\pi}{6}\right) = \cos\frac{\pi}{6} = x = \frac{\sqrt{3}}{2}$

(c) $\tan\left(-\frac{\pi}{6}\right) = -\tan\frac{\pi}{6} = -\frac{y}{x} = -\frac{1/2}{\sqrt{3}/2} = -\frac{1}{\sqrt{3}} = -\frac{\sqrt{3}}{3}$

17. Evaluate the given trigonometric function.

(a) $\sin\frac{11\pi}{6}$ (b) $\cos\frac{11\pi}{6}$ (c) $\tan\frac{11\pi}{6}$

Solution:

$t = \frac{11\pi}{6}$ corresponds to the point $\left(\frac{\sqrt{3}}{2}, -\frac{1}{2}\right)$.

(a) $\sin\frac{11\pi}{6} = y = -\frac{1}{2}$

(b) $\cos\frac{11\pi}{6} = x = \frac{\sqrt{3}}{2}$

(c) $\tan\frac{11\pi}{6} = \frac{y}{x} = \frac{-1/2}{\sqrt{3}/2} = -\frac{1}{\sqrt{3}} = -\frac{\sqrt{3}}{3}$

21. Evaluate the six trigonometric functions for $t = \pi/4$.

Solution:

$t = \frac{\pi}{4}$ corresponds to the point $\left(\frac{\sqrt{2}}{2}, \frac{\sqrt{2}}{2}\right)$.

$$\sin\frac{\pi}{4} = y = \frac{\sqrt{2}}{2}$$

$$\cos\frac{\pi}{4} = x = \frac{\sqrt{2}}{2}$$

$$\tan\frac{\pi}{4} = \frac{y}{x} = 1$$

$$\cot\frac{\pi}{4} = \frac{x}{y} = 1$$

$$\sec\frac{\pi}{4} = \frac{1}{x} = \sqrt{2}$$

$$\csc\frac{\pi}{4} = \frac{1}{y} = \sqrt{2}$$

27. Use the periodic nature of the sine and cosine to evaluate $\sin 3\pi$.

Solution:

$$\sin 3\pi = \sin(2\pi + \pi) = \sin \pi = 0$$

31. Use the periodic nature of the sine and cosine to evaluate $\cos \frac{19\pi}{6}$.

Solution:

$$\cos \frac{19\pi}{6} = \cos\left(2\pi + \frac{7\pi}{6}\right) = \cos \frac{7\pi}{6} = -\frac{\sqrt{3}}{2}$$

35. Use $\sin t = \frac{1}{3}$ to evaluate the indicated functions.

(a) $\sin(-t)$

(b) $\csc(-t)$

Solution:

(a) $\sin(-t) = -\sin t = -\dfrac{1}{3}$

(b) $\csc(-t) = -\csc t = -\dfrac{1}{\sin t} = -\dfrac{1}{1/3} = -3$

41. Use a calculator to evaluate $\cos(-3)$. [Set your calculator in radian mode and round your answer to four decimal places.]

Solution:

$$\cos(-3) = \cos 3 \approx -0.9899925 \approx -0.9900$$

43. Use a calculator to evaluate $\tan(\pi/10)$. [Set your calculator in radian mode and round your answer to four decimal places.]

Solution:

$$\tan \frac{\pi}{10} \approx 0.3249197 \approx 0.3249$$

47. Use a calculator to evaluate $\csc 0.8$. [Set your calculator in radian mode and round your answer to four decimal places.]

Solution:

$$\csc 0.8 = \frac{1}{\sin 0.8} \approx 1.3940078 \approx 1.3940$$

SECTION 5.3

Trigonometric Functions of an Acute Angle

- You should know the right triangle definition of trigonometric functions.

(a) $\sin\theta = \dfrac{\text{opp}}{\text{hyp}}$ (b) $\cos\theta = \dfrac{\text{adj}}{\text{hyp}}$

(c) $\tan\theta = \dfrac{\text{opp}}{\text{adj}}$ (d) $\cot\theta = \dfrac{\text{adj}}{\text{opp}}$

(e) $\sec\theta = \dfrac{\text{hyp}}{\text{adj}}$ (f) $\csc\theta = \dfrac{\text{hyp}}{\text{opp}}$

- You should know the sine, cosine, and tangent of the special angles 30°, 45°, and 60°.

(a) For 45°, use the triangle (b) For 30° and 60°, use the triangle

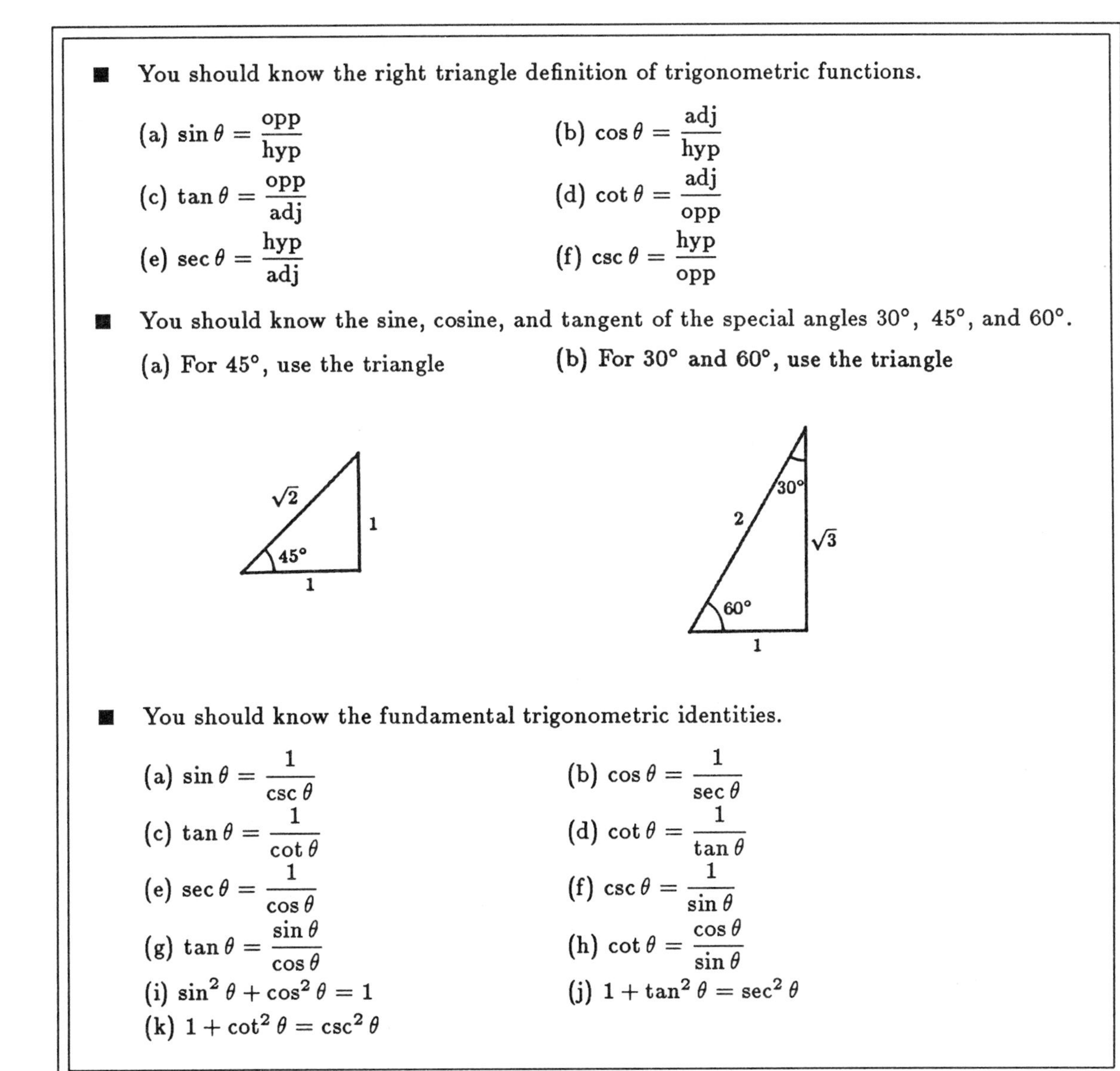

- You should know the fundamental trigonometric identities.

(a) $\sin\theta = \dfrac{1}{\csc\theta}$ (b) $\cos\theta = \dfrac{1}{\sec\theta}$

(c) $\tan\theta = \dfrac{1}{\cot\theta}$ (d) $\cot\theta = \dfrac{1}{\tan\theta}$

(e) $\sec\theta = \dfrac{1}{\cos\theta}$ (f) $\csc\theta = \dfrac{1}{\sin\theta}$

(g) $\tan\theta = \dfrac{\sin\theta}{\cos\theta}$ (h) $\cot\theta = \dfrac{\cos\theta}{\sin\theta}$

(i) $\sin^2\theta + \cos^2\theta = 1$ (j) $1 + \tan^2\theta = \sec^2\theta$

(k) $1 + \cot^2\theta = \csc^2\theta$

Solutions to Selected Exercises

3. Find the exact value of the six trigonometric functions of the angle θ given in the figure. (Use the Pythagorean Theorem to find the third side of the triangle.)

Solution:

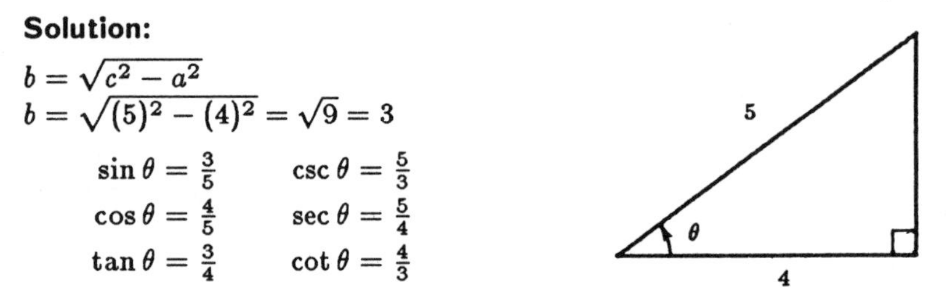

$b = \sqrt{c^2 - a^2}$

$b = \sqrt{(5)^2 - (4)^2} = \sqrt{9} = 3$

$\sin\theta = \frac{3}{5} \qquad \csc\theta = \frac{5}{3}$

$\cos\theta = \frac{4}{5} \qquad \sec\theta = \frac{5}{4}$

$\tan\theta = \frac{3}{4} \qquad \cot\theta = \frac{4}{3}$

9. Sketch a right triangle corresponding to $\sin\theta = \frac{2}{3}$, and find the other five trigonometric functions of θ.

Solution:

$\sin\theta = \dfrac{2}{3}$

Adjacent side $= \sqrt{9-4} = \sqrt{5}$

$$\cos\theta = \frac{\sqrt{5}}{3} \qquad \sec\theta = \frac{3}{\sqrt{5}} = \frac{3\sqrt{5}}{5}$$

$$\tan\theta = \frac{2}{\sqrt{5}} = \frac{2\sqrt{5}}{5} \qquad \csc\theta = \frac{3}{2}$$

$$\cot\theta = \frac{\sqrt{5}}{2}$$

13. Sketch a right triangle corresponding to $\tan\theta = 3$, and find the other five trigonometric functions of θ.

Solution:

$\tan\theta = 3 = \dfrac{3}{1}$

Hypotenuse $= \sqrt{1+9} = \sqrt{10}$

$$\sin\theta = \frac{3}{\sqrt{10}} = \frac{3\sqrt{10}}{10} \qquad \sec\theta = \sqrt{10}$$

$$\cos\theta = \frac{1}{\sqrt{10}} = \frac{\sqrt{10}}{10} \qquad \csc\theta = \frac{\sqrt{10}}{3}$$

$$\cot\theta = \frac{1}{3}$$

19. Use $\csc\theta = 3$ and $\sec\theta = \frac{3\sqrt{2}}{4}$ to find the following trigonometric functions.

(a) $\sin\theta$ (b) $\cos\theta$

(c) $\tan\theta$ (d) $\sec(90^\circ - \theta)$

Solution:

(a) $\sin\theta = \dfrac{1}{\csc\theta} = \dfrac{1}{3}$

(b) $\cos\theta = \dfrac{1}{\sec\theta} = \dfrac{4}{3\sqrt{2}} = \dfrac{2\sqrt{2}}{3}$

(c) $\tan\theta = \dfrac{\sin\theta}{\cos\theta} = \dfrac{1/3}{(2\sqrt{2})/3} = \dfrac{1}{2\sqrt{2}} = \dfrac{\sqrt{2}}{4}$

(d) $\sec(90° - \theta) = \csc\theta = 3$

21. Evaluate (a) $\cos 60°$ and (b) $\tan 30°$ by memory or by constructing an appropriate triangle.

Solution:

(a) $\cos 60° = \dfrac{1}{2}$

(b) $\tan 30° = \dfrac{1}{\sqrt{3}} = \dfrac{\sqrt{3}}{3}$

25. Use a calculator to evaluate (a) $\sin 10°$ and (b) $\cos 80°$. Round your answers to four decimal places. (Be sure the calculator is in the correct mode.)

Solution:

(a) $\sin 10° \approx 0.1736$

(b) $\cos 80° = \sin 10° \approx 0.1736$

31. Use a calculator to evaluate (a) $\cot(\pi/16)$ and (b) $\tan(\pi/16)$. Round your answers to four decimal places.

Solution:

Make sure that your calculator is in radian mode.

(a) $\cot\dfrac{\pi}{16} = \dfrac{1}{\tan(\pi/16)} \approx 5.0273$

(b) $\tan\dfrac{\pi}{16} \approx 0.1989$

35. Find the value of θ in degrees $(0° < \theta < 90°)$ and radians $(0 < \theta < \pi/2)$ for (a) $\sin\theta = 1/2$ and (b) $\csc\theta = 2$ without a calculator.

Solution:

(a) $\sin\theta = \dfrac{1}{2}$

$\theta = 30° = \dfrac{\pi}{6}$

(b) $\csc\theta = 2 \Rightarrow \sin\theta = \dfrac{1}{2}$ Same as (a)

$\theta = 30° = \dfrac{\pi}{6}$

39. Find the value of θ in degrees $(0° < \theta < 90°)$ and radians $(0 < \theta < \pi/2)$ for (a) $\csc\theta = (2\sqrt{3})/3$ and (b) $\sin\theta = \sqrt{2}/2$ without a calculator.

Solution:

(a) $$\csc\theta = \frac{2\sqrt{3}}{3} = \frac{2}{\sqrt{3}}$$

$$\theta = 60° = \frac{\pi}{3}$$

(b) $$\sin\theta = \frac{\sqrt{2}}{2} = \frac{1}{\sqrt{2}}$$

$$\theta = 45° = \frac{\pi}{4}$$

47. Solve for x.

Solution:

$$\tan 60° = \frac{25}{x}$$

$$\sqrt{3} = \frac{25}{x}$$

$$x = \frac{25}{\sqrt{3}} = \frac{25\sqrt{3}}{3}$$

51. Solve for y.

Solution:

$$\sin 50° = \frac{y}{12}$$

$$0.7660 = \frac{y}{12}$$

$$y = 9.19$$

53. A six-foot person standing 12 feet from a streetlight casts an eight-foot shadow, as shown in the figure. What is the height of the streetlight?

Solution:

$$\frac{h}{6} = \frac{20}{8}$$

$$h = \frac{120}{8} = 15 \text{ ft}$$

57. From a 150-foot observation tower on the coast, a Coast Guard officer sights a boat in difficulty. The angle of depression of the boat is 4°, as shown in the figure. How far is the boat from the shoreline?

Solution:

Let $x =$ distance from the boat to the shoreline.

$$\tan 4^\circ = \frac{150}{x}$$

$$x = \frac{150}{\tan 4^\circ} \approx 2145.10 \text{ ft}$$

150 4°

61. Determine whether the statement is true or false, and give reasons.

$$\sin 45^\circ + \cos 45^\circ = 1.$$

Solution:

False; $\sin 45^\circ + \cos 45^\circ = \frac{\sqrt{2}}{2} + \frac{\sqrt{2}}{2} = \sqrt{2} \neq 1$

SECTION 5.4

Trigonometric Functions of Any Angle

- You should know the trigonometric functions of any angle θ in standard position with (x, y) on the terminal side of θ and $r = \sqrt{x^2 + y^2}$.

  (a) $\sin\theta = \dfrac{y}{r}$ (b) $\cos\theta = \dfrac{x}{r}$

  (c) $\tan\theta = \dfrac{y}{x}, \quad x \neq 0$ (d) $\cot\theta = \dfrac{x}{y}, \quad y \neq 0$

  (e) $\sec\theta = \dfrac{r}{x}, \quad x \neq 0$ (f) $\csc\theta = \dfrac{r}{y}, \quad y \neq 0$

- You should know the signs of the trigonometric functions in the four quadrants.

- You should be able to find the trigonometric functions of the quadrant angles (if they exist).

  (a) For 0, use (1, 0).
  (b) For $\pi/2$, use (0, 1).
  (c) For π, use $(-1, 0)$.
  (d) For $3\pi/2$, use $(0, -1)$.

- You should be able to use reference angles with the special angles to find trigonometric values.

Solutions to Selected Exercises

3. Determine the exact value of the six trigonometric functions of the given angle θ.

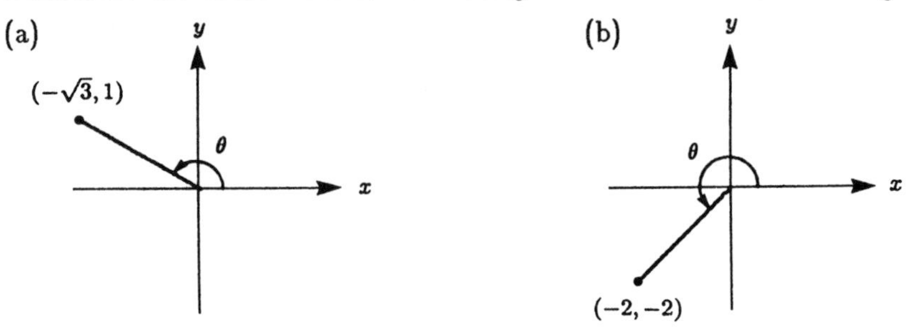

Solution:

(a) $x = -\sqrt{3},\ y = 1,\ r = \sqrt{3+1} = 2$

$$\sin\theta = \frac{y}{r} = \frac{1}{2} \qquad \csc\theta = \frac{r}{y} = 2$$

$$\cos\theta = \frac{x}{r} = -\frac{\sqrt{3}}{2} \qquad \sec\theta = \frac{r}{x} = -\frac{2\sqrt{3}}{3}$$

$$\tan\theta = \frac{y}{x} = -\frac{1}{\sqrt{3}} = -\frac{\sqrt{3}}{3} \qquad \cot\theta = \frac{x}{y} = -\sqrt{3}$$

(b) $x = -2,\ y = -2,\ r = \sqrt{4+4} = 2\sqrt{2}$

$$\sin\theta = \frac{y}{r} = -\frac{2}{2\sqrt{2}} = -\frac{\sqrt{2}}{2} \qquad \csc\theta = \frac{r}{y} = \frac{2\sqrt{2}}{-2} = -\sqrt{2}$$

$$\cos\theta = \frac{x}{r} = -\frac{2}{2\sqrt{2}} = -\frac{\sqrt{2}}{2} \qquad \sec\theta = \frac{r}{x} = \frac{2\sqrt{2}}{-2} = -\sqrt{2}$$

$$\tan\theta = \frac{y}{x} = \frac{-2}{-2} = 1 \qquad \cot\theta = \frac{x}{y} = \frac{-2}{-2} = 1$$

7. The given point is on the terminal side of an angle in standard position. Determine the exact value of the six trigonometric functions of the angle.

(a) $(-4,\ 10)$ (b) $(3,\ -5)$

Solution:

(a) $x = -4,\ y = 10,\ r = \sqrt{16+100} = 2\sqrt{29}$

$$\sin\theta = \frac{10}{2\sqrt{29}} = \frac{5\sqrt{29}}{29} \qquad \csc\theta = \frac{2\sqrt{29}}{10} = \frac{\sqrt{29}}{5}$$

$$\cos\theta = \frac{-4}{2\sqrt{29}} = -\frac{2\sqrt{29}}{29} \qquad \sec\theta = \frac{2\sqrt{29}}{-4} = -\frac{\sqrt{29}}{2}$$

$$\tan\theta = \frac{10}{-4} = -\frac{5}{2} \qquad \cot\theta = \frac{-4}{10} = -\frac{2}{5}$$

(b) $x = 3,\ y = -5,\ r = \sqrt{9+25} = \sqrt{34}$

$$\sin\theta = \frac{-5}{\sqrt{34}} = -\frac{5\sqrt{34}}{34} \qquad \csc\theta = -\frac{\sqrt{34}}{5}$$

$$\cos\theta = \frac{3}{\sqrt{34}} = \frac{3\sqrt{34}}{34} \qquad \sec\theta = \frac{\sqrt{34}}{3}$$

$$\tan\theta = -\frac{5}{3} \qquad \cot\theta = -\frac{3}{5}$$

9. Use the two similar triangles in the figure to find (a) the unknown sides of the triangles and (b) the six trigonometric functions of the angles α_1 and α_2.

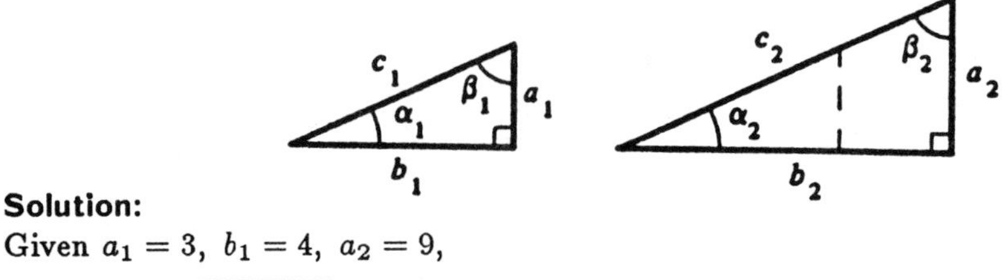

Solution:

Given $a_1 = 3,\ b_1 = 4,\ a_2 = 9,$

(a) $c_1 = \sqrt{a_1{}^2 + b_1{}^2} = \sqrt{9 + 16} = 5$

$$\frac{a_1}{a_2} = \frac{b_1}{b_2}$$

$$\frac{3}{9} = \frac{4}{b_2}$$

$$b_2 = 12$$

$$c_2 = \sqrt{a_2{}^2 + b_2{}^2} = \sqrt{81 + 144} = 15$$

(b) $\sin\alpha_1 = \sin\alpha_2 = \dfrac{3}{5}$ $\qquad \csc\alpha_1 = \csc\alpha_2 = \dfrac{5}{3}$

$\cos\alpha_1 = \cos\alpha_2 = \dfrac{4}{5}$ $\qquad \sec\alpha_1 = \sec\alpha_2 = \dfrac{5}{4}$

$\tan\alpha_1 = \tan\alpha_2 = \dfrac{3}{4}$ $\qquad \cot\alpha_1 = \cot\alpha_2 = \dfrac{4}{3}$

13. Determine the quadrant in which θ lies.

(a) $\sin\theta < 0$ and $\cos\theta < 0$
(b) $\sin\theta > 0$ and $\cos\theta < 0$

Solution:

(a) $\sin\theta < 0 \Rightarrow \theta$ lies in Quadrant III or in Quadrant IV.
$\cos\theta < 0 \Rightarrow \theta$ lies in Quadrant II or in Quadrant III.
$\sin\theta < 0$ *and* $\cos\theta < 0 \Rightarrow \theta$ lies in Quadrant III.

(b) $\sin\theta > 0 \Rightarrow \theta$ lies in Quadrant I or in Quadrant II.
$\cos\theta < 0 \Rightarrow \theta$ lies in Quadrant II or in Quadrant III.
$\sin\theta > 0$ *and* $\cos\theta < 0 \Rightarrow \theta$ lies in Quadrant II.

17. Find the exact value of the six trigonometric functions of θ, given θ lies in Quadrant II and $\sin\theta = \frac{3}{5}$.

Solution:

$y = 3,\ r = 5,\ x = -\sqrt{25 - 9} = -4,$ $\ x$ is negative since θ lies in Quadrant II.

$\sin\theta = \frac{3}{5}$ $\qquad \csc\theta = \frac{5}{3}$

$\cos\theta = -\frac{4}{5}$ $\qquad \sec\theta = -\frac{5}{4}$

$\tan\theta = -\frac{3}{4}$ $\qquad \cot\theta = -\frac{4}{3}$

21. Find the exact value of the six trigonometric functions of θ, given $\sin\theta > 0$ and $\sec\theta = -2$.

Solution:

θ is in Quadrant II.

$\sec\theta = \dfrac{2}{-1}$, $r = 2$, $x = -1$, $y = \sqrt{4-1} = \sqrt{3}$

$$\sin\theta = \frac{\sqrt{3}}{2} \qquad \csc\theta = \frac{2\sqrt{3}}{3}$$

$$\cos\theta = -\frac{1}{2} \qquad \sec\theta = -2$$

$$\tan\theta = -\sqrt{3} \qquad \cot\theta = -\frac{\sqrt{3}}{3}$$

25. Find the exact value of the six trigonometric functions of θ, given the terminal side of θ is in Quadrant III and lies on the line $y = 2x$.

Solution:

To find a point on the terminal side of θ, use any point on the line $y = 2x$ that lies in Quadrant III. $(-1, -2)$ is one such point.

$x = -1$, $y = -2$, $r = \sqrt{5}$

$$\sin\theta = -\frac{2}{\sqrt{5}} = -\frac{2\sqrt{5}}{5} \qquad \csc\theta = \frac{\sqrt{5}}{-2} = -\frac{\sqrt{5}}{2}$$

$$\cos\theta = -\frac{1}{\sqrt{5}} = -\frac{\sqrt{5}}{5} \qquad \sec\theta = \frac{\sqrt{5}}{-1} = -\sqrt{5}$$

$$\tan\theta = \frac{-2}{-1} = 2 \qquad \cot\theta = \frac{-1}{-2} = \frac{1}{2}$$

29. Find the reference angle θ', and draw a sketch for (a) $\theta = -245°$ and (b) $\theta = -72°$.

Solution:

(a) $\theta = -245°$

$\theta' = 245° - 180°$

$\theta' = 65°$

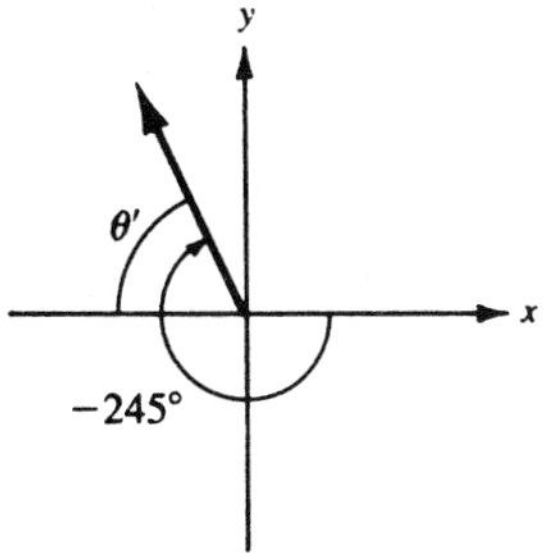

(b) $\theta = -72°$

$\theta' = |-72°| = 72°$

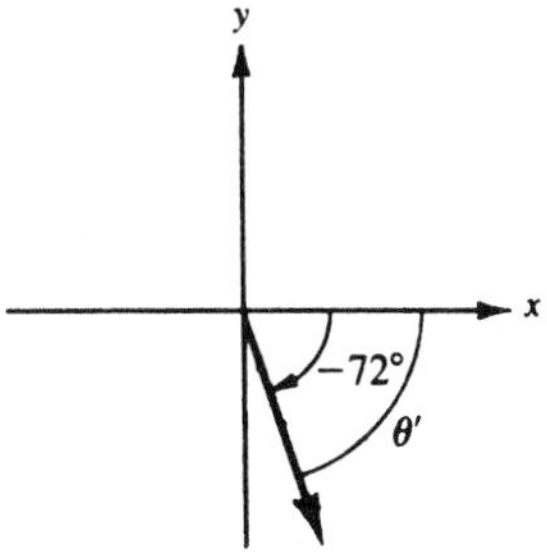

33. Find the reference angle θ' and draw a sketch for (a) $\theta = 3.5$ and (b) $\theta = 5.8$.

Solution:

(a) $\theta = 3.5$

$\theta' = 3.5 - \pi$

$\theta' \approx 0.3584$

(b) $\theta = 5.8$

$\theta' = 2\pi - 5.8$

$\theta' \approx 0.4832$

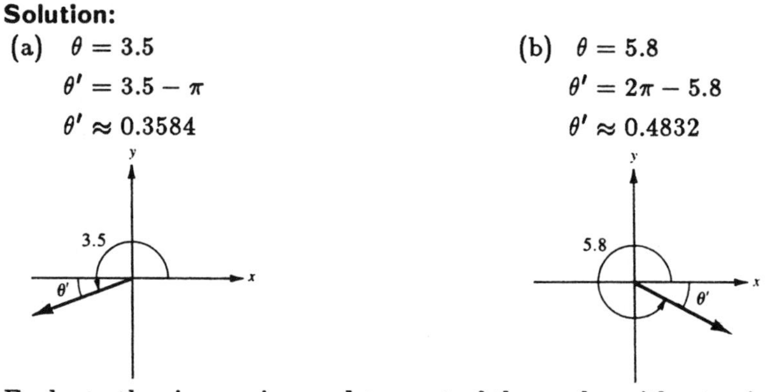

35. Evaluate the sine, cosine, and tangent of the angles without using a calculator.

(a) $225°$

(b) $-225°$

Solution:

(a) The reference angle of $225°$ is $45°$, and $225°$ lies in Quadrant III.

$$\sin 225° = -\sin 45° = -\frac{\sqrt{2}}{2}$$

$$\cos 225° = -\cos 45° = -\frac{\sqrt{2}}{2}$$

$$\tan 225° = \tan 45° = 1$$

(b) The reference angle is $45°$, and $-225°$ lies in Quadrant II.

$$\sin(-225°) = \sin 45° = \frac{\sqrt{2}}{2}$$

$$\cos(-225°) = -\cos 45° = -\frac{\sqrt{2}}{2}$$

$$\tan(-225°) = -\tan 45° = -1$$

39. Evaluate the sine, cosine, and tangent of the angles without using a calculator.

(a) $\dfrac{4\pi}{3}$

(b) $\dfrac{2\pi}{3}$

Solution:

(a) The reference angle of $4\pi/3$ is $\pi/3$, and $4\pi/3$ lies in Quadrant III.

$$\sin \frac{4\pi}{3} = -\sin \frac{\pi}{3} = -\frac{\sqrt{3}}{2}$$

$$\cos \frac{4\pi}{3} = -\cos \frac{\pi}{3} = -\frac{1}{2}$$

$$\tan \frac{4\pi}{3} = \tan \frac{\pi}{3} = \sqrt{3}$$

(b) The reference angle of $2\pi/3$ is $\pi/3$, and $2\pi/3$ lies in Quadrant II.

$$\sin\frac{2\pi}{3} = \sin\frac{\pi}{3} = \frac{\sqrt{3}}{2}$$
$$\cos\frac{2\pi}{3} = -\cos\frac{\pi}{3} = -\frac{1}{2}$$
$$\tan\frac{2\pi}{3} = -\tan\frac{\pi}{3} = -\sqrt{3}$$

45. Use a calculator to evaluate (a) $\sin 10°$ and (b) $\csc 10°$ to four decimal places. (Be sure the calculator is set in the correct mode.)

Solution:

(a) $\sin 10° \approx 0.1736$

(b) $\csc 10° = \dfrac{1}{\sin 10°} \approx 5.7588$

49. Use a calculator to evaluate (a) $\cos(-110°)$ and (b) $\cos 250°$ to four decimal places. (Be sure the calculator is set in the correct mode.)

Solution:

(a) $\cos(-110°) = \cos 110° \approx -0.3420$

(b) $\cos 250° \approx -0.3420$

53. Find two values of θ that satisfy (a) $\sin\theta = \frac{1}{2}$ and (b) $\sin\theta = -\frac{1}{2}$. List your answers in degrees $(0° \le \theta < 360°)$ and radians $(0 \le \theta < 2\pi)$. Do not use a calculator.

Solution:

(a) $\sin\theta = \dfrac{1}{2} > 0$

θ is in either Quadrant I or Quadrant II.

$$\theta = 30° = \frac{\pi}{6} \text{ or } \theta = 150° = \frac{5\pi}{6}$$

(b) $\sin\theta = -\dfrac{1}{2} < 0$

θ is in either Quadrant III or Quadrant IV.

$$\theta = 210° = \frac{7\pi}{6} \text{ or } \theta = 330° = \frac{11\pi}{6}$$

57. Find two values of θ that satisfy (a) $\tan\theta = 1$ and (b) $\cot\theta = -\sqrt{3}$. List your answers in degrees $(0° \le \theta < 360°)$ and radians $(0 \le \theta < 2\pi)$. Do not use a calculator.

Solution:

(a) $\tan\theta = 1$, θ lies in either Quadrant I or Quadrant III.

$$\theta = 45° = \frac{\pi}{4} \text{ or } \theta = 225° = \frac{5\pi}{4}$$

(b) $\cot\theta = -\sqrt{3}$, θ lies in either Quadrant II or Quadrant IV.

$$\theta = 150° = \frac{5\pi}{6} \text{ or } \theta = 330° = \frac{11\pi}{6}$$

65. Evaluate $\sin^2 2 + \cos^2 2$ without using a calculator.

Solution:

Since $\sin^2\theta + \cos^2\theta = 1$, we have $\sin^2 2 + \cos^2 2 = 1$.

69. The average daily temperature (in degrees Fahrenheit) for a certain city is given by

$$T = 45 - 23\cos\left[\frac{2\pi}{365}(t - 32)\right]$$

where t is the time in days with $t = 1$ corresponding to January 1. Find the average temperature on (a) January 1, (b) July 4 ($t = 185$), and (c) October 18 ($t = 291$).

Solution:

(a) $t = 1$

$$T = 45 - 23\cos\left[\frac{2\pi}{365}(1 - 32)\right] \approx 25.2^\circ\text{F}$$

(b) $t = 185$

$$T = 45 - 23\cos\left[\frac{2\pi}{365}(185 - 32)\right] \approx 65.1^\circ\text{F}$$

(c) $t = 291$

$$T = 45 - 23\cos\left[\frac{2\pi}{365}(291 - 32)\right] \approx 50.8^\circ\text{F}$$

SECTION 5.5

Graphs of Sine and Cosine

- You should be able to graph $y = a\sin(bx - c)$ and $y = a\cos(bx - c)$.
- Amplitude: $|a|$
- Period: $\dfrac{2\pi}{|b|}$
- Shift: Solve $bx - c = 0$ and $bx - c = 2\pi$.
- Key Increments: $\dfrac{1}{4}$ (period)

Solutions to Selected Exercises

5. Determine the period and amplitude of $y = \frac{1}{2}\sin \pi x$.

Solution:

$y = \dfrac{1}{2}\sin \pi x;\quad a = \dfrac{1}{2},\ b = \pi,\ c = 0$

Period: $\dfrac{2\pi}{|b|} = \dfrac{2\pi}{\pi} = 2$

Amplitude: $|a| = \left|\dfrac{1}{2}\right| = \dfrac{1}{2}$

9. Determine the period and amplitude of $y = -2\sin 10x$.

Solution:

$y = -2\sin 10x;\quad a = -2,\ b = 10,\ c = 0$

Period: $\dfrac{2\pi}{|b|} = \dfrac{2\pi}{10} = \dfrac{\pi}{5}$

Amplitude: $|a| = |-2| = 2$

11. Determine the period and amplitude of

$$y = \frac{1}{2}\cos\frac{2x}{3}.$$

Solution:

$y = \frac{1}{2}\cos\frac{2x}{3}$; $a = \frac{1}{2}$, $b = \frac{2}{3}$, $c = 0$

Period: $\frac{2\pi}{|b|} = \frac{2\pi}{2/3} = 3\pi$

Amplitude: $|a| = \left|\frac{1}{2}\right| = \frac{1}{2}$

15. Describe the relationship between the graphs of $f(x) = \sin x$ and $g(x) = \sin(x - \pi)$.

Solution:

$f(x) = \sin x$ and $g(x) = \sin(x - \pi)$ both have a period of 2π and an amplitude of 1. However, the graph of $g(x) = \sin(x - \pi)$ is the graph of $f(x) = \sin x$ shifted to the right π units.

19. Describe the relationship between the graphs of $f(x) = \cos x$ and $g(x) = \cos 2x$.

Solution:

$f(x) = \cos x$ and $g(x) = \cos 2x$ both have an amplitude of 1. However, $f(x) = \cos x$ has a period of 2π, whereas $g(x) = \cos 2x$ has a period of π.

23. Sketch the graphs of $f(x) = -2\sin x$ and $g(x) = 4\sin x$ on the same coordinate plane. (Include two full periods.)

Solution:

$f(x) = -2\sin x$
Period: 2π
Amplitude: 2

$g(x) = 4\sin x$
Period: 2π
Amplitude: 4

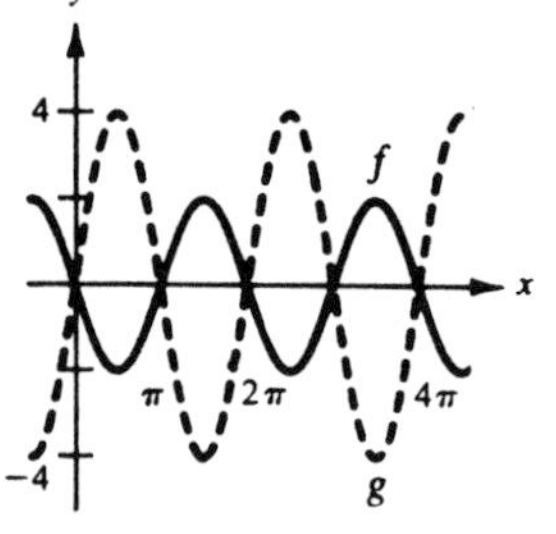

27. Sketch the graphs of the following on the same coordinate plane. (Include two full periods.)

$$f(x) = -\frac{1}{2}\sin\frac{x}{2} \quad \text{and} \quad g(x) = 3 - \frac{1}{2}\sin\frac{x}{2}$$

Solution:

$f(x) = -\frac{1}{2}\sin\frac{x}{2}$
Period: 4π
Amplitude: $\frac{1}{2}$

$g(x) = 3 - \frac{1}{2}\sin\frac{x}{2}$ is the graph of $f(x)$ shifted vertically three units upward.

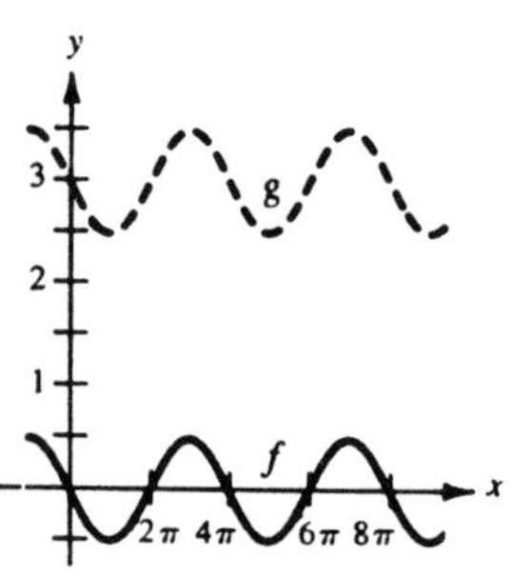

31. Sketch the graph of $y = -2\sin 6x$. (Include two full periods.)

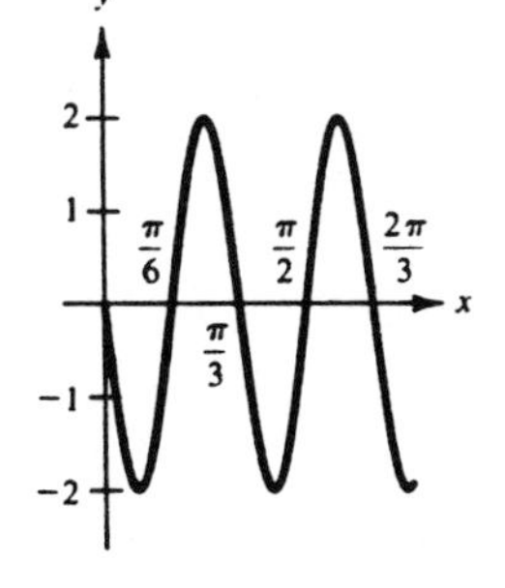

Solution:

$y = -2\sin 6x;\quad a = -2,\ b = 6,\ c = 0$

Period: $\dfrac{2\pi}{6} = \dfrac{\pi}{3}$

Amplitude: $|-2| = 2$

Key points: $(0,\ 0),\ \left(\frac{\pi}{12},\ -2\right),\ \left(\frac{\pi}{6},\ 0\right),\ \left(\frac{\pi}{4},\ 2\right),\ \left(\frac{\pi}{3},\ 0\right)$

35. Sketch the graph of the following. (Include two full periods.)

$$y = -\sin\frac{2\pi x}{3}$$

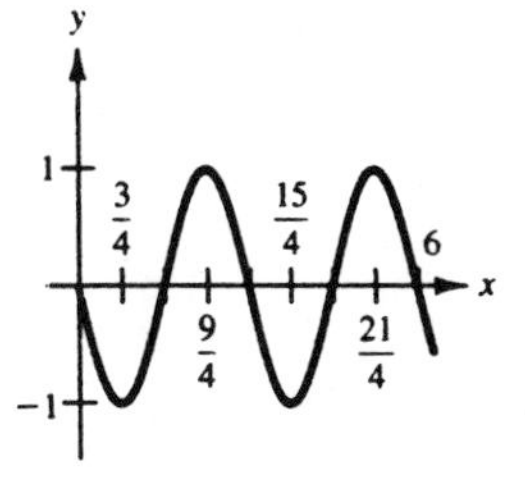

Solution:

$y = -\sin\dfrac{2\pi x}{3};\quad a = -1,\ b = \dfrac{2\pi}{3},\ c = 0$

Period: $\dfrac{2\pi}{2\pi/3} = 3$

Amplitude: 1

Key points: $(0,\ 0),\ \left(\frac{3}{4},\ -1\right),\ \left(\frac{3}{2},\ 0\right),\ \left(\frac{9}{4},\ 1\right),\ (3,\ 0)$

39. Sketch the graph of the following. (Include two full periods.)

$$y = \sin\left(x - \frac{\pi}{4}\right)$$

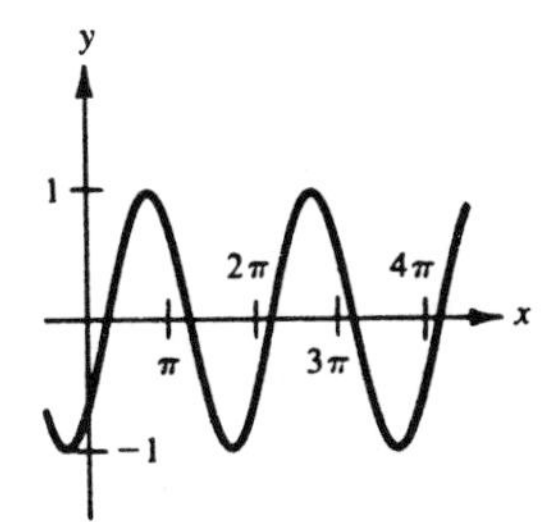

Solution:

$y = \sin\left(x - \dfrac{\pi}{4}\right);\quad a = 1,\ b = 1,\ c = \dfrac{\pi}{4}$

Period: 2π

Amplitude: 1

Shift: Set $x - \dfrac{\pi}{4} = 0$ and $x - \dfrac{\pi}{4} = 2\pi$

$x = \dfrac{\pi}{4}$ $\qquad x = \dfrac{9\pi}{4}$

Key points: $\left(\frac{\pi}{4},\ 0\right),\ \left(\frac{3\pi}{4},\ 1\right),\ \left(\frac{5\pi}{4},\ 0\right),\ \left(\frac{7\pi}{4},\ -1\right),\ \left(\frac{9\pi}{4},\ 0\right)$

45. Sketch the graph of the following. (Include two full periods.)

$$y = \frac{2}{3}\cos\left(\frac{x}{2} - \frac{\pi}{4}\right)$$

Solution:

$y = \frac{2}{3}\cos\left(\frac{x}{2} - \frac{\pi}{4}\right);\quad a = \frac{2}{3},\ b = \frac{1}{2},\ c = \frac{\pi}{4}$

Period: 4π

Amplitude: $\frac{2}{3}$

Shift: Set $\frac{x}{2} - \frac{\pi}{4} = 0$ and $\frac{x}{2} - \frac{\pi}{4} = 2\pi$

$x = \frac{\pi}{2}\qquad\qquad x = \frac{9\pi}{2}$

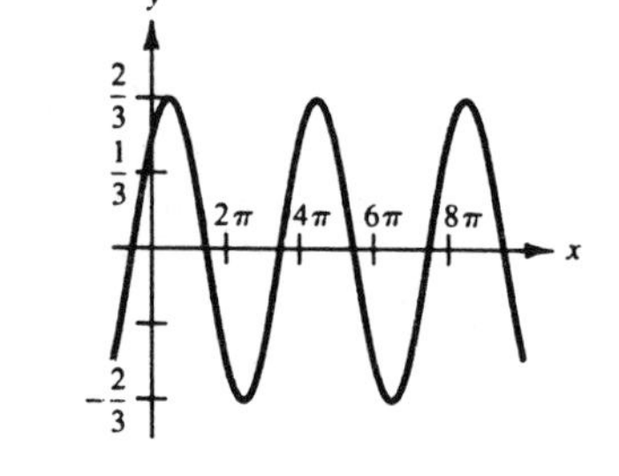

Key points: $\left(\frac{\pi}{2}, \frac{2}{3}\right), \left(\frac{3\pi}{2}, 0\right), \left(\frac{5\pi}{2}, \frac{-2}{3}\right), \left(\frac{7\pi}{2}, 0\right), \left(\frac{9\pi}{2}, \frac{2}{3}\right)$

49. Sketch the graph of the following. (Include two full periods.)

$$y = \cos\left(2\pi x - \frac{\pi}{2}\right) + 1$$

Solution:

$y = \cos\left(2\pi x - \frac{\pi}{2}\right) + 1;\quad a = 1,\ b = 2\pi,\ c = \frac{\pi}{2}$

Period: 1

Amplitude: 1

Shift: Set $2\pi x - \frac{\pi}{2} = 0$ and $2\pi x - \frac{\pi}{2} = 2\pi$

$x = \frac{1}{4}\qquad\qquad x = \frac{5}{4}$

Key points: $\left(\frac{1}{4}, 2\right), \left(\frac{1}{2}, 1\right), \left(\frac{3}{4}, 0\right), (1, 1), \left(\frac{5}{4}, 2\right)$

Vertical shift: One unit upward

51. Sketch the graph of the following. (Include two full periods.)

$$y = -0.1\sin\left(\frac{\pi x}{10} + \pi\right)$$

Solution:

$y = -0.1\sin\left(\dfrac{\pi x}{10} + \pi\right)$; $a = -0.1$, $b = \dfrac{\pi}{10}$, $c = -\pi$

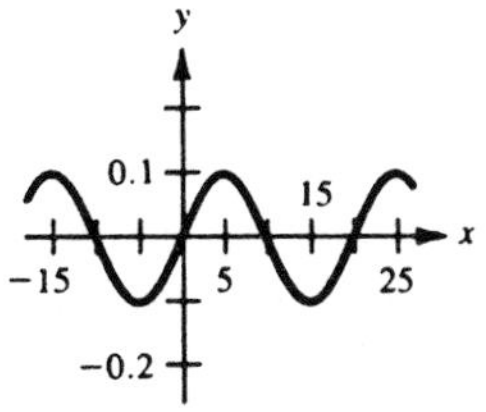

Period: $\dfrac{2\pi}{\pi/10} = 20$

Amplitude: $|-0.1| = 0.1$

Shift: Set $\dfrac{\pi x}{10} + \pi = 0$ and $\dfrac{\pi x}{10} + \pi = 2\pi$

$x = -10$ $\qquad$ $x = 10$

Key points: $(-10, 0)$, $(-5, -0.1)$, $(0, 0)$, $(5, 0.1)$, $(10, 0)$

55. Sketch the graph of $y = \frac{1}{10}\cos 60\pi x$. (Include two full periods.)

Solution:

$y = \dfrac{1}{10}\cos(60\pi x)$; $a = \dfrac{1}{10}$, $b = 60\pi$, $c = 0$

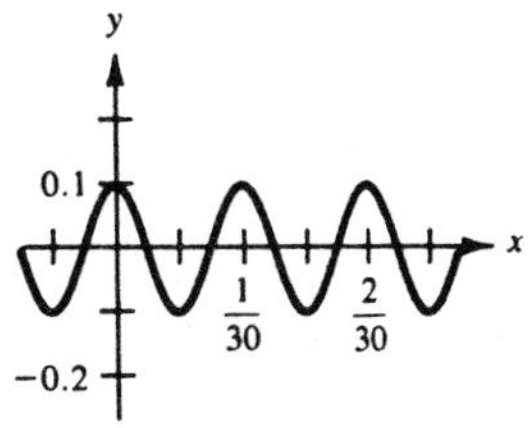

Period: $\dfrac{2\pi}{60\pi} = \dfrac{1}{30}$

Amplitude: $\dfrac{1}{10}$

Key points: $\left(0, \dfrac{1}{10}\right)$, $\left(\dfrac{1}{120}, 0\right)$, $\left(\dfrac{1}{60}, -\dfrac{1}{10}\right)$, $\left(\dfrac{1}{40}, 0\right)$, $\left(\dfrac{1}{30}, \dfrac{1}{10}\right)$

59. Find a, b, and c so that the graph of the function matches the graph in the figure.

Solution:

$y = a\cos(bx - c)$

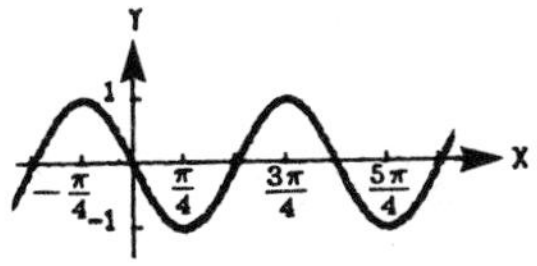

Amplitude: $1 \Rightarrow a = 1$

Period: $\dfrac{2\pi}{b} = \pi \Rightarrow b = 2$

Shift: The graph begins at $-\dfrac{\pi}{4}$.

$$2\left(-\frac{\pi}{4}\right) - c = 0 \Rightarrow c = -\frac{\pi}{2}$$

Thus, $y = \cos\left(2x + \dfrac{\pi}{2}\right)$.

65. For a person at rest, the velocity v (in liters per second) of air flow during a respiratory cycle is

$$v = 0.85\sin\frac{\pi t}{3}$$

where t is the time in seconds. (Inhalation occurs when $v > 0$, and exhalation occurs when $v < 0$.)

(a) Find the time for one full respiratory cycle.
(b) Find the number of cycles per minute.
(c) Sketch the graph of the velocity function.

Solution:

(a) Time for one cycle = period = $\dfrac{2\pi}{\pi/3} = 6$ sec

(b) Cycles per min = $\dfrac{60}{6} = 10$ cycles per min

(c) Amplitude: 0.85
Period: 6
Key points: $(0, 0)$, $\left(\frac{3}{2}, 0.85\right)$, $(3, 0)$, $\left(\frac{9}{2}, -0.85\right)$, $(6, 0)$

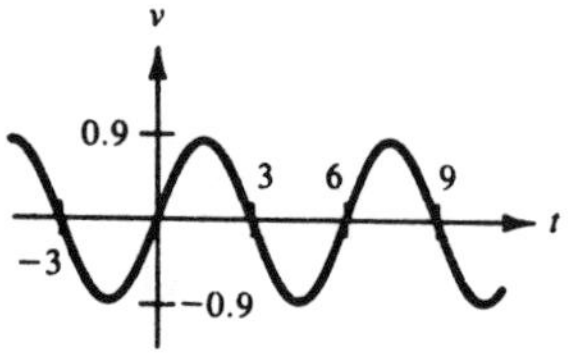

v
0.9
3 6 9
t
−3
−0.9

67. When tuning a piano, a technician strikes a tuning fork for the A above middle C and sets up wave motion that can be approximated by

$$y = 0.001 \sin 880\pi t$$

where t is the time in seconds.

(a) What is the period p of this function?
(b) The frequency f is given by $f = 1/p$. What is the frequency of this note?
(c) Sketch the graph of this function.

Solution:

(a) Period: $\dfrac{2\pi}{880\pi} = \dfrac{1}{440}$

(b) $f = \dfrac{1}{p} = 440$

(c) Amplitude: 0.001
Period: $\dfrac{1}{440}$
Key points: $(0, 0)$, $\left(\frac{1}{1760}, 0.001\right)$, $\left(\frac{1}{880}, 0\right)$, $\left(\frac{3}{1760}, -0.001\right)$, $\left(\frac{1}{440}, 0\right)$

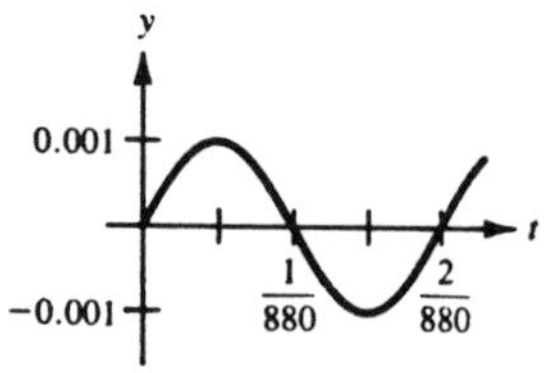

y
0.001
−0.001
$\frac{1}{880}$
$\frac{2}{880}$
t

SECTION 5.6

Graphs of Other Trigonometric Functions

- You should be able to graph

$$y = a\tan(bx - c) \qquad y = a\cot(bx - c)$$
$$y = a\sec(bx - c) \qquad y = a\csc(bx - c)$$

- When graphing

$$y = a\sec(bx - c) \quad \text{or} \quad y = a\csc(bx - c)$$

you should know to first graph

$$y = a\cos(bx - c) \quad \text{or} \quad y = a\sin(bx - c)$$

since

(a) The intercepts of sine and cosine are vertical asymptotes of cosecant and secant.

(b) The maximum points of sine and cosine are local minimums of cosecant and secant.

(c) The minimum points of sine and cosine are local maximums of cosecant and secant.

Solutions to Selected Exercises

5. Match $y = \cot \pi x$ with the correct graph and give the period of the function.

Solution:

Period: $\dfrac{\pi}{\pi} = 1$

Matches graph (d)

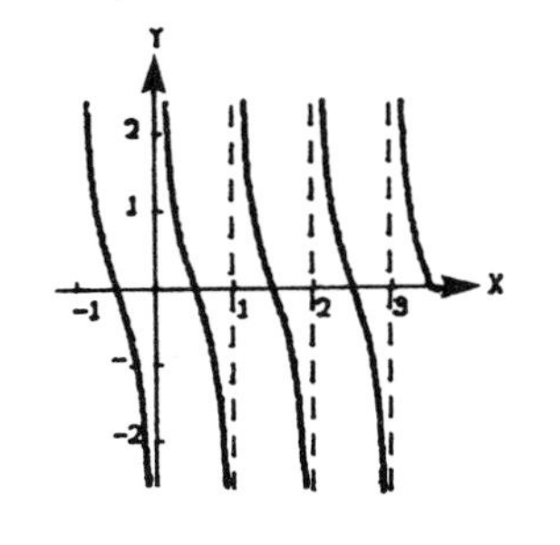

9. Sketch the graph of $y = \tan 2x$ through two periods.

Solution:

Period: $\dfrac{\pi}{2}$

One cycle: $-\dfrac{\pi}{4}$ to $\dfrac{\pi}{4}$

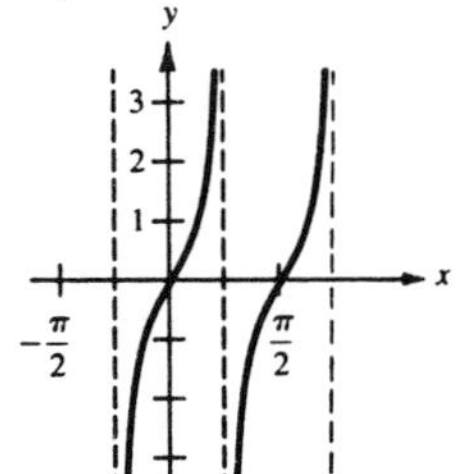

13. Sketch the graph of $y = -2\sec 4x$ through two periods.

Solution:

Period: $\dfrac{2\pi}{4} = \dfrac{\pi}{2}$

One cycle: 0 to $\dfrac{\pi}{2}$

19. Sketch the graph of the following through two periods.

$$y = \csc \frac{x}{2}$$

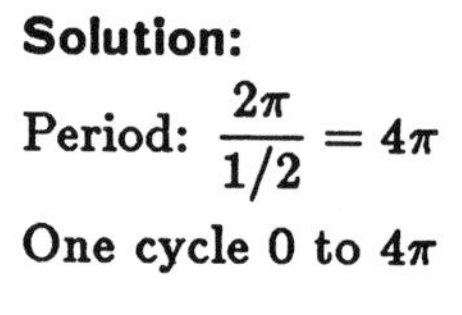

Solution:

Period: $\dfrac{2\pi}{1/2} = 4\pi$

One cycle 0 to 4π

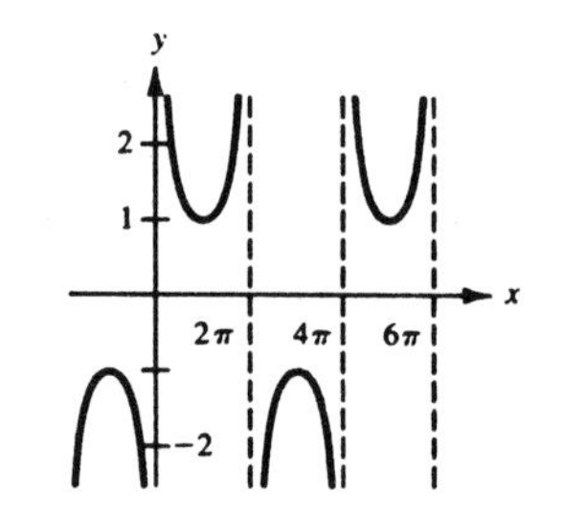

21. Sketch the graph of the following through two periods.

$$y = \cot \frac{x}{2}$$

Solution:

Period: $\dfrac{\pi}{1/2} = 2\pi$

One cycle: 0 to 2π

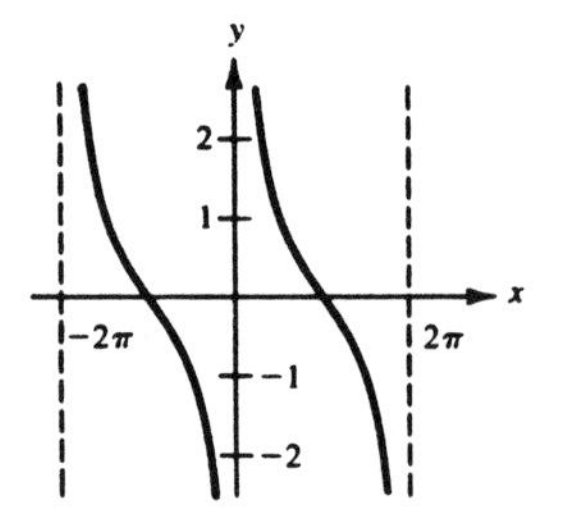

25. Sketch the graph of the following through two periods.

$$y = \tan\left(x - \frac{\pi}{4}\right)$$

Solution:

Period: π

Shift: Set $x - \dfrac{\pi}{4} = -\dfrac{\pi}{2}$ and $x - \dfrac{\pi}{4} = \dfrac{\pi}{2}$

$x = -\dfrac{\pi}{4}$ to $x = \dfrac{3\pi}{4}$

29. Sketch the graph of the following through two periods.

$$y = \frac{1}{4}\cot\left(x - \frac{\pi}{2}\right)$$

Solution:

Period: π

Shift: Set $x - \dfrac{\pi}{2} = 0$ and $x - \dfrac{\pi}{2} = \pi$

$x = \dfrac{\pi}{2}$ to $x = \dfrac{3\pi}{2}$

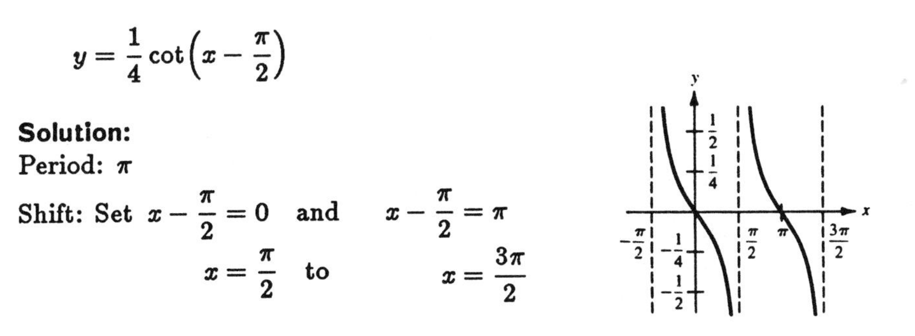

31. Sketch the graph of $y = 2\sec(2x - \pi)$ through two periods.

Solution:

$y = 2\sec(2x - \pi)$

Period: π

Shift: Set $2x - \pi = 0$ and $2x - \pi = 2\pi$

$x = \dfrac{\pi}{2}$ to $x = \dfrac{3\pi}{2}$

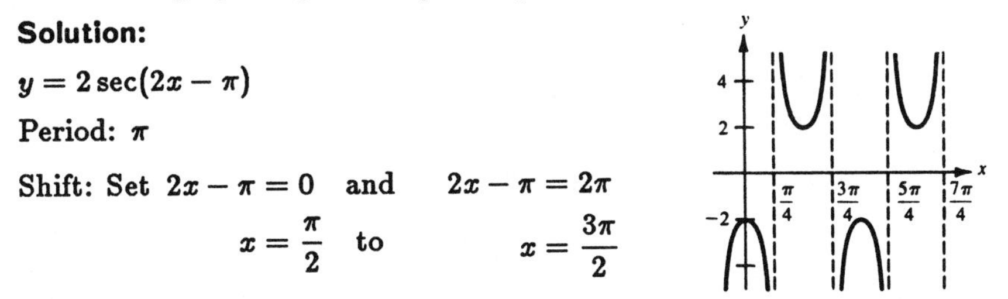

35. Sketch the graph of $y = \csc(\pi - x)$ through two periods.

Solution:

$y = \csc(\pi - x)$

Period: 2π

Shift: Set $\pi - x = 0$ and $\pi - x = 2\pi$

$x = \pi$ to $x = -\pi$

37. A plane flying at an altitude of 6 miles over level ground will pass directly over a radar antenna, as shown in the figure. Let d be the ground distance from the antenna to the point directly under the plane and let x be the angle of elevation to the plane from the antenna. Write d as a function of x, $0 < x < \pi/2$.

Solution:

$$\tan x = \frac{6}{d}$$

$$d = \frac{6}{\tan x} = 6\cot x$$

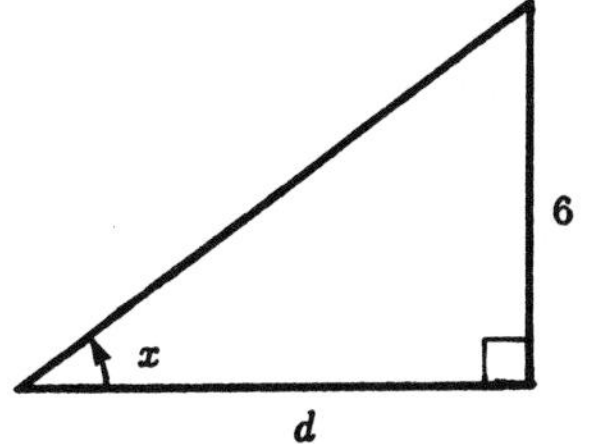

SECTION 5.7

Additional Graphing Techniques

- You should be able to graph by addition of ordinates.
- You should be able to graph vertical translations.
- You should be able to graph using a damping factor.

Solutions to Selected Exercises

1. Use addition of ordinates to sketch the graph of

$$y = 2 - 2\sin\frac{x}{2}.$$

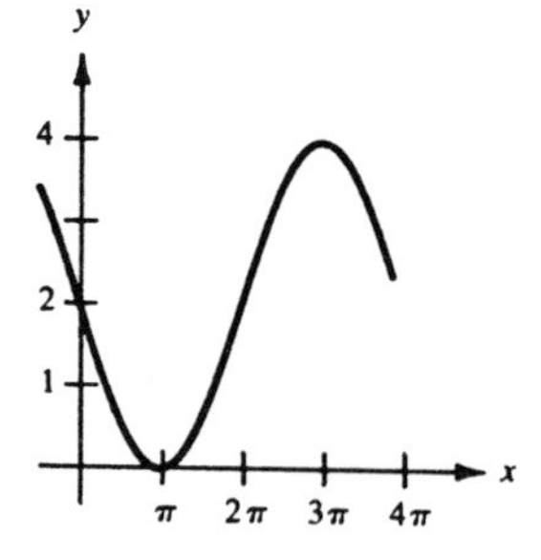

Solution:
Vertical translation of the graph of
$y = -2\sin\frac{x}{2}$ by two units

7. Use addition of ordinates to sketch the graph of $y = 1 + \csc x$.

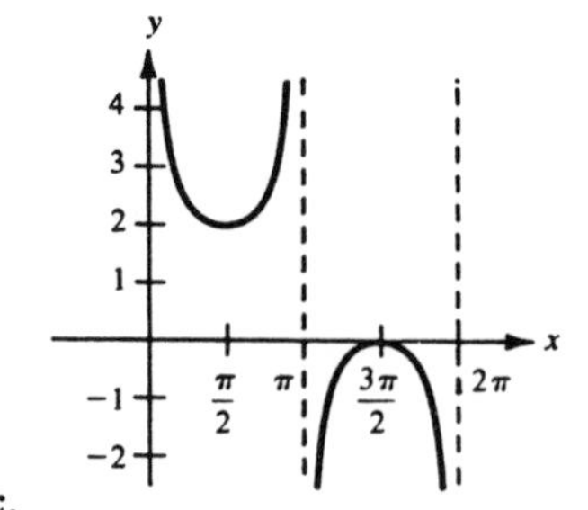

Solution:
Vertical translation of the graph of
$y = 1 + \csc x$ by one unit

11. Use addition of ordinates to sketch the graph of $y = \frac{1}{2}x - 2\cos x$.

Solution:

$$y = \frac{1}{2}x - 2\cos x$$

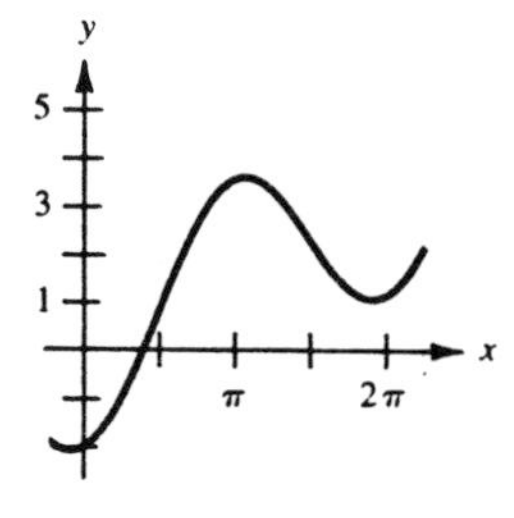

15. Use addition of ordinates to sketch the graph of $y = 2\sin x + \sin 2x$.

Solution:

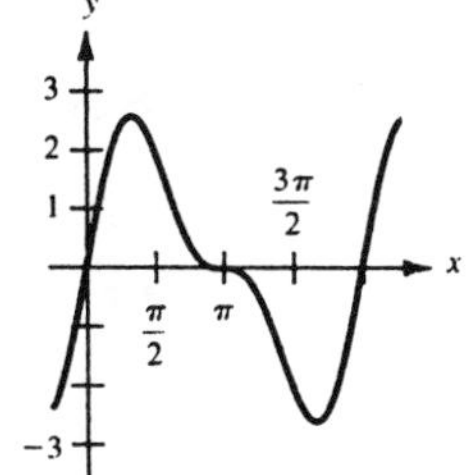

$$y = 2\sin x + \sin 2x$$

21. Use addition of ordinates to sketch the graph of $y = -3 + \cos x + 2\sin 2x$.

Solution:

$$y = -3 + \cos x + 2\sin 2x$$

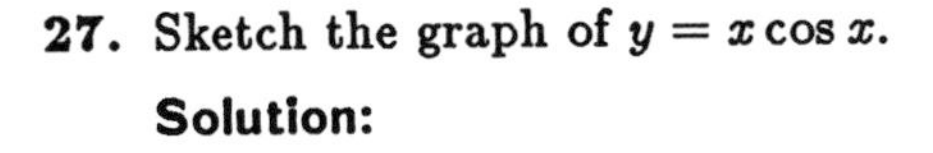

27. Sketch the graph of $y = x\cos x$.

Solution:

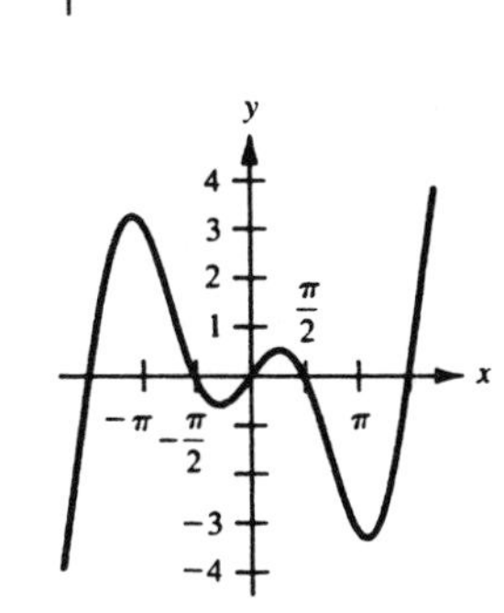

$$y = x\cos x$$

$$|x\cos x| = |x||\cos x| \le |x|$$

Thus, the graph lies between the lines $y = -x$ and $y = x$.

31. Sketch the graph of $y = e^{-x^2/2}\sin x$.

Solution:

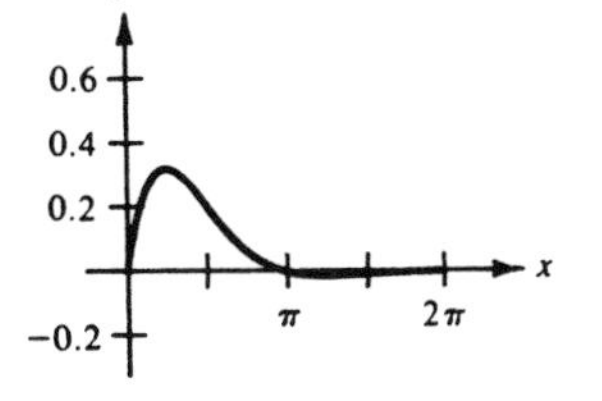

$$y = e^{-x^2/2}\sin x$$

$$|e^{-x^2/2}\sin x| = |e^{-x^2/2}||\sin x| \le |e^{-x^2/2}|$$

Thus, $-e^{-x^2/2} \le y \le e^{-x^2/2}$.

35. The monthly sales S (in thousands of units) of a seasonal product is approximated by $S = 74.50 + 43.75\sin(\pi t/6)$ where t is the time in months with $t = 1$ corresponding to January. Sketch the graph of this sales function over one year.

Solution:

$$S = 74.50 + 43.75 \sin \frac{\pi t}{6}$$

Vertical translation of the graph of $y = 43.75 \sin \frac{\pi t}{6}$ by 74.50 units.

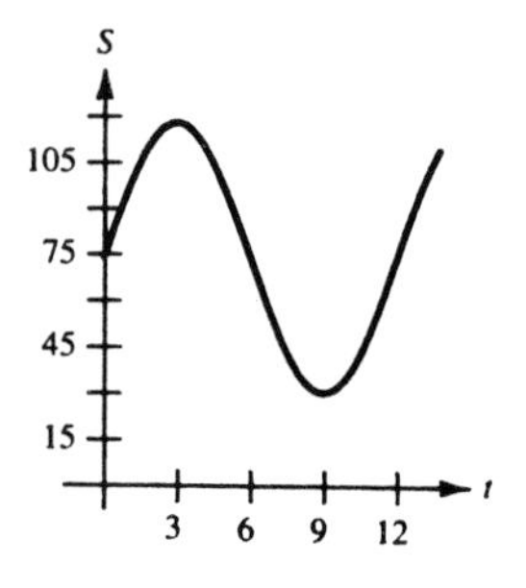

39. Use a calculator to evaluate the function

$$f(x) = \frac{1 - \cos x}{x}$$

at several points in the interval $[-1, 1]$, and then use these points to sketch the graph of f. This function is undefined when $x = 0$. From your graph, estimate the value that $f(x)$ is approaching as x approaches 0.

Solution:
Make sure your calculator is in radian mode.

x	-0.5	-0.4	-0.3	-0.2	-0.1	0.1	0.2	0.3	0.4	0.5
$\frac{1-\cos x}{x}$	-0.245	-0.197	-0.149	-0.100	-0.050	0.050	0.100	0.149	0.197	0.245

As $x \to 0$, $f(x) = \frac{1 - \cos x}{x} \to 0$.

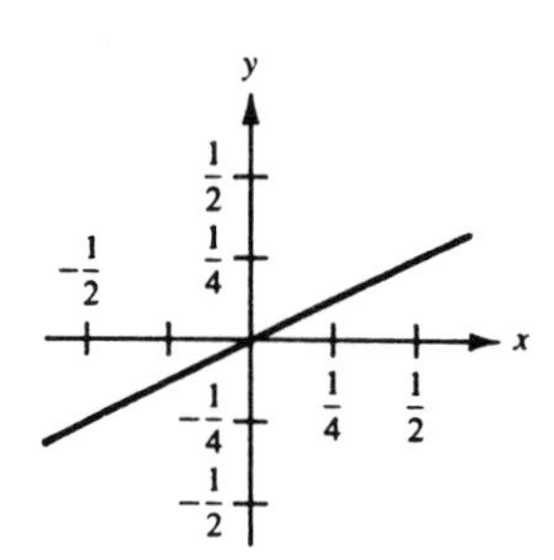

SECTION 5.8

Inverse Trigonometric Functions

- You should know the definitions, domains, and ranges of $y = \arcsin x$, $y = \arccos x$, and $y = \arctan x$.
- You should know the inverse properties of the inverse trigonometric functions.
- You should be able to use the triangle technique to convert trigonometric expressions into algebraic expressions.

Solutions to Selected Exercises

3. Evaluate $\arccos \frac{1}{2}$ without using a calculator.

Solution:

$$\arccos \frac{1}{2} = \theta$$
$$\cos \theta = \frac{1}{2}$$
$$\theta = \frac{\pi}{3}$$

7. Evaluate $\arccos\left(-\frac{\sqrt{3}}{2}\right)$ without using a calculator.

Solution:

$$\arccos\left(-\frac{\sqrt{3}}{2}\right) = \theta$$
$$\cos \theta = -\frac{\sqrt{3}}{2}, \quad \frac{\pi}{2} < \theta < \pi$$
$$\theta = \frac{5\pi}{6}$$

9. Evaluate $\arctan(-\sqrt{3})$ without using a calculator.

Solution:

$$\arctan(-\sqrt{3}) = \theta$$
$$\tan\theta = -\sqrt{3}, \quad -\frac{\pi}{2} < \theta < 0$$
$$\theta = -\frac{\pi}{3}$$

13. Evaluate $\arcsin \frac{\sqrt{3}}{2}$ without using a calculator.

Solution:

$$\arcsin \frac{\sqrt{3}}{2} = \theta$$
$$\sin\theta = \frac{\sqrt{3}}{2}$$
$$\theta = \frac{\pi}{3}$$

17. Use a calculator to approximate $\arccos 0.28$. (Round your answer to two decimal places.)

Solution:
Make sure that your calculator is in radian mode.

$$\arccos 0.28 \approx 1.29$$

21. Use a calculator to approximate $\arctan(-2)$. (Round your answer to two decimal places.)

Solution:

$$\arctan(-2) \approx -1.11$$

27. Use a calculator to approximate $\arctan 0.92$. (Round your answer to two decimal places.)

Solution:

$$\arctan 0.92 \approx 0.74$$

31. Use the properties of inverse trigonometric functions to evaluate $\cos[\arccos(-0.1)]$.

Solution:

$$\cos[\arccos(-0.1)] = -0.1$$

35. Find the exact value of $\sin(\arctan \frac{3}{4})$ without using a calculator. [*Hint:* Make a sketch of a right triangle, as illustrated in Example 6.]

Solution:

Let $y = \arctan \frac{3}{4}$. Then,

$$\tan y = \frac{3}{4}, \quad 0 < y < \frac{\pi}{2}$$

and $\sin y = \frac{3}{5}$.

41. Find the exact value of $\sec[\arctan(-\frac{3}{5})]$ without using a calculator. [*Hint:* Make a sketch of a right triangle, as illustrated in Example 6.]

Solution:

Let $y = \arctan\left(-\frac{3}{5}\right)$. Then,

$$\tan y = -\frac{3}{5}, \quad -\frac{\pi}{2} < y < 0$$

and $\sec y = \frac{\sqrt{34}}{5}$.

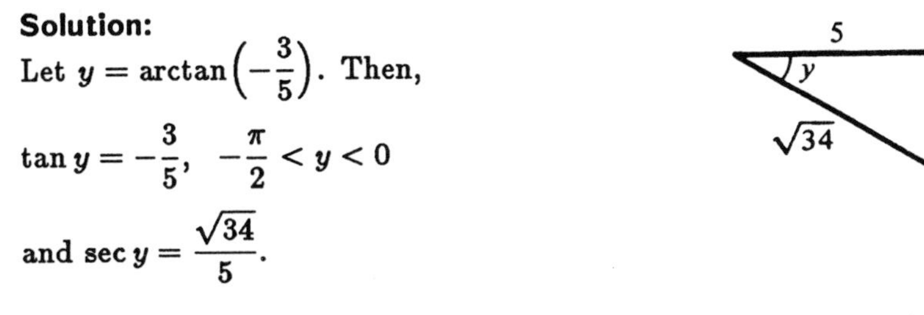

45. Write an algebraic expression that is equivalent to $\cos(\arcsin 2x)$. [*Hint:* Sketch a right triangle, as demonstrated in Example 7.]

Solution:

Let $y = \arcsin(2x)$.

Then, $\sin y = 2x = \frac{2x}{1}$

and $\cos y = \sqrt{1 - 4x^2}$.

49. Write an algebraic expression that is equivalent to $\tan[\arccos(x/3)]$. [*Hint:* Sketch a right triangle, as demonstrated in Example 7.]

Solution:

Let $y = \arccos \frac{x}{3}$.

Then, $\cos y = \frac{x}{3}$

and $\tan y = \frac{\sqrt{9 - x^2}}{x}$.

53. Fill in the blank.

$$\arctan \frac{9}{x} = \arcsin(\_\_\_\_\_\_)$$

Solution:

$$\arctan \frac{9}{x} = \arcsin \frac{9}{\sqrt{x^2 + 81}}$$

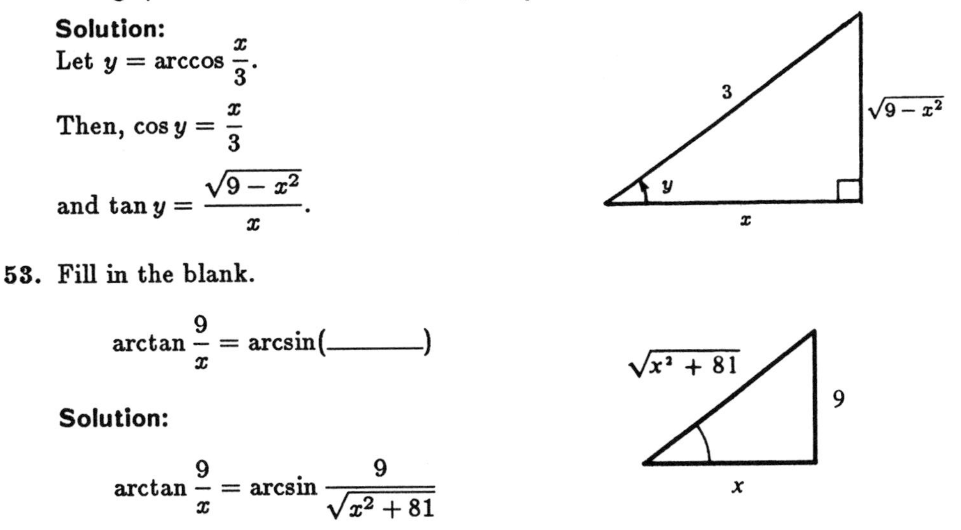

57. Sketch the graph of $f(x) = \arcsin(x - 1)$.

Solution:

The graph of $f(x) = \arcsin(x - 1)$ is a horizontal translation of the graph of $y = \arcsin x$ by one unit.

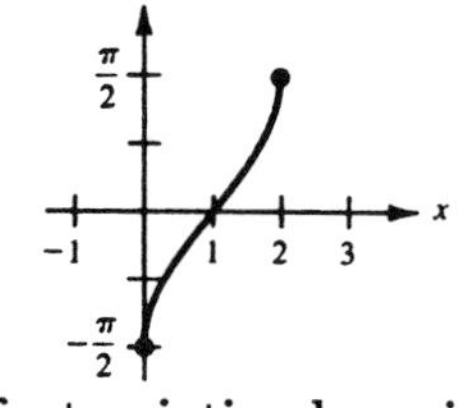

61. A photographer is taking a picture of a four-foot painting hung in an art gallery. The camera lens is one foot below the lower edge of the painting, as shown in the figure. The angle β subtended by the camera lens x feet from the painting is given by

$$\beta = \arctan \frac{4x}{x^2 + 5}.$$

Find β when (a) $x = 3$ feet and (b) $x = 6$ feet.

Solution:

(a) When $x = 3$,

$$\beta = \arctan\left(\frac{12}{9 + 5}\right)$$
$$\approx 0.7086 \text{ radians or } 40.6°$$

(b) When $x = 6$,

$$\beta = \arctan\left(\frac{24}{36 + 5}\right)$$
$$\approx 0.5296 \text{ radians or } 30.3°$$

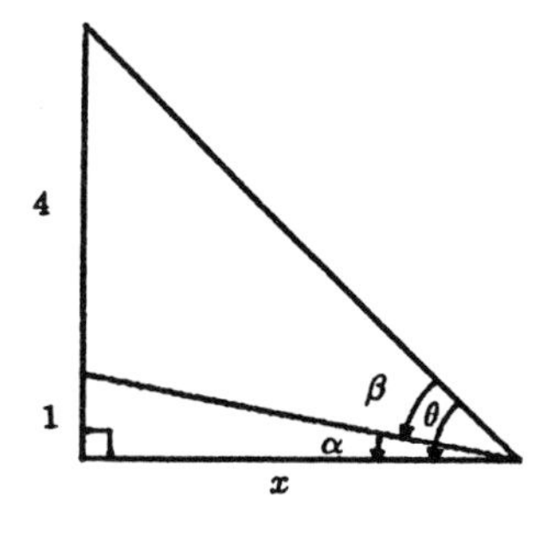

67. Prove the identity of $\arcsin(-x) = -\arcsin x$.

Solution:

Let $y = \arcsin(-x)$. Then,

$$\sin y = -x$$
$$-\sin y = x$$
$$\sin(-y) = x$$
$$-y = \arcsin x$$
$$y = -\arcsin x.$$

Therefore, $\arcsin(-x) = -\arcsin x$.

SECTION 5.9

Applications of Trigonometry

- You should be able to solve right triangles.
- You should be able to solve right triangle applications.

Solutions to Selected Exercises

5. Solve the right triangle, given $A = 12°15'$ and $c = 430.5$. (Round your answers to two decimal places.)

Solution:

$A = 12°15', \quad c = 430.5$

$B = 90° - 12°15' = 77°45'$

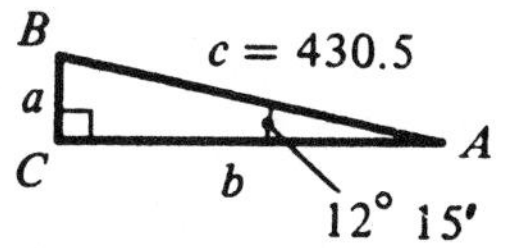

$$\sin 12°15' = \frac{a}{430.5}$$

$$a = 430.5 \sin 12°15' \approx 91.34$$

$$\cos 12°15' = \frac{b}{430.5}$$

$$b = 430.5 \cos 12°15' \approx 420.70$$

9. Solve the right triangle, given $b = 16$ and $c = 52$. (Round your answers to two decimal places.)

Solution:

$b = 16, \quad c = 52$

$a = \sqrt{52^2 - 16^2} = \sqrt{2448} = 12\sqrt{17}$

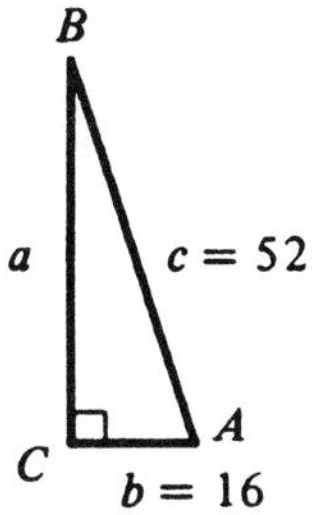

$$\cos A = \frac{16}{52}$$

$$A = \arccos \frac{16}{52} \approx 72.08°$$

$$B = 90 - 72.08 \approx 17.92°$$

13. An isosceles triangle has two angles of 52°, as shown in the figure. The base of the triangle is 4 inches. Find the altitude of the triangle.

Solution:

Divide the triangle in half. Then

$$\tan 52^\circ = \frac{h}{2}$$
$$h = 2\tan 52^\circ$$
$$\approx 2.56 \text{ inches}$$

15. An amateur radio operator erects a 75-foot vertical tower for his antenna. Find the angle of elevation to the top of the tower at a point on level ground 50 feet from the base.

Solution:

$$\tan\theta = \frac{75}{50}$$
$$\theta = \arctan\frac{3}{2} = 56.3^\circ$$

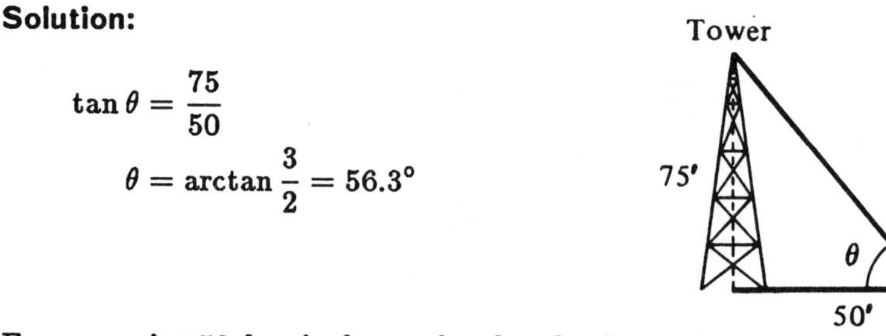

19. From a point 50 feet in front of a church, the angles of elevation to the base of the steeple and the top of the steeple are 35° and 47°40′, respectively, as shown in the figure. Find the height of the steeple.

Solution:

Let Height of the church $= x$.
Height of the church and steeple $= y$.

Then, $\tan 35^\circ = \dfrac{x}{50}$ and $\tan 47^\circ 40' = \dfrac{y}{50}$

$x = 35.01$ $\qquad$ $y = 54.88$

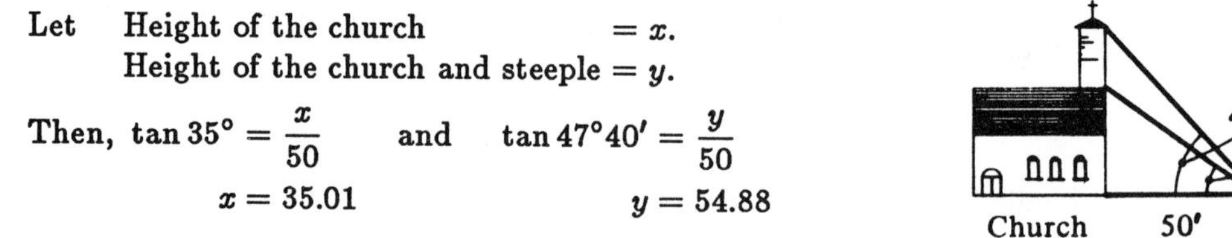

Height of the steeple $= y - x = 19.9$ feet

23. A ship is 45 miles east and 30 miles south of port. If the captain wants to travel directly to port, what bearing should be taken?

Solution:

$$\tan A = \frac{30}{45}$$
$$A = \arctan\frac{2}{3}$$
$$A = 30.69^\circ$$

Bearing $= 90 - 30.69 = 56.3^\circ =$ N 56.3° W

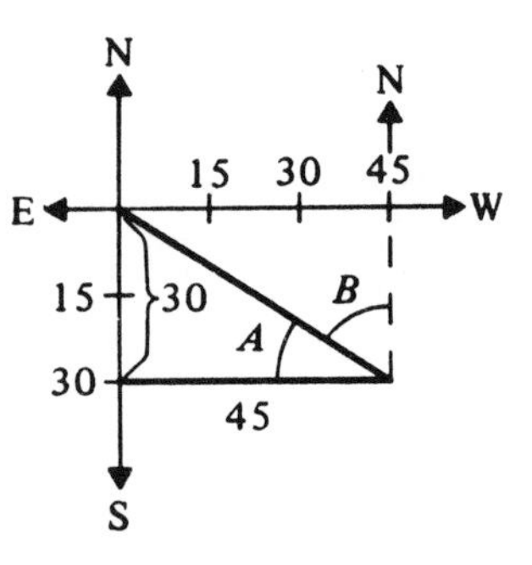

27. An observer in a lighthouse 300 feet above sea level spots two ships directly offshore. The angles of depression to the ships are 4° and 6.5°, as shown in the figure. How far apart are the ships?

Solution:

$$\tan 4° = \frac{300}{y}$$

$$y = \frac{300}{\tan 4°} \approx 4290.20$$

$$\tan 6.5° = \frac{x}{300}$$

$$x = \frac{300}{\tan 6.5} \approx 2633.07$$

Distance between ships = 1657.13 feet

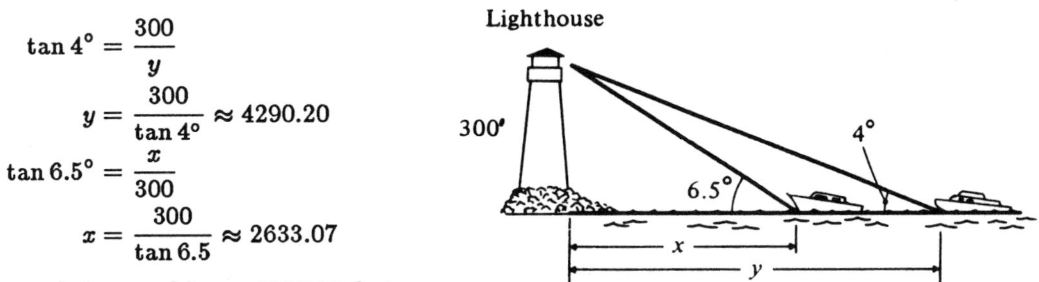

33. Use the figure to find the distance y across the flat sides of the hexagonal nut as a function of r.

Solution:

$$\sin 60° = \frac{\text{opp.}}{\text{hyp.}} = \frac{\frac{1}{2}y}{r}$$

$$\frac{\sqrt{3}}{2} = \frac{y}{2r}$$

$$y = \sqrt{3}r$$

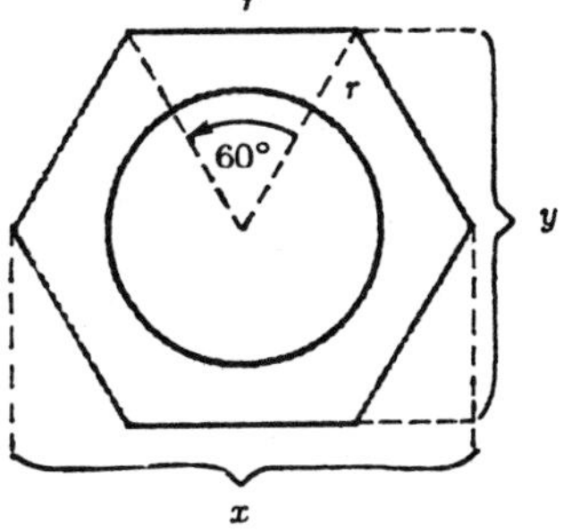

37. For the simple harmonic motion described by $d = 4\cos 8\pi t$, find (a) the maximum displacement, (b) the frequency, and (c) the least positive value of t for which $d = 0$.

Solution:

(a) Maximum displacement = amplitude = 4

(b) Frequency $= \dfrac{\omega}{\text{period}} = \dfrac{8\pi}{2\pi} = 4$ cycles per unit of time

(c) $d = 0$ when $8\pi t = \dfrac{\pi}{2}$ or $t = \dfrac{1}{16}$.

41. A point on the end of a tuning fork moves in simple harmonic motion described by $d = a\sin\omega t$. Find ω given that the tuning fork for middle C has a frequency of 264 vibrations per second.

Solution:

$$\text{Frequency} = \frac{\omega}{2\pi} = 264$$

$$\omega = 528\pi$$

REVIEW EXERCISES FOR CHAPTER 5

Solutions to Selected Exercises

3. Sketch the angle $-110°$ in standard position, and list one positive and one negative coterminal angle.

Solution:

$\theta = -110°$

Coterminal angles:

$$\theta_1 = 360° + (-110°) = 250°$$
$$\theta_2 = -360° + (-110°) = -470°$$

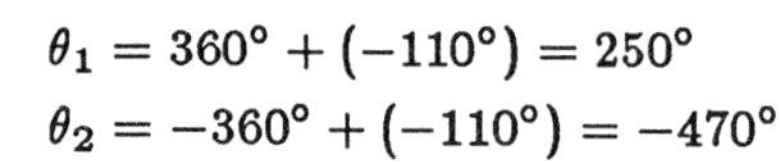

7. Convert $5°22'53''$ to decimal form. Round your answer to two decimal places.

Solution:

$$5°22'53'' = 5 + \frac{22}{60} + \frac{53}{3600} \approx 5.38°$$

11. Convert $-85.15°$ to D° M′ S″ form.

Solution:

$$-85.15° = -[85° + .15(60)'] = -85°9'$$

15. Convert -3.5 from radians to degrees. Round your answer to two decimal places.

Solution:

$$-3.5 = -3.5\left(\frac{180}{\pi}\right)^{\circ} \approx -200.54°$$

19. Convert $-33°45'$ from degrees to radians. Round your answer to four decimal places.

Solution:

$$\begin{aligned} -33°45' &= -\left[33 + \frac{45}{60}\right]^{\circ} \\ &= -33.75° \\ &= -33.75\left(\frac{\pi}{180}\right) \\ &\approx -0.5890 \text{ radians} \end{aligned}$$

23. Find the reference angle for 252°.

Solution:
252° is in Quadrant III. The reference angle is

$$\theta = 252° - 180° = 72°.$$

27. Find the six trigonometric functions of the angle θ (in standard position) whose terminal side passes through the point $(-4, -6)$.

Solution:
$x = -4,\ y = -6,\ r = \sqrt{(-4)^2 + (-6)^2} = 2\sqrt{13}$

$$\sin\theta = \frac{y}{r} = \frac{-6}{2\sqrt{13}} = -\frac{3\sqrt{13}}{13} \qquad \csc\theta = \frac{r}{y} = \frac{2\sqrt{13}}{-6} = -\frac{\sqrt{13}}{3}$$

$$\cos\theta = \frac{x}{r} = \frac{-4}{2\sqrt{13}} = -\frac{2\sqrt{13}}{13} \qquad \sec\theta = \frac{r}{x} = \frac{2\sqrt{13}}{-4} = -\frac{\sqrt{13}}{2}$$

$$\tan\theta = \frac{y}{x} = \frac{-6}{-4} = \frac{3}{2} \qquad \cot\theta = \frac{x}{y} = \frac{-4}{-6} = \frac{2}{3}$$

31. Use a right triangle to find the remaining five trigonometric functions of θ, given $\sin\theta = \frac{3}{8}$ and $\cos\theta < 0$.

Solution:
$\sin\theta = \frac{3}{8},\ \ \cos\theta < 0,\ \theta$ is in Quadrant II
$y = 3,\ r = 8,\ x = -\sqrt{64 - 9} = -\sqrt{55}$

$$\cos\theta = -\frac{\sqrt{55}}{8} \qquad \sec\theta = \frac{8}{-\sqrt{55}} = -\frac{8\sqrt{55}}{55}$$

$$\tan\theta = \frac{3}{-\sqrt{55}} = -\frac{3\sqrt{55}}{55} \qquad \cot\theta = -\frac{\sqrt{55}}{3}$$

$$\csc\theta = \frac{8}{3}$$

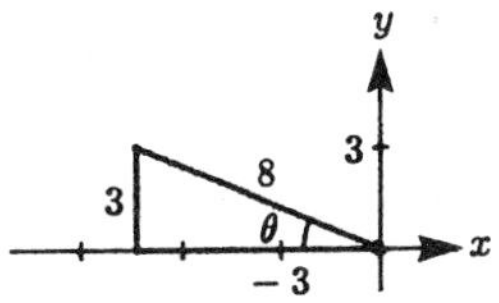

y
3
8
3
θ
x
−3

35. Evaluate $\cos 495°$ without the use of a calculator.

Solution:
The reference angle for 495° is 45° and 495° is in Quadrant II. Therefore,

$$\cos 495° = -\cos 45° = -\tfrac{\sqrt{2}}{2}.$$

39. Use a calculator to evaluate $\sec(12\pi/5)$. Round your answer to two decimal places.

Solution:
Make sure that your calculator is in radian mode.

$$\sec\left(\frac{12\pi}{5}\right) = \frac{1}{\cos(12\pi/5)} \approx 3.24$$

43. Given $\csc\theta = -2$, find two values of θ in degrees $(0° \le \theta < 360°)$ and in radians $(0 \le \theta < 2\pi)$ without using a calculator.

Solution:

Since $\csc\theta < 0$, we know that θ is in either Quadrant III or in Quadrant IV. Also, since $\csc 30° = 2$, we know that $30°$ is the reference angle.

$$\text{In Quadrant III: } \theta = 180° + 30° = 210° = \frac{7\pi}{6}$$

$$\text{In Quadrant IV: } \theta = 360° - 30° = 330° = \frac{11\pi}{6}$$

47. Given $\sec\theta = -1.0353$, find two values of θ in degrees $(0° \le \theta < 360°)$ and in radians $(0 \le \theta < 2\pi)$ by using a calculator.

Solution:

Since $\sec\theta < 0$, we know that θ is in either Quadrant II or in Quadrant III. To find the reference angle θ', use

$$\cos\theta' = \frac{1}{\sec\theta'} = \frac{1}{1.0353}$$

$$\theta' = \arccos\left(\frac{1}{1.0353}\right)$$

The reference angle θ' is $15°$.

$$\text{In Quadrant II: } \theta = 180° - 15° = 165° \approx 2.8798$$

$$\text{In Quadrant III: } \theta = 180° + 15° = 195° \approx 3.4034$$

51. Use a right triangle to write an algebraic expression for

$$\sin\left(\arccos\frac{x^2}{4-x^2}\right).$$

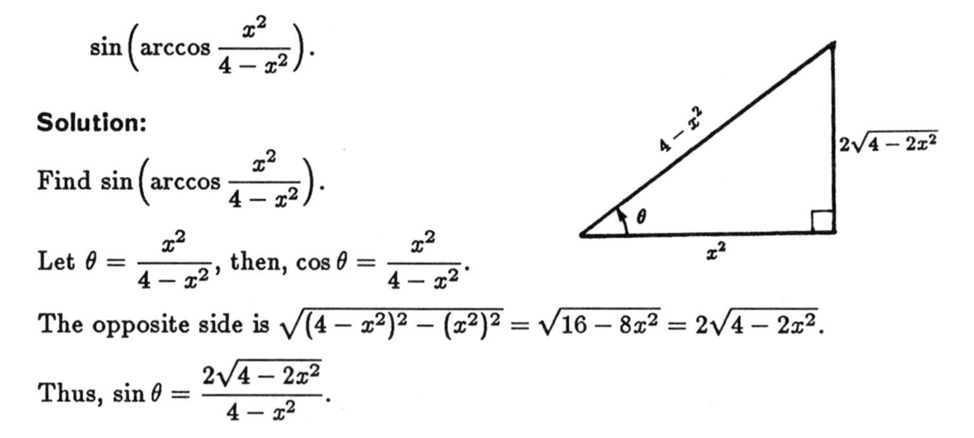

Solution:

Find $\sin\left(\arccos\dfrac{x^2}{4-x^2}\right)$.

Let $\theta = \dfrac{x^2}{4-x^2}$, then, $\cos\theta = \dfrac{x^2}{4-x^2}$.

The opposite side is $\sqrt{(4-x^2)^2-(x^2)^2} = \sqrt{16-8x^2} = 2\sqrt{4-2x^2}$.

Thus, $\sin\theta = \dfrac{2\sqrt{4-2x^2}}{4-x^2}$.

55. Sketch the graph of

$$f(x) = -\frac{1}{4}\cos\frac{\pi x}{4}.$$

Solution:

Amplitude: $\left|-\frac{1}{4}\right| = \frac{1}{4}$

Period: $\dfrac{2\pi}{\pi/4} = 8$

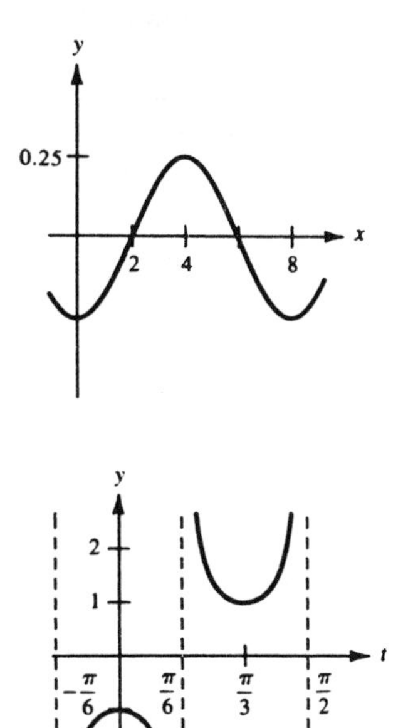

59. Sketch the graph of

$$h(t) = \csc\left(3t - \frac{\pi}{2}\right).$$

Solution:

Period: $\dfrac{2\pi}{3}$

Shift: $3t - \dfrac{\pi}{2} = 0$ and $3t - \dfrac{\pi}{2} = 2\pi$

$t = \dfrac{\pi}{6}$ $\qquad t = \dfrac{5\pi}{6}$

63. Sketch the graph of

$$f(x) = \frac{x}{4} - \sin x.$$

Solution:

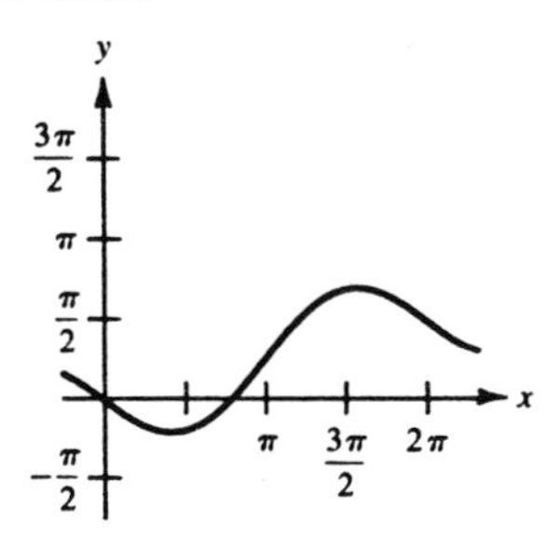

67. Sketch the graph of

$$f(x) = \arcsin\frac{x}{2}.$$

Solution:

Domain: $-2 \le x \le 2$

Range: $-\dfrac{\pi}{2} \le f(x) \le \dfrac{\pi}{2}$

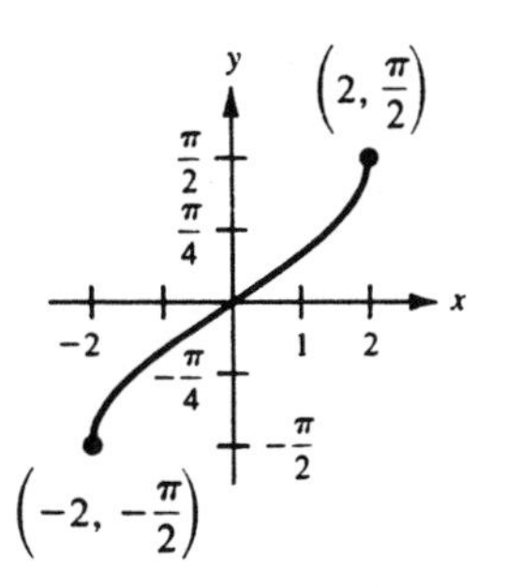

71. An observer 2.5 miles from the launch pad of a space shuttle measures the angle of elevation to the base of the vehicle to be 28° soon after liftoff (see figure). How high is the shuttle at that instant? Assume that the shuttle is still moving vertically.

Solution:

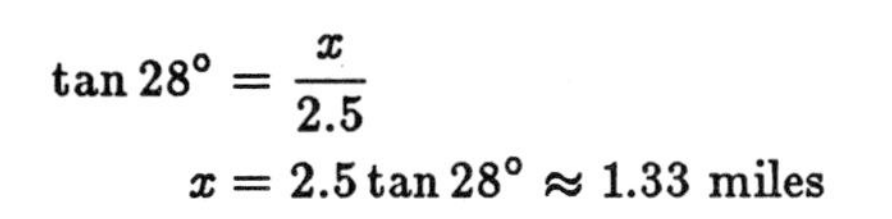

$$\tan 28^\circ = \frac{x}{2.5}$$

$$x = 2.5 \tan 28^\circ \approx 1.33 \text{ miles}$$

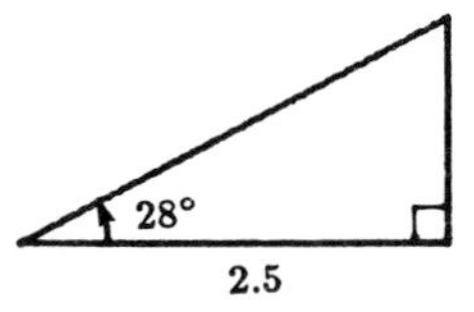

Practice Test for Chapter 5

1. (a) Express 350° in radian measure. (b) Express $\dfrac{5\pi}{9}$ in degree measure.

2. (a) Convert $135°14'12''$ to decimal form. (b) Convert $-22.569°$ to D° M′ S″ form.

3. Use the unit circle to evaluate

(a) $\sin\dfrac{5\pi}{6}$ (b) $\tan\dfrac{5\pi}{4}$

4. Use the unit circle and the periodic nature of sine and cosine to evaluate

(a) $\sin 7\pi$ (b) $\cos\left(-\dfrac{13\pi}{3}\right)$

5. If $\cos\theta = \frac{2}{3}$, use the trigonometric identities to find $\tan\theta$.

6. Find θ given $\sin\theta = 0.9063$.

7. Solve for x.

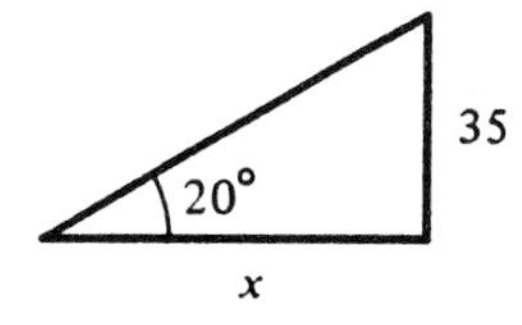
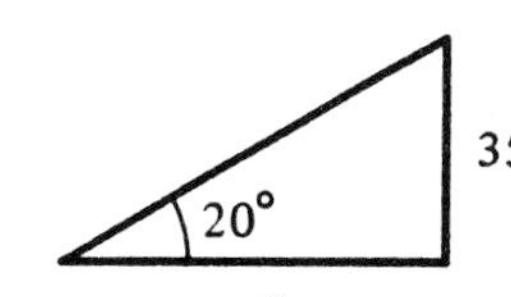

8. Find the magnitude of the reference angle for $\theta = \dfrac{6\pi}{5}$.

9. Evaluate $\csc 3.92$.

10. Find $\sec\theta$ given that θ lies in Quadrant III and $\tan\theta = 6$.

11. Graph $y = 3\sin\dfrac{x}{2}$.

12. Graph $y = -2\cos(x-\pi)$.

13. Graph $y = \tan 2x$.

14. Graph $y = -\csc\left(x + \dfrac{\pi}{4}\right)$.

15. Graph $y = 2x + \sin x$.

16. Graph $y = 3x\cos x$.

17. Evaluate $\arcsin 1$.

18. Evaluate $\arctan(-3)$.

19. Evaluate $\sin\left(\arccos\dfrac{4}{\sqrt{35}}\right)$.

20. Write an algebraic expression for $\cos\left(\arcsin\dfrac{x}{4}\right)$.

For Exercises 21–23, solve the right triangle.

21. $A = 40°$, $c = 12$

22. $B = 6.84°$, $a = 21.3$

23. $a = 5$, $b = 9$

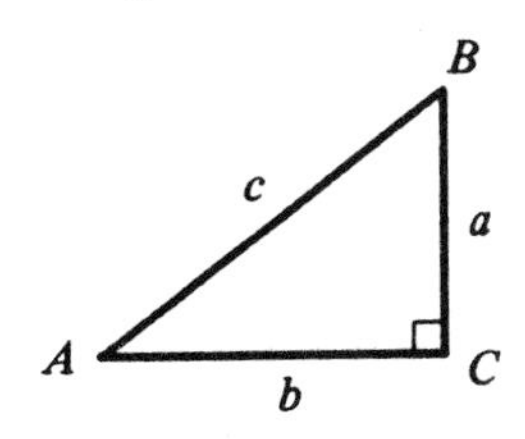

24. A 20-foot ladder leans against the side of a barn. Find the height of the top of the ladder if the angle of elevation of the ladder is $67°$.

25. An observer in a lighthouse 250 feet above sea level spots a ship off the shore. If the angle of depression to the ship is $5°$, how far out is the ship?

CHAPTER 6
Analytic Trigonometry

SECTION 6.1

Applications of Fundamental Identities

- You should know the following identities.
  (a) Reciprocal Identities
  (b) Tangent and Cotangent Identities
  (c) Pythagorean Identities
  (d) Cofunction Identities
  (e) Negative Angle Identities
- You should be able to use these fundamental identities to find function values.
- You should be able to convert trigonometric expressions to equivalent forms by using the fundamental identities.

Solutions to Selected Exercises

3. Given $\sec\theta = \sqrt{2}$ and $\sin\theta = -\sqrt{2}/2$, use the fundamental identities to find the values of the other four trigonometric functions.

Solution:

$$\cos\theta = \frac{1}{\sec\theta} = \frac{1}{\sqrt{2}} = \frac{\sqrt{2}}{2}$$

$$\tan\theta = \frac{\sin\theta}{\cos\theta} = \frac{-\sqrt{2}/2}{\sqrt{2}/2} = -1$$

$$\cot\theta = \frac{1}{\tan\theta} = \frac{1}{-1} = -1$$

$$\csc\theta = \frac{1}{\sin\theta} = \frac{1}{-\sqrt{2}/2} = -\sqrt{2}$$

7. Given $\sec\phi = -1$ and $\sin\phi = 0$, use the fundamental identities to find the values of the other four trigonometric functions.

Solution:

$$\cos\phi = \frac{1}{\sec\phi} = \frac{1}{-1} = -1$$
$$\tan\phi = \frac{\sin\phi}{\cos\phi} = \frac{0}{-1} = 0$$
$$\cot\phi = \frac{1}{\tan\phi} = \frac{1}{0} \text{ undefined}$$
$$\csc\phi = \frac{1}{\sin\phi} = \frac{1}{0} \text{ undefined}$$

11. Given $\tan\theta = 2$ and $\sin\theta < 0$, use the fundamental identities to find the values of the other four trigonometric functions.

Solution:

θ is in Quadrant III.

$$\sec\theta = -\sqrt{1+\tan^2\theta} = -\sqrt{1+(2)^2} = -\sqrt{5}$$
$$\cot\theta = \frac{1}{\tan\theta} = \frac{1}{2}$$
$$\csc\theta = -\sqrt{1+\cot^2\theta} = -\sqrt{1+(1/2)^2} = -\sqrt{1+(1/4)} = -\frac{\sqrt{5}}{2}$$
$$\sin\theta = \frac{1}{\csc\theta} = -\frac{2}{\sqrt{5}} = -\frac{2\sqrt{5}}{5}$$
$$\cos\theta = \frac{1}{\sec\theta} = -\frac{1}{\sqrt{5}} = -\frac{\sqrt{5}}{5}$$

17. Match $\tan^2 x - \sec^2 x$ with one of the following.

(a) -1 (b) $\cos x$ (c) $\cot x$
(d) 1 (e) $-\tan x$ (f) $\sin x$

Solution:

$$\tan^2 x - \sec^2 x = (\sec^2 x - 1) - \sec^2 x = -1$$

Matches (a).

21. Match $\sin x \sec x$ with one of the following.

(a) $\csc x$ (b) $\tan x$ (c) $\sin^2 x$
(d) $\sin x \tan x$ (e) $\sec^2 x$ (f) $\sec^2 x + \tan^2 x$

Solution:

$$\sin x \sec x = \sin x\left(\frac{1}{\cos x}\right) = \frac{\sin x}{\cos x} = \tan x$$

Matches (b).

25. Match $\sec^4 x - \tan^4 x$ with one of the following.

(a) $\csc x$ (b) $\tan x$ (c) $\sin^2 x$
(d) $\sin x \tan x$ (e) $\sec^2 x$ (f) $\sec^2 x + \tan^2 x$

Solution:

$$\begin{aligned}\sec^4 x - \tan^4 x &= (\sec^2 x + \tan^2 x)(\sec^2 x - \tan^2 x)\\ &= (\sec^2 x + \tan^2 x)[(1 + \tan^2 x) - \tan^2 x]\\ &= \sec^2 x + \tan^2 x\end{aligned}$$

Matches (f).

29. Use the fundamental identities to simplify $\cos\beta \tan\beta$.

Solution:

$$\cos\beta \tan\beta = \cos\beta\left(\frac{\sin\beta}{\cos\beta}\right) = \sin\beta$$

33. Use the fundamental identities to simplify $\sec^2 x(1 - \sin^2 x)$.

Solution:

$$\sec^2 x(1 - \sin^2 x) = \sec^2 x(\cos^2 x) = \frac{1}{\cos^2 x}(\cos^2 x) = 1$$

39. Use the fundamental identities to simplify

$$\frac{\cos^2 y}{1 - \sin y}.$$

Solution:

$$\frac{\cos^2 y}{1 - \sin y} = \frac{1 - \sin^2 y}{1 - \sin y} = \frac{(1 + \sin y)(1 - \sin y)}{1 - \sin y} = 1 + \sin y$$

43. Factor $\sin^2 x \sec^2 x - \sin^2 x$ and use the fundamental identities to simplify the result.

Solution:

$$\sin^2 x \sec^2 x - \sin^2 x = \sin^2 x(\sec^2 x - 1) = \sin^2 x \tan^2 x$$

47. Factor $\sin^4 x - \cos^4 x$ and use the fundamental identities to simplify the result.

Solution:

$$\sin^4 x - \cos^4 x = (\sin^2 x + \cos^2 x)(\sin^2 x - \cos^2 x) = 1(\sin^2 x - \cos^2 x) = \sin^2 x - \cos^2 x$$

51. Multiply $(\sec x + 1)(\sec x - 1)$ and use the fundamental identities to simplify the result.

Solution:

$$(\sec x + 1)(\sec x - 1) = \sec^2 x - 1 = \tan^2 x$$

55. Add the following and use the fundamental identities to simplify the result.

$$\frac{\cos x}{1 + \sin x} + \frac{1 + \sin x}{\cos x}$$

Solution:

$$\begin{aligned}
\frac{\cos x}{1 + \sin x} + \frac{1 + \sin x}{\cos x} &= \frac{(\cos x)(\cos x) + (1 + \sin x)(1 + \sin x)}{\cos x(1 + \sin x)} \\
&= \frac{\cos^2 x + 1 + 2\sin x + \sin^2 x}{\cos x(1 + \sin x)} \\
&= \frac{(\sin^2 x + \cos^2 x) + 1 + 2\sin x}{\cos x(1 + \sin x)} \\
&= \frac{1 + 1 + 2\sin x}{\cos x(1 + \sin x)} \\
&= \frac{2\cancel{(1 + \sin x)}}{\cos x\cancel{(1 + \sin x)}} \\
&= \frac{2}{\cos x} \\
&= 2\left(\frac{1}{\cos x}\right) \\
&= 2\sec x
\end{aligned}$$

59. Rewrite the following expression so that it is *not* in fractional form.

$$\frac{3}{\sec x - \tan x}$$

Solution:

$$\begin{aligned}
\frac{3}{\sec x - \tan x} &= \frac{3}{\sec x - \tan x} \cdot \frac{\sec x + \tan x}{\sec x + \tan x} = \frac{3(\sec x + \tan x)}{\sec^2 x - \tan^2 x} \\
&= \frac{3(\sec x + \tan x)}{(1 + \tan^2 x) - \tan^2 x} = 3(\sec x + \tan x)
\end{aligned}$$

63. Use $x = 3\sec\theta$ to write $\sqrt{x^2 - 9}$ as a trigonometric function involving θ, where $0 < \theta < \pi/2$.

Solution:

Since $x = 3\sec\theta$,

$$\begin{aligned}\sqrt{x^2 - 9} &= \sqrt{(3\sec\theta)^2 - 9}\\ &= \sqrt{9\sec^2\theta - 9}\\ &= \sqrt{9(\sec^2\theta - 1)}\\ &= \sqrt{9\tan^2\theta}\\ &= 3\tan\theta\end{aligned}$$

69. Use $x = 3\tan\theta$ to write $\sqrt{(9 + x^2)^3}$ as a trigonometric function involving θ, where $0 < \theta < \pi/2$.

Solution:

Since $x = 3\tan\theta$,

$$\begin{aligned}\sqrt{(9 + x^2)^3} &= \sqrt{[9 + (3\tan\theta)^2]^3}\\ &= \sqrt{(9 + 9\tan^2\theta)^3}\\ &= \sqrt{[9(1 + \tan^2\theta)]^3}\\ &= \sqrt{(9\sec^2\theta)^3}\\ &= (\sqrt{9\sec^2\theta})^3\\ &= (3\sec\theta)^3\\ &= 27\sec^3\theta\end{aligned}$$

73. Rewrite $\ln|\cos\theta| - \ln|\sin\theta|$ as a single logarithm and simplify.

Solution:

$$\ln|\cos\theta| - \ln|\sin\theta| = \ln\frac{|\cos\theta|}{|\sin\theta|} = \ln|\cot\theta|$$

75. Determine whether the following statement is true or false, and give a reason for your answer.

$$\frac{\sin k\theta}{\cos k\theta} = \tan\theta, \quad k \text{ is constant}$$

Solution:

$\dfrac{\sin k\theta}{\cos k\theta} = \tan\theta$ is false, since $\dfrac{\sin k\theta}{\cos k\theta} = \tan k\theta$.

79. Use a calculator to demonstrate that $\csc^2\theta - \cot^2\theta = 1$ is true for

(a) $\theta = 132°$ (b) $\theta = 2\pi/7$.

Solution:

(a) $\theta = 132°$

$\csc 132° \approx 1.3456$

$\cot 132° \approx -0.9004$

Therefore, $\csc^2 132° - \cot^2 132° \approx 1.8107 - 0.8107 = 1$.

(b) $\theta = \dfrac{2\pi}{7}$

$\csc \dfrac{2\pi}{7} \approx 1.2790$

$\cot \dfrac{2\pi}{7} \approx 0.7975$

Therefore, $\csc^2 \dfrac{2\pi}{7} - \cot^2 \dfrac{2\pi}{7} \approx 1.6360 - 0.6360 = 1$.

83. Express each of the other trigonometric functions of θ in terms of $\sin\theta$.

Solution:

Since $\sin^2\theta + \cos^2\theta = 1$ and $\cos^2\theta = 1 - \sin^2\theta$,

$$\cos\theta = \pm\sqrt{1-\sin^2\theta}$$

$$\tan\theta = \frac{\sin\theta}{\cos\theta} = \frac{\pm\sin\theta}{\sqrt{1-\sin^2\theta}}$$

$$\cot\theta = \frac{1}{\tan\theta} = \frac{\pm\sqrt{1-\sin^2\theta}}{\sin\theta}$$

$$\sec\theta = \frac{1}{\cos\theta} = \frac{\pm 1}{\sqrt{1-\sin^2\theta}}$$

$$\csc\theta = \frac{1}{\sin\theta}$$

SECTION 6.2

Verifying Trigonometric Identities

- You should know the difference between an expression, an equation, and an identity.
- You should be able to solve trigonometric identities, using the following techniques.
  (a) Work with *one* side at a time. Do not "cross" the equal sign.
  (b) Use algebraic techniques such as combining fractions, factoring expressions, rationalizing denominators, and squaring binomials.
  (c) Use the fundamental identities.
  (d) Convert all the terms into sines and cosines.

Solutions to Selected Exercises

5. Verify $\cos^2\beta - \sin^2\beta = 1 - 2\sin^2\beta$.

Solution:

$$\cos^2\beta - \sin^2\beta = (1 - \sin^2\beta) - \sin^2\beta = 1 - 2\sin^2\beta$$

9. Verify $\sin^2\alpha - \sin^4\alpha = \cos^2\alpha - \cos^4\alpha$.

Solution:

$$\sin^2\alpha - \sin^4\alpha = \sin^2\alpha(1 - \sin^2\alpha) = (1 - \cos^2\alpha)(\cos^2\alpha) = \cos^2\alpha - \cos^4\alpha$$

13. Verify

$$\frac{\cot^2 t}{\csc t} = \csc t - \sin t.$$

Solution:

$$\frac{\cot^2 t}{\csc t} = \frac{\csc^2 t - 1}{\csc t} = \frac{\csc^2 t}{\csc t} - \frac{1}{\csc t} = \csc t - \sin t$$

17. Verify

$$\frac{1}{\sec x \tan x} = \csc x - \sin x.$$

Solution:

$$\begin{aligned}\frac{1}{\sec x \tan x} &= \frac{1}{\sec x} \cdot \frac{1}{\tan x} \\ &= \cos x \cot x \\ &= \cos x \left(\frac{\cos x}{\sin x}\right) \\ &= \frac{\cos^2 x}{\sin x} \\ &= \frac{1 - \sin^2 x}{\sin x} \\ &= \frac{1}{\sin x} - \frac{\sin^2 x}{\sin x} \\ &= \csc x - \sin x\end{aligned}$$

21. Verify $\csc x - \sin x = \cos x \cot x$.

Solution:

$$\csc x - \sin x = \frac{1}{\sin x} - \frac{\sin^2 x}{\sin x} = \frac{1 - \sin^2 x}{\sin x} = \frac{\cos^2 x}{\sin x} = \cos x \left(\frac{\cos x}{\sin x}\right) = \cos x \cot x$$

25. Verify

$$\frac{\cos \theta \cot \theta}{1 - \sin \theta} - 1 = \csc \theta.$$

Solution:

$$\begin{aligned}\frac{\cos \theta \cot \theta}{1 - \sin \theta} - 1 &= \frac{\cos \theta \left(\frac{\cos \theta}{\sin \theta}\right)}{1 - \sin \theta} \cdot \frac{1 + \sin \theta}{1 + \sin \theta} - 1 \\ &= \frac{\cos^2 \theta (1 + \sin \theta)}{\sin \theta (1 - \sin^2 \theta)} - 1 \\ &= \frac{\cancel{\cos^2 \theta}(1 + \sin \theta)}{\sin \theta \cancel{\cos^2 \theta}} - 1 \\ &= \frac{1 + \sin \theta}{\sin \theta} - 1 \\ &= \frac{1}{\sin \theta} + \frac{\sin \theta}{\sin \theta} - 1 \\ &= \csc \theta + 1 - 1 \\ &= \csc \theta\end{aligned}$$

29. Verify $2\sec^2 x - 2\sec^2 x\sin^2 x - \sin^2 x - \cos^2 x = 1$.

Solution:

$$\begin{aligned}2\sec^2 x - 2\sec^2 x\sin^2 x - \sin^2 x - \cos^2 x &= 2\sec^2 x(1-\sin^2 x) - (\sin^2 x + \cos^2 x)\\ &= 2\sec^2 x(\cos^2 x) - 1\\ &= 2\left(\frac{1}{\cos^2 x}\right)(\cos^2 x) - 1\\ &= 2 - 1 = 1\end{aligned}$$

33. Verify $\csc^4 x - 2\csc^2 x + 1 = \cot^4 x$.

Solution:

$$\csc^4 x - 2\csc^2 x + 1 = (\csc^2 x - 1)^2 = (\cot^2 x)^2 = \cot^4 x$$

37. Verify

$$\frac{\sin\beta}{1-\cos\beta} = \frac{1+\cos\beta}{\sin\beta}.$$

Solution:

$$\frac{\sin\beta}{1-\cos\beta} = \frac{\sin\beta}{1-\cos\beta}\cdot\frac{1+\cos\beta}{1+\cos\beta} = \frac{\sin\beta(1+\cos\beta)}{1-\cos^2\beta} = \frac{\sin\beta(1+\cos\beta)}{\sin^2\beta} = \frac{1+\cos\beta}{\sin\beta}$$

41. Verify

$$\cos\left(\frac{\pi}{2} - x\right)\csc x = 1.$$

Solution:

$$\cos\left(\frac{\pi}{2} - x\right)\csc x = \sin x\csc x = \sin x\left(\frac{1}{\sin x}\right) = 1$$

45. Verify

$$\frac{\cos(-\theta)}{1+\sin(-\theta)} = \sec\theta + \tan\theta.$$

Solution:

$$\begin{aligned}\frac{\cos(-\theta)}{1+\sin(-\theta)} &= \frac{\cos\theta}{1-\sin\theta}\\ &= \frac{\cos\theta}{1-\sin\theta}\cdot\frac{1+\sin\theta}{1+\sin\theta}\\ &= \frac{\cos\theta(1+\sin\theta)}{1-\sin^2\theta}\\ &= \frac{\cos\theta(1+\sin\theta)}{\cos^2\theta}\\ &= \frac{1+\sin\theta}{\cos\theta}\\ &= \frac{1}{\cos\theta}+\frac{\sin\theta}{\cos\theta}\\ &= \sec\theta+\tan\theta\end{aligned}$$

49. Verify

$$\frac{\tan x+\cot y}{\tan x\cot y} = \tan y+\cot x.$$

Solution:

$$\frac{\tan x+\cot y}{\tan x\cot y} = \frac{\tan x}{\tan x\cot y}+\frac{\cot y}{\tan x\cot y} = \frac{1}{\cot y}+\frac{1}{\tan x} = \tan y+\cot x$$

53. Verify $\ln|\tan\theta| = \ln|\sin\theta| - \ln|\cos\theta|$.

Solution:

$$\ln|\tan\theta| = \ln\left|\frac{\sin\theta}{\cos\theta}\right| = \ln\frac{|\sin\theta|}{|\cos\theta|} = \ln|\sin\theta| - \ln|\cos\theta|$$

57. Verify $\sin^2 x+\sin^2\left(\frac{\pi}{2}-x\right) = 1.$

Solution:

$$\sin^2 x+\sin^2 x\left(\frac{\pi}{2}-x\right) = \sin^2 x+\cos^2 x = 1$$

63. Explain why $\sqrt{\tan^2 x} = \tan x$ is *not* an identity and find one value of the variable for which the equation is not true.

Solution:

$$\sqrt{\tan^2 x} = |\tan x|$$

To show that $\sqrt{\tan^2 x} = \tan x$ is not true, pick any value of x whose tangent is negative. For example, $\sqrt{\tan^2 135^\circ} = 1$, whereas, $\tan 135^\circ = -1$.

SECTION 6.3

Solving Trigonometric Equations

- You should be able to identify and solve trigonometric equations.
- A trigonometric equation is a conditional equation. It is true for a specific set of values.
- To solve trigonometric equations, use algebraic techniques such as collecting like terms, taking square roots, factoring, squaring, converting to quadratic form, and using formulas.

Solutions to Selected Exercises

5. Verify that the given values of x are solutions of the equation $2\sin^2 x - \sin x - 1 = 0$.

(a) $x = \dfrac{\pi}{2}$ (b) $x = \dfrac{7\pi}{6}$

Solution:

(a) $x = \dfrac{\pi}{2}$

$$\sin\frac{\pi}{2} = 1$$

$$2\sin^2\frac{\pi}{2} - \sin\frac{\pi}{2} - 1 = 2(1)^2 - 1 - 1 = 0$$

(b) $x = \dfrac{7\pi}{6}$

$$\sin\frac{7\pi}{6} = -\frac{1}{2}$$

$$\begin{aligned} 2\sin^2\frac{7\pi}{6} - \sin\frac{7\pi}{6} - 1 &= 2\left(-\frac{1}{2}\right)^2 - \left(-\frac{1}{2}\right) - 1 \\ &= 2\left(\frac{1}{4}\right) + \frac{1}{2} - 1 \\ &= \frac{1}{2} + \frac{1}{2} - 1 \\ &= 0 \end{aligned}$$

9. Find all solutions of $\sqrt{3}\csc x - 2 = 0$. (Do not use a calculator.)

Solution:

$$\sqrt{3}\csc x - 2 = 0$$

$$\csc x = \frac{2}{\sqrt{3}}$$

$$x = \frac{\pi}{3} \text{ or } \frac{2\pi}{3} \text{ in } [0,\ 2\pi)$$

In general form, $x = \frac{\pi}{3} + 2n\pi$ or $x = \frac{2\pi}{3} + 2n\pi$ where n is an integer.

13. Find all solutions of $3\sec^2 x - 4 = 0$. (Do not use a calculator.)

Solution:

$$3\sec^2 x - 4 = 0$$

$$\sec^2 x = \frac{4}{3}$$

$$\sec x = \pm\sqrt{\frac{4}{3}} = \pm\frac{2}{\sqrt{3}} = \pm\frac{2\sqrt{3}}{3}$$

$$\sec x = \frac{2\sqrt{3}}{3} \qquad \text{OR} \qquad \sec x = -\frac{2\sqrt{3}}{3}$$

$$x = \frac{\pi}{6}, \frac{11\pi}{6} \qquad\qquad x = \frac{5\pi}{6}, \frac{7\pi}{6} \text{ in } [0,\ 2\pi)$$

In general form, the solutions are

$$x = \frac{\pi}{6} + n\pi \text{ or } x = \frac{5\pi}{6} + n\pi$$

where n is an integer.

17. Find all solutions of $\sin x(\sin x + 1) = 0$. (Do not use a calculator.)

Solution:

$$\sin x = 0 \qquad \text{OR} \qquad \sin x = -1$$

$$x = 0,\ \pi \qquad\qquad x = \frac{3\pi}{2} \text{ in } [0,\ 2\pi)$$

In general form, the solutions are

$$x = n\pi \quad \text{and} \quad x = \frac{3\pi}{2} + 2n\pi,$$

where n is an integer.

21. Find all solutions of $\sec x \csc x - 2\csc x = 0$ in the interval $[0, 2\pi)$. (Do not use a calculator.)

Solution:

$$\sec x \csc x - 2\csc x = 0$$
$$\csc x(\sec x - 2) = 0$$

$\csc x = 0$ OR $\sec x = 2$

Not possible $\qquad x = \frac{\pi}{3}, \frac{5\pi}{3}$

25. Find all solutions of $\cos^3 x = \cos x$ in the interval $[0, 2\pi)$. (Do not use a calculator.)

Solution:

$$\cos^3 x = \cos x$$
$$\cos^3 x - \cos x = 0$$
$$\cos x(\cos^2 x - 1) = 0$$

$\cos x = 0$ OR $\cos^2 x - 1 = 0$

$x = \frac{\pi}{2}, \frac{3\pi}{2} \qquad \cos x = \pm 1$

$x = 0, \pi$

$$x = 0, \frac{\pi}{2}, \pi, \frac{3\pi}{2}$$

29. Find all solutions of $2\sec^2 x + \tan^2 x - 3 = 0$ in the interval $[0, 2\pi)$. (Do not use a calculator.)

Solution:

$$2\sec^2 x + \tan^2 x - 3 = 0$$
$$2\sec^2 x + (\sec^2 x - 1) - 3 = 0$$
$$3\sec^2 x - 4 = 0$$
$$\sec^2 x = \frac{4}{3}$$
$$\sec x = \pm\frac{2}{\sqrt{3}}$$

$\sec x = \frac{2}{\sqrt{3}}$ OR $\sec x = -\frac{2}{\sqrt{3}}$

$x = \frac{\pi}{6}, \frac{11\pi}{6} \qquad x = \frac{5\pi}{6}, \frac{7\pi}{6}$

$$x = \frac{\pi}{6}, \frac{5\pi}{6}, \frac{7\pi}{6}, \frac{11\pi}{6}$$

33. Find all solutions of $\sin 2x = -\sqrt{3}/2$ in the interval $[0,\ 2\pi)$. (Do not use a calculator.)

Solution:

$$2x = \frac{4\pi}{3}, \quad 2x = \frac{5\pi}{3}, \quad 2x = \frac{10\pi}{3}, \quad 2x = \frac{11\pi}{3}$$
$$x = \frac{2\pi}{3}, \quad x = \frac{5\pi}{6}, \quad x = \frac{5\pi}{3}, \quad x = \frac{11\pi}{6}$$

39. Find all solutions of

$$\frac{1+\sin x}{\cos x} + \frac{\cos x}{1+\sin x} = 4$$

in the interval $[0,\ 2\pi)$. (Do not use a calculator.)

Solution:

$$\frac{1+\sin x}{\cos x} + \frac{\cos x}{1+\sin x} = 4$$
$$(1+\sin x)^2 + (\cos x)^2 = 4\cos x(1+\sin x)$$
$$1 + 2\sin x + \sin^2 x + \cos^2 x = 4\cos x(1+\sin x)$$
$$1 + 2\sin x + 1 = 4\cos x(1+\sin x)$$
$$2(1+\sin x) - 4\cos(1+\sin x) = 0$$
$$2(1+\sin x)(1-2\cos x) = 0$$

$\sin x = -1$ OR $\cos x = \dfrac{1}{2}$

$x = \dfrac{3\pi}{2}$ $\qquad x = \dfrac{\pi}{3}, \dfrac{5\pi}{3}$

Extraneous

The only solutions are $x = \dfrac{\pi}{3}, \dfrac{5\pi}{3}$.

43. Use a calculator to find all solutions of $12\sin^2 x - 13\sin x + 3 = 0$ in the interval $[0,\ 2\pi)$.

Solution:

$$12\sin^2 x - 13\sin x + 3 = 0$$
$$(4\sin x - 3)(3\sin x - 1) = 0$$

$\sin x = \dfrac{3}{4}$ OR $\sin x = \dfrac{1}{3}$

$x = 0.8481,\ 2.2935$ $\qquad x = 0.3398,\ 2.8018$

47. Use a calculator to find all solutions of $\tan^2 x - 8\tan x + 13 = 0$ in the interval $[0, 2\pi)$.

Solution:

$$\tan^2 x - 8\tan x + 13 = 0$$

$$\tan x = \frac{8 \pm \sqrt{64 - 52}}{2}$$

$$= \frac{8 \pm 2\sqrt{3}}{2}$$

$$\tan x = 4 + \sqrt{3} \qquad \text{OR} \qquad \tan x = 4 - \sqrt{3}$$

$$x = 1.3981,\ 4.5397 \qquad\qquad x = 1.1555,\ 4.2971$$

53. A 5-pound weight is oscillating on the end of a spring, and the position of the weight relative to the point of equilibrium is given by

$$h(t) = \frac{1}{4}(\cos 8t - 3\sin 8t)$$

where t is the time in seconds. Find the times when the weight is at the point of equilibrium $[h(t) = 0]$ for $0 \le t \le 1$.

Solution:

$$\frac{1}{4}(\cos 8t - 3\sin 8t) = 0$$

$$\cos 8t = 3\sin 8t$$

$$\frac{1}{3} = \tan 8t$$

$$8t = 0.32175 + n\pi$$

$$t = 0.04 + \frac{n\pi}{8}$$

In the interval $0 \le t \le 1$, we have $t = 0.04,\ 0.43$, and 0.83.

SECTION 6.4

Sum and Difference Formulas

- You should memorize the sum and difference formulas given at the beginning of the section.
- You should be able to use these formulas to find the values of the trigonometric functions of angles whose sums or differences are special angles.
- You should be able to use these formulas to solve trigonometric equations.

Solutions to Selected Exercises

5. Determine the exact value of the sine, cosine, and tangent of the angle $195^\circ = 225^\circ - 30^\circ$.

Solution:

$$\begin{aligned}
\sin 195^\circ &= \sin(225^\circ - 30^\circ) \\
&= \sin 225^\circ \cos 30^\circ - \cos 225^\circ \sin 30^\circ \\
&= (-\sin 45^\circ)\cos 30^\circ - (-\cos 45^\circ)\sin 30^\circ \\
&= -\frac{\sqrt{2}}{2}\cdot\frac{\sqrt{3}}{2} + \frac{\sqrt{2}}{2}\cdot\frac{1}{2} = \frac{\sqrt{2}}{4}(1-\sqrt{3})
\end{aligned}$$

$$\begin{aligned}
\cos 195^\circ &= \cos(225^\circ - 30^\circ) \\
&= \cos 225^\circ \cos 30^\circ + \sin 225^\circ \sin 30^\circ \\
&= (-\cos 45^\circ)\cos 30^\circ + (-\sin 45^\circ)\sin 30^\circ \\
&= -\frac{\sqrt{2}}{2}\cdot\frac{\sqrt{3}}{2} - \frac{\sqrt{2}}{2}\cdot\frac{1}{2} = -\frac{\sqrt{2}}{4}(1+\sqrt{3})
\end{aligned}$$

$$\begin{aligned}
\tan 195^\circ &= \tan(225^\circ - 30^\circ) = \frac{\tan 225^\circ - \tan 30^\circ}{1 + \tan 225^\circ \tan 30^\circ} = \frac{\tan 45^\circ - \tan 30^\circ}{1 + \tan 45^\circ \tan 30^\circ} \\
&= \frac{1 - \sqrt{3}/3}{1 + (1)(\sqrt{3}/3)}\cdot\frac{3}{3} = \frac{3-\sqrt{3}}{3+\sqrt{3}}\cdot\frac{3-\sqrt{3}}{3-\sqrt{3}} \\
&= \frac{(3-\sqrt{3})^2}{9-3} = \frac{[\sqrt{3}(\sqrt{3}-1)]^2}{6} \\
&= \frac{3(\sqrt{3}-1)^2}{6} = \frac{1}{2}(\sqrt{3}-1)^2
\end{aligned}$$

11. Simplify $\cos 25° \cos 15° - \sin 25° \sin 15°$.

Solution:

$$\cos 25° \cos 15° - \sin 25° \sin 15° = \cos(25° + 15°) = \cos 40°$$

15. Simplify

$$\frac{\tan 325° - \tan 86°}{1 + \tan 325° \tan 86°}.$$

Solution:

$$\frac{\tan 325° - \tan 86°}{1 + \tan 325° \tan 86°} = \tan(325° - 86°) = \tan 239°$$

19. Simplify

$$\frac{\tan 2x + \tan x}{1 - \tan 2x \tan x}.$$

Solution:

$$\frac{\tan 2x + \tan x}{1 - \tan 2x \tan x} = \tan(2x + x) = \tan 3x$$

23. Find the exact value of $\cos(v + u)$ given that $\sin u = \frac{5}{13}$, $0 < u < \pi/2$, and $\cos v = -\frac{3}{5}$, $\pi/2 < v < \pi$.

Solution:

$$\sin u = \frac{5}{13},\ 0 < u < \frac{\pi}{2} \Rightarrow \cos u = \frac{12}{13}$$
$$\cos v = -\frac{3}{5},\ \frac{\pi}{2} < v < \pi \Rightarrow \sin v = \frac{4}{5}$$

$$\cos(v + u) = \cos v \cos u - \sin v \sin u = \left(-\frac{3}{5}\right)\left(\frac{12}{13}\right) - \left(\frac{4}{5}\right)\left(\frac{5}{13}\right) = -\frac{56}{65}$$

27. Find the exact value of $\sin(v - u)$ given that $\sin u = \frac{7}{25}$, $\pi/2 < u < \pi$ and $\cos v = \frac{4}{5}$, $3\pi/2 < v < 2\pi$.

Solution:

$$\sin u = \frac{7}{25},\ \frac{\pi}{2} < u < \pi \Rightarrow \cos u = -\frac{24}{25}$$
$$\cos v = \frac{4}{5},\ \frac{3\pi}{2} < v < 2\pi \Rightarrow \sin v = -\frac{3}{5}$$

$$\sin(v - u) = \sin v \cos u - \cos v \sin u = \left(-\frac{3}{5}\right)\left(-\frac{24}{25}\right) - \left(\frac{4}{5}\right)\left(\frac{7}{25}\right) = \frac{44}{125}$$

31. Verify $\cos\left(\frac{3\pi}{2} - x\right) = -\sin x$.

Solution:

$$\cos\left(\frac{3\pi}{2} - x\right) = \cos\frac{3\pi}{2}\cos x + \sin\frac{3\pi}{2}\sin x = (0)(\cos x) + (-1)(\sin x) = -\sin x$$

35. Verify $\cos(\pi - \theta) + \sin\left(\frac{\pi}{2} + \theta\right) = 0$.

Solution:

$$\begin{aligned}\cos(\pi - \theta) + \sin\left(\frac{\pi}{2} + \theta\right) &= (\cos\pi\cos\theta + \sin\pi\sin\theta) + \left(\sin\frac{\pi}{2}\cos\theta + \cos\frac{\pi}{2}\sin\theta\right)\\ &= (-1)\cos\theta + (0)\sin\theta + (1)\cos\theta + (0)\sin\theta\\ &= -\cos\theta + 0 + \cos\theta + 0\\ &= 0\end{aligned}$$

39. Verify $\cos(x + y)\cos(x - y) = \cos^2 x - \sin^2 y$.

Solution:

$$\begin{aligned}\cos(x+y)\cos(x-y) &= (\cos x\cos y - \sin x\sin y)(\cos x\cos y + \sin x\sin y)\\ &= \cos^2 x\cos^2 y - \sin^2 x\sin^2 y\\ &= \cos^2 x(1 - \sin^2 y) - (1 - \cos^2 x)\sin^2 y\\ &= \cos^2 x - \cos^2 x\sin^2 y - \sin^2 y + \cos^2 x\sin^2 y\\ &= \cos^2 x - \sin^2 y\end{aligned}$$

43. Verify

$$\sin(x + y + z) = \sin x\cos y\cos z + \sin y\cos x\cos z + \sin z\cos x\cos y - \sin x\sin y\sin z.$$

Solution:

$$\begin{aligned}\sin(x+y+z) &= \sin[x + (y+z)]\\ &= \sin x\cos(y+z) + \cos x\sin(y+z)\\ &= \sin x(\cos y\cos z - \sin y\sin z) + \cos x(\sin y\cos z + \cos y\sin z)\\ &= \sin x\cos y\cos z - \sin x\sin y\sin z + \sin y\cos x\cos z + \sin z\cos x\cos y\\ &= \sin x\cos y\cos z + \sin y\cos x\cos z + \sin z\cos x\cos y - \sin x\sin y\sin z\end{aligned}$$

47. Verify $a\sin B\theta + b\cos B\theta = \sqrt{a^2 + b^2}\sin(B\theta + C)$, where $C = \arctan(b/a)$.

Solution:

$$\sqrt{a^2+b^2}\sin(B\theta+C) = \sqrt{a^2+b^2}(\sin B\theta\cos C + \cos B\theta\sin C)$$
$$= \sqrt{a^2+b^2}\left[\sin B\theta\left(\frac{a}{\sqrt{a^2+b^2}}\right) + \cos B\theta\left(\frac{b}{\sqrt{a^2+b^2}}\right)\right]$$
$$= \frac{\sqrt{a^2+b^2}}{\sqrt{a^2+b^2}}(a\sin B\theta + b\cos B\theta)$$
$$= a\sin B\theta + b\cos B\theta$$

$\sqrt{a^2+b^2}$

b

C

a

51. Use the formulas given in Exercises 47 and 48 to write $12\sin 3\theta + 5\cos 3\theta$ in the following forms.
(a) $\sqrt{a^2+b^2}\sin(B\theta+C)$ (b) $\sqrt{a^2+b^2}\cos(B\theta-C)$

Solution:
$a = 12,\ b = 5,\ B = 3,\ C = \arctan\frac{5}{12} \approx 0.3948$
$\sqrt{12^2+5^2} = \sqrt{169} = 13$

(a) Thus, $\sqrt{a^2+b^2}\sin(B\theta+C) = 13\sin(3\theta+0.3948)$, and
(b) $\sqrt{a^2+b^2}\cos(B\theta-C) = 13\cos(3\theta-0.3948)$.

55. Write $\sin(\arcsin x + \arccos x)$ as an algebraic expression in x. [*Hint:* See Examples 6 and 7 in Section 7.7.]

Solution:

$$\sin(\arcsin x + \arccos x) = \sin(\arcsin x)\cos(\arccos x) + \cos(\arcsin x)\sin(\arccos x)$$
$$= (x)(x) + \frac{\sqrt{1-x^2}}{1}\cdot\frac{\sqrt{1-x^2}}{1} = x^2 + 1 - x^2 = 1$$

59. Find all solutions in the interval $[0,\ 2\pi)$ of $\cos\left(x+\frac{\pi}{4}\right) + \cos\left(x-\frac{\pi}{4}\right) = 1$.

Solution:

$$\cos\left(x+\frac{\pi}{4}\right) + \cos\left(x-\frac{\pi}{4}\right) = 1$$
$$\cos x\cos\frac{\pi}{4} - \sin x\sin\frac{\pi}{4} + \cos x\cos\frac{\pi}{4} + \sin x\sin\frac{\pi}{4} = 1$$
$$\frac{\sqrt{2}}{2}\cos x + \frac{\sqrt{2}}{2}\cos x = 1$$
$$\sqrt{2}\cos x = 1$$
$$\cos x = \frac{1}{\sqrt{2}}$$
$$x = \frac{\pi}{4},\ \frac{7\pi}{4}$$

63. Show that

$$\frac{\cos(x+h)-\cos x}{h}=\cos x\left(\frac{\cos h-1}{h}\right)-\sin x\left(\frac{\sin h}{h}\right).$$

Solution:

$$\begin{aligned}\frac{\cos(x+h)-\cos x}{h}&=\frac{\cos x\cos h-\sin x\sin h-\cos x}{h}\\&=\frac{\cos x(\cos h-1)-\sin x\sin h}{h}\\&=\frac{\cos x(\cos h-1)}{h}-\frac{\sin x\sin h}{h}\\&=\cos x\left(\frac{\cos h-1}{h}\right)-\sin x\left(\frac{\sin h}{h}\right)\end{aligned}$$

SECTION 6.5

Multiple-Angle Formulas and Product-Sum Formulas

- You should know the following double-angle formulas.

  (a) $\sin 2u = 2 \sin u \cos u$

  (b) $\cos 2u = \cos^2 u - \sin^2 u$

  $= 2\cos^2 u - 1$

  $= 1 - 2\sin^2 u$

  (c) $\tan 2u = \dfrac{2 \tan u}{1 - \tan^2 u}$

- You should be able to reduce the power of a trigonometric function.

  (a) $\sin^2 u = \dfrac{1 - \cos 2u}{2}$

  (b) $\cos^2 u = \dfrac{1 + \cos 2u}{2}$

  (c) $\tan^2 u = \dfrac{1 - \cos 2u}{1 + \cos 2u}$

- You should be able to use the half-angle formulas.

  (a) $\sin^2 \dfrac{u}{2} = \dfrac{1 - \cos u}{2}$

  (b) $\cos^2 \dfrac{u}{2} = \dfrac{1 + \cos u}{2}$

  (c) $\tan^2 \dfrac{u}{2} = \dfrac{1 - \cos u}{1 + \cos u}$

- You should be able to use the following.

  Product-Sum Formulas

  (a) $\sin u \sin v = \dfrac{1}{2}[\cos(u - v) - \cos(u + v)]$

  (b) $\cos u \cos v = \dfrac{1}{2}[\cos(u - v) + \cos(u + v)]$

  (c) $\sin u \cos v = \dfrac{1}{2}[\sin(u + v) + \sin(u - v)]$

  (d) $\cos u \sin v = \dfrac{1}{2}[\sin(u + v) - \sin(u - v)]$

- You should be able to use the following.

  Sum-Product Formulas

  (a) $\sin x + \sin y = 2\sin\left(\frac{x+y}{2}\right)\cos\left(\frac{x-y}{2}\right)$

  (b) $\sin x - \sin y = 2\cos\left(\frac{x+y}{2}\right)\sin\left(\frac{x-y}{2}\right)$

  (c) $\cos x + \cos y = 2\cos\left(\frac{x+y}{2}\right)\cos\left(\frac{x-y}{2}\right)$

  (d) $\cos x - \cos y = -2\sin\left(\frac{x+y}{2}\right)\sin\left(\frac{x-y}{2}\right)$

Solutions to Selected Exercises

3. Use a double-angle formula to determine the exact values of the sine, cosine, and tangent of the angle $60° = 2(30°)$.

Solution:

$$\sin 60° = 2\sin 30° \cos 30° = 2\left(\frac{1}{2}\right)\left(\frac{\sqrt{3}}{2}\right) = \frac{\sqrt{3}}{2}$$

$$\cos 60° = \cos^2 30° - \sin^2 30° = \left(\frac{\sqrt{3}}{2}\right)^2 - \left(\frac{1}{2}\right)^2 = \frac{1}{2}$$

$$\tan 60° = \frac{2\tan 30°}{1 - \tan^2 30°} = \frac{2(1/\sqrt{3})}{1-(1/\sqrt{3})^2} = \frac{2/\sqrt{3}}{2/3} = \frac{3}{\sqrt{3}} = \sqrt{3}$$

9. Find the exact values of $\sin 2u$, $\cos 2u$, and $\tan 2u$, given $\tan u = \frac{1}{2}$, $\pi < u < 3\pi/2$.

Solution:

$$\tan u = \frac{1}{2},\ u \text{ lies in Quadrant III}$$

$$\sin u = -\frac{1}{5} \text{ and } \cos u = -\frac{2}{\sqrt{5}}$$

$$\sin 2u = 2\left(-\frac{1}{\sqrt{5}}\right)\left(-\frac{2}{\sqrt{5}}\right) = \frac{4}{5}$$

$$\cos 2u = \left(-\frac{2}{\sqrt{5}}\right)^2 - \left(-\frac{1}{\sqrt{5}}\right)^2 = \frac{3}{5}$$

$$\tan 2u = \frac{\sin 2u}{\cos 2u} = \frac{4}{3}$$

15. Use half-angle formulas to determine the exact values of the sine, cosine, and tangent of the angle $112°30' = \frac{1}{2}(225°)$.

Solution:

$$\sin 112°30' = +\sqrt{\frac{1-\cos 225°}{2}} = \sqrt{\frac{1-(-\sqrt{2}/2)}{2}} = \sqrt{\frac{2+\sqrt{2}}{4}} = \frac{\sqrt{2+\sqrt{2}}}{2}$$

$$\cos 112°30' = -\sqrt{\frac{1+\cos 225°}{2}} = -\sqrt{\frac{1+(-\sqrt{2}/2)}{2}} = -\sqrt{\frac{2-\sqrt{2}}{4}} = -\frac{\sqrt{2-\sqrt{2}}}{2}$$

$$\tan 112°30' = -\sqrt{\frac{1-\cos 225°}{1+\cos 225°}} = -\sqrt{\frac{1+\sqrt{2}/2}{1-\sqrt{2}/2}} = -\sqrt{\frac{2+\sqrt{2}}{2-\sqrt{2}}} \cdot \sqrt{\frac{2+\sqrt{2}}{2+\sqrt{2}}}$$

$$= -\sqrt{\frac{(2+\sqrt{2})^2}{4-2}} = -\frac{2+\sqrt{2}}{\sqrt{2}} = -(\sqrt{2}+1) = -1-\sqrt{2}$$

21. Find the exact values of $\sin(u/2)$, $\cos(u/2)$, and $\tan(u/2)$ by using the half-angle formulas, given $\tan u = -\frac{5}{8}$, $3\pi/2 < u < 2\pi$.

Solution:

$$\tan u = -\frac{5}{8}, \ u \text{ lies in Quadrant IV}$$

$$\sin u = -\frac{5}{\sqrt{89}} \text{ and } \cos u = \frac{8}{\sqrt{89}}$$

$$\sin\frac{u}{2} = \sqrt{\frac{1-8/\sqrt{89}}{2}} = \sqrt{\frac{\sqrt{89}-8}{2\sqrt{89}}}$$

$$\cos\frac{u}{2} = -\sqrt{\frac{1+8/\sqrt{89}}{2}} = -\sqrt{\frac{\sqrt{89}+8}{2\sqrt{89}}}$$

$$\tan\frac{u}{2} = -\sqrt{\frac{1-8/\sqrt{89}}{1+8/\sqrt{89}}} = -\sqrt{\frac{\sqrt{89}-8}{\sqrt{89}+8}} \cdot \sqrt{\frac{\sqrt{89}-8}{\sqrt{89}-8}}$$

$$= \sqrt{\frac{(\sqrt{89}-8)^2}{89-64}} = -\left(\frac{\sqrt{89}-8}{5}\right) = \frac{1}{5}(8-\sqrt{89})$$

25. Use the half-angle formulas to simplify

$$\sqrt{\frac{1-\cos 6x}{2}}.$$

Solution:

$$\sqrt{\frac{1-\cos 6x}{2}} = \sqrt{\frac{1-\cos 2(3x)}{2}} = \sin 3x$$

29. Use the power-reducing formulas to write each expression in terms of the first power of the cosine.

(a) $\cos^4 x$ (b) $\sin^2 x \cos^4 x$

Solution:

(a)
$$\begin{aligned}
\cos^4 x &= (\cos^2 x)^2 \\
&= \left(\frac{1+\cos 2x}{2}\right)^2 \\
&= \frac{1}{4}(1 + 2\cos 2x + \cos^2 2x) \\
&= \frac{1}{4}\left(1 + 2\cos 2x + \frac{1+\cos 4x}{2}\right) \\
&= \frac{1}{4}\left(\frac{3}{2} + 2\cos 2x + \frac{1}{2}\cos 4x\right) \\
&= \frac{1}{8}(3 + 4\cos 2x + \cos 4x)
\end{aligned}$$

(b)
$$\begin{aligned}
\sin^2 x \cos^4 x &= \left(\frac{1-\cos 2x}{2}\right)\cos^4 x \\
&= \frac{1}{2}(1-\cos 2x)\frac{1}{8}(3 + 4\cos 2x + \cos 4x) \text{ from part (a)} \\
&= \frac{1}{16}(1-\cos 2x)(3 + 4\cos 2x + \cos 4x)
\end{aligned}$$

33. Rewrite $\sin 5\theta \cos 3\theta$ as a sum.

Solution:

$$\sin 5\theta \cos 3\theta = \tfrac{1}{2}[\sin(5\theta + 3\theta) + \sin(5\theta - 3\theta)] = \tfrac{1}{2}(\sin 8\theta + \sin 2\theta)$$

37. Rewrite $\sin(x+y)\sin(x-y)$ as a sum.

Solution:

$$\begin{aligned}
\sin(x+y)\sin(x-y) &= \tfrac{1}{2}\{\cos[(x+y)-(x-y)] - \cos[(x+y)+(x-y)]\} \\
&= \tfrac{1}{2}\{\cos 2y - \cos 2x\}
\end{aligned}$$

41. Express $\sin 60° + \sin 30°$ as a product.

Solution:

$$\sin 60° + \sin 30° = 2\sin\left(\frac{60° + 30°}{2}\right)\cos\left(\frac{60° - 30°}{2}\right) = 2\sin 45°\cos 15°$$

45. Express $\cos 6x + \cos 2x$ as a product.

Solution:

$$\cos 6x + \cos 2x = 2\cos\left(\frac{6x + 2x}{2}\right)\cos\left(\frac{6x - 2x}{2}\right) = 2\cos 4x\cos 2x$$

49. Express $\cos(\phi + 2\pi) + \cos\phi$ as a product.

Solution:

$$\cos(\phi + 2\pi) + \cos\phi = 2\cos\left(\frac{\phi + 2\pi + \phi}{2}\right)\cos\left(\frac{\phi + 2\pi - \phi}{2}\right) = 2\cos(\phi + \pi)\cos\pi$$

51. Verify

$$\csc 2\theta = \frac{\csc\theta}{2\cos\theta}.$$

Solution:

$$\csc 2\theta = \frac{1}{\sin 2\theta} = \frac{1}{2\sin\theta(\cos\theta)} = \frac{1}{\sin\theta}\cdot\frac{1}{2\cos\theta} = \frac{\csc\theta}{2\cos\theta}$$

55. Verify $(\sin x + \cos x)^2 = 1 + \sin 2x$.

Solution:

$$\begin{aligned}(\sin x + \cos x)^2 &= \sin^2 x + 2\sin x\cos x + \cos^2 x\\ &= (\sin^2 x + \cos^2 x) + 2\sin x\cos x\\ &= 1 + \sin 2x\end{aligned}$$

59. Verify $1 + \cos 10y = 2\cos^2 5y$.

Solution:

$$2\cos^2 5y = 2\left[\frac{1 + \cos 10y}{2}\right] = 1 + \cos 10y$$

65. Verify

$$\frac{\cos 4x - \cos 2x}{2 \sin 3x} = -\sin x.$$

Solution:

$$\frac{\cos 4x - \cos 2x}{2 \sin 3x} = \frac{-2 \sin\left(\frac{4x + 2x}{2}\right) \sin\left(\frac{4x - 2x}{2}\right)}{2 \sin 3x} = \frac{-2 \sin 3x \sin x}{2 \sin 3x} = -\sin x$$

69. Verify

$$\frac{\cos t + \cos 3t}{\sin 3t - \sin t} = \cot t.$$

Solution:

$$\frac{\cos t + \cos 3t}{\sin 3t - \sin t} = \frac{2 \cos\left(\frac{t + 3t}{2}\right) \cos\left(\frac{t - 3t}{2}\right)}{2 \cos\left(\frac{3t + t}{2}\right) \sin\left(\frac{3t - t}{2}\right)}$$

$$= \frac{2 \cos 2t \cos(-t)}{2 \cos 2t \sin t} = \frac{\cos t}{\sin t} = \cot t$$

73. Find all the solutions of $4 \sin x \cos x = 1$ in the interval $[0, 2\pi)$.

Solution:

$$\begin{aligned} 4 \sin x(\cos x) &= 1 \\ 2[2(\sin x) \cos x] &= 1 \\ 2 \sin 2x &= 1 \\ \sin 2x &= \frac{1}{2} \end{aligned}$$

$$2x = \frac{\pi}{6}, \quad 2x = \frac{5\pi}{6}, \quad 2x = \frac{13\pi}{6}, \quad 2x = \frac{17\pi}{6}$$

$$x = \frac{\pi}{12}, \quad x = \frac{5\pi}{12}, \quad x = \frac{13\pi}{12}, \quad x = \frac{17\pi}{12}$$

77. Find all solutions of $\sin 4x + 2\sin 2x = 0$ in the interval $[0,\ 2\pi)$.

Solution:

$$\begin{aligned} \sin 4x + 2\sin 2x &= 0 \\ 2\sin 2x\cos 2x + 2\sin 2x &= 0 \\ 2\sin 2x(\cos 2x + 1) &= 0 \end{aligned}$$

$$\sin 2x = 0 \qquad \text{or} \qquad \cos 2x = -1$$

$$2x = 0,\ \pi,\ 2\pi,\ 3\pi \qquad\qquad 2x = \pi,\ 3\pi$$

$$x = 0,\ \frac{\pi}{2},\ \pi,\ \frac{3\pi}{2} \qquad\qquad x = \frac{\pi}{2},\ \frac{3\pi}{2}$$

81. Find all solutions of $\sin 6x + \sin 2x = 0$ in the interval $[0,\ 2\pi)$.

Solution:

$$\begin{aligned} \sin 6x + \sin 2x &= 0 \\ 2\sin\left(\frac{6x+2x}{2}\right)\cos\left(\frac{6x-2x}{2}\right) &= 0 \\ 2(\sin 4x)\cos 2x &= 0 \end{aligned}$$

$$\sin 4x = 0 \qquad \text{or} \qquad \cos 2x = 0$$

$$4x = n\pi \qquad\qquad 2x = \frac{\pi}{2} + n\pi$$

$$x = \frac{n\pi}{4} \qquad\qquad x = \frac{\pi}{4} + \frac{n\pi}{2}$$

In the interval $[0,\ 2\pi)$, we have

$$x = 0,\ \frac{\pi}{4},\ \frac{\pi}{2},\ \frac{3\pi}{4},\ \pi,\ \frac{5\pi}{4},\ \frac{3\pi}{2},\ \frac{7\pi}{4}.$$

85. Sketch the graph of $f(x) = \sin^2 x$ by using the power-reducing formulas.

Solution:

$$\begin{aligned} f(x) &= \sin^2 x \\ &= \frac{1-\cos 2x}{2} \\ &= \frac{1}{2} - \frac{1}{2}\cos 2x \end{aligned}$$

Period: π

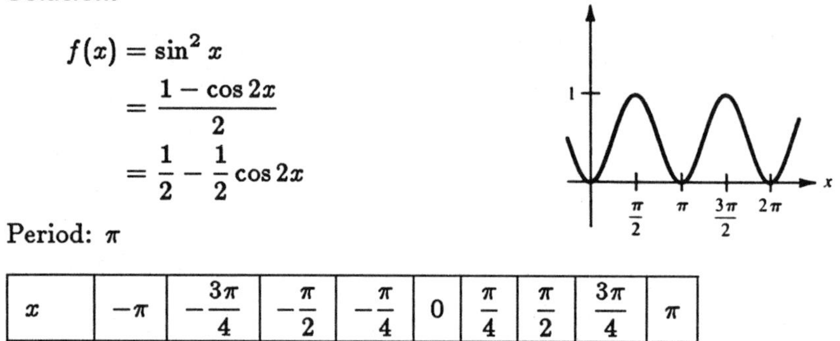

x	$-\pi$	$-\frac{3\pi}{4}$	$-\frac{\pi}{2}$	$-\frac{\pi}{4}$	0	$\frac{\pi}{4}$	$\frac{\pi}{2}$	$\frac{3\pi}{4}$	π
$f(x)$	0	$\frac{1}{2}$	1	$\frac{1}{2}$	0	$\frac{1}{2}$	1	$\frac{1}{2}$	0

89. Verify $\sin(\phi - \theta) = \cos 2\theta$ for complementary angles ϕ and θ.

Solution:

Since ϕ and θ are complementary, $\phi = 90° - \theta$.

$$\begin{aligned}
\sin(\phi - \theta) &= \sin[(90° - \theta) - \theta] \\
&= \sin[90° - 2\theta] \\
&= \sin 90° \cos 2\theta - \cos 90° \sin 2\theta \\
&= (1)\cos 2\theta - (0)\sin 2\theta \\
&= \cos 2\theta
\end{aligned}$$

93. Prove $\cos u \sin v = \frac{1}{2}[\sin(u + v) - \sin(u - v)]$.

Solution:

$$\begin{aligned}
\tfrac{1}{2}[\sin(u + v) - \sin(u - v)] &= \tfrac{1}{2}\{(\sin u \cos v + \cos u \sin v) - (\sin u \cos v - \cos u \sin v)\} \\
&= \tfrac{1}{2}\{2 \cos u \sin v\} \\
&= \cos u \sin v
\end{aligned}$$

REVIEW EXERCISES FOR CHAPTER 6

Solutions to Selected Exercises

3. Simplify

$$\frac{\sin^2\alpha - \cos^2\alpha}{\sin^2\alpha - \sin\alpha\cos\alpha}.$$

Solution:

$$\begin{aligned}\frac{\sin^2\alpha - \cos^2\alpha}{\sin^2\alpha - \sin\alpha\cos\alpha} &= \frac{(\sin\alpha + \cos\alpha)(\sin\alpha - \cos\alpha)}{\sin\alpha(\sin\alpha - \cos\alpha)}\\ &= \frac{\sin\alpha + \cos\alpha}{\sin\alpha} = \frac{\sin\alpha}{\sin\alpha} + \frac{\cos\alpha}{\sin\alpha} = 1 + \cot\alpha\end{aligned}$$

7. Simplify $\tan^2\theta(\csc^2\theta - 1)$.

Solution:

$$\tan^2\theta(\csc^2\theta - 1) = \tan^2\theta(\cot^2\theta) = \tan^2\theta\left(\frac{1}{\tan^2\theta}\right) = 1$$

11. Verify $\tan x(1 - \sin^2 x) = \frac{1}{2}\sin 2x$.

Solution:

$$\tan x(1 - \sin^2 x) = \frac{\sin x}{\cos x}(\cos^2 x) = (\sin x)\cos x = \frac{1}{2}(2(\sin x)\cos x) = \frac{1}{2}\sin 2x$$

15. Verify $\sin^2 x\cos^4 x = \frac{1}{16}(1 - \cos 4x + 2\sin^2 2x\cos 2x)$.

Solution:

$$\begin{aligned}\tfrac{1}{16}(1 - \cos 4x + 2\sin^2 2x\cos 2x) &= \tfrac{1}{16}[1 - (\cos^2 2x - \sin^2 2x) + 2(\sin^2 2x)\cos 2x]\\ &= \tfrac{1}{16}[2\sin^2 2x + 2(\sin^2 2x)\cos 2x]\\ &= \tfrac{1}{8}[\sin^2 2x(1 + \cos 2x)]\\ &= \tfrac{1}{8}[(2(\sin x)\cos x)^2(1 + \cos^2 x - \sin^2 x)]\\ &= \tfrac{1}{8}[4(\sin^2 x)\cos^2 x(2\cos^2 x)]\\ &= \sin^2 x(\cos^4 x)\end{aligned}$$

19. Verify $\sqrt{\dfrac{1 - \sin\theta}{1 + \sin\theta}} = \dfrac{1 - \sin\theta}{|\cos\theta|}$.

Solution:

$$\sqrt{\frac{1-\sin\theta}{1+\sin\theta}} = \sqrt{\frac{1-\sin\theta}{1+\sin\theta}\cdot\frac{1-\sin\theta}{1-\sin\theta}} = \sqrt{\frac{(1-\sin\theta)^2}{\cos^2\theta}} = \frac{1-\sin\theta}{|\cos\theta|}$$

23. Verify $\sin\left(x-\frac{3\pi}{2}\right) = \cos x$.

Solution:

$$\sin\left(x-\frac{3\pi}{2}\right) = (\sin x)\cos\frac{3\pi}{2} - (\cos x)\sin\frac{3\pi}{2} = (\sin x)(0) - \cos x(-1) = \cos x$$

27. Verify $\dfrac{\cos 3x - \cos x}{\sin 3x - \sin x} = -\tan 2x$.

Solution:

$$\frac{\cos 3x - \cos x}{\sin 3x - \sin x} = \frac{-2\sin\left(\frac{3x+x}{2}\right)\sin\left(\frac{3x-x}{2}\right)}{2\cos\left(\frac{3x+x}{2}\right)\sin\left(\frac{3x-x}{2}\right)} = \frac{-2(\sin 2x)\sin x}{2(\cos 2x)\sin x} = -\frac{\sin 2x}{\cos 2x} = -\tan 2x$$

31. Verify $2\sin y\cos y\sec 2y = \tan 2y$.

Solution:

$$2\sin y\cos y\sec 2y = \sin 2y\sec 2y = \sin 2y\left(\frac{1}{\cos 2y}\right) = \frac{\sin 2y}{\cos 2y} = \tan 2y$$

35. Verify $\tan^2 x = \dfrac{1-\cos 2x}{1+\cos 2x}$.

Solution:

$$\frac{1-\cos 2x}{1+\cos 2x} = \frac{1-(1-2\sin^2 x)}{1+(2\cos^2 x - 1)} = \frac{2\sin^2 x}{2\cos^2 x} = \tan^2 x$$

39. Verify $1+\cos 2x+\cos 4x+\cos 6x = 4\cos x\cos 2x\cos 3x$.

Solution:

$$\begin{aligned}4\cos x\cos 2x\cos 3x &= 4\cos x\left[\tfrac{1}{2}(\cos(-x)+\cos(5x))\right] = 2\cos x(\cos x+\cos 5x)\\ &= 2\cos^2 x + 2\cos x\cos 5x = 2\cos^2 x + 2\left[\tfrac{1}{2}(\cos(-4x)+\cos 6x)\right]\\ &= 2\cos^2 x + \cos(-4x) + \cos 6x = 1+\cos 2x+\cos 4x+\cos 6x\end{aligned}$$

43. Using the sum, difference, or half-angle formulas, find the exact value of

$$\cos(157°30') = \cos\frac{315°}{2}.$$

Solution:

$$\cos(157°30') = \cos\frac{315°}{2} = -\sqrt{\frac{1+\cos 315°}{2}} = -\sqrt{\frac{1+\cos 45°}{2}}$$
$$= -\sqrt{\frac{1+\frac{\sqrt{2}}{2}}{2}} = -\sqrt{\frac{2+\sqrt{2}}{4}} = -\frac{\sqrt{2+\sqrt{2}}}{2}$$

47. Find the exact value of $\cos(u-v)$, given that $\sin u = \frac{3}{4}$, $\cos v = -\frac{5}{13}$, (u and v are in Quadrant II).

Solution:
Since u and v are in Quadrant II,

$$\sin u = \frac{3}{4} \Rightarrow \cos u = -\frac{\sqrt{7}}{4}$$
$$\cos v = -\frac{5}{13} \Rightarrow \sin v = \frac{12}{13}.$$

$$\cos(u-v) = \cos u \cos v + \sin u \sin v = \left(-\frac{\sqrt{7}}{4}\right)\left(-\frac{5}{13}\right) + \left(\frac{3}{4}\right)\left(\frac{12}{13}\right) = \frac{5\sqrt{7}+36}{52}$$

51. Find all solutions of $\sin x - \tan x = 0$ in the interval $[0,\ 2\pi)$.

Solution:

$$\sin x - \tan x = 0$$
$$\sin x - \frac{\sin x}{\cos x} = 0$$
$$(\sin x)\cos x - \sin x = 0$$
$$\sin x(\cos x - 1) = 0$$

$$\sin x = 0 \qquad \text{OR} \qquad \cos x = 1$$
$$x = 0,\ \pi \qquad\qquad x = 0$$

55. Find all solutions of $\sin 2x + \sqrt{2}\sin x = 0$ in the interval $[0,\ 2\pi)$.

Solution:

$$\sin 2x + \sqrt{2}\sin x = 0$$
$$2(\sin x)\cos x + \sqrt{2}\sin x = 0$$
$$\sin x(2\cos x + \sqrt{2}) = 0$$

$$\sin x = 0 \qquad \text{OR} \qquad \cos x = -\frac{\sqrt{2}}{2}$$
$$x = 0,\ \pi \qquad\qquad x = \frac{3\pi}{4},\ \frac{5\pi}{4}$$

59. Find all solutions of $\tan^3 x - \tan^2 x + 3\tan x - 3 = 0$ in the interval $[0,\ 2\pi)$.

Solution:

$$\tan^3 x - \tan^2 x + 3\tan x - 3 = 0$$
$$(\tan x - 1)(\tan^2 x + 3) = 0$$
$$\tan x = 1 \qquad \text{OR} \qquad \tan^2 x = -3$$
$$x = \frac{\pi}{4},\ \frac{5\pi}{4} \qquad \text{No real solutions}$$

63. Write $\sin 3\alpha \sin 2\alpha$ as a sum or difference.

Solution:

$$\sin 3\alpha \sin 2\alpha = \tfrac{1}{2}[\cos(3\alpha - 2\alpha) - \cos(3\alpha + 2\alpha)] = \tfrac{1}{2}[\cos\alpha - \cos 5\alpha]$$

67. A standing wave on a string of given length is modeled by the equation

$$y = A\left(\cos\left[2\pi\left(\frac{t}{T} - \frac{x}{\lambda}\right)\right] + \cos\left[2\pi\left(\frac{t}{T} + \frac{x}{\lambda}\right)\right]\right).$$

Use the trigonometric identities for the cosine of the sum and difference of two angles to verify that the following equation is an equivalent model for the standing wave.

$$y = 2A\cos\frac{2\pi t}{T}\cos\frac{2\pi x}{\lambda}$$

Solution:

$$\begin{aligned}
y &= A\left(\cos\left[2\pi\left(\frac{t}{T} - \frac{x}{\lambda}\right)\right] + \cos\left[2\pi\left(\frac{t}{T} + \frac{x}{\lambda}\right)\right]\right)\\
&= A\left(2\cos\left[\frac{2\pi\left(\frac{t}{T} - \frac{x}{\lambda}\right) + 2\pi\left(\frac{t}{T} + \frac{x}{\lambda}\right)}{2}\right]\cos\left[\frac{2\pi\left(\frac{t}{T} - \frac{x}{\lambda}\right) - 2\pi\left(\frac{t}{T} + \frac{x}{\lambda}\right)}{2}\right]\right)\\
&= 2A\cos\left(\frac{4\pi\left(\frac{t}{T}\right)}{2}\right)\cos\left(\frac{-4\pi\left(\frac{x}{\lambda}\right)}{2}\right)\\
&= 2A\cos\left(\frac{2\pi t}{T}\right)\cos\left(-\frac{2\pi x}{\lambda}\right)\\
&= 2A\cos\left(\frac{2\pi t}{T}\right)\cos\left(\frac{2\pi x}{\lambda}\right) \quad \text{since} \quad \cos(-\theta) = \cos\theta.
\end{aligned}$$

Practice Test for Chapter 6

1. Find the value of the other five trigonometric functions, given $\tan x = \frac{4}{11}$, $\sec x < 0$.

2. Simplify $\dfrac{\sec^2 x + \csc^2 x}{\csc^2 x(1 + \tan^2 x)}$.

3. Rewrite as a single logarithm and simplify $\ln|\tan\theta| - \ln|\cot\theta|$.

4. True or False: $\cos\left(\dfrac{\pi}{2} - x\right) = \dfrac{1}{\csc x}$?

5. Factor and simplify $\sin^4 x + (\sin^2 x)\cos^2 x$.

6. Multiply and simplify $(\csc x + 1)(\csc x - 1)$.

7. Rationalize the denominator and simplify $\dfrac{\cos^2 x}{1 - \sin x}$.

8. Verify $\dfrac{1 + \cos\theta}{\sin\theta} + \dfrac{\sin\theta}{1 + \cos\theta} = 2\csc\theta$.

9. Verify $\tan^4 x + 2\tan^2 x + 1 = \sec^4 x$.

10. Use the sum or difference formulas to determine:

(a) $\sin 105°$ (b) $\tan 15°$

11. Simplify $(\sin 42°)\cos 38° - (\cos 42°)\sin 38°$.

12. Verify $\tan\left(\theta + \dfrac{\pi}{4}\right) = \dfrac{1 + \tan\theta}{1 - \tan\theta}$.

13. Write $\sin(\arcsin x - \arccos x)$ as an algebraic expression in x.

14. Use the double-angle formulas to determine:

(a) $\cos 120°$ (b) $\tan 300°$

15. Use the half-angle formulas to determine:

(a) $\sin 22.5°$ (b) $\tan \dfrac{\pi}{12}$

16. Given $\sin = \dfrac{4}{5}$, θ lies in Quadrant II, find $\cos\dfrac{\theta}{2}$.

17. Use the power-reducing identities to write $(\sin^2 x)\cos^2 x$ in terms of the first power of cosine.

18. Rewrite as a sum $6(\sin 5\theta)\cos 2\theta$.

19. Rewrite as a product $\sin(x+\pi)+\sin(x-\pi)$.

20. Verify $\dfrac{\sin 9x+\sin 5x}{\cos 9x-\cos 5x} = -\cot 2x$.

21. Verify $(\cos u)\sin v = \frac{1}{2}[\sin(u+v)-\sin(u-v)]$.

22. Find all solutions in the interval $[0,\ 2\pi)$ $4\sin^2 x = 1$.

23. Find all solutions in the interval $[0,\ 2\pi)$ $\tan^2\theta + (\sqrt{3}-1)\tan\theta - \sqrt{3} = 0$.

24. Find all solutions in the interval $[0,\ 2\pi)$ $\sin 2x = \cos x$.

25. Use the quadratic formula to find all solutions in the interval $[0,\ 2\pi)$ $\tan^2 x - 6\tan + 4 = 0$.

CHAPTER 7

Additional Applications of Trigonometry

SECTION 7.1

Law of Sines

- If ABC is any oblique triangle with sides a, b, and c, then

$$\frac{a}{\sin A} = \frac{b}{\sin B} = \frac{c}{\sin C}.$$

- You should be able to use the Law of Sines to solve an oblique triangle for the remaining three parts, given:
  (a) Two angles and any side (AAS or ASA)
  (b) Two sides and an angle opposite one of them (SSA)
    1. If A is acute and:
       (a) $a < h$, no triangle is possible.
       (b) $a = h$ or $a > b$, one triangle is possible.
       (c) $h < a < b$, two triangles are possible.
    2. If A is obtuse and:
       (a) $a \leq b$, no triangle is possible.
       (b) $a > b$, one triangle is possible.

- The area of any triangle equals one-half the product of the lengths of two sides times the sine of their included angle.

$$A = \frac{1}{2}ab \sin C = \frac{1}{2}ac \sin B = \frac{1}{2}bc \sin A$$

Solutions to Selected Exercises

3. Find the remaining sides and angles of the triangle.

Solution:

$A = 10°$, $B = 60°$, $a = 4.5$

$C = 180° - (10° + 60°) = 110°$

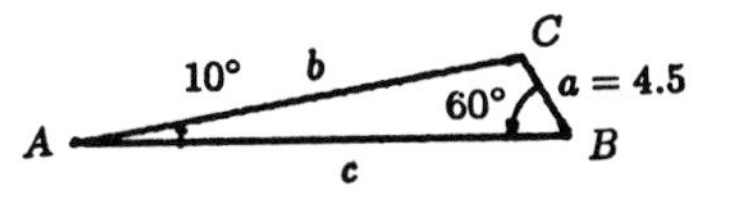

$$\frac{4.5}{\sin 10°} = \frac{b}{\sin 60°}$$

$$b = \sin 60° \left(\frac{4.5}{\sin 10°}\right) \approx 22.4$$

$$\frac{4.5}{\sin 10°} = \frac{c}{\sin 110°}$$

$$c = \sin 110° \left(\frac{4.5}{\sin 10°}\right) \approx 24.4$$

7. Find the remaining sides and angles of the triangle, given $A = 150°$, $C = 20°$, $a = 200$.

Solution:

$A = 150°$, $C = 20°$, $a = 200$

$B = 180° - (150° + 20°) = 10°$

$$\frac{200}{\sin 150°} = \frac{b}{\sin 10°}$$

$$b = \sin 10° \left(\frac{200}{\sin 150°}\right) \approx 69.5$$

$$\frac{200}{\sin 150°} = \frac{c}{\sin 20°}$$

$$c = \sin 20° \left(\frac{200}{\sin 150°}\right) \approx 136.8$$

11. Find the remaining sides and angles of the triangle, given $B = 15°30'$, $a = 4.5$, $b = 6.8$.

Solution:

$B = 15°30'$, $a = 4.5$, $b = 6.8$

$B = 15°30' = 15.5°$

$$\frac{6.8}{\sin 15.5°} = \frac{4.5}{\sin A}$$

$$\sin A = \frac{4.5 \sin 15.5}{6.8}$$

$$\sin A = 0.1768$$

$$A \approx 10.19° \approx 10°11'$$

$$C = 180° - (10°11' + 15°30') = 154°19'$$

$$\frac{6.8}{\sin 15.5°} = \frac{c}{\sin 154.32°}$$

$$c \approx 11$$

15. Find the remaining sides and angles of the triangle, given $A = 110°15'$, $a = 48$, $b = 16$.

Solution:

$A = 100°15'$, $a = 48$, $b = 16$

$$\frac{48}{\sin 100.25°} = \frac{16}{\sin B}$$

$$\sin B = 0.3127$$

$$B \approx 18.22° = 18°13'$$

$$C = 180° - (110°15' + 18°13') = 51°32'$$

$$\frac{48}{\sin 110.25°} = \frac{c}{\sin 51.53°}$$

$$c \approx 40.1$$

19. Solve the triangle: $A = 58°$, $a = 4.5$, $b = 5$, (if possible). If two solutions exist, find both.

Solution:

$A = 58°$, $a = 4.5$, $b = 5$

$h = b \sin A = 5 \sin 58° = 4.24$

A is acute and $h < a < b$, so there are two possible solutions.

$$\frac{4.5}{\sin 58°} = \frac{5}{\sin B}$$
$$\sin B = 0.9423$$

$B \approx 70.4°$ or $B \approx 109.6°$

$C = 51.6°$ $\quad$ $C = 12.4°$

$$\frac{4.5}{\sin 58°} = \frac{c}{\sin 51.6°} \qquad \frac{4.5}{\sin 58°} = \frac{c}{\sin 12.4°}$$

$c \approx 4.16$ $\quad$ $c \approx 1.14$

25. Find the area of the triangle with $C = 120°$, $a = 4$, and $b = 6$.

Solution:

$$\text{Area} = \tfrac{1}{2}ab \sin C = \tfrac{1}{2}(4)(6)\sin 120° = 10.4 \text{ square units}$$

31. Find the length d of the brace required to support the streetlight shown in the figure.

Solution:

$$\frac{3}{\sin 30°} = \frac{5}{\sin \alpha}$$
$$\sin \alpha = 0.8333$$
$$\alpha \approx 56.44°$$
$$\beta = 93.56°$$
$$\frac{3}{\sin 30°} = \frac{d}{\sin 93.56°}$$
$$d \approx 6 \text{ feet}$$

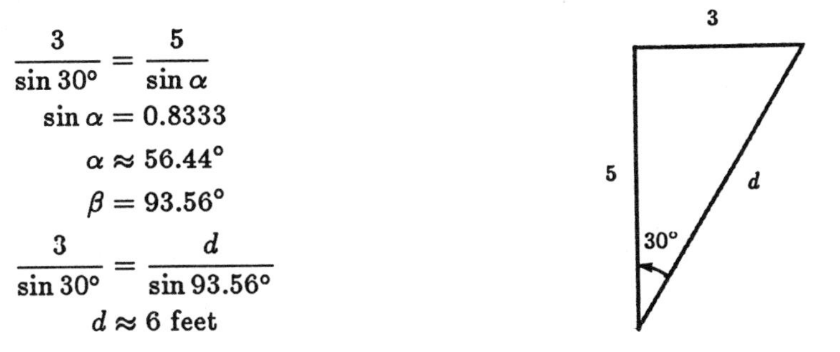

37. Two fire towers A and B are 18.5 miles apart. The bearing from A to B is N 65° E. A fire is spotted by the ranger in each tower, and its bearings from A and B are N 28° E and N 16.5° W, respectively (see figure). Find the distance of the fire from each tower.

Solution:

$A = 37°,\ B = 98.5°,\ C = 44.5°,\ c = 18.5$

$$\frac{b}{\sin 98.5°} = \frac{18.5}{\sin 44.5°}$$

$$b = 26.1 \text{ mi}$$

$$\frac{a}{\sin 37°} = \frac{18.5}{\sin 44.5°}$$

$$a = 15.9 \text{ mi}$$

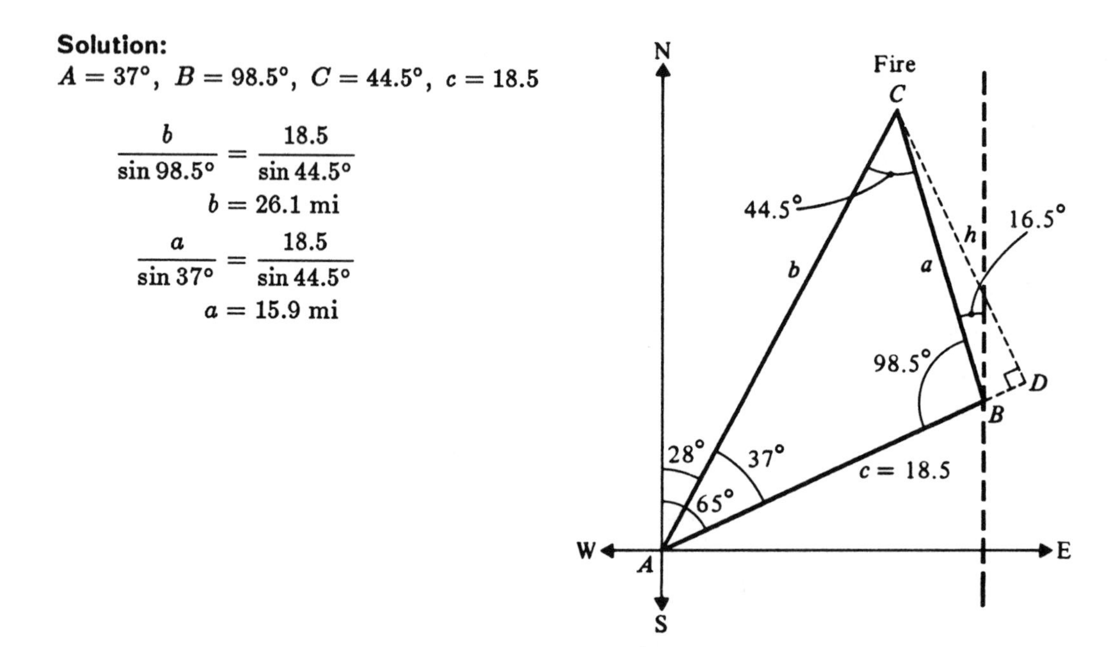

41. The following information about a triangular parcel of land is given at a zoning board meeting: "One side is 450 feet long and another is 120 feet long. The angle opposite the shorter side is 30°." Could this information be correct?

Solution:

$$h = b \sin A = 450 \sin 30° = 225$$

$$a = 120$$

Since $a < h$, no such triangle is possible.

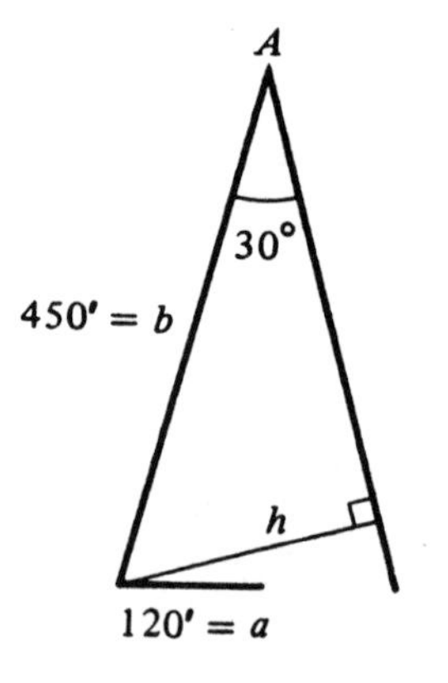

SECTION 7.2

Law of Cosines

- If ABC is any oblique triangle with sides a, b, and c, then the following equations are valid.

  (a) $a^2 = b^2 + c^2 - 2bc\cos A$ or $\cos A = \dfrac{b^2 + c^2 - a^2}{2bc}$

  (b) $b^2 = a^2 + c^2 - 2ac\cos B$ or $\cos B = \dfrac{a^2 + c^2 - b^2}{2ac}$

  (c) $c^2 = a^2 + b^2 - 2ab\cos C$ or $\cos C = \dfrac{a^2 + b^2 - c^2}{2ab}$

- You should be able to use the Law of Cosines to solve an oblique triangle for the remaining three parts, given:

  (a) Three sides (SSS)
  (b) Two sides and their included angle (SAS)

- Given any triangle with sides of length a, b, and c, then the area of the triangle is

$$\text{Area} = \sqrt{s(s-a)(s-b)(s-c)}, \text{ where } s = \frac{a+b+c}{2}.$$

Solutions to Selected Exercises

5. Use the Law of Cosines to solve the triangle: $a = 9,\ b = 12,\ c = 15.$

Solution:

$$\cos A = \frac{144 + 225 - 81}{360} = 0.8$$

$$A \approx 36.9^\circ$$

$$\cos B = \frac{81 + 225 - 144}{270} = 0.6$$

$$B \approx 53.1^\circ$$

$$C = 180^\circ - (36.9^\circ + 53.1^\circ)$$

$$C = 90^\circ$$

7. Use the Law of Cosines to solve the triangle: $a = 75.4,\ b = 52,\ c = 52$.

Solution:

$$\cos A = \frac{(52)^2 + (52)^2 - (75.4)^2}{2(52)(52)} = -0.05125$$

$$A \approx 92.9^\circ$$

Since $b = c$, the triangle is isosceles and $B = C$.

$$2B = 180^\circ - 92.9^\circ$$

$$B = C = 43.55^\circ$$

13. Use the Law of Cosines to solve the triangle: $C = 125^\circ 40',\ a = 32,\ b = 32$.

Solution:
Since $a = b$, the triangle is isosceles and $A = B$.

$$2A = 180^\circ - 125^\circ 40'$$

$$A = B = 27^\circ 10'$$

$$c^2 = (32)^2 + (32)^2 - 2(32)(32)\cos 125^\circ 40'$$

$$c \approx 56.9$$

17. Use Heron's Formula to find the area of the triangle: $a = 12,\ b = 15,\ c = 9$.

Solution:

$$s = \tfrac{1}{2}(12 + 15 + 9) = 18$$

$$\text{Area} = \sqrt{18(18-12)(18-15)(18-9)}$$

$$= \sqrt{2916}$$

$$= 54 \text{ square units}$$

23. Two ships leave a port at 9 A. M. One travels at a bearing of N 53° W at 12 miles per hour and the other at a bearing of S 67° W at 16 miles per hour. Approximately how far apart are they at noon of that day?

Solution:
By noon, the first ship has traveled $3(12) = 36$ miles, and the second ship has traveled $3(16) = 48$ miles.

$$d^2 = 36^2 + 48^2 - 2(36)(48)\cos 60^\circ$$

$$d^2 = 1872$$

$$d \approx 43.3 \text{ mi}$$

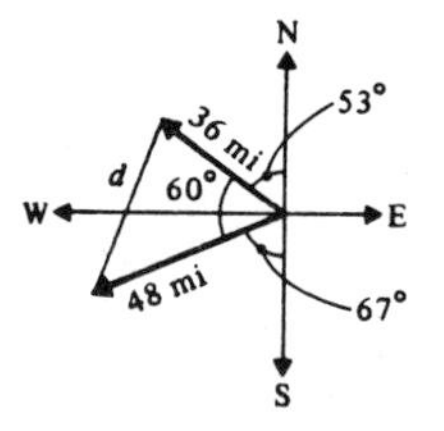

27. Determine the angle θ as shown on the streetlight in the figure.

Solution:

$$\cos\theta = \frac{3^2 + 2^2 - 4.25^2}{2(3)(2)} = -0.421875$$

$$\theta \approx 114.95^\circ$$

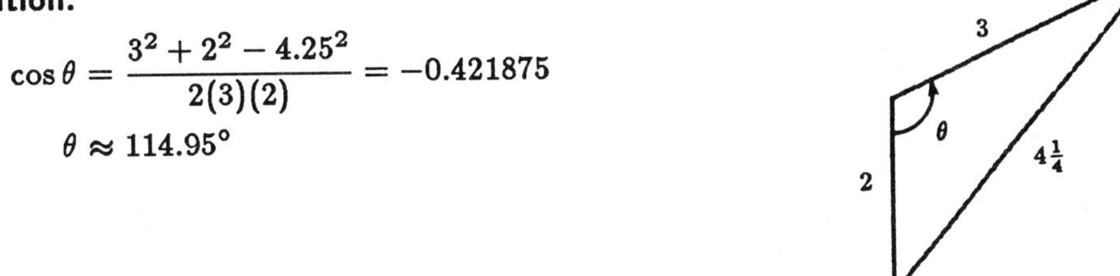

31. In a (square) baseball diamond with 90-foot sides the pitcher's mound is 60 feet from home plate.

(a) How far is it from the pitcher's mound to third base?

(b) When a runner is halfway from second to third, how far is the runner from the pitcher's mound?

Solution:

(a) $x^2 = 90^2 + 60^2 - 2(90)(60)\cos 45^\circ$

$x \approx 63.7$ feet

(b) $\dfrac{60}{\sin\alpha} = \dfrac{63.7}{\sin 45^\circ}$

$\sin\alpha = \dfrac{60\sin 45^\circ}{63.7}$

$\alpha \approx 41.73^\circ$

$\beta = 90^\circ - \alpha = 48.27^\circ$

$y^2 = 45^2 + 63.7^2 - 2(45)(63.7)\cos 48.27^\circ$

$y \approx 47.6$ feet

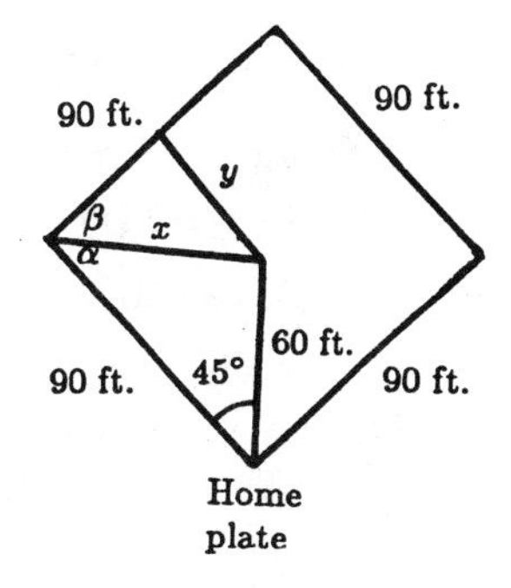

37. Use the Law of Cosines to prove that $\frac{1}{2}bc(1 + \cos A) = \frac{a+b+c}{2} \cdot \frac{-a+b+c}{2}$.

Solution:

$$\begin{aligned}
\frac{1}{2}bc(1 + \cos A) &= \frac{1}{2}bc\left[1 + \frac{b^2 + c^2 - a^2}{2bc}\right] \\
&= \frac{1}{2}bc\left[\frac{2bc + b^2 + c^2 - a^2}{2bc}\right] \\
&= \frac{1}{4}[(b+c)^2 - a^2] \\
&= \frac{1}{4}[(b+c) + a][(b+c) - a] \\
&= \frac{b+c+a}{2} \cdot \frac{b+c-a}{2} \\
&= \frac{a+b+c}{2} \cdot \frac{-a+b+c}{2}
\end{aligned}$$

SECTION 7.3

Vectors

- A vector $\mathbf{v}$ is the collection of all directed line segments that are equivalent to a given directed line segment $\overrightarrow{PQ}$.
- You should be able to *geometrically* perform the operations of vector addition and scalar multiplication.
- The component form of the vector with initial point $P = (p_1, p_2)$ and terminal point $Q = (q_1, q_2)$ is

  $$\overrightarrow{PQ} = \langle q_1 - p_1, q_2 - p_2 \rangle = \langle v_1, v_2 \rangle = \mathbf{v}.$$

- The magnitude of $\mathbf{v} = \langle v_1, v_2 \rangle$ is given by $\|\mathbf{v}\| = \sqrt{{v_1}^2 + {v_2}^2}$.
- You should be able to perform the operations of scalar multiplication and vector addition in component form.
- You should know the following properties of vector addition and scalar multiplication.
  (a) $\mathbf{u} + \mathbf{v} = \mathbf{v} + \mathbf{u}$
  (b) $(\mathbf{u} + \mathbf{v}) + \mathbf{w} = \mathbf{u} + (\mathbf{v} + \mathbf{w})$
  (c) $\mathbf{u} + \mathbf{O} = \mathbf{u}$
  (d) $\mathbf{u} + (-\mathbf{u}) = \mathbf{O}$
  (e) $c(d\mathbf{u}) = (cd)\mathbf{u}$
  (f) $(c + d)\mathbf{u} = c\mathbf{u} + d\mathbf{u}$
  (g) $c(\mathbf{u} + \mathbf{v}) = c\mathbf{u} + c\mathbf{v}$
  (h) $1(\mathbf{u}) = \mathbf{u}$, $0\mathbf{u} = \mathbf{O}$
  (i) $\|c\mathbf{v}\| = |c|\,\|\mathbf{v}\|$
- A unit vector in the direction of $\mathbf{v}$ is given by $\mathbf{u} = \dfrac{\mathbf{v}}{\|\mathbf{v}\|}$.
- The standard unit vectors are $\mathbf{i} = \langle 1, 0 \rangle$ and $\mathbf{j} = \langle 0, 1 \rangle$. $\mathbf{v} = \langle v_1, v_2 \rangle$ can be written as $\mathbf{v} = v_1\mathbf{i} + v_2\mathbf{j}$.
- A vector $\mathbf{v}$ with magnitude $\|\mathbf{v}\|$ and direction θ can be written as $\mathbf{v} = a\mathbf{i} + b\mathbf{j} = \|\mathbf{v}\|(\cos\theta)\mathbf{i} + \|\mathbf{v}\|(\sin\theta)\mathbf{j}$ where $\tan\theta = b/a$.

Solutions to Selected Exercises

3. Use the figure to sketch a graph of $\mathbf{u} + \mathbf{v}$.

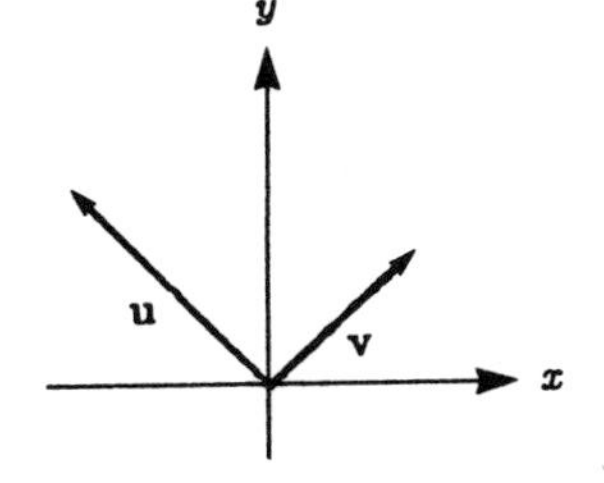

Solution:
Move the initial point of $\mathbf{v}$ to the terminal point of $\mathbf{u}$.

9. Find the component form and the magnitude of the vector $\mathbf{v}$.

Solution:

$$\mathbf{v} = \langle -1 - 2,\ 3 - 1 \rangle = \langle -3,\ 2 \rangle$$
$$\|\mathbf{v}\| = \sqrt{(-3)^2 + (2)^2} = \sqrt{13}$$

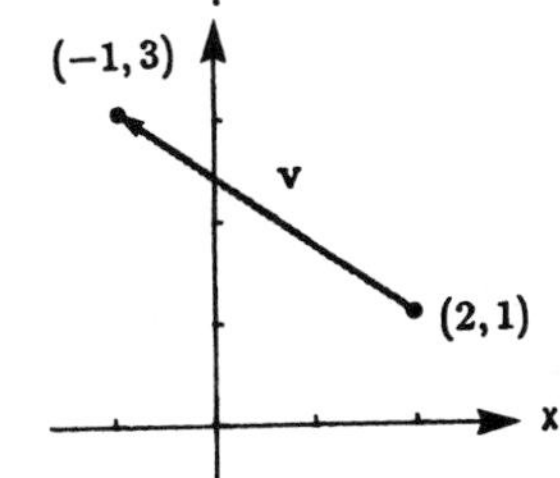

13. Find the component form and the magnitude of the vector $\mathbf{v}$ with initial point $(-1,\ 5)$ and terminal point $(15,\ 2)$.

Solution:

$$\mathbf{v} = \langle 15 - (-1),\ 2 - 5 \rangle = \langle 16,\ -3 \rangle$$
$$\|\mathbf{v}\| = \sqrt{(16)^2 + (-3)^2} = \sqrt{265}$$

19. Find (a) $\mathbf{u} + \mathbf{v}$, (b) $\mathbf{u} - \mathbf{v}$, and (c) $2\mathbf{u} - 3\mathbf{v}$ for $\mathbf{u} = \langle -2,\ 3 \rangle$, $\mathbf{v} = \langle -2,\ 1 \rangle$.

Solution:

(a) $\mathbf{u} + \mathbf{v} = \langle -2 + (-2),\ 3 + 1 \rangle = \langle -4,\ 4 \rangle$
(b) $\mathbf{u} - \mathbf{v} = \langle -2 - (-2),\ 3 - 1 \rangle = \langle 0,\ 2 \rangle$
(c) $2\mathbf{u} - 3\mathbf{v} = \langle 2(-2) - 3(-2),\ 2(3) - 3(1) \rangle = \langle 2,\ 3 \rangle$

25. Find (a) $\mathbf{u}+\mathbf{v}$, (b) $\mathbf{u}-\mathbf{v}$, and (c) $2\mathbf{u}-3\mathbf{v}$ for $\mathbf{u}=2\mathbf{i}$, $\mathbf{v}=\mathbf{j}$.

Solution:

(a) $\mathbf{u}+\mathbf{v}=2\mathbf{i}+\mathbf{j}$
(b) $\mathbf{u}-\mathbf{v}=2\mathbf{i}-\mathbf{j}$
(c) $2\mathbf{u}-3\mathbf{v}=4\mathbf{i}-3\mathbf{j}$

29. Sketch $\mathbf{v}$ and find its component form given $\|\mathbf{v}\| = 1$, $\theta = 150°$. (Assume θ is measured counterclockwise from the x-axis to the vector.)

Solution:

$$\begin{aligned}\mathbf{v} &= 1\cos 150°\mathbf{i} + 1\sin 150°\mathbf{j}\\ &= -\frac{\sqrt{3}}{2}\mathbf{i} + \frac{1}{2}\mathbf{j}\\ &= \left\langle -\frac{\sqrt{3}}{2}, \frac{1}{2}\right\rangle\end{aligned}$$

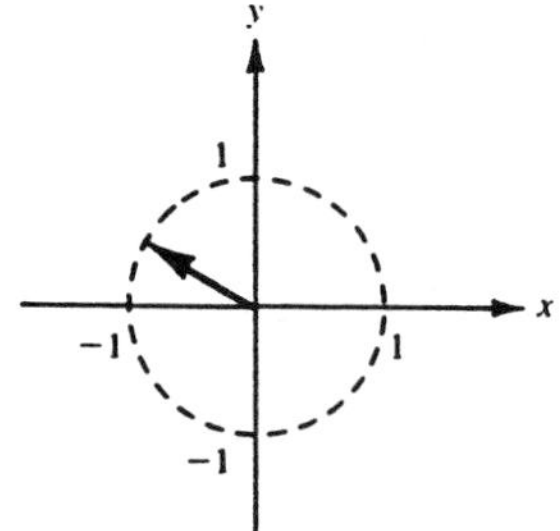

33. Sketch $\mathbf{v}$ and find its component form given $\|\mathbf{v}\| = 2$, and $\mathbf{v}$ is in the direction of $\mathbf{i}+3\mathbf{j}$. (Assume θ is measured counterclockwise from the x-axis to the vector.)

Solution:

$$\tan\theta = \frac{3}{1} \Rightarrow \sin\theta = \frac{3\sqrt{10}}{10} \text{ and } \cos\theta = \frac{\sqrt{10}}{10}$$

$$\begin{aligned}\mathbf{v} &= 2\left(\frac{\sqrt{10}}{10}\right)\mathbf{i} + 2\left(\frac{3\sqrt{10}}{10}\right)\mathbf{j}\\ &= \left\langle \frac{\sqrt{10}}{5}, \frac{3\sqrt{10}}{5}\right\rangle\end{aligned}$$

37. Find the component form of $\mathbf{v}=\mathbf{u}+2\mathbf{w}$ and sketch the indicated vector operations geometrically, where $\mathbf{u}=2\mathbf{i}-\mathbf{j}$ and $\mathbf{w}=\mathbf{i}+2\mathbf{j}$.

Solution:

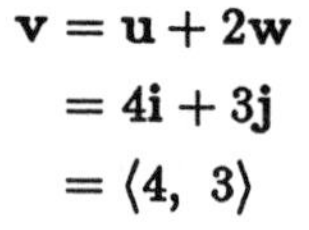

$$\begin{aligned}\mathbf{v} &= \mathbf{u}+2\mathbf{w}\\ &= 4\mathbf{i}+3\mathbf{j}\\ &= \langle 4, 3\rangle\end{aligned}$$

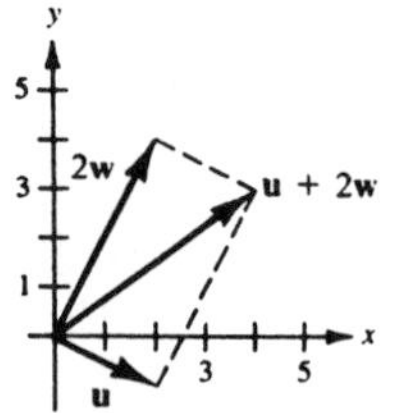

41. Find the component form of the sum of the vectors $\mathbf{u}$ and $\mathbf{v}$ with direction angles $\theta_{\mathbf{u}}$ and $\theta_{\mathbf{v}}$, respectively, given $\|\mathbf{u}\| = 5$, $\theta_{\mathbf{u}} = 0°$, and $\|\mathbf{v}\| = 5$, $\theta_{\mathbf{v}} = 90°$.

Solution:

$$\begin{aligned}\|\mathbf{u}\| = 5,\ \theta_{\mathbf{u}} = 0° &\Rightarrow \mathbf{u} = 5\mathbf{i}\\ \|\mathbf{v}\| = 5,\ \theta_{\mathbf{v}} = 90° &\Rightarrow \mathbf{v} = 5\mathbf{j}\end{aligned}$$

$$\mathbf{u}+\mathbf{v} = 5\mathbf{i}+5\mathbf{j} = \langle 5, 5\rangle$$

45. Find a unit vector in the direction of $\mathbf{v} = 4\mathbf{i} - 3\mathbf{j}$.

Solution:

$$\|\mathbf{v}\| = \sqrt{(4)^2 + (-3)^2} = 5$$

$$\frac{\mathbf{v}}{\|\mathbf{v}\|} = \frac{4\mathbf{i} - 3\mathbf{j}}{5} = \left\langle \frac{4}{5}, -\frac{3}{5} \right\rangle$$

49. Use the Law of Cosines to find the angle α between the vectors $\mathbf{v} = \mathbf{i} + \mathbf{j}$ and $\mathbf{w} = 2(\mathbf{i} - \mathbf{j})$. (Assume $0° \le \alpha \le 180°$.)

Solution:

$\mathbf{v} = \mathbf{i} + \mathbf{j}, \quad \mathbf{w} = 2(\mathbf{i} - \mathbf{j})$

$\mathbf{u} = \mathbf{w} - \mathbf{v} = \mathbf{i} - 3\mathbf{j}$

$$\|\mathbf{v}\| = \sqrt{2}$$

$$\|\mathbf{w}\| = 2\sqrt{2}$$

$$\|\mathbf{u}\| = \sqrt{10}$$

$$\cos\theta = \frac{2 + 8 - 10}{2(\sqrt{2})(2\sqrt{2})} = 0$$

$$\theta = 90°$$

53. Two forces, one of 35 pounds and the other of 50 pounds, act on the same object. The angle between the forces is 30°. Find the magnitude of the resultant (vector sum) of these forces.

Solution:

$$\mathbf{u} = 35\mathbf{i}$$

$$\mathbf{v} = 50(\cos 30°\mathbf{i} + \sin 30°\mathbf{j}) = 50\left(\frac{\sqrt{3}}{2}\mathbf{i} + \frac{1}{2}\mathbf{j}\right)$$

$$\mathbf{r} = \mathbf{u} + \mathbf{v} = (35 + 25\sqrt{3})\mathbf{i} + 25\mathbf{j}$$

$$\|\mathbf{r}\| \approx 82.2 \text{ lb}$$

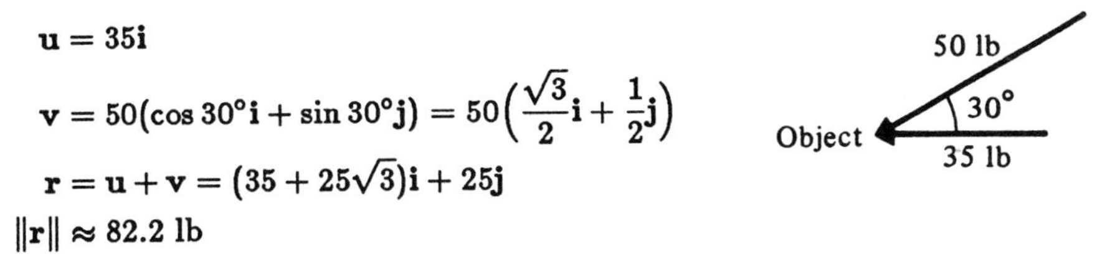

57. A heavy implement is dragged 10 feet across the floor, using a force of 85 pounds. Find the work done if the direction of the force is 60° above the horizontal (see figure). (Use the formula for work, $W = FD$, where F is the horizontal component of force and D is the horizontal distance.)

Solution:

The horizontal component of the force is $85 \cos 60° = \frac{85}{2}$.

$$W = FD = \frac{85}{2}(10) = 425 \text{ ft-lb}$$

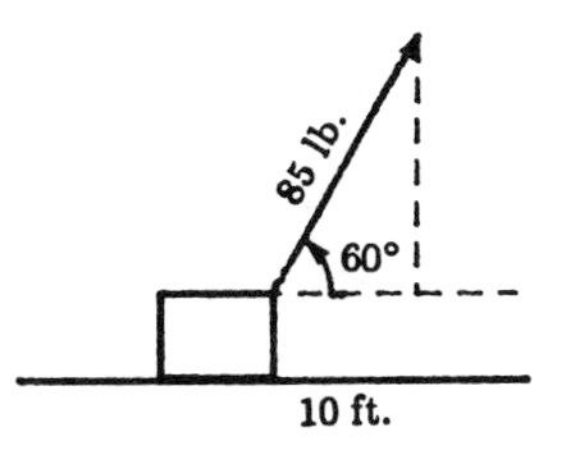

61. An airplane is flying in the direction S 32° E, with an airspeed of 540 miles per hour. Because of the wind, its groundspeed and direction are 500 miles per hour and S 40° E, respectively. Find the direction and speed of the wind.

Solution:

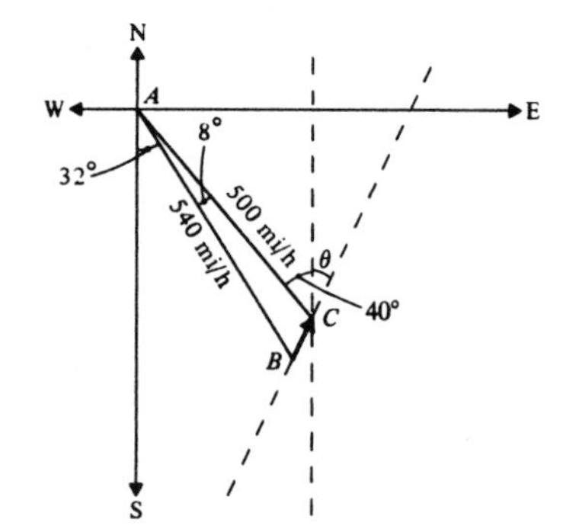

$$a = |\overrightarrow{BC}| = \text{speed of wind}$$

$$a^2 = 500^2 + 540^2 - 2(500)(540)\cos 8°$$

$$a \approx 82.8 \text{ mi/hr}$$

$$\cos C \approx \frac{82.8^2 + 500^2 - 540^2}{2(82.8)(500)}$$

$$C \approx 114.8°$$

$$C + \theta + 40 \approx 180°$$

$$\theta \approx \text{N } 25.2° \text{ E}$$

SECTION 7.4

Trigonometric Form of Complex Numbers

- You should be able to graphically represent complex numbers and know the following facts about them.
- The absolute value of the complex number $z = a + bi$ is $|z| = \sqrt{a^2 + b^2}$.
- The trigonometric form of the complex number $z = a + bi$ is $z = r(\cos\theta + i\sin\theta)$ where
  (a) $a = r\cos\theta$
  (b) $b = r\sin\theta$
  (c) $r = \sqrt{a^2 + b^2}$
  (d) $\tan\theta = b/a$
- Given $z_1 = r_1(\cos\theta_1 + i\sin\theta_1)$ and $z_2 = r_2(\cos\theta_2 + i\sin\theta_2)$:
  (a) $z_1 z_2 = r_1 r_2[\cos(\theta_1 + \theta_2) + i\sin(\theta_1 + \theta_2)]$
  (b) $\dfrac{z_1}{z_2} = \dfrac{r_1}{r_2}[\cos(\theta_1 - \theta_2) + i\sin(\theta_1 - \theta_2)],\quad z \neq 0$

Solutions to Selected Exercises

3. Express the complex number $-3 - 3i$ in trigonometric form.

Solution:

$$z = -3 - 3i$$
$$r = \sqrt{(-3)^2 + (-3)^2} = \sqrt{18} = 3\sqrt{2}$$
$$\tan\theta = \frac{-3}{-3} = 1,\ \theta \text{ is in Quadrant III}$$
$$\theta = 225^\circ \text{ or } \frac{5\pi}{4}$$
$$z = 3\sqrt{2}\left(\cos\frac{5\pi}{4} + i\sin\frac{5\pi}{4}\right)$$

7. Represent $\sqrt{3} + i$ graphically, and find the trigonometric form of the number.

Solution:

$$z = \sqrt{3} + i$$

$$r = \sqrt{(\sqrt{3})^2 + 1^2} = 2$$

$$\tan\theta = \frac{1}{\sqrt{3}}$$

$$\theta = 30^\circ \text{ or } \frac{\pi}{6}$$

$$z = 2\left(\cos\frac{\pi}{6} + i\sin\frac{\pi}{6}\right)$$

11. Represent $6i$ graphically, and find the trigonometric form of the number.

Solution:

$$z = 6i$$

$$r = \sqrt{0^2 + 6^2} = 6$$

$$\tan\theta = \frac{6}{0}, \text{ undefined}$$

$$\theta = \frac{\pi}{2}$$

$$z = 6\left(\cos\frac{\pi}{2} + i\sin\frac{\pi}{2}\right)$$

17. Represent $1 + 6i$ graphically, and find the trigonometric form of the number.

Solution:

$$z = 1 + 6i$$

$$r = \sqrt{37}$$

$$\tan\theta = 6$$

$$\theta \approx 1.41 \text{ rad}$$

$$z = \sqrt{37}[\cos(1.41) + i\sin(1.41)]$$

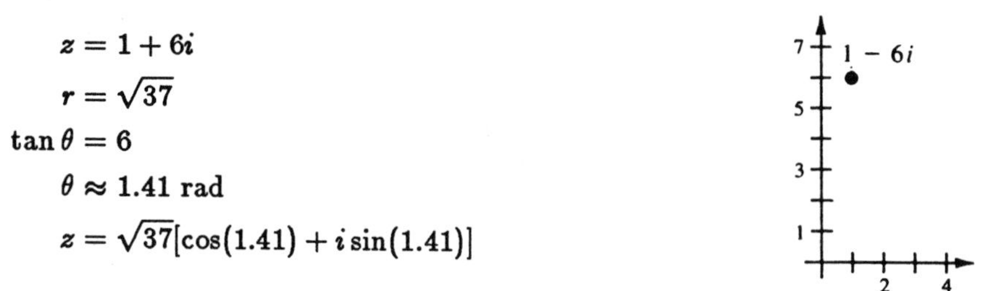

21. Represent $2(\cos 150^\circ + i\sin 150^\circ)$ graphically, and find the standard form of the number.

Solution:

$$z = 2(\cos 150^\circ + i\sin 150^\circ)$$

$$a = 2\cos 150^\circ = -\sqrt{3}$$

$$b = 2\sin 150^\circ = 1$$

$$z = -\sqrt{3} + i$$

29. Represent $3[\cos(18°45') + i\sin(18°45')]$ graphically, and find the standard form of the number.

Solution:

$$z = 3(\cos 18°45' + i \sin 18°45')$$
$$a = 3\cos 18°45' = 3\cos 18.75° = 2.8408$$
$$b = 3\sin 18°45' = 3\sin 18.75° = 0.9643$$
$$z = 2.8408 + 0.9643i$$

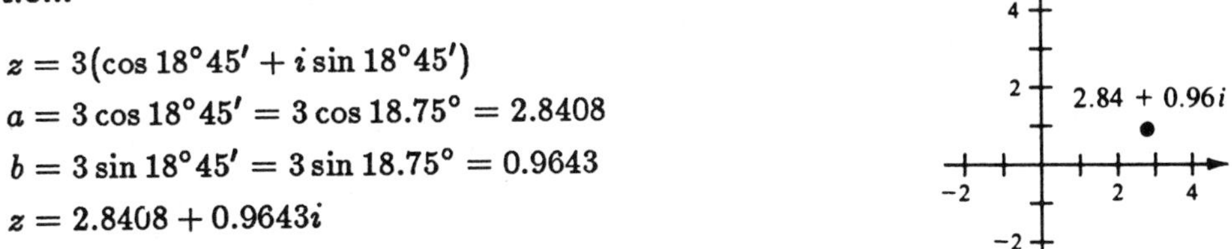

33. Perform the indicated operation and leave the result in trigonometric form.

$$\left[\tfrac{5}{3}(\cos 140° + i\sin 140°)\right]\left[\tfrac{2}{3}(\cos 60° + i\sin 60°)\right]$$

Solution:

$$\left[\tfrac{5}{3}(\cos 140° + i\sin 140°)\right]\left[\tfrac{2}{3}(\cos 60° + i\sin 60°)\right]$$
$$= \left(\tfrac{5}{3}\right)\left(\tfrac{2}{3}\right)[\cos(140° + 60°) + i\sin(140° + 60°)] = \tfrac{10}{9}[\cos 200° + i\sin 200°]$$

39. Perform the indicated operation and leave the result in trigonometric form.

$$\frac{\cos(5\pi/3) + i\sin(5\pi/3)}{\cos\pi + i\sin\pi}$$

Solution:

$$\frac{\cos(5\pi/3) + i\sin(5\pi/3)}{\cos\pi + i\sin\pi} = \cos\left(\frac{5\pi}{3} - \pi\right) + i\sin\left(\frac{5\pi}{3} - \pi\right) = \cos\frac{2\pi}{3} + i\sin\frac{2\pi}{3}$$

43. For $(2 + 2i)(1 - i)$, (a) give the trigonometric form of the complex numbers, (b) perform the indicated operation using the trigonometric form, and (c) perform the indicated operation using the standard form and check your result with the answer in part (b).

Solution:

(a) Trigonometric form:

$$\left[2\sqrt{2}\left(\cos\frac{\pi}{4} + i\sin\frac{\pi}{4}\right)\right]\left[\sqrt{2}\left(\cos\frac{7\pi}{4} + i\sin\frac{7\pi}{4}\right)\right]$$

(b) Operation in trigonometric form:

$$(2\sqrt{2})(\sqrt{2})\left[\cos\left(\frac{\pi}{4} + \frac{7\pi}{4}\right) + i\sin\left(\frac{\pi}{4} + \frac{7\pi}{4}\right)\right] = 4[\cos(2\pi) + i\sin(2\pi)]$$
$$= 4[\cos(0°) + i\sin(0°)]$$
$$= 4$$

(c) $(2 + 2i)(1 - i) = 2 - 2i + 2i - 2i^2 = 4$

47. For $5/(2+3i)$, (a) give the trigonometric form of the complex numbers, (b) perform the indicated operation using the trigonometric form, and (c) perform the indicated operation using the standard form and check your result with the answer in part (b).

Solution:

(a) Trigonometric form:

$$\frac{5[\cos 0^\circ + i \sin 0^\circ]}{\sqrt{13}[\cos 56.3^\circ + i \sin 56.3^\circ]}$$

(b) Operation in trigonometric form:

$$\frac{5}{\sqrt{13}}[\cos(-56.3^\circ) + i \sin(-56.3^\circ)] = \frac{5\sqrt{13}}{13}[\cos(-56.3^\circ) + i \sin(-56.3^\circ)] \approx \frac{5}{13}(2-3i)$$

(c) $\dfrac{5}{2+3i} \cdot \dfrac{2-3i}{2-3i} = \dfrac{5(2-3i)}{13} = \dfrac{5}{13}(2-3i)$

51. Use the trigonometric form $z = r(\cos\theta + i\sin\theta)$ and $\overline{z} = r[\cos(-\theta) + i\sin(-\theta)]$ to find (a) $z\overline{z}$ and (b) $z/\overline{z}$, $z \neq 0$.

Solution:

(a)
$$\begin{aligned} z\overline{z} &= [r(\cos\theta + i\sin\theta)][r\cos(-\theta) + i\sin(-\theta)] \\ &= r^2[\cos(\theta-\theta) + i\sin(\theta-\theta)] \\ &= r^2 \end{aligned}$$

(b)
$$\begin{aligned} \frac{z}{\overline{z}} &= \frac{r}{r}[\cos(\theta-(-\theta)) + i\sin(\theta-(-\theta))] \\ &= \cos 2\theta + i\sin 2\theta \end{aligned}$$

SECTION 7.5

DeMoivre's Theorem and nth Roots

- You should know DeMoivre's Theorem: If $z = r(\cos\theta + i\sin\theta)$, then for any positive integer n,

$$z^n = r^n(\cos n\theta + i\sin n\theta).$$

- You should know that for any positive integer n, $z = r(\cos\theta + i\sin\theta)$ has n distinct nth roots given by

$$\sqrt[n]{r}\left[\cos\left(\frac{\theta + 2\pi k}{n}\right) + i\sin\left(\frac{\theta + 2\pi k}{n}\right)\right]$$

where $k = 0,\ 1,\ 2,\ \ldots,\ n-1$.

Solutions to Selected Exercises

3. Use DeMoivre's Theorem to find $(-1+i)^{10}$. Express the result in standard form.

Solution:

$$\begin{aligned}(-1+i)^{10} &= \left[\sqrt{2}\left(\cos\frac{3\pi}{4} + i\sin\frac{3\pi}{4}\right)\right]^{10}\\ &= (\sqrt{2})^{10}\left[\cos 10\left(\frac{3\pi}{4}\right) + i\sin 10\left(\frac{3\pi}{4}\right)\right]\\ &= 32\left[\cos\frac{15\pi}{2} + i\sin\frac{15\pi}{2}\right]\\ &= 32[0 - i]\\ &= -32i\end{aligned}$$

7. Use DeMoivre's Theorem to find $[5(\cos 20^\circ + i\sin 20^\circ)]^3$. Express the result in standard form.

Solution:

$$\begin{aligned}[5(\cos 20^\circ + i\sin 20^\circ)]^3 &= 5^3[\cos 60^\circ + i\sin 60^\circ]\\ &= 125\left(\frac{1}{2} + \frac{\sqrt{3}}{2}i\right)\\ &= \frac{125}{2} + \frac{125\sqrt{3}}{2}i\end{aligned}$$

11. Use DeMoivre's Theorem to find $[5(\cos 3.2 + i\sin 3.2)]^4$. Express the result in standard form.

Solution:

$$\begin{aligned}[5(\cos 3.2 + i\sin 3.2)]^4 &= 5^4[\cos 12.8 + i\sin 12.8]\\ &\approx 625(0.97283 + 0.2315i)\\ &\approx 608.02 + 144.69i\end{aligned}$$

15. (a) Use DeMoivre's Theorem to find the fourth roots of

$$16\left(\cos\frac{4\pi}{3} + i\sin\frac{4\pi}{3}\right),$$

(b) represent each of the roots graphically, and (c) express each of the roots in standard form.

Solution:

(a) & (c) $n = 4$

$$k = 0:\ 2\left(\cos\frac{\pi}{3} + i\sin\frac{\pi}{3}\right) = 1 + \sqrt{3}i$$

$$k = 1:\ 2\left(\cos\frac{5\pi}{6} + i\sin\frac{5\pi}{6}\right) = -\sqrt{3} + i$$

$$k = 2:\ 2\left(\cos\frac{4\pi}{3} + i\sin\frac{4\pi}{3}\right) = -1 - \sqrt{3}i$$

$$k = 3:\ 2\left(\cos\frac{11\pi}{6} + i\sin\frac{11\pi}{6}\right) = \sqrt{3} - i$$

(b)

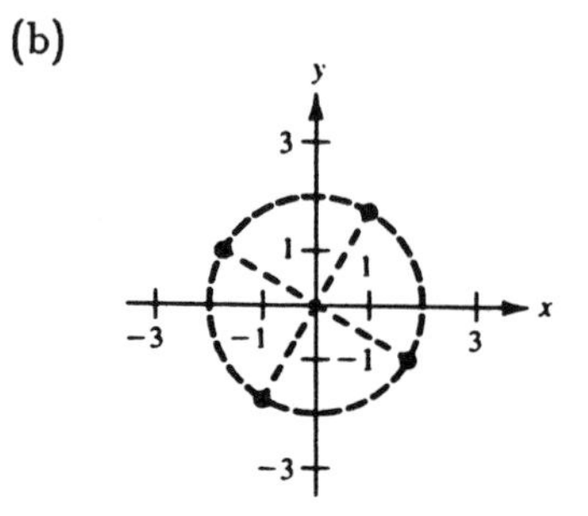

19. (a) Use DeMoivre's Theorem to find the cube roots of

$$-\frac{125}{2}(1 + \sqrt{3}i),$$

(b) represent each of the roots graphically, and (c) express each of the roots in standard form.

Solution:

$$-\frac{125}{2}(1 + \sqrt{3}i) = 125\left(\cos\frac{4\pi}{3} + i\sin\frac{4\pi}{3}\right)$$

(a) & (c) $n = 3$

$$k = 0:\ 5\left(\cos\frac{4\pi}{9} + i\sin\frac{4\pi}{9}\right) \approx 0.8682 + 4.9240i$$

$$k = 1:\ 5\left(\cos\frac{10\pi}{9} + i\sin\frac{10\pi}{9}\right) \approx -4.6985 - 1.7101i$$

$$k = 2:\ 5\left(\cos\frac{16\pi}{9} + i\sin\frac{16\pi}{9}\right) \approx 3.8302 - 3.2139i$$

(b)

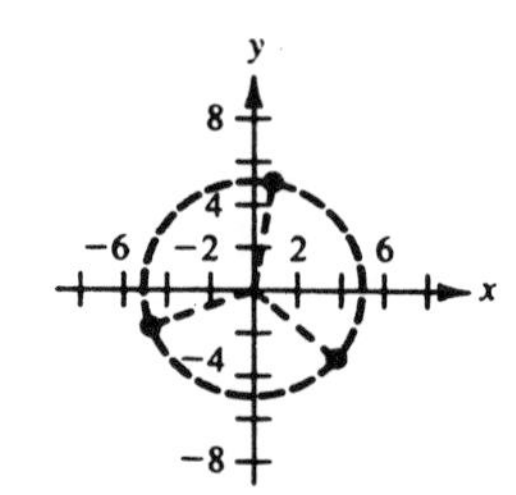

25. Find all the solutions of $x^4 - i = 0$ and represent your solutions graphically.

Solution:

$$x^4 - i = 0$$
$$x^4 = i$$

Find the fourth roots of $i = \cos\frac{\pi}{2} + i\sin\frac{\pi}{2}$.

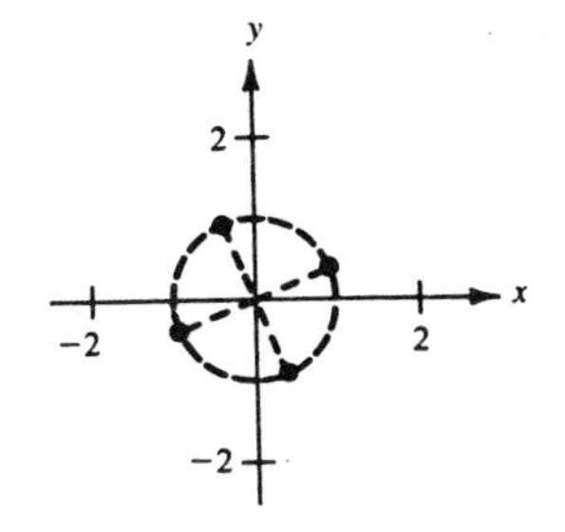

$n = 4$

$k = 0:\ \cos\frac{\pi}{8} + i\sin\frac{\pi}{8} \approx 0.9239 + 0.3827i$

$k = 1:\ \cos\frac{5\pi}{8} + i\sin\frac{5\pi}{8} \approx -0.3827 + 0.9239i$

$k = 2:\ \cos\frac{9\pi}{8} + i\sin\frac{9\pi}{8} \approx -0.9239 - 0.3827i$

$k = 3:\ \cos\frac{13\pi}{8} + i\sin\frac{13\pi}{8} \approx 0.3827 - 0.9239i$

29. Find all the solutions of $x^3 + 64i = 0$ and represent your solutions graphically.

Solution:

$$x^3 + 64i = 0$$
$$x^3 = -64i$$

Find the cube roots of $-64i = 64\left(\cos\frac{3\pi}{2} + i\sin\frac{3\pi}{2}\right)$.

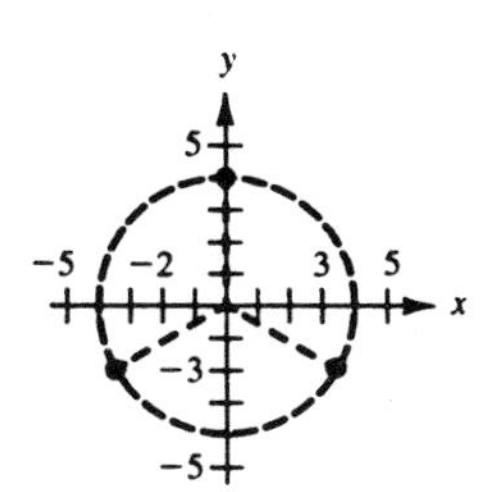

$n = 3$

$k = 0:\ 4\left(\cos\frac{\pi}{2} + i\sin\frac{\pi}{2}\right) = 4i$

$k = 1:\ 4\left(\cos\frac{7\pi}{6} + i\sin\frac{7\pi}{6}\right) = -2\sqrt{3} - 2i$

$k = 2:\ 4\left(\cos\frac{11\pi}{6} + i\sin\frac{11\pi}{6}\right) = 2\sqrt{3} - 2i$

REVIEW EXERCISES FOR CHAPTER 7

Solutions to Selected Exercises

5. Solve the triangle, given $B = 110°$, $a = 4$, $c = 4$.

Solution:
Since the triangle is isosceles,

$$A = C = \frac{1}{2}(180 - 110) = 35°.$$

By the Law of Sines:

$$\frac{4}{\sin 35°} = \frac{b}{\sin 110°}$$
$$b \approx 6.6$$

9. Solve the triangle, given $B = 115°$, $a = 7$, $b = 14.5$.

Solution:
By the Law of Sines:

$$\frac{\sin A}{7} = \frac{\sin 115°}{14.5}$$
$$\sin A \approx 0.4375$$
$$A \approx 25.9°$$
$$C = 180 - (115 + 25.9)$$
$$C = 39.1°$$
$$\frac{c}{\sin 39.1°} = \frac{14.5}{\sin 115°}$$
$$c \approx 10.1$$

13. Solve the triangle, given $B = 150°$, $a = 10$, $c = 20$.

Solution:
By the Law of Cosines:

$$b^2 = 10^2 + 20^2 - 2(10)(20)\cos 150°$$
$$b \approx 29.1$$

By the Law of Sines:

$$\frac{\sin A}{10} = \frac{\sin 150^\circ}{29.1}$$
$$\sin A \approx 0.1719$$
$$A \approx 9.9^\circ$$
$$C = 180 - (150 + 9.9)$$
$$C = 20.1^\circ$$

17. Find the area of the triangle with $a = 4$, $b = 5$, and $c = 7$.

Solution:

$$s = \frac{4+5+7}{2} = 8$$
$$A = \sqrt{8(8-4)(8-5)(8-7)} = \sqrt{96} = 4\sqrt{6} \text{ square units}$$

21. Find the height of a tree that stands on a hillside of slope 32° (from the horizontal) if from a point 75 feet downhill the angle of elevation to the top of the tree is 48° (see figure).

Solution:
Let $x =$ the height of the hillside.

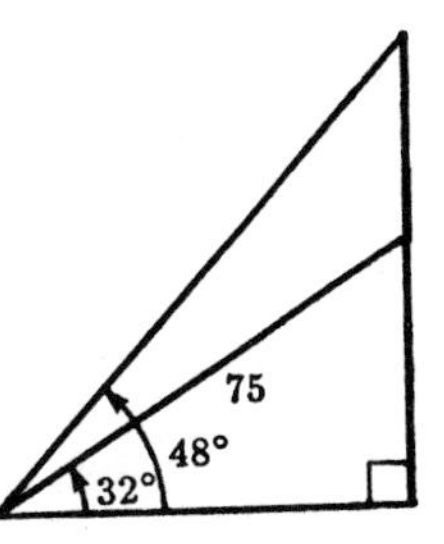

$$\sin 32^\circ = \frac{x}{75}$$
$$x = 75 \sin 32 \approx 39.7439 \text{ feet}$$

Let $y =$ the horizontal distance.

$$y = \sqrt{75^2 - x^2} = \sqrt{75^2 - 39.7439^2} \approx 63.6036 \text{ feet}$$

Let $h =$ the height of the tree.

$$\tan 48^\circ = \frac{x+h}{y}$$
$$\tan 48^\circ = \frac{39.7439 + h}{63.6036}$$
$$h = 63.6036 \tan 48^\circ - 39.7439$$
$$h \approx 31 \text{ feet}$$

23. From a certain distance, the angle of elevation of the top of a building is 17°. At a point 50 meters closer to the building, the angle of elevation is 31°. Approximate the height of the building.

Solution:

$$\tan 17^\circ = \frac{h}{50+y} \Rightarrow h = (50+y)\tan 17^\circ$$

$$\tan 31^\circ = \frac{h}{y} \Rightarrow h = y\tan 31^\circ$$

$$(50+y)\tan 17^\circ = y\tan 31^\circ$$

$$50\tan 17^\circ + y\tan 17^\circ = y\tan 31^\circ$$

$$y(\tan 17^\circ - \tan 31^\circ) = -50\tan 17^\circ$$

$$y = \frac{-50\tan 17^\circ}{\tan 17^\circ - \tan 31^\circ} \approx 51.7959 \text{ m}$$

$$h = y\tan 31^\circ = 51.7959\tan 31^\circ \approx 31.1 \text{ m}$$

27. Find the component form of the vector $\mathbf{v}$ with initial point (0, 10), and terminal point (7, 3).

Solution:

$$\mathbf{v} = \langle 7-0,\ 3-10\rangle = \langle 7,\ -7\rangle$$

33. Find the component form of $4\mathbf{u}-5\mathbf{v}$ and sketch its graph given that $\mathbf{u} = 6\mathbf{i}-5\mathbf{j}$ and $\mathbf{v} = 10\mathbf{i}+3\mathbf{j}$.

Solution:

$$\begin{aligned} 4\mathbf{u}-5\mathbf{v} &= (24\mathbf{i}-20\mathbf{j})-(50\mathbf{i}+15\mathbf{j}) \\ &= -26\mathbf{i}-35\mathbf{j} \\ &= \langle -26,\ -35\rangle \end{aligned}$$

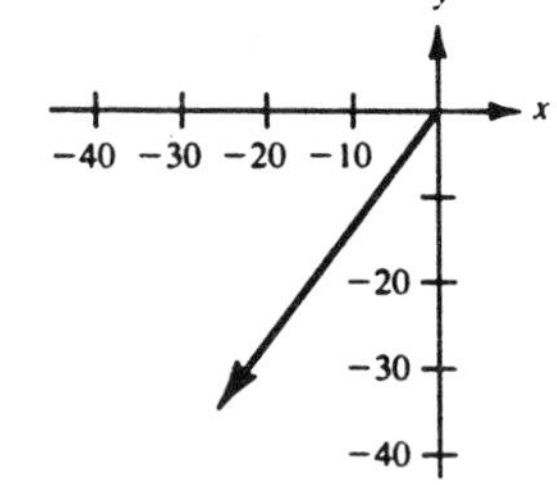

37. Find the direction and magnitude of the resultant of the three forces shown in the figure.

Solution:

$$\|\mathbf{u}\| = 13$$

$$\|\mathbf{v}\| = 5$$

$$\mathbf{u} = 300\left(\frac{5}{13}\mathbf{i} + \frac{12}{13}\mathbf{j}\right)$$

$$\mathbf{v} = 150\left(-\frac{4}{5}\mathbf{i} + \frac{3}{5}\mathbf{j}\right)$$

$$\mathbf{w} = 250(0\mathbf{i} - \mathbf{j})$$

$\tan\beta = \frac{3}{4}$ $\quad\tan\alpha = \frac{12}{5}$

$$\mathbf{r} = \mathbf{u}+\mathbf{v}+\mathbf{w} = \left(\frac{1500}{13} - 120 + 0\right)\mathbf{i} + \left(\frac{3600}{13} + 90 - 250\right)\mathbf{j} = \frac{-60}{13}\mathbf{i} + \frac{1520}{13}\mathbf{j}$$

$$\|\mathbf{r}\| = \sqrt{\left(\frac{-60}{13}\right)^2 + \left(\frac{1520}{13}\right)^2} \approx 117.0 \text{ lb}$$

$$\theta = 180^\circ - \arctan\frac{1520}{60} \approx 92.3^\circ$$

41. Find the trigonometric form of $5 - 5i$.

Solution:

$$
\begin{aligned}
z &= 5 - 5i \\
|z| &= \sqrt{5^2 + (-5)^2} = 5\sqrt{2} \\
\tan\theta &= -\frac{5}{5} = -1, \ \theta \text{ is in Quadrant IV} \\
\theta &= 315^\circ \\
z &= 5\sqrt{2}(\cos 315^\circ + i\sin 315^\circ)
\end{aligned}
$$

45. Write $100(\cos 240^\circ + i\sin 240^\circ)$ in standard form.

Solution:

$$z = 100(\cos 240^\circ + i\sin 240^\circ) = 100\left(-\frac{1}{2} - \frac{\sqrt{3}}{2}i\right) = -50 - 50\sqrt{3}i$$

51. Use DeMoivre's Theorem to find the indicated power of the following complex number. Express the result in standard form.

$$\left[5\left(\cos\frac{\pi}{12} + i\sin\frac{\pi}{12}\right)\right]^4$$

Solution:

$$\left[5\left(\cos\frac{\pi}{12} + i\sin\frac{\pi}{12}\right)\right]^4 = 625\left(\cos\frac{\pi}{3} + i\sin\frac{\pi}{3}\right) = \frac{625}{2} + \frac{625\sqrt{3}}{2}i$$

55. Use DeMoivre's Theorem to find the sixth roots of $-729i$.

Solution:

Find the sixth roots of $-729i = 729\left(\cos\frac{3\pi}{2} + i\sin\frac{3\pi}{2}\right)$.

$$
\begin{aligned}
&n = 6 \\
&k = 0: \ 3\left(\cos\frac{\pi}{4} + i\sin\frac{\pi}{4}\right) \\
&k = 1: \ 3\left(\cos\frac{7\pi}{12} + i\sin\frac{7\pi}{12}\right) \\
&k = 2: \ 3\left(\cos\frac{11\pi}{12} + i\sin\frac{11\pi}{12}\right) \\
&k = 3: \ 3\left(\cos\frac{5\pi}{4} + i\sin\frac{5\pi}{4}\right) \\
&k = 4: \ 3\left(\cos\frac{19\pi}{12} + i\sin\frac{19\pi}{12}\right) \\
&k = 5: \ 3\left(\cos\frac{23\pi}{12} + i\sin\frac{23\pi}{12}\right)
\end{aligned}
$$

Practice Test for Chapter 7

For Exercises 1 and 2, use the Law of Sines to find the remaining sides and angles of the triangle.

1. $A = 40°,\ B = 12°,\ b = 100$

2. $C = 150°,\ a = 5,\ c = 20$

3. Find the area of the triangle: $a = 3,\ b = 5,\ C = 130°$.

4. Determine the number of solutions to the triangle: $a = 10,\ b = 35,\ A = 22.5°$.

For Exercises 5 and 6, use the Law of Cosines to find the remaining sides and angles of the triangle.

5. $a = 49,\ b = 53,\ c = 38$

6. $C = 29°,\ a = 100,\ b = 300$

7. Use Heron's Formula to find the area of the triangle: $a = 4.1,\ b = 6.8,\ c = 5.5$.

8. A ship travels 40 miles due east, then adjusts its course 12° southward. After traveling 70 miles in that direction, how far is the ship from its point of departure?

9. $\mathbf{w}$ is $4\mathbf{u} - 7\mathbf{v}$ where $\mathbf{u} = 3\mathbf{i} + \mathbf{j}$ and $\mathbf{v} = -\mathbf{i} + 2\mathbf{j}$. Find $\mathbf{w}$.

10. Find a unit vector in the direction of $\mathbf{v} = 5\mathbf{i} - 3\mathbf{j}$.

11. Find the angle between $\mathbf{u} = 6\mathbf{i} + 5\mathbf{j}$ and $\mathbf{v} = 2\mathbf{i} - 3\mathbf{j}$.

12. $\mathbf{v}$ is a vector of magnitude 4 making an angle of 30° with the positive x-axis. Find $\mathbf{v}$.

13. Give the trigonometric form of $z = 5 - 5i$.

14. Give the standard form of $z = 6(\cos 225° + i \sin 225°)$.

15. Multiply $[7(\cos 23° + i \sin 23°)][4(\cos 7° + i \sin 7°)]$.

16. Divide $\dfrac{9\left(\cos \frac{5\pi}{4} + i \sin \frac{5\pi}{4}\right)}{3(\cos \pi + i \sin \pi)}$.

17. Find $(2 + 2i)^8$.

18. Find the cube roots of $8\left(\cos \frac{\pi}{3} + i \sin \frac{\pi}{3}\right)$.

19. Find all the solutions to $x^3 + 125 = 0$.

20. Find all the solutions to $x^4 + i = 0$.

CHAPTER 8

Systems of Equations and Inequalities

SECTION 8.1

Systems of Equations

- You should be able to solve systems of equations by the method of substitution.
- You should be able to solve systems of equations by graphically finding points of intersection.

Solutions to Selected Exercises

5. Solve the following system by the method of substitution.

$$\begin{aligned} x + 3y &= 15 \\ x^2 + y^2 &= 25 \end{aligned}$$

Solution:

$x + 3y = 15 \Rightarrow x = 15 - 3y$

$$\begin{aligned} x^2 + y^2 &= 25 \\ (15 - 3y)^2 + y^2 &= 25 \\ 225 - 90y + 10y^2 &= 25 \\ 10y^2 - 90y + 200 &= 0 \\ 10(y^2 - 9y + 20) &= 0 \\ 10(y - 4)(y - 5) &= 0 \end{aligned}$$

$$\begin{aligned} y &= 4 \quad \text{or} \quad y = 5 \\ x &= 3 \qquad\quad\;\; x = 0 \end{aligned}$$

Solutions: (3, 4), (0, 5)

9. Solve the following system by the method of substitution.

$$\begin{aligned} x - 3y &= -4 \\ x^2 - y^3 &= 0 \end{aligned}$$

Solution:

$x - 3y = -4 \Rightarrow x = 3y - 4$

$$\begin{aligned} x^2 - y^3 &= 0 \\ (3y-4)^2 - y^3 &= 0 \\ y^3 - 9y^2 + 24y - 16 &= 0 \\ (y-1)(y-4)^2 &= 0 \end{aligned}$$

$y = 1$ or $y = 4$

$x = -1$ $\quad$ $x = 8$

Solutions: $(-1, 1)$, $(8, 4)$

13. Solve the following system by the method of substitution.

$$\begin{aligned} 2x - y + 2 &= 0 \\ 4x + y - 5 &= 0 \end{aligned}$$

Solution:

$2x - y + 2 = 0 \Rightarrow y = 2x + 2$

$$\begin{aligned} 4x + y - 5 &= 0 \\ 4x + (2x + 2) - 5 &= 0 \\ 6x &= 3 \\ x &= \tfrac{1}{2} \\ y &= 3 \end{aligned}$$

Solution: $(\frac{1}{2}, 3)$

17. Solve the following system by the method of substitution.

$$\begin{aligned} \tfrac{1}{5}x + \tfrac{1}{2}y &= 8 \\ x + \quad y &= 20 \end{aligned}$$

Solution:

$\frac{1}{5}x + \frac{1}{2}y = 8 \Rightarrow 2x + 5y = 80$

$x + y = 20 \Rightarrow y = 20 - x$

$$\begin{aligned} 2x + 5(20 - x) &= 80 \\ -3x &= -20 \\ x &= \tfrac{20}{3} \\ y &= \tfrac{40}{3} \end{aligned}$$

Solution: $(\frac{20}{3}, \frac{40}{3})$

23. Solve the following system by the method of substitution.

$$3x - 7y + 6 = 0$$
$$x^2 - y^2 = 4$$

Solution:

$3x - 7y + 6 = 0 \Rightarrow x = \dfrac{7y - 6}{3}$

$$x^2 - y^2 = 4$$
$$\left(\frac{7y-6}{3}\right)^2 - y^2 = 4$$
$$\frac{49y^2 - 84y + 36}{9} - y^2 = 4$$
$$49y^2 - 84y + 36 - 9y^2 = 36$$
$$40y^2 - 84y = 0$$
$$4y(10y - 21) = 0$$

$y = 0$ or $y = \dfrac{21}{10}$

$x = -2$ $\qquad x = \dfrac{29}{10}$

Solutions: $(-2,\ 0)$, $\left(\dfrac{29}{10},\ \dfrac{21}{10}\right)$

27. Solve the following system by the method of substitution.

$$y = x^4 - 2x^2 + 1$$
$$y = 1 - x^2$$

Solution:

$$x^4 - 2x^2 + 1 = 1 - x^2$$
$$x^4 - x^2 = 0$$
$$x^2(x^2 - 1) = 0$$
$$x^2(x + 1)(x - 1) = 0$$

$x = 0$ or $x = -1$ or $x = 1$

$y = 1$ $\qquad y = 0$ $\qquad y = 0$

Solutions: $(0,\ 1)$, $(\pm 1,\ 0)$

29. Solve the following system by the method of substitution.

$$xy - 1 = 0$$
$$2x - 4y + 7 = 0$$

Solution:

$$xy - 1 = 0$$
$$2x - 4y + 7 = 0 \Rightarrow x = \frac{4y - 7}{2}$$

$$\left(\frac{4y-7}{2}\right)y - 1 = 0$$
$$4y^2 - 7y - 2 = 0$$
$$(4y + 1)(y - 2) = 0$$

$$y = -\frac{1}{4} \quad \text{or} \quad y = 2$$
$$x = -4 \qquad\qquad x = \frac{1}{2}$$

Solutions: $\left(-4, -\frac{1}{4}\right)$, $\left(\frac{1}{2}, 2\right)$

33. Find all points of intersection of the graphs of the given pair of equations. [*Hint:* A graphical approach, as demonstrated in Example 5, may be helpful.]

$$2x - y + 3 = 0$$
$$x^2 + y^2 - 4x = 0$$

Solution:

$$2x - y + 3 = 0 \Rightarrow y = 2x + 3$$
$$x^2 + y^2 - 4x = 0 \Rightarrow (x - 2)^2 + y^2 = 4$$

No points of intersection

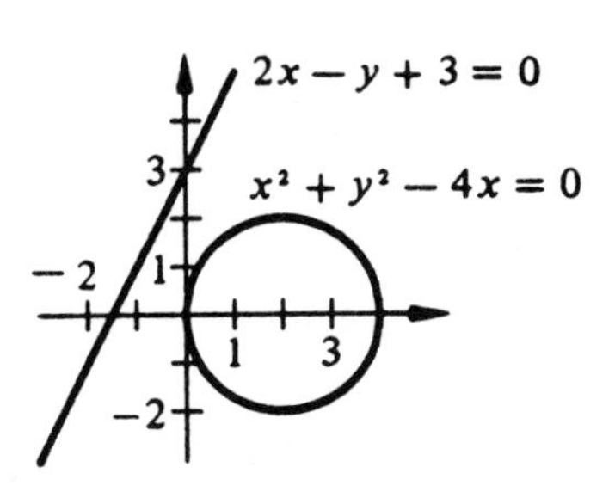

39. Find x, y, and λ satisfying the given system. These systems arise in certain optimization problems in calculus, and λ is called a Lagrange Multiplier. [*Hint:* You can reduce each system to a system of two equations in two variables by solving for λ in the first equation and substituting into the second equation.]

$$y + \lambda = 0$$
$$x + \lambda = 0$$
$$x + y - 10 = 0$$

Solution:

$$\begin{aligned} y + \lambda = 0 &\Rightarrow y = -\lambda \\ x + \lambda = 0 &\Rightarrow x = -\lambda \\ x + y - 10 = 0 &\Rightarrow -\lambda - \lambda - 10 = 0 \\ &\qquad -2\lambda = 10 \\ &\qquad \lambda = -5 \\ &\qquad x = y = 5 \end{aligned}$$

Solution: $x = 5,\ y = 5,\ \lambda = -5$

45. A person is setting up a small business and has invested \$16,000 to produce an item that will sell for \$5.95. If each unit can be produced for \$3.45, how many units must be sold to break even?

Solution:

Let $x =$ the number of units

$$C = 3.45x + 16{,}000$$

$$R = 5.95x$$

To break even: $R = C$

$$\begin{aligned} 5.95x &= 3.45x + 16{,}000 \\ 2.5x &= 16{,}000 \\ x &= 6400 \text{ units} \end{aligned}$$

49. What are the dimensions of a rectangle if its perimeter is 40 miles and its area is 96 square miles?

Solution:

Let $l =$ the length of the rectangle and $w =$ the width of the rectangle.

$$\begin{aligned} \text{Perimeter: } 2l + 2w &= 40 \quad \Rightarrow w = 20 - l \\ \text{Area: } lw &= 96 \quad \Rightarrow l(20 - l) = 96 \\ 20l - l^2 &= 96 \\ 0 &= l^2 - 20l + 96 \\ 0 &= (l - 8)(l - 12) \end{aligned}$$

$l = 8$ or $l = 12$

$w = 12$ $\quad w = 8$

The dimensions are 12 miles by 8 miles.

51. Find the initial velocity v_0 and the time t of travel of a projectile that is thrown from the point (0, 0) at an angle of inclination $\theta = 45°$ and that lands at the point (144, 0). The path of the projectile is described by the model $x = (v_0 \cos\theta)t$ and $y = (v_0 \sin\theta)t - 16t^2$ where t is time in seconds, x and y are distances in feet.

Solution:

$x = (v_0 \cos\theta)t$ and $y = (v_0 \sin\theta)t - 16t^2$

$\theta = 45°,\ x = 144,\ y = 0$

Find v_0 and t:

$$144 = v_0\left(\frac{\sqrt{2}}{2}\right)t$$

and

$$0 = v_0\left(\frac{\sqrt{2}}{2}\right)t - 16t^2$$

$$0 = 144 - 16t^2$$

$$0 = 16(9 - t^2)$$

$$t = 3 \text{ sec}$$

$$v_0 = 144\left(\frac{2}{\sqrt{2}}\right)\frac{1}{3} = 48\sqrt{2} \text{ ft/sec}$$

SECTION 8.2

Systems of Linear Equations in Two Variables

- You should be able to solve a linear system by the method of elimination.
- You should know that for a system of two linear equations, one of the following is true.
  (a) There are infinitely many solutions; the lines are identical.
  (b) There is no solution; the lines are parallel.
  (c) There is one solution; the lines intersect at one point.

Solutions to Selected Exercises

5. Solve the linear system by elimination. Identify and label each line with the appropriate equation.

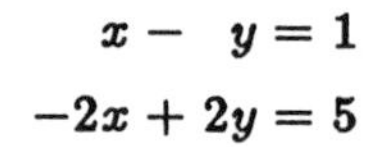

$$\begin{aligned} x - \quad y &= 1 \\ -2x + 2y &= 5 \end{aligned}$$

Solution:

$$\begin{aligned} x - \quad y = 1 &\Rightarrow \quad 2x - 2y = 2 \\ -2x + 2y = 5 &\Rightarrow \underline{-2x + 2y = 5} \\ & \qquad\qquad 0 = 7 \end{aligned}$$

Inconsistent; no solution

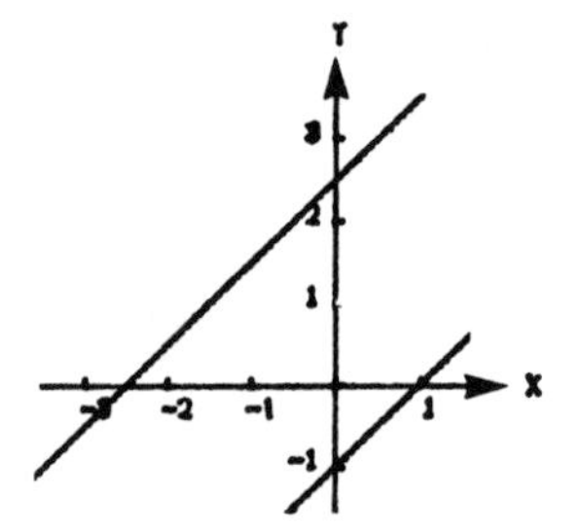

9. Solve the linear system by elimination. Identify and label each line with the appropriate equation.

$$\begin{aligned} 9x - 3y &= -1 \\ 3x + 6y &= -5 \end{aligned}$$

Solution:

$$\begin{aligned} 9x - 3y = -1 &\Rightarrow 18x - 6y = -2 \\ 3x + 6y = -5 &\Rightarrow \underline{\ \ 3x + 6y = -5} \\ & \quad\ 21x \qquad = -7 \\ & \qquad\qquad\ x = -\tfrac{1}{3} \\ & \qquad\qquad\ y = -\tfrac{2}{3} \end{aligned}$$

Consistent; one solution

Solution: $\left(-\frac{1}{3}, -\frac{2}{3}\right)$

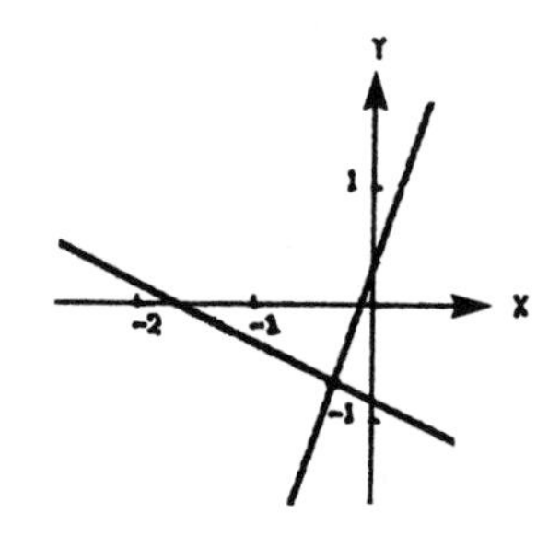

13. Solve the system by elimination.

$$\begin{aligned} 2x + 3y &= 18 \\ 5x - y &= 11 \end{aligned}$$

Solution:

$$\begin{aligned} 2x + 3y = 18 &\Rightarrow 2x + 3y = 18 \\ 5x - y = 11 &\Rightarrow \underline{15x - 3y = 33} \\ & \quad\; 17x \qquad\;\; = 51 \\ & \qquad\qquad\quad x = 3 \\ & \qquad\qquad\quad y = 4 \end{aligned}$$

Consistent; one solution

Solution: (3, 4)

17. Solve the system by elimination.

$$\begin{aligned} 2u + v &= 120 \\ u + 2v &= 120 \end{aligned}$$

Solution:

$$\begin{aligned} 2u + v = 120 &\Rightarrow -4u - 2v = -240 \\ u + 2v = 120 &\Rightarrow \underline{\qquad u + 2v = 120} \\ & \quad\; -3u \qquad\;\; = -120 \\ & \qquad\qquad\quad u = 40 \\ & \qquad\qquad\quad v = 40 \end{aligned}$$

Consistent; one solution

Solution: (40, 40)

23. Solve the system by elimination.

$$\begin{aligned} \frac{x+3}{4} + \frac{y-1}{3} &= 1 \\ 2x - y &= 12 \end{aligned}$$

Solution:

$$\frac{x+3}{4}+\frac{y-1}{3}=1 \Rightarrow 3(x+3)+4(y-1)=12 \Rightarrow \begin{aligned} 3x+4y &= 7 \end{aligned}$$

$$2x-y=12 \Rightarrow \quad \begin{aligned} 8x-4y &= 48 \\ \hline 11x \quad &= 55 \\ x &= 5 \\ y &= -2 \end{aligned}$$

Consistent; one solution

Solution: $(5,\ -2)$

25. Solve the system by elimination.

$$\begin{aligned} 2.5x - \ \ 3y &= 1.5 \\ 10x - 12y &= 6 \end{aligned}$$

Solution:

$$\begin{aligned} 2.5x - \ \ 3y = 1.5 &\Rightarrow 25x - 30y = 15 \Rightarrow & 5x - 6y &= 3 \\ 10x - 12y = 6 &\Rightarrow 5x - \ \ 6y = \ \ 3 \Rightarrow & -5x + 6y &= -3 \\ \hline & & 0 &= 0 \end{aligned}$$

Consistent; infinite solutions of the form $\left(a,\ \frac{5}{6}a - \frac{1}{2}\right)$

29. Solve the system by elimination.

$$\begin{aligned} 4b + \ \ 3m &= 3 \\ 3b + 11m &= 13 \end{aligned}$$

Solution:

$$\begin{aligned} 4b + \ \ 3m = \ \ 3 &\Rightarrow & 44b + 33m &= 33 \\ 3b + 11m = 13 &\Rightarrow & -9b - 33m &= -39 \\ \hline & & 35b \qquad &= -6 \\ & & b &= -\tfrac{6}{35} \\ & & m &= \tfrac{43}{35} \end{aligned}$$

Consistent; one solution

Solution: $\left(-\frac{6}{35},\ \frac{43}{35}\right)$

33. Find two numbers whose sum is 20 and difference is 2.

Solution:

Let $x =$ one number and $y =$ the other number.

$$\begin{array}{rl} x + y &= 20 \\ x - y &= 2 \\ \hline 2x \quad &= 22 \\ x &= 11 \\ y &= 9 \end{array}$$

The two numbers are 11 and 9.

37. Ten gallons of a 30% acid solution are obtained by mixing a 20% solution with a 50% solution. How much of each must be used?

Solution:

Let $x =$ amount of 20% solution and $y =$ amount of 50% solution.

$$\begin{array}{rrl} x + \quad y = \quad 10 \Rightarrow & -2x - 2y &= -20 \\ 0.2x + 0.5y = 0.3(10) \Rightarrow & 2x + 5y &= 30 \\ \hline & 3y &= 10 \\ & y &= \frac{10}{3} \\ & x &= \frac{20}{3} \end{array}$$

Solution: $\frac{20}{3}$ gal 20% solution, $\frac{10}{3}$ gal 50% solution

41. The perimeter of a rectangle is 40 feet and the length is 4 feet greater than the width. Find the dimensions of the rectangle.

Solution:

$$\begin{array}{rrl} 2l + 2w = 40 \quad \Rightarrow & l + w &= 20 \\ l = w + 4 \Rightarrow & l - w &= 4 \\ \hline & 2l \quad &= 24 \\ & l &= 12 \\ & w &= 8 \end{array}$$

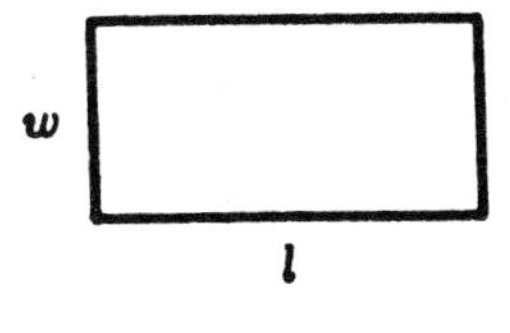

Solution: Length = 12 ft, width = 8 ft

45. (a) Find the least squares regression line, and (b) plot the given points and sketch the least squares regression line on the same axes. The *least squares regression line*, $y = ax + b$, for the points (x_1, y_1), (x_2, y_2), ..., (x_n, y_n) is obtained by solving the following system of linear equations for a and b.

$$nb + \left(\sum_{i=1}^{n} x_i\right)a = \sum_{i=1}^{n} y_i$$

$$\left(\sum_{i=1}^{n} x_i\right)b + \left(\sum_{i=1}^{n} x_i^2\right)a = \sum_{i=1}^{n} x_i y_i$$

Use the points (0, 4), (1, 3), (1, 1), (2, 0).

Solution:

(a) $n = 4$

$$\sum_{i=1}^{4} x_i = 4$$

$$\sum_{i=1}^{4} x_i^2 = 6$$

$$\sum_{i=1}^{4} y_i = 8$$

$$\sum_{i=1}^{4} x_i y_i = 4$$

$$\begin{aligned} 4b + 4a &= 8 \\ -(4b + 6a &= 4) \\ \hline -2a &= 4 \\ a &= -2 \\ b &= 4 \end{aligned}$$

$y = -2x + 4$

(b)

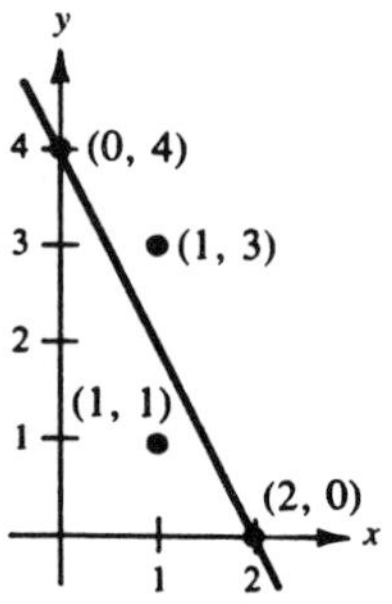

SECTION 8.3

Systems of Linear Equations in More Than Two Variables

- You should know the operations that lead to equivalent systems of equations:
  (a) Interchange any two equations.
  (b) Multiply all terms of an equation by a nonzero constant.
  (c) Replace an equation by the sum of itself and a constant multiple of any other equation in the system.
- You should be able to use the method of elimination.

Solutions to Selected Exercises

3. Solve the following system of linear equations.

$$\begin{aligned} 4x + y - 3z &= 11 \\ 2x - 3y + 2z &= 9 \\ x + y + z &= -3 \end{aligned}$$

Solution:

$$\begin{array}{r} 4x + y - 3z = 11 \\ \underline{-4x + 6y - 4z = -18} \\ 7y - 7z = -7 \\ y - z = -1 \end{array} \qquad \begin{array}{r} 2x - 3y + 2z = 9 \\ \underline{-2x - 2y - 2z = 6} \\ -5y = 15 \\ y = -3 \\ z = y + 1 \\ z = -2 \\ x = 2 \end{array}$$

Solution: $(2, -3, -2)$

7. Solve the following system of linear equations.

$$\begin{aligned} 3x - 2y + 4z &= 1 \\ x + y - 2z &= 3 \\ 2x - 3y + 6z &= 8 \end{aligned}$$

Solution:

$$\begin{array}{r} 3x - 2y + 4z = 1 \\ -3x - 3y + 6z = -9 \\ \hline -5y + 10z = -8 \end{array} \qquad \begin{array}{r} -2x - 2y + 4z = -6 \\ 2x - 3y + 6z = 8 \\ \hline -5y + 10z = 2 \end{array} \qquad \begin{array}{r} -5y + 10z = -8 \\ 5y - 10z = -2 \\ \hline 0 = -10 \end{array}$$

Inconsistent; no solution

11. Solve the following system of linear equations.

$$\begin{aligned} x + 2y - 7z &= -4 \\ 2x + y + z &= 13 \\ 3x + 9y - 36z &= -33 \end{aligned}$$

Solution:

$$\begin{array}{r} 2x + 4y - 14z = -8 \\ -2x - y - z = -13 \\ \hline 3y - 15z = -21 \\ y - 5z = -7 \end{array} \qquad \begin{array}{r} 3x + 6y - 21z = -12 \\ -3x - 9y + 36z = 33 \\ \hline -3y + 15z = 21 \\ y - 5z = -7 \end{array}$$

Let $z = a$. Then

$$\begin{aligned} y &= 5a - 7 \\ x &= -4 - 2(5a - 7) + 7a = -3a + 10 \end{aligned}$$

Solution: $(-3a + 10,\ 5a - 7,\ a)$, a is any real number

15. Solve the following system of linear equations.

$$\begin{aligned} x - 2y + 5z &= 2 \\ 3x + 2y - z &= -2 \end{aligned}$$

Solution:

$$\begin{array}{r} x - 2y + 5z = 2 \\ 3x + 2y - z = -2 \\ \hline 4x \quad\quad + 4z = 0 \end{array} \Rightarrow x = -z$$

Let $z = a$. Then

$$\begin{aligned} x &= -a \\ y &= \tfrac{1}{2}[-a + 5a - 2] = 2a - 1 \end{aligned}$$

Solution: $(-a,\ 2a - 1,\ a)$, a is any real number.

21. Solve the following system of linear equations.

$$\begin{aligned} x \qquad + 4z &= 1 \\ x + y + 10z &= 10 \\ 2x - y + 2z &= -5 \end{aligned}$$

Solution:

$$\begin{array}{r} -x \qquad - 4z = -1 \\ x + y + 10z = 10 \\ \hline y + 6z = 9 \end{array} \qquad \begin{array}{r} -2x \qquad - 8z = -2 \\ 2x - y + 2z = -5 \\ \hline - y - 6z = -7 \end{array} \qquad \begin{array}{r} y + 6z = 9 \\ -y - 6z = -7 \\ \hline 0 = 2 \end{array}$$

Inconsistent; no solution

23. Solve the following system of linear equations.

$$\begin{aligned} 4x + 3y + 17z &= 0 \\ 5x + 4y + 22z &= 0 \\ 4x + 2y + 19z &= 0 \end{aligned}$$

Solution:

$$\begin{array}{r} 20x + 15y + 85z = 0 \\ -20x - 16y - 88z = 0 \\ \hline -y - 3z = 0 \end{array} \qquad \begin{array}{r} 4x + 3y + 17z = 0 \\ -4x - 2y - 19z = 0 \\ \hline y + 2z = 0 \\ -y - 3z = 0 \\ \hline -z = 0 \end{array}$$

$z = 0, \; y = 0, \; x = 0$

Solution: $(0, 0, 0)$

27. Find the equation of the parabola $y = ax^2 + bx + c$ that passes through the points $(0, -4)$, $(1, 1)$, and $(2, 10)$.

Solution:

$$\begin{aligned} -4 &= a(0)^2 + b(0) + c \Rightarrow -4 = c \\ 1 &= a(1)^2 + b(1) + c \Rightarrow 1 = a + b + c \\ 10 &= a(2)^2 + b(2) + c \Rightarrow 10 = 4a + 2b + c \end{aligned}$$

$$\begin{aligned} 1 &= a + b - 4 \Rightarrow a + b = 5 \Rightarrow -a - b = -5 \\ 10 &= 4a + 2b - 4 \Rightarrow 4a + 2b = 14 \Rightarrow \underline{2a + b = 7} \\ & \qquad a = 2 \\ & \qquad b = 3 \end{aligned}$$

Thus, $y = 2x^2 + 3x - 4$.

31. Find the equation of the circle $x^2 + y^2 + Dx + Ey + F = 0$ that passes through the points $(0, 0)$, $(2, -2)$, and $(4, 0)$.

Solution:

$$\begin{aligned} (0)^2 + (0)^2 + D(0) + E(0) + F = 0 &\Rightarrow F = 0 \\ (2)^2 + (-2)^2 + D(2) + E(-2) + F = 0 &\Rightarrow 2D - 2E + F = -8 \\ (4)^2 + (0)^2 + D(4) + E(0) + F = 0 &\Rightarrow 4D + F = -16 \end{aligned}$$

$$\begin{aligned} 2D - 2E + 0 &= -8 \Rightarrow D - E = -4 \\ 4D + 0 &= -16 \Rightarrow D = -4 \Rightarrow E = 0 \end{aligned}$$

Thus, $x^2 + y^2 - 4x + 0y + 0 = 0$

$x^2 + y^2 - 4x = 0.$

35. Find a, v_0, and s_0 in the position equation $s = \frac{1}{2}at^2 + v_0t + s_0$.

At $t = 1$ second, $s = 128$ feet.
At $t = 2$ seconds, $s = 80$ feet.
At $t = 3$ seconds, $s = 0$ feet.

Solution:

$s = \frac{1}{2}at^2 + v_0t + s_0$

$(1, 128)$, $(2, 80)$, $(3, 0)$

$$\begin{aligned} 128 &= \tfrac{1}{2}a + v_0 + s_0 \Rightarrow a + 2v_0 + 2s_0 = 256 \\ 80 &= 2a + 2v_0 + s_0 \Rightarrow 2a + 2v_0 + s_0 = 80 \\ 0 &= \tfrac{9}{2}a + 3v_0 + s_0 \Rightarrow 9a + 6v_0 + 2s_0 = 0 \end{aligned}$$

$$\begin{aligned} 2a + 4v_0 + 4s_0 &= 512 \\ -2a - 2v_0 - s_0 &= -80 \\ \hline 2v_0 + 3s_0 &= 432 \end{aligned}$$

$$\begin{aligned} 18a + 18v_0 + 9s_0 &= 720 \\ -18a - 12v_0 - 4s_0 &= 0 \\ \hline 6v_0 + 5s_0 &= 720 \\ -6v_0 - 9s_0 &= -1296 \\ \hline -4s_0 &= -576 \\ s_0 &= 144 \\ v_0 &= 0 \\ a &= -32 \end{aligned}$$

Thus, $s = \frac{1}{2}(-32)t^2 + (0)t + 144$
$= -16t^2 + 144.$

39. Use a system of linear equations to decompose the following rational fraction into partial fractions. (See Example 9 and Section 5.2.)

$$\frac{1}{x^3 - x} = \frac{A}{x} + \frac{B}{x-1} + \frac{C}{x+1}$$

Solution:

$$\begin{aligned} \frac{1}{x^3 - x} &= \frac{A}{x} + \frac{B}{x-1} + \frac{C}{x+1} \\ 1 &= A(x+1)(x-1) + Bx(x+1) + Cx(x-1) \\ 1 &= Ax^2 - A + Bx^2 + Bx + Cx^2 - Cx \\ 1 &= (A+B+C)x^2 + (B-C)x - A \end{aligned}$$

By equating coefficients, we have

$$\begin{aligned} 0 &= A + B + C \\ 0 &= B - C \\ 1 &= -A \end{aligned} \quad \Rightarrow \quad \begin{aligned} A &= -1 \\ B + C &= 1 \\ B - C &= 0 \\ \hline 2B &= 1 \Rightarrow B = \tfrac{1}{2} \\ C &= \tfrac{1}{2} \end{aligned}$$

$$\begin{aligned} \frac{A}{x} + \frac{B}{x-1} + \frac{C}{x+1} &= \frac{-1}{x} + \frac{1/2}{x-1} + \frac{1/2}{x+1} \\ &= \frac{1}{2}\left(\frac{-2}{x} + \frac{1}{x-1} + \frac{1}{x+1}\right) \end{aligned}$$

43. A small company that manufactures products A and B has an order for 15 units of product A and 16 units of product B. The company has trucks of three different sizes that can haul the products, as shown in the following table.

	Product	
Truck	A	B
Large	6	3
Medium	4	4
Small	0	3

How many trucks of each size are needed to deliver the order? (Give *two* possible solutions.)

Solution:

Possible solutions:

(1) 4 medium trucks
(2) 2 large trucks, 1 medium truck, 2 small trucks
(3) 3 large trucks, 1 medium truck, 1 small truck
(4) 3 large trucks, 3 small trucks

47. (a) Find the least squares regression parabola, then (b) plot the given points and sketch the least squares parabola on the same axes. The least squares regression parabola, $y = ax^2 + bx + c$, for the points (x_1, y_1), (x_2, y_2), $\ldots$, (x_n, y_n) is obtained by solving the following system of linear equations for a, b, and c.

$$\begin{aligned}
nc + \left(\sum_{i=1}^{n} x_i\right)b + \left(\sum_{i=1}^{n} x_i^2\right)a &= \sum_{i=1}^{n} y_1 \\
\left(\sum_{i=1}^{n} x_i\right)c + \left(\sum_{i=1}^{n} x_i^2\right)b + \left(\sum_{i=1}^{n} x_i^3\right)a &= \sum_{i=1}^{n} x_i y_i \\
\left(\sum_{i=1}^{n} x_i^2\right)c + \left(\sum_{i=1}^{n} x_i^3\right)b + \left(\sum_{i=1}^{n} x_i^4\right)a &= \sum_{i=1}^{n} x_i^2 y_i
\end{aligned}$$

Use the points (0, 0), (2, 2), (3, 6), (4, 12).

Solution:

(a)

$$\begin{aligned}
n &= 4 & \sum y_i &= 20 \\
\sum x_i &= 9 & \sum x_i^3 &= 99 \\
\sum x_i^2 &= 29 & \sum x_i y_i &= 70 \\
\sum x_i^4 &= 353 & & \\
\sum x_i^2 y_i &= 254 & &
\end{aligned}$$

$$\begin{aligned}
353a + 99b + 29c &= 254 \\
99a + 29b + 9c &= 70 \\
29a + 9b + 4c &= 20
\end{aligned}$$

By solving, we get $a = 1$, $b = -1$, $c = 0$. Thus, $y = x^2 - x$.

(b)

SECTION 8.4

Systems of Inequalities

- You should be able to sketch the graph of an inequality in two variables:
  (a) Replace the inequality with an equal sign and graph the equation. Use a dashed line for $<$ or $>$, a solid line for $\leq$ or $\geq$.
  (b) Test a point in each region formed by the graph. If the point satisfies the inequality, shade the whole region.

Solutions to Selected Exercises

5. Match $x^2 + y^2 < 4$ with its graph.

Solution:
Since $x^2 + y^2 = 4$ is a circle with center $(0, 0)$ and radius $r = 2$, it matches graph a.

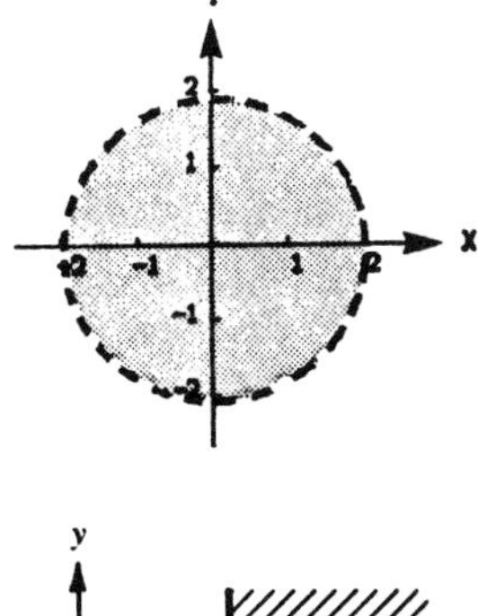

9. Sketch the graph of $x \geq 2$.

Solution:
Using a solid line, sketch the graph of the vertical line $x = 2$. Test point $(3, 0)$. Shade the half-plane to the right of $x = 2$.

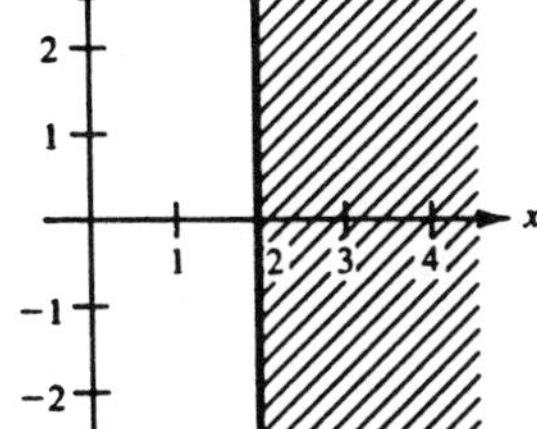

13. Sketch the graph of $y < 2 - x$.

Solution:
Using a dashed line, graph $x + y = 2$, and then shade the half-plane below the line. (Use $(0, 0)$ as a test point.)

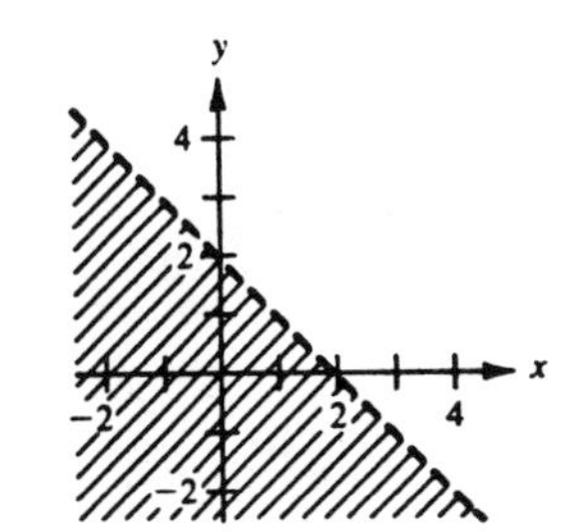

17. Sketch the graph of $(x+1)^2 + (y-2)^2 < 9$.

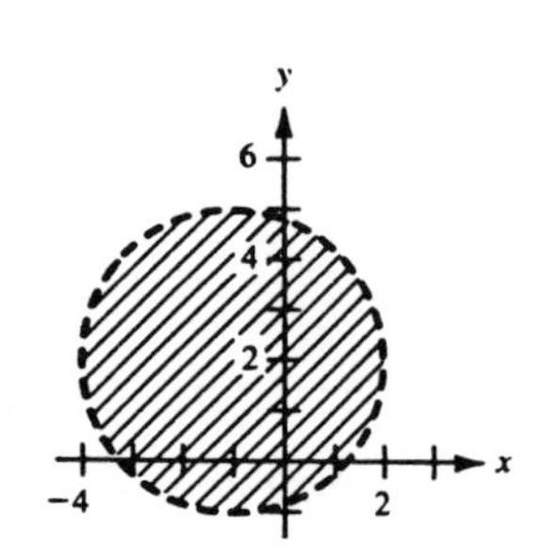

Solution:

Using a dashed line, sketch the circle $(x+1)^2 + (y-2)^2 = 9$.
Center: $(-1,\ 2)$
Radius: 3
Test Point: $(0,\ 0)$. Shade inside of circle.

21. Sketch the graph of the solution of the system of inequalities.

$$\begin{aligned} x + y &\le 1 \\ -x + y &\le 1 \\ y &\ge 0 \end{aligned}$$

Solution:

First, find the points of intersection of each pair of equations.

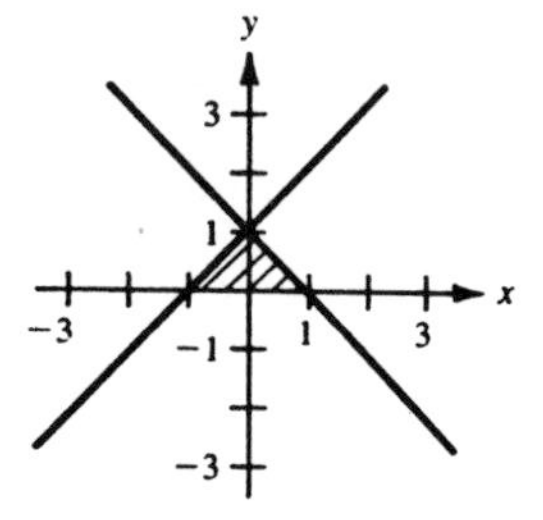

Vertex A	Vertex B	Vertex C
$x + y = 1$	$x + y = 1$	$-x + y = 1$
$-x + y = 1$	$y = 0$	$y = 0$
$(0,\ 1)$	$(1,\ 0)$	$(-1,\ 0)$

25. Sketch the graph of the solution of the system of inequalities.

$$\begin{aligned} -3x + 2y &< 6 \\ x + 4y &> -2 \\ 2x + y &< 3 \end{aligned}$$

Solution:

First, find the points of intersection of each pair of equations.

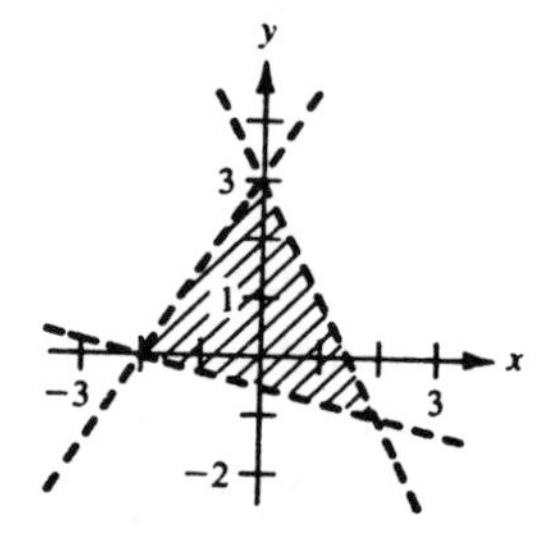

Vertex A	Vertex B	Vertex C
$-3x + 2y = 6$	$-3x + 2y = 6$	$x + 4y = -2$
$x + 4y = -2$	$2x + y = 3$	$2x + y = 3$
$(-2,\ 0)$	$(0,\ 3)$	$(2,\ -1)$

29. Sketch the graph of the solution of the system of inequalities.

$$\begin{aligned} x &\ge 1 \\ x - 2y &\le 3 \\ 3x + 2y &\ge 9 \\ x + y &\le 6 \end{aligned}$$

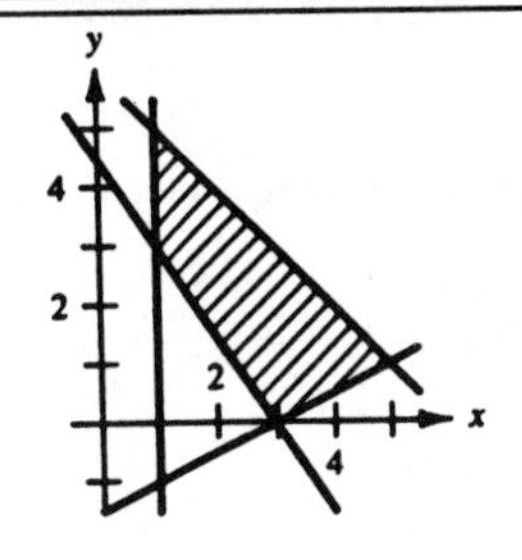

Solution:
The vertices of the region are
(1, 5), (1, 3), (3, 0), and (5, 1).

33. Sketch the graph of the solution of the system of inequalities.

$$x > y^2$$
$$x < y + 2$$

Solution:
Points of intersection:

$$y^2 = y + 2$$
$$y^2 - y - 2 = 0$$
$$(y + 1)(y - 2) = 0$$
$$y = -1, \ y = 2$$
$$(1, -1), \ (4, 2)$$

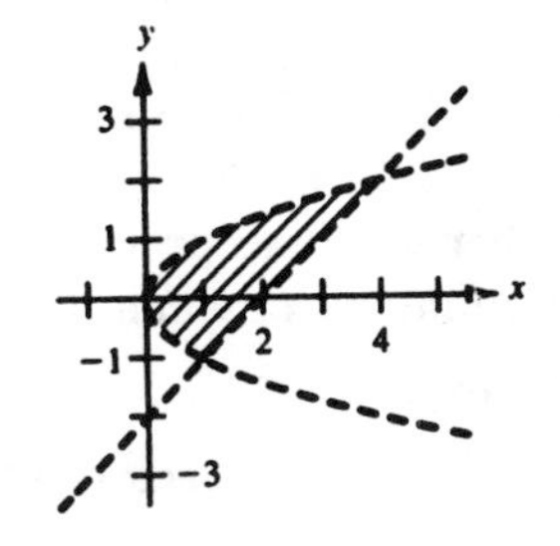

37. Sketch the graph of the solution of the system of inequalities.

$$y < x^3 - 2x + 1$$
$$y > -2x$$
$$x \leq 1$$

Solution:
Points of intersection:

$$x^3 - 2x + 1 = -2x$$
$$x^3 + 1 = 0$$
$$x = -1$$
$$(-1, 2)$$

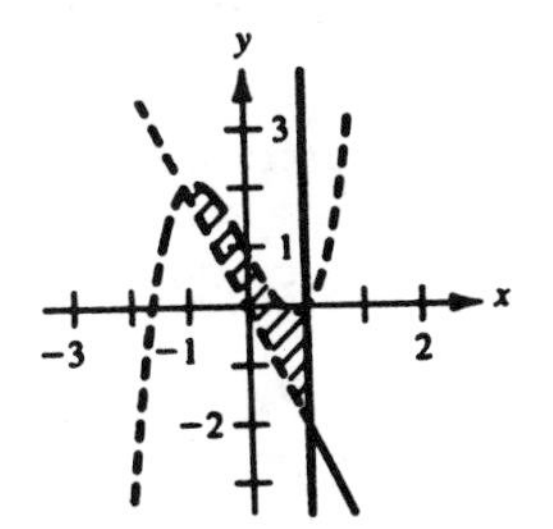

$x = 1$	$x = 1$
$y = x^3 - 2x + 1$	$y = -2x$
$(1, 0)$	$(1, -2)$

41. Derive a set of inequalities to describe the rectangular region with vertices at (2, 1), (5, 1), (5, 7), and (2, 7).

Solution:

$x \geq 2$

$x \leq 5$

$y \geq 1$

$y \leq 7$

Thus, $2 \leq x \leq 5,\ 1 \leq y \leq 7$.

47. A furniture company can sell all the tables and chairs it produces. Each table requires 1 hour in the assembly center and $1\frac{1}{3}$ hours in the finishing center. Each chair requires $1\frac{1}{2}$ hours in the assembly center and $1\frac{1}{2}$ hours in the finishing center. The company's assembly center is available 12 hours per day, and its finishing center is available 15 hours per day. If x is the number of tables produced per day and y is the number of chairs, find a system of inequalities describing all possible production levels. Sketch the graph of the system.

Solution:

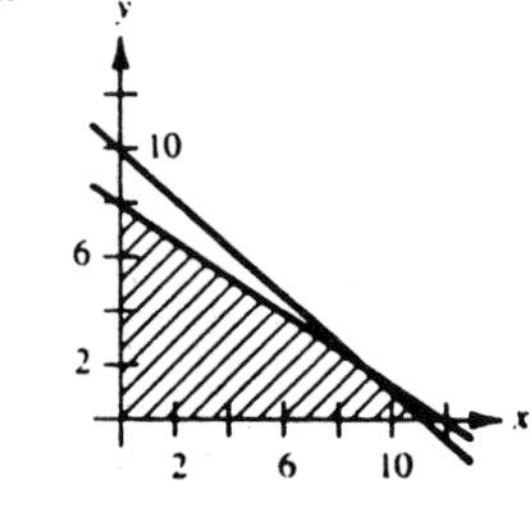

Assembly center constraint: $x + \frac{3}{2}y \leq 12$

Finishing center constraint: $\frac{4}{3}x + \frac{3}{2}y \leq 15$

Physical constraints: $x \geq 0$ and $y \geq 0$

49. A person plans to invest a total of at most \$20,000 in two different interest bearing accounts. Each account is to contain at least \$5000. Moreover, one account should have at least twice the amount that is in the other account. Find a system of inequalities to describe the various amounts that can be deposited in each account, and sketch the graph of the system.

Solution:

Account constraints:

$x \geq 5,000$

$y \geq 5,000$

$2x \leq y$

$x + y \leq 20,000$

SECTION 8.5

Linear Programming

■ To solve a linear programming problem:

1. Sketch the solution set for the system of constraints.
2. Find the vertices of the region.
3. Test the objective function at each of the vertices.

Solutions to Selected Exercises

1. Find the minimum and maximum values of the objective function $C = 3x + 2y$, subject to the constraints:

$$x \geq 0$$
$$y \geq 0$$
$$x + 3y \leq 15$$
$$4x + y \leq 16.$$

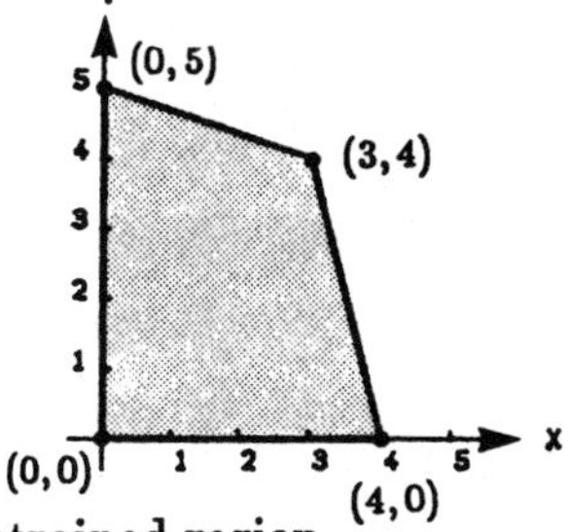

Solution:
The minimum and maximum values occur at the vertices of the constrained region.

Vertex	Value of $C = 3x + 2y$
(0, 0)	$C = 0$, minimum value
(4, 0)	$C = 12$
(0, 5)	$C = 10$
(3, 4)	$C = 17$, maximum value

7. Find the minimum and maximum values of the objective function $C = 25x + 30y$, subject to the constraints:

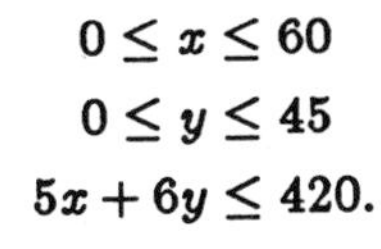

$$0 \leq x \leq 60$$
$$0 \leq y \leq 45$$
$$5x + 6y \leq 420.$$

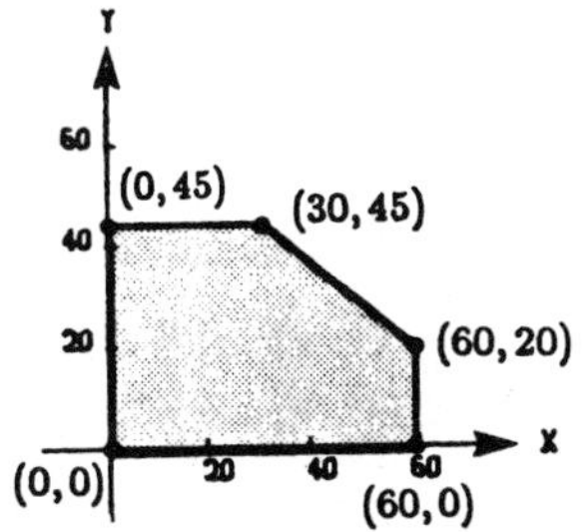

Solution:

Vertex	Value of $C = 25x + 30y$
(0, 0)	$C = 0$, minimum value
(60, 0)	$C = 1500$
(60, 20)	$C = 2100$, maximum value
(30, 45)	$C = 2100$, maximum value
(0, 45)	$C = 1350$

13. Find the minimum and maximum values of the objective function $C = 4x + y$, subject to the constraints:

$$x \geq 0$$
$$y \geq 0$$
$$x + 2y \leq 40$$
$$x + y \geq 30$$
$$2x + 3y \geq 72.$$

Solution:

Vertex	Value of $C = 4x + y$
(36, 0)	$C = 144$
(40, 0)	$C = 160$, maximum value
(24, 8)	$C = 104$, minimum value

17. A merchant plans to sell two models of home computers at costs of \$250 and \$400, respectively. The \$250 model yields a profit of \$45 and the \$400 model yields a profit of \$50. The merchant estimates that the total monthly demand will not exceed 250 units. Find the number of units of each model that should be stocked in order to maximize profit. Assume that the merchant does not want to invest more than \$70,000 in computer inventory.

Solution:

Objective function:

Maximize $P = 45x + 50y$ subject to the constraints:

$$x \geq 0$$
$$y \geq 0$$
$$x + y \leq 250$$
$$250x + 400y \leq 70,000$$

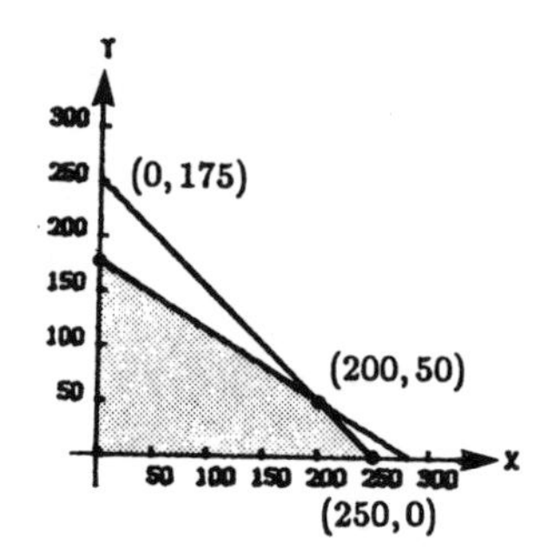

Testing the vertices shows that the profit is maximized when $x = 200$ units and $y = 50$ units.

19. A farmer mixes two brands of cattle feed. Brand X costs \$25 per bag and contains 2 units of nutritional element A, 2 units of element B, and 2 units of element C. Brand Y costs \$20 per bag and contains 1 unit of nutritional element A, 9 units of element B, and 3 units of element C. Find the number of bags of each brand that should be mixed to produce a mixture having a minimum cost per bag. The minimum requirements of nutrients A, B, and C are 12 units, 36 units, and 24 units, respectively.

Solution:

Objective function:

Minimize $C = 25x + 20y$ subject to the constraints:

$$2x + y \geq 12$$
$$2x + 9y \geq 36$$
$$2x + 3y \geq 24$$
$$x \geq 0,\ y \geq 0$$

Vertex	Value of $C = 25x + 20y$
(0, 12)	$C = 240$
(3, 6)	$C = 195$, minimum value
(9, 2)	$C = 265$
(18, 0)	$C = 450$

The cost is a minimum when three bags of Brand X and six bags of Brand Y are used.

23. Sketch a graph of the solution region for the given linear programming problem and describe its unusual characteristic. (The objective function is to be maximized.)

Objective function: $C = -x + 2y$

Constraints: $x \geq 0,\ y \geq 0$

$x \leq 10,\ x + y \leq 7$

Solution:

The constraint $x \leq 10$ is extraneous.

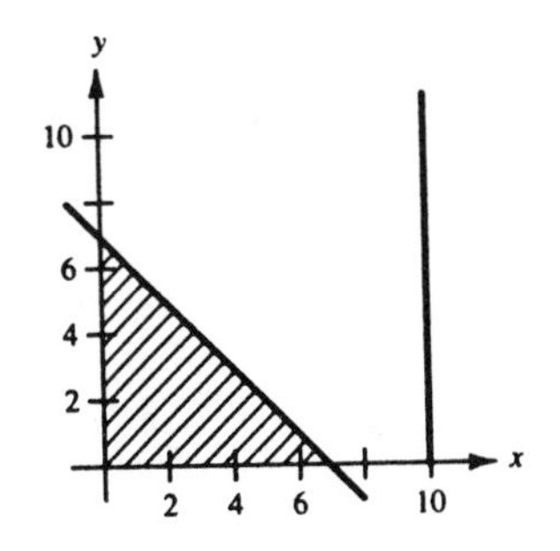

Vertex	Value of $C = -x + 2y$
(0, 0)	$C = 0$
(7, 0)	$C = -7$
(0, 7)	$C = 14$, maximum value

REVIEW EXERCISES FOR CHAPTER 8

Solutions to Selected Exercises

3. Solve the system of equations.

$$\begin{aligned} x^2 - y^2 &= 9 \\ x - y &= 1 \end{aligned}$$

Solution:

$$\begin{aligned} x^2 - y^2 &= 9 \\ x - y &= 1 \Rightarrow x = y + 1 \end{aligned}$$

$$\begin{aligned} (y+1)^2 - y^2 &= 9 \\ 2y + 1 &= 9 \\ y &= 4 \\ x &= 5 \end{aligned}$$

Solution: $(5, 4)$

7. Solve the system of equations.

$$\begin{aligned} y^2 - 2y + x &= 0 \\ x + y &= 0 \end{aligned}$$

Solution:

$$\begin{aligned} y^2 - 2y + x &= 0 \\ x + y &= 0 \Rightarrow x = -y \end{aligned}$$

$$\begin{aligned} y^2 - 2y - y &= 0 \\ y(y - 3) &= 0 \end{aligned}$$

$$\begin{aligned} y &= 0, \quad y = 3 \\ x &= 0, \quad x = -3 \end{aligned}$$

Solutions: $(0, 0)$, $(-3, 3)$

11. Solve the system of equations.

$$\begin{aligned} 0.2x + 0.3y &= 0.14 \\ 0.4x + 0.5y &= 0.20 \end{aligned}$$

Solution:

$$\begin{aligned}
0.2x + 0.3y = 0.14 &\Rightarrow 20x + 30y = 14 \Rightarrow && 20x + 30y = 14 \\
0.4x + 0.5y = 0.20 &\Rightarrow 4x + 5y = 2 \Rightarrow && -20x - 25y = -10 \\
& && 5y = 4 \\
& && y = \tfrac{4}{5} \\
& && x = -\tfrac{1}{2}
\end{aligned}$$

Solution: $(-\frac{1}{2}, \frac{4}{5})$ or $(-0.5, 0.8)$

15. Solve the system of equations.

$$\begin{aligned}
x + 2y + 6z &= 4 \\
-3x + 2y - z &= -4 \\
4x + 2z &= 16
\end{aligned}$$

Solution:

$$\begin{aligned}
3x + 6y + 18z &= 12 \\
-3x + 2y - z &= -4 \\
\hline
8y + 17z &= 8
\end{aligned}
\qquad
\begin{aligned}
2x + 4y + 12z &= 8 \\
-2x - z &= -8 \\
\hline
4y + 11z &= 0
\end{aligned}
\qquad
\begin{aligned}
8y + 17z &= 8 \\
-8y - 22z &= 0 \\
\hline
-5z &= 8 \\
z &= -\tfrac{8}{5} = -1.6 \\
y &= \tfrac{1}{8}[8 - 17(-1.6)] = 4.4 \\
x &= \tfrac{1}{2}[8 - (-1.6)] = 4.8
\end{aligned}$$

Solution: $(4.8, 4.4, -1.6)$

19. Solve the system of equations.

$$\begin{aligned}
2x + 5y - 19z &= 34 \\
3x + 8y - 31z &= 54
\end{aligned}$$

Solution:

$$\begin{aligned}
2x + 5y - 19z = 34 &\Rightarrow && 6x + 15y - 57z = 102 \\
3x + 8y - 31z = 54 &\Rightarrow && -6x - 16y + 62z = -108 \\
& && -y + 5z = -6
\end{aligned}$$

Let $z = a$. Then,

$$\begin{aligned}
y &= 5a + 6 \\
x &= \tfrac{1}{2}[34 - 5(5a + 6) + 19a] = -3a + 2
\end{aligned}$$

Solution: $(-3a + 2, 5a + 6, a)$

23. A mixture of 6 parts of chemical A, 8 parts of chemical B, and 13 parts of chemical C is required to kill a certain destructive crop insect. Commercial spray X contains 1, 2, and 2 parts, respectively, of these chemicals. Commercial spray Y contains only chemical C. Commercial spray Z contains chemicals A, B, and C in equal amounts. How much of each type of commercial spray is needed to get the desired mixture?

Solution:

From the following chart we obtain our system of equations.

	A	B	C
Mixture X	$\frac{1}{5}$	$\frac{2}{5}$	$\frac{2}{5}$
Mixture Y	0	0	1
Mixture Z	$\frac{1}{3}$	$\frac{1}{3}$	$\frac{1}{3}$
Desired Mixture	$\frac{6}{27}$	$\frac{8}{27}$	$\frac{13}{27}$

$$\left.\begin{aligned}\tfrac{1}{5}x + \tfrac{1}{3}z &= \tfrac{6}{27}\\ \tfrac{2}{5}x + \tfrac{1}{3}z &= \tfrac{8}{27}\end{aligned}\right\} x = \tfrac{10}{27},\ z = \tfrac{12}{27}$$

$$\tfrac{2}{5}x + y + \tfrac{1}{3}z = \tfrac{13}{27} \Rightarrow y = \tfrac{5}{27}$$

To obtain the desired mixture, the commercial sprays X, Y, and Z should be combined in a ratio of 10, 5, 12, respectively.

27. Sketch the graph of the solution of the system of inequalities.

$$\begin{aligned} 3x + 2y &\geq 24 \\ x + 2y &\geq 12 \\ 2 \leq x &\leq 15 \\ y &\leq 15 \end{aligned}$$

Solution:

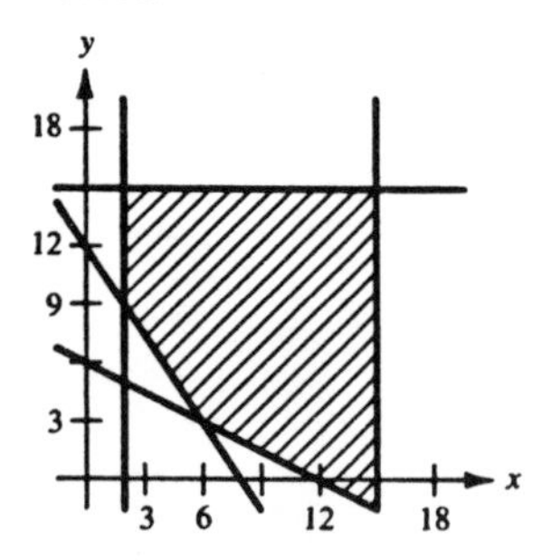

33. A Pennsylvania fruit grower has 1500 bushels of apples that are to be divided between markets in Harrisburg and Philadelphia. These two markets need at least 400 bushels and 600 bushels, respectively. Determine a system of inequalities and sketch a graph of the solution of the system.

Solution:
Let x = number of bushels for Harrisburg and y = number of bushels for Philadelphia.

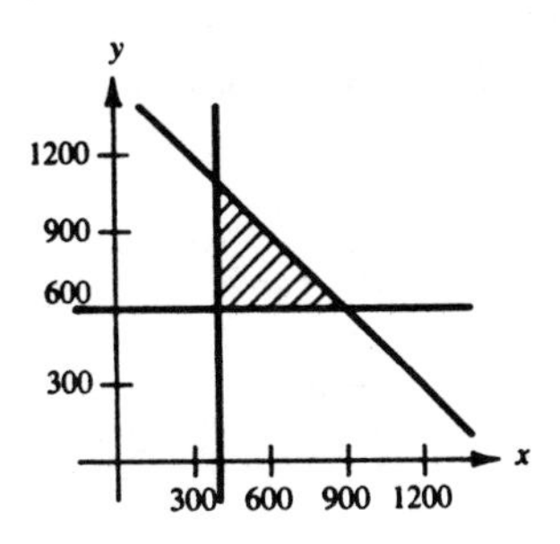

$$x \geq 400$$
$$y \geq 600$$
$$x + y \leq 1500$$

37. Minimize $C = 1.75x + 2.25y$ subject to the constraints:

$$2x + y \geq 25$$
$$3x + 2y \geq 45$$
$$x \geq 0$$
$$y \geq 0.$$

Solution:

Vertex	Value of $C = 1.75x + 2.25y$
(0, 25)	$C = 56.25$
(5, 15)	$C = 42.5$
(15, 0)	$C = 26.25$ minimum value

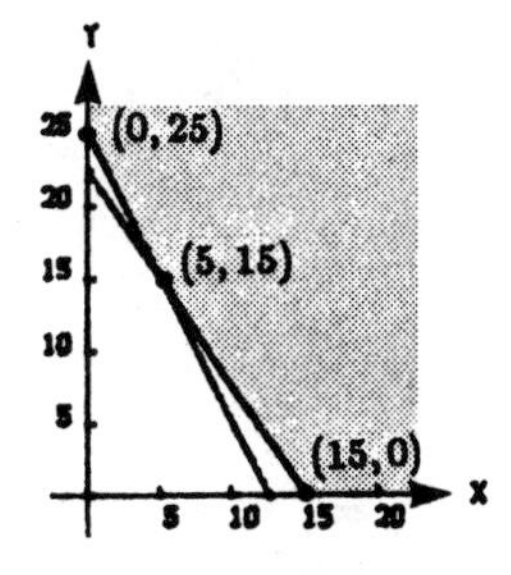

Practice Test for Chapter 8

For Exercises 1–3, solve the given system by the method of substitution.

1. $$\begin{aligned} x + y &= 1 \\ 3x - y &= 15 \end{aligned}$$

2. $$\begin{aligned} x - 3y &= -3 \\ x^2 + 6y &= 5 \end{aligned}$$

3. $$\begin{aligned} x + y + z &= 6 \\ 2x - y + 3z &= 0 \\ 5x + 2y - z &= -3 \end{aligned}$$

4. Find two numbers whose sum is 110 and product is 2800.

5. Find the dimensions of a rectangle if its perimeter is 170 feet and its area is 2800 square feet.

For Exercises 6–8, solve the linear system by elimination.

6. $$\begin{aligned} 2x + 15y &= 4 \\ x - 3y &= 23 \end{aligned}$$

7. $$\begin{aligned} x + y &= 2 \\ 38x - 19y &= 7 \end{aligned}$$

8. $$\begin{aligned} 0.4x + 0.5y &= 0.112 \\ 0.3x - 0.7y &= -0.131 \end{aligned}$$

9. Herbert invests $17,000 in two funds that pay 11% and 13% simple interest, respectively. If he receives $2080 in yearly interest, how much is invested in each fund?

10. Find the least squares regression line for the points (4, 3), (1, 1), (−1, −2), and (−2, −1).

For Exercises 11–13, solve the system of equations.

11. $$\begin{aligned} x + y \phantom{{}+z} &= -2 \\ 2x - y + z &= 11 \\ 4y - 3z &= -20 \end{aligned}$$

12. $$\begin{aligned} 4x - y + 5z &= 4 \\ 2x + y - z &= 0 \\ 2x + 4y + 8z &= 0 \end{aligned}$$

13. $$\begin{aligned} 3x + 2y - z &= 5 \\ 6x - y + 5z &= 2 \end{aligned}$$

14. Find the equation of the parabola $y = ax^2 + bx + c$ passing through the points $(0, -1)$, $(1, 4)$ and $(2, 13)$.

15. Find the position equation $s = (1/2)at^2 + v_0t + s_0$ given that $s = 12$ feet after 1 second, $s = 5$ feet after 2 seconds, and $s = 4$ feet after 3 seconds.

16. Graph $x^2 + y^2 \geq 9$.

17. Graph the solution of the system.

$$\begin{aligned} x + y &\leq 6 \\ x &\geq 2 \\ y &\geq 0 \end{aligned}$$

18. Derive a set of inequalities to describe the triangle with vertices $(0, 0)$, $(0, 7)$, and $(2, 3)$.

19. Find the maximum value of the objective function $C = 30x + 26y$ subject to the following constraints:

$$\begin{aligned} x &\geq 0 \\ y &\geq 0 \\ 2x + 3y &\leq 21 \\ 5x + 3y &\leq 30. \end{aligned}$$

20. Graph the system of inequalities.

$$\begin{aligned} x^2 + y^2 &\leq 4 \\ (x-2)^2 + y^2 &\geq 4 \end{aligned}$$

CHAPTER 9

Matrices and Determinants

SECTION 9.1

Matrices and Systems of Linear Equations

- You should be able to use elementary row operations to produce a triangular form of an augmented matrix.
- You should be able to transform a matrix into reduced row-echelon form. This is called Gauss-Jordan elimination.

Solutions to Selected Exercises

3. Determine the order of the matrix.

$$\begin{bmatrix} -9 \\ 2 \\ 36 \\ 11 \\ 3 \end{bmatrix}$$

Solution:
Since the matrix has five rows and one column, its order is 5×1.

9. Determine whether the following matrix is in row-echelon form.

$$\begin{bmatrix} 2 & 0 & 4 & 0 \\ 0 & -1 & 3 & 6 \\ 0 & 0 & 1 & 5 \end{bmatrix}$$

Solution:
Since the first nonzero entries in rows one and two are not 1, the matrix is *not* in row-echelon form.

15. Write the following matrix in row-echelon form. Remember that the row-echelon form of a matrix is not unique.

$$\begin{bmatrix} 1 & -1 & -1 & 1 \\ 5 & -4 & 1 & 8 \\ -6 & 8 & 18 & 0 \end{bmatrix}$$

Solution:

$$\begin{bmatrix} 1 & -1 & -1 & 1 \\ 5 & -4 & 1 & 8 \\ -6 & 8 & 18 & 0 \end{bmatrix} \quad \begin{matrix} \\ -5R_1 + R_2 \rightarrow \\ 6R_1 + R_3 \rightarrow \end{matrix} \begin{bmatrix} 1 & -1 & -1 & 1 \\ 0 & 1 & 6 & 3 \\ 0 & 2 & 12 & 6 \end{bmatrix}$$

$$\begin{matrix} \\ \\ -2R_2 + R_3 \rightarrow \end{matrix} \begin{bmatrix} 1 & -1 & -1 & 1 \\ 0 & 1 & 6 & 3 \\ 0 & 0 & 0 & 0 \end{bmatrix}$$

19. Write the following matrix in *reduced* row-echelon form.

$$\begin{bmatrix} 1 & 2 & 3 & -5 \\ 1 & 2 & 4 & -9 \\ -2 & -4 & -4 & 3 \\ 4 & 8 & 11 & -14 \end{bmatrix}$$

Solution:

$$\begin{bmatrix} 1 & 2 & 3 & -5 \\ 1 & 2 & 4 & -9 \\ -2 & -4 & -4 & 3 \\ 4 & 8 & 11 & -14 \end{bmatrix} \quad \begin{matrix} \\ -R_1 + R_2 \rightarrow \\ 2R_1 + R_3 \rightarrow \\ -4R_1 + R_4 \rightarrow \end{matrix} \begin{bmatrix} 1 & 2 & 3 & -5 \\ 0 & 0 & 1 & -4 \\ 0 & 0 & 2 & -7 \\ 0 & 0 & -1 & 6 \end{bmatrix}$$

$$\begin{matrix} -3R_2 + R_1 \rightarrow \\ \\ -2R_2 + R_3 \rightarrow \\ R_2 + R_4 \rightarrow \end{matrix} \begin{bmatrix} 1 & 2 & 0 & 7 \\ 0 & 0 & 1 & -4 \\ 0 & 0 & 0 & 1 \\ 0 & 0 & 0 & 2 \end{bmatrix}$$

$$\begin{matrix} -7R_3 + R_1 \rightarrow \\ 4R_3 + R_2 \rightarrow \\ \\ -2R_3 + R_4 \rightarrow \end{matrix} \begin{bmatrix} 1 & 2 & 0 & 0 \\ 0 & 0 & 1 & 0 \\ 0 & 0 & 0 & 1 \\ 0 & 0 & 0 & 0 \end{bmatrix}$$

23. Write the system of linear equations represented by the augmented matrix.

$$\left[\begin{array}{ccc|c} 1 & 0 & 2 & -10 \\ 0 & 3 & -1 & 5 \\ 4 & 2 & 0 & 3 \end{array}\right]$$

Solution:

Row 1: $1x + 0y + 2z = -10 \Rightarrow x + 2z = -10$

Row 2: $0x + 3y - 1z = 5 \Rightarrow 3y - z = 5$

Row 3: $4x + 2y + 0z = 3 \Rightarrow 4x + 2y = 3$

27. Determine the augmented matrix for the given system of linear equations.

$$\begin{aligned} x + 10y - 3z &= 2 \\ 5x - 3y + 4z &= 0 \\ 2x + 4y &= 6 \end{aligned}$$

Solution:
The augmented matrix for this system is

$$\begin{bmatrix} 1 & 10 & -3 & \vdots & 2 \\ 5 & -3 & 4 & \vdots & 0 \\ 2 & 4 & 0 & \vdots & 6 \end{bmatrix}.$$

31. Solve the system of equations. Use Gaussian elimination with back-substitution or Gauss-Jordan elimination.

$$\begin{aligned} -3x + 5y &= -22 \\ 3x + 4y &= 4 \\ 4x - 8y &= 32 \end{aligned}$$

Solution:

$$\begin{bmatrix} -3 & 5 & \vdots & -22 \\ 3 & 4 & \vdots & 4 \\ 4 & -8 & \vdots & 32 \end{bmatrix} \quad \begin{matrix} -\frac{1}{3}R_1 \rightarrow \\ -3R_1 + R_2 \rightarrow \\ -4R_1 + R_3 \rightarrow \end{matrix} \begin{bmatrix} 1 & -\frac{5}{3} & \vdots & \frac{22}{3} \\ 0 & 9 & \vdots & -18 \\ 0 & -\frac{4}{3} & \vdots & \frac{8}{3} \end{bmatrix}$$

$$\begin{matrix} \frac{5}{3}R_2 + R_1 \rightarrow \\ \frac{1}{9}R_2 \rightarrow \\ \frac{4}{3}R_2 + R_3 \rightarrow \end{matrix} \begin{bmatrix} 1 & 0 & \vdots & 4 \\ 0 & 1 & \vdots & -2 \\ 0 & 0 & \vdots & 0 \end{bmatrix}$$

Solution: $(4, -2)$

35. Solve the system of equations. Use Gaussian elimination with back-substitution or Gauss-Jordan elimination.

$$\begin{aligned} -x + 2y &= 1.5 \\ 2x - 4y &= 3 \end{aligned}$$

Solution:

$$\begin{bmatrix} -1 & 2 & \vdots & 1.5 \\ 2 & -4 & \vdots & 3 \end{bmatrix} \quad \begin{matrix} -R_1 \rightarrow \\ -2R_1 + R_2 \rightarrow \end{matrix} \begin{bmatrix} 1 & -2 & \vdots & -1.5 \\ 0 & 0 & \vdots & 6 \end{bmatrix}$$

The second line says $0 = 6$. This is inconsistent.

39. Solve the system of equations. Use Gaussian elimination with back-substitution or Gauss-Jordan elimination.

$$\begin{aligned} x + y - 5z &= 3 \\ x \qquad - 2z &= 1 \\ 2x - y - \ z &= 0 \end{aligned}$$

Solution:

$$\begin{bmatrix} 1 & 1 & -5 & \vdots & 3 \\ 1 & 0 & -2 & \vdots & 1 \\ 2 & -1 & -1 & \vdots & 0 \end{bmatrix} \quad \begin{matrix} \\ -R_1 + R_2 \rightarrow \\ -2R_1 + R_3 \rightarrow \end{matrix} \begin{bmatrix} 1 & 1 & -5 & \vdots & 3 \\ 0 & -1 & 3 & \vdots & -2 \\ 0 & -3 & 9 & \vdots & -6 \end{bmatrix}$$

$$\begin{matrix} R_2 + R_1 \rightarrow \\ -R_2 \rightarrow \\ 3R_2 + R_3 \rightarrow \end{matrix} \begin{bmatrix} 1 & 0 & -2 & \vdots & 1 \\ 0 & 1 & -3 & \vdots & 2 \\ 0 & 0 & 0 & \vdots & 0 \end{bmatrix}$$

Thus, $x - 2z = 1$ and $y - 3z = 2$. By letting $z = a$, we have $y = 3a + 2$ and $x = 2a + 1$.

Solution: $(2a + 1,\ 3a + 2,\ a)$

45. Solve the system of equations. Use Gaussian elimination with back-substitution or Gauss-Jordan elimination.

$$\begin{aligned} 2x + \ y - \ z + 2w &= -6 \\ 3x + 4y \qquad + \ w &= \ \ 1 \\ x + 5y + 2z + 6w &= -3 \\ 5x + 2y - \ z - \ w &= \ \ 3 \end{aligned}$$

Solution:

$$\begin{bmatrix} 2 & 1 & -1 & 2 & \vdots & -6 \\ 3 & 4 & 0 & 1 & \vdots & 1 \\ 1 & 5 & 2 & 6 & \vdots & -3 \\ 5 & 2 & -1 & -1 & \vdots & 3 \end{bmatrix} \quad \begin{matrix} R_3 \rightarrow \\ \\ R_1 \rightarrow \\ \\ \end{matrix} \begin{bmatrix} 1 & 5 & 2 & 6 & \vdots & -3 \\ 3 & 4 & 0 & 1 & \vdots & 1 \\ 2 & 1 & -1 & 2 & \vdots & -6 \\ 5 & 2 & -1 & -1 & \vdots & 3 \end{bmatrix}$$

$$\begin{matrix} \\ -3R_1 + R_2 \rightarrow \\ -2R_1 + R_3 \rightarrow \\ -5R_1 + R_4 \rightarrow \end{matrix} \begin{bmatrix} 1 & 5 & 2 & 6 & \vdots & -3 \\ 0 & -11 & -6 & -17 & \vdots & 10 \\ 0 & -9 & -5 & -10 & \vdots & 0 \\ 0 & -23 & -11 & -31 & \vdots & 18 \end{bmatrix}$$

$$\begin{matrix} 5R_4 + R_1 \rightarrow \\ -11R_4 + R_2 \rightarrow \\ -9R_4 + R_3 \rightarrow \\ -2R_2 + R_4 \rightarrow \end{matrix} \begin{bmatrix} 1 & 0 & 7 & 21 & \vdots & -13 \\ 0 & 0 & -17 & -50 & \vdots & 32 \\ 0 & 0 & -14 & -37 & \vdots & 18 \\ 0 & -1 & 1 & 3 & \vdots & -2 \end{bmatrix}$$

$$\begin{matrix} -7R_3 + R_1 \rightarrow \\ 17R_3 + R_2 \rightarrow \\ -\frac{1}{14}R_3 \rightarrow \\ -R_4 \rightarrow \end{matrix} \begin{bmatrix} 1 & 0 & 0 & \frac{5}{2} & \vdots & -4 \\ 0 & 0 & 0 & -\frac{71}{14} & \vdots & \frac{71}{7} \\ 0 & 0 & 1 & \frac{37}{14} & \vdots & -\frac{9}{7} \\ 0 & 1 & -1 & -3 & \vdots & 2 \end{bmatrix}$$

$$\begin{matrix} \\ -\frac{14}{71}R_2 \rightarrow \\ \\ \\ \end{matrix} \begin{bmatrix} 1 & 0 & 0 & \frac{5}{2} & \vdots & -4 \\ 0 & 0 & 0 & 1 & \vdots & -2 \\ 0 & 0 & 1 & \frac{37}{14} & \vdots & -\frac{9}{7} \\ 0 & 1 & -1 & -3 & \vdots & 2 \end{bmatrix}$$

$$\begin{aligned} x \qquad + \tfrac{5}{2}w &= -4 \\ w &= -2 \\ z + \tfrac{37}{14}w &= -\tfrac{9}{7} \\ y - z - \ 3w &= \ 2 \end{aligned}$$

Thus,

$$\begin{aligned} &w = -2 \\ &x = -4 - \tfrac{5}{2}(-2) = 1 \\ &z = -\tfrac{9}{7} - \tfrac{37}{14}(-2) = 4 \\ &y = 2 + 4 + 3(-2) = 0. \end{aligned}$$

Solution: $(1,\ 0,\ 4,\ -2)$

49. Solve the system of equations. Use Gaussian elimination with back-substitution or Gauss-Jordan elimination.

$$\begin{aligned} x + y + z &= 0 \\ 2x + 3y + z &= 0 \\ 3x + 5y + z &= 0 \end{aligned}$$

Solution:

$$\begin{bmatrix} 1 & 1 & 1 & \vdots & 0 \\ 2 & 3 & 1 & \vdots & 0 \\ 3 & 5 & 1 & \vdots & 0 \end{bmatrix} \quad \begin{matrix} \\ -2R_1 + R_2 \rightarrow \\ -3R_1 + R_3 \rightarrow \end{matrix} \begin{bmatrix} 1 & 1 & 1 & \vdots & 0 \\ 0 & 1 & -1 & \vdots & 0 \\ 0 & 2 & -2 & \vdots & 0 \end{bmatrix}$$

$$\begin{matrix} -R_2 + R_1 \rightarrow \\ \\ -2R_2 + R_3 \rightarrow \end{matrix} \begin{bmatrix} 1 & 0 & 2 & \vdots & 0 \\ 0 & 1 & -1 & \vdots & 0 \\ 0 & 0 & 0 & \vdots & 0 \end{bmatrix}$$

Thus, $x + 2z = 0$ and $y - z = 0$. By letting $z = a$, we have $x = -2a$ and $y = a$.

Solution: $(-2a,\ a,\ a)$

51. A small corporation borrowed $1,500,000 to expand its product line. Some of the money was borrowed at 8%, some at 9%, and some at 12%. How much was borrowed at each rate if the annual interest was $133,000 and the amount borrowed at 8% was 4 times the amount borrowed at 12%?

Solution:

Let $x =$ 8% amount
$y =$ 9% amount
$z =$ 12% amount

$$\begin{aligned} x + y + z &= 1,500,000 \\ 0.08x + 0.09y + 0.12z &= 133,000 \\ x - 4z &= 0 \end{aligned}$$

$$\begin{bmatrix} 1 & 1 & 1 & \vdots & 1,500,000 \\ 8 & 9 & 12 & \vdots & 13,300,000 \\ 1 & 0 & -4 & \vdots & 0 \end{bmatrix}$$

$$\begin{matrix} \\ -8R_1 + R_2 \to \\ -R_1 + R_3 \to \end{matrix} \begin{bmatrix} 1 & 1 & 1 & \vdots & 1,500,000 \\ 0 & 1 & 4 & \vdots & 1,300,000 \\ 0 & -1 & -5 & \vdots & -1,500,000 \end{bmatrix}$$

$$\begin{matrix} -R_2 + R_1 \to \\ \\ R_2 + R_3 \to \end{matrix} \begin{bmatrix} 1 & 0 & -3 & \vdots & 200,000 \\ 0 & 1 & 4 & \vdots & 1,300,000 \\ 0 & 0 & -1 & \vdots & -200,000 \end{bmatrix}$$

$$\begin{matrix} 3R_3 + R_1 \to \\ -4R_3 + R_2 \to \\ -R_3 \to \end{matrix} \begin{bmatrix} 1 & 0 & 0 & \vdots & 800,000 \\ 0 & 1 & 0 & \vdots & 500,000 \\ 0 & 0 & 1 & \vdots & 200,000 \end{bmatrix}$$

Thus,

$$\begin{aligned} x &= \$800,000 \\ y &= \$500,000 \\ z &= \$200,000. \end{aligned}$$

55. Find D, E, and F such that (1, 1), (3, 3), and (4, 2) are solution points of the equation $x^2 + y^2 + Dx + Ey + F = 0$. [*Hint:* See Example 8 in Section 8.3.]

Solution:

At (1, 1): $(1)^2 + (1)^2 + D(1) + E(1) + F = 0 \Rightarrow D + E + F = -2$

At (3, 3): $(3)^2 + (3)^2 + D(3) + E(3) + F = 0 \Rightarrow 3D + 3E + F = -18$

At (4, 2): $(4)^2 + (2)^2 + D(4) + E(2) + F = 0 \Rightarrow 4D + 2E + F = -20$

$$\begin{bmatrix} 1 & 1 & 1 & -2 \\ 3 & 3 & 1 & -18 \\ 4 & 2 & 1 & -20 \end{bmatrix} \quad \begin{matrix} \\ -3R_1 + R_2 \to \\ -4R_1 + R_3 \to \end{matrix} \begin{bmatrix} 1 & 1 & 1 & \vdots & -2 \\ 0 & 0 & -2 & \vdots & -12 \\ 0 & -2 & -3 & \vdots & -12 \end{bmatrix}$$

$$\begin{matrix} -R_2 + R_1 \to \\ -\frac{1}{2}R_2 \to \\ 3R_2 + R_3 \to \end{matrix} \begin{bmatrix} 1 & 1 & 0 & \vdots & -8 \\ 0 & 0 & 1 & \vdots & 6 \\ 0 & -2 & 0 & \vdots & 6 \end{bmatrix}$$

$$\begin{aligned} D + E &= -8 \Rightarrow D = -5 \\ F &= 6 \Rightarrow F = 6 \\ -2E &= 6 \Rightarrow E = -3 \end{aligned}$$

The equation of the circle is $x^2 + y^2 - 5x - 3y + 6 = 0$.

SECTION 9.2

Operations with Matrices

- $A = B$ if and only if they have the same order and $a_{ij} = b_{ij}$.
- You should be able to perform the operations of matrix addition, scalar multiplication, and matrix multiplication.
- Some properties of matrix addition, scalar multiplication, and matrix multiplication are:

  (a) $A + B = B + A$
  (b) $A + (B + C) = (A + B) + C$
  (c) $(cd)A = c(dA)$
  (d) $1A = A$
  (e) $c(A + B) = cA + cB$
  (f) $(c + d)A = cA + dA$
  (g) $A(BC) = (AB)C$
  (h) $A(B + C) = AB + AC$
  (i) $(A + B)C = AC + BC$
  (j) $c(AB) = (cA)B = A(cB)$
- You should remember that $AB \neq BA$ in general.

Solutions to Selected Exercises

3. Find x and y.

$$\begin{bmatrix} 16 & 4 & 5 & 4 \\ -3 & 13 & 15 & 6 \\ 0 & 2 & 4 & 0 \end{bmatrix} = \begin{bmatrix} 16 & 4 & 2x+1 & 4 \\ -3 & 13 & 15 & 3x \\ 0 & 2 & 3y-5 & 0 \end{bmatrix}$$

Solution:

$$\left.\begin{array}{l} 5 = 2x + 1 \\ 6 = 3x \end{array}\right\} \Rightarrow x = 2$$

$$4 = 3y - 5 \quad \Rightarrow y = 3$$

Solution: $x = 2, \ y = 3$

9. Find (a) $A+B$, (b) $A-B$, (c) $3A$, and (d) $3A-2B$.

$$A = \begin{bmatrix} 2 & 2 & -1 & 0 & 1 \\ 1 & 1 & -2 & 0 & -1 \end{bmatrix}, \quad B = \begin{bmatrix} 1 & 1 & -1 & 1 & 0 \\ -3 & 4 & 9 & -6 & -7 \end{bmatrix}$$

Solution:

(a) $A+B = \begin{bmatrix} 3 & 3 & -2 & 1 & 1 \\ -2 & 5 & 7 & -6 & -8 \end{bmatrix}$

(b) $A-B = \begin{bmatrix} 1 & 1 & 0 & -1 & 1 \\ 4 & -3 & -11 & 6 & 6 \end{bmatrix}$

(c) $3A = \begin{bmatrix} 6 & 6 & -3 & 0 & 3 \\ 3 & 3 & -6 & 0 & -3 \end{bmatrix}$

(d) $3A-2B = \begin{bmatrix} 4 & 4 & -1 & -2 & 3 \\ 9 & -5 & -24 & 12 & 11 \end{bmatrix}$

13. Find (a) AB, (b) BA, and if possible (c) A^2.

$$A = \begin{bmatrix} 3 & -1 \\ 1 & 3 \end{bmatrix}, \quad B = \begin{bmatrix} 1 & -3 \\ 3 & 1 \end{bmatrix}$$

Solution:

(a) $AB = \begin{bmatrix} 0 & -10 \\ 10 & 0 \end{bmatrix}$

(b) $BA = \begin{bmatrix} 0 & -10 \\ 10 & 0 \end{bmatrix}$

(c) $A^2 = \begin{bmatrix} 8 & -6 \\ 6 & 8 \end{bmatrix}$

17. Find AB, if possible.

$$A = \begin{bmatrix} 2 & 1 \\ -3 & 4 \\ 1 & 6 \end{bmatrix}, \quad B = \begin{bmatrix} 0 & -1 & 0 \\ 4 & 0 & 2 \\ 8 & -1 & 7 \end{bmatrix}$$

Solution:
A is 3×2 and B is 3×3. Since the number of columns of A does not equal the number of rows of B, the multiplication is not possible.

Note: BA is possible.

21. Find AB, if possible.

$$A = \begin{bmatrix} 5 & 0 & 0 \\ 0 & -8 & 0 \\ 0 & 0 & 7 \end{bmatrix}, \quad B = \begin{bmatrix} \frac{1}{5} & 0 & 0 \\ 0 & -\frac{1}{8} & 0 \\ 0 & 0 & \frac{1}{2} \end{bmatrix}$$

Solution:

$$AB = \begin{bmatrix} 5 & 0 & 0 \\ 0 & -8 & 0 \\ 0 & 0 & 7 \end{bmatrix} \begin{bmatrix} \frac{1}{5} & 0 & 0 \\ 0 & -\frac{1}{8} & 0 \\ 0 & 0 & \frac{1}{2} \end{bmatrix} = \begin{bmatrix} 1 & 0 & 0 \\ 0 & 1 & 0 \\ 0 & 0 & \frac{7}{2} \end{bmatrix}$$

25. Solve for X in $X = 3A - 2B$, given

$$A = \begin{bmatrix} -2 & -1 \\ 1 & 0 \\ 3 & -4 \end{bmatrix} \quad \text{and} \quad B = \begin{bmatrix} 0 & 3 \\ 2 & 0 \\ -4 & -1 \end{bmatrix}.$$

Solution:

$$X = 3A - 2B = \begin{bmatrix} -6 & -3 \\ 3 & 0 \\ 9 & -12 \end{bmatrix} - \begin{bmatrix} 0 & 6 \\ 4 & 0 \\ -8 & -2 \end{bmatrix} = \begin{bmatrix} -6 & -9 \\ -1 & 0 \\ 17 & 10 \end{bmatrix}$$

29. Find matrices A, X, and B such that the given system of linear equations can be written as the matrix equation $AX = B$. Solve the system of equations.

$$-x + y = 4$$
$$-2x + y = 0$$

Solution:

$$A = \begin{bmatrix} -1 & 1 \\ -2 & 1 \end{bmatrix}, \quad X = \begin{bmatrix} x \\ y \end{bmatrix}, \quad B = \begin{bmatrix} 4 \\ 0 \end{bmatrix}$$

By Gauss-Jordan elimination on

$$\begin{bmatrix} -1 & 1 & \vdots & 4 \\ -2 & 1 & \vdots & 0 \end{bmatrix} \quad \begin{matrix} -R_1 \rightarrow \\ 2R_1 + R_2 \rightarrow \end{matrix} \begin{bmatrix} 1 & -1 & \vdots & -4 \\ 0 & -1 & \vdots & -8 \end{bmatrix}$$

$$\begin{matrix} R_2 + R_1 \rightarrow \\ -R_2 \rightarrow \end{matrix} \begin{bmatrix} 1 & 0 & \vdots & 4 \\ 0 & 1 & \vdots & 8 \end{bmatrix},$$

we have $x = 4$ and $y = 8$.

35. If A and B are real numbers, then the following equations are true. If A and B are $n \times n$ matrices, are they true? Give reasons for your answers.

(a) $(A+B)(A-B) = A^2 - B^2$
(b) $(A+B)(A+B) = A^2 + 2AB + B^2$

Solution:
Since $AB \neq BA$ in general, neither equation is true.

(a) $(A+B)(A-B) = A^2 + BA - AB - B^2$
(b) $(A+B)(A+B) = A^2 + BA + AB + B^2$

39. Find $f(A)$, given

$$f(x) = x^2 - 5x + 2 \quad \text{and} \quad A = \begin{bmatrix} 2 & 0 \\ 4 & 5 \end{bmatrix}.$$

Solution:

$$\begin{aligned} f(A) &= \begin{bmatrix} 2 & 0 \\ 4 & 5 \end{bmatrix}\begin{bmatrix} 2 & 0 \\ 4 & 5 \end{bmatrix} - 5\begin{bmatrix} 2 & 0 \\ 4 & 5 \end{bmatrix} + 2\begin{bmatrix} 1 & 0 \\ 0 & 1 \end{bmatrix} \\ &= \begin{bmatrix} 4 & 0 \\ 28 & 25 \end{bmatrix} - \begin{bmatrix} 10 & 0 \\ 20 & 25 \end{bmatrix} + \begin{bmatrix} 2 & 0 \\ 0 & 2 \end{bmatrix} \\ &= \begin{bmatrix} -4 & 0 \\ 8 & 2 \end{bmatrix} \end{aligned}$$

SECTION 9.3

The Inverse of a Matrix

- You should be able to find the inverse, if it exists, of a matrix.
- You should be able to use inverse matrices to solve systems of equations.

Solutions to Selected Exercises

7. Find the inverse of the following matrix.

$$\begin{bmatrix} 1 & -2 \\ 2 & -3 \end{bmatrix}$$

Solution:

$$\left[\begin{array}{cc:cc} 1 & -2 & 1 & 0 \\ 2 & -3 & 0 & 1 \end{array}\right] \begin{array}{r} \\ -2R_1 + R_2 \rightarrow \end{array} \left[\begin{array}{cc:cc} 1 & -2 & 1 & 0 \\ 0 & 1 & -2 & 1 \end{array}\right]$$

$$2R_2 + R_1 \rightarrow \left[\begin{array}{cc:cc} 1 & 0 & -3 & 2 \\ 0 & 1 & -2 & 1 \end{array}\right]$$

$$A^{-1} = \begin{bmatrix} -3 & 2 \\ -2 & 1 \end{bmatrix}$$

11. Find the inverse of the following matrix, (if it exists).

$$\begin{bmatrix} 2 & 4 \\ 4 & 8 \end{bmatrix}$$

Solution:

$$\left[\begin{array}{cc:cc} 2 & 4 & 1 & 0 \\ 4 & 8 & 0 & 1 \end{array}\right] \begin{array}{r} \frac{1}{2}R_1 \rightarrow \\ -4R_1 + \quad R_2 \rightarrow \end{array} \left[\begin{array}{cc:cc} 1 & 2 & \frac{1}{2} & 0 \\ 0 & 0 & -2 & 1 \end{array}\right]$$

Since the left side does not reduce to I_2, the inverse does not exist.

15. Find the inverse of the following matrix.

$$\begin{bmatrix} 1 & 1 & 1 \\ 3 & 5 & 4 \\ 3 & 6 & 5 \end{bmatrix}$$

Solution:

$$\left[\begin{array}{ccc|ccc} 1 & 1 & 1 & 1 & 0 & 0 \\ 3 & 5 & 4 & 0 & 1 & 0 \\ 3 & 6 & 5 & 0 & 0 & 1 \end{array}\right] \begin{array}{r} \\ -3R_1 + R_2 \rightarrow \\ -3R_1 + R_3 \rightarrow \end{array} \left[\begin{array}{ccc|ccc} 1 & 1 & 1 & 1 & 0 & 0 \\ 0 & 2 & 1 & -3 & 1 & 0 \\ 0 & 3 & 2 & -3 & 0 & 1 \end{array}\right]$$

$$\begin{array}{r} -R_2 + R_1 \rightarrow \\ \frac{1}{2}R_2 \rightarrow \\ -3R_2 + R_3 \rightarrow \end{array} \left[\begin{array}{ccc|ccc} 1 & 0 & \frac{1}{2} & -\frac{5}{2} & -\frac{1}{2} & 0 \\ 0 & 1 & \frac{1}{2} & -\frac{3}{2} & \frac{1}{2} & 0 \\ 0 & 0 & \frac{1}{2} & \frac{3}{2} & -\frac{3}{2} & 1 \end{array}\right]$$

$$\begin{array}{r} -R_3 + R_1 \rightarrow \\ -R_3 + R_2 \rightarrow \\ 2R_3 \rightarrow \end{array} \left[\begin{array}{ccc|ccc} 1 & 0 & 0 & 1 & 1 & -1 \\ 0 & 1 & 0 & -3 & 2 & -1 \\ 0 & 0 & 1 & 3 & -3 & 2 \end{array}\right]$$

$$A^{-1} = \begin{bmatrix} 1 & 1 & -1 \\ -3 & 2 & -1 \\ 3 & -3 & 2 \end{bmatrix}$$

19. Find the inverse of the following matrix.

$$\begin{bmatrix} 1 & -2 & -1 & -2 \\ 3 & -5 & -2 & -3 \\ 2 & -5 & -2 & -5 \\ -1 & 4 & 4 & 11 \end{bmatrix}$$

Solution:

$$\left[\begin{array}{rrrr|rrrr} 1 & -2 & -1 & -2 & 1 & 0 & 0 & 0 \\ 3 & -5 & -2 & -3 & 0 & 1 & 0 & 0 \\ 2 & -5 & -2 & -5 & 0 & 0 & 1 & 0 \\ -1 & 4 & 4 & 11 & 0 & 0 & 0 & 1 \end{array}\right]$$

$$\begin{array}{r} \\ -3R_1 + R_2 \rightarrow \\ -2R_1 + R_3 \rightarrow \\ R_1 + R_4 \rightarrow \end{array} \left[\begin{array}{rrrr|rrrr} 1 & -2 & -1 & -2 & 1 & 0 & 0 & 0 \\ 0 & 1 & 1 & 3 & -3 & 1 & 0 & 0 \\ 0 & -1 & 0 & -1 & -2 & 0 & 1 & 0 \\ 0 & 2 & 3 & 9 & 1 & 0 & 0 & 1 \end{array}\right]$$

$$\begin{array}{r} 2R_2 + R_1 \rightarrow \\ \\ R_2 + R_3 \rightarrow \\ -2R_2 + R_4 \rightarrow \end{array} \left[\begin{array}{rrrr|rrrr} 1 & 0 & 1 & 4 & -5 & 2 & 0 & 0 \\ 0 & 1 & 1 & 3 & -3 & 1 & 0 & 0 \\ 0 & 0 & 1 & 2 & -5 & 1 & 1 & 0 \\ 0 & 0 & 1 & 3 & 7 & -2 & 0 & 1 \end{array}\right]$$

$$\begin{array}{r} -R_3 + R_1 \rightarrow \\ -R_3 + R_2 \rightarrow \\ \\ -R_3 + R_4 \rightarrow \end{array} \left[\begin{array}{rrrr|rrrr} 1 & 0 & 0 & 2 & 0 & 1 & -1 & 0 \\ 0 & 1 & 0 & 1 & 2 & 0 & -1 & 0 \\ 0 & 0 & 1 & 2 & -5 & 1 & 1 & 0 \\ 0 & 0 & 0 & 1 & 12 & -3 & -1 & 1 \end{array}\right]$$

$$\begin{array}{r} -2R_4 + R_1 \rightarrow \\ -R_4 + R_2 \rightarrow \\ -2R_4 + R_3 \rightarrow \\ \\ \end{array} \left[\begin{array}{rrrr|rrrr} 1 & 0 & 0 & 0 & -24 & 7 & 1 & -2 \\ 0 & 1 & 0 & 0 & -10 & 3 & 0 & -1 \\ 0 & 0 & 1 & 0 & -29 & 7 & 3 & -2 \\ 0 & 0 & 0 & 1 & 12 & -3 & -1 & 1 \end{array}\right]$$

$$A^{-1} = \begin{bmatrix} -24 & 7 & 1 & -2 \\ -10 & 3 & 0 & -1 \\ -29 & 7 & 3 & -2 \\ 12 & -3 & -1 & 1 \end{bmatrix}$$

23. Find the inverse of the following matrix.

$$\begin{bmatrix} 0.1 & 0.2 & 0.3 \\ -0.3 & 0.2 & 0.2 \\ 0.5 & 0.4 & 0.4 \end{bmatrix}$$

Solution:

$$\left[\begin{array}{rrr|rrr} 0.1 & 0.2 & 0.3 & 1 & 0 & 0 \\ -0.3 & 0.2 & 0.2 & 0 & 1 & 0 \\ 0.5 & 0.4 & 0.4 & 0 & 0 & 1 \end{array}\right] \begin{array}{r} 10R_1 \rightarrow \\ 10R_2 \rightarrow \\ 10R_3 \rightarrow \end{array} \left[\begin{array}{rrr|rrr} 1 & 2 & 3 & 10 & 0 & 0 \\ -3 & 2 & 2 & 0 & 10 & 0 \\ 5 & 4 & 4 & 0 & 0 & 10 \end{array}\right]$$

$$\begin{array}{r} \\ 3R_1 + R_2 \rightarrow \\ -5R_1 + R_3 \rightarrow \end{array} \left[\begin{array}{rrr|rrr} 1 & 2 & 3 & 10 & 0 & 0 \\ 0 & 8 & 11 & 30 & 10 & 0 \\ 0 & -6 & -11 & -50 & 0 & 10 \end{array}\right]$$

$$\begin{array}{r} \\ R_3 + R_2 \rightarrow \\ 3R_2 + R_3 \rightarrow \end{array} \left[\begin{array}{rrr|rrr} 1 & 2 & 3 & 10 & 0 & 0 \\ 0 & 2 & 0 & -20 & 10 & 10 \\ 0 & 0 & -11 & -110 & 30 & 40 \end{array}\right]$$

$$\begin{array}{r} -R_2 + R_1 \rightarrow \\ \frac{1}{2}R_2 \rightarrow \\ -\frac{1}{11}R_3 \rightarrow \end{array} \left[\begin{array}{rrr|rrr} 1 & 0 & 3 & 30 & -10 & -10 \\ 0 & 1 & 0 & -10 & 5 & 5 \\ 0 & 0 & 1 & 10 & -\frac{30}{11} & -\frac{40}{11} \end{array}\right]$$

$$\begin{array}{r} \\ -3R_3 + R_1 \rightarrow \\ \\ \end{array} \left[\begin{array}{rrr|rrr} 1 & 0 & 0 & 0 & -\frac{20}{11} & \frac{10}{11} \\ 0 & 1 & 0 & -10 & 5 & 5 \\ 0 & 0 & 1 & 10 & -\frac{30}{11} & -\frac{40}{11} \end{array}\right]$$

$$A^{-1} = \begin{bmatrix} 0 & -\frac{20}{11} & \frac{10}{11} \\ -10 & 5 & 5 \\ 10 & -\frac{30}{11} & -\frac{40}{11} \end{bmatrix} = \frac{5}{11}\begin{bmatrix} 0 & -4 & 2 \\ -22 & 11 & 11 \\ 22 & -6 & -8 \end{bmatrix}$$

27. Find the inverse of the following matrix.

$$\begin{bmatrix} 1 & 0 & 0 \\ 3 & 4 & 0 \\ 2 & 5 & 5 \end{bmatrix}$$

Solution:

$$\left[\begin{array}{ccc:ccc} 1 & 0 & 0 & 1 & 0 & 0 \\ 3 & 4 & 0 & 0 & 1 & 0 \\ 2 & 5 & 5 & 0 & 0 & 1 \end{array}\right] \begin{array}{l} \\ -3R_1 + R_2 \rightarrow \\ -2R_1 + R_3 \rightarrow \end{array} \left[\begin{array}{ccc:ccc} 1 & 0 & 0 & 1 & 0 & 0 \\ 0 & 4 & 0 & -3 & 1 & 0 \\ 0 & 5 & 5 & -2 & 0 & 1 \end{array}\right]$$

$$\begin{array}{r} \\ \frac{1}{4}R_2 \rightarrow \\ -5R_2 + R_3 \rightarrow \end{array} \left[\begin{array}{ccc:ccc} 1 & 0 & 0 & 1 & 0 & 0 \\ 0 & 1 & 0 & -0.75 & 0.25 & 0 \\ 0 & 0 & 5 & 1.75 & -1.25 & 1 \end{array}\right]$$

$$\begin{array}{r} \\ \\ \frac{1}{5}R_3 \rightarrow \end{array} \left[\begin{array}{ccc:ccc} 1 & 0 & 0 & 1 & 0 & 0 \\ 0 & 1 & 0 & -0.75 & 0.25 & 0 \\ 0 & 0 & 1 & 0.35 & -0.25 & 0.2 \end{array}\right]$$

$$A^{-1} = \begin{bmatrix} 1 & 0 & 0 \\ -0.75 & 0.25 & 0 \\ 0.35 & -0.25 & 0.2 \end{bmatrix}$$

33. Use an inverse matrix to solve the following systems. (See Exercise 19.)

(a)
$$\begin{aligned} x_1 - 2x_2 - x_3 - 2x_4 &= 0 \\ 3x_1 - 5x_2 - 2x_3 - 3x_4 &= 1 \\ 2x_1 - 5x_2 - 2x_3 - 5x_4 &= -1 \\ -x_1 + 4x_2 + 4x_3 + 11x_4 &= 2 \end{aligned}$$

(b)
$$\begin{aligned} x_1 - 2x_2 - x_3 - 2x_4 &= 1 \\ 3x_1 - 5x_2 - 2x_3 - 3x_4 &= -2 \\ 2x_1 - 5x_2 - 2x_3 - 5x_4 &= 0 \\ -x_1 + 4x_2 + 4x_3 + 11x_4 &= -3 \end{aligned}$$

Solution:
From Exercise 19, we have that the inverse of

$$\begin{bmatrix} 1 & -2 & -1 & -2 \\ 3 & -5 & -2 & -3 \\ 2 & -5 & -2 & -5 \\ -1 & 4 & 4 & 11 \end{bmatrix} \text{ is } A^{-1} = \begin{bmatrix} -24 & 7 & 1 & -2 \\ -10 & 3 & 0 & -1 \\ -29 & 7 & 3 & -2 \\ 12 & -3 & -1 & 1 \end{bmatrix}.$$

(a)
$$\begin{bmatrix} x_1 \\ x_2 \\ x_3 \\ x_4 \end{bmatrix} = \begin{bmatrix} -24 & 7 & 1 & -2 \\ -10 & 3 & 0 & -1 \\ -29 & 7 & 3 & -2 \\ 12 & -3 & -1 & 1 \end{bmatrix} \begin{bmatrix} 0 \\ 1 \\ -1 \\ 2 \end{bmatrix} = \begin{bmatrix} 2 \\ 1 \\ 0 \\ 0 \end{bmatrix}$$

Solution: (2, 1, 0, 0)

(b)
$$\begin{bmatrix} x_1 \\ x_2 \\ x_3 \\ x_4 \end{bmatrix} = \begin{bmatrix} -24 & 7 & 1 & -2 \\ -10 & 3 & 0 & -1 \\ -29 & 7 & 3 & -2 \\ 12 & -3 & -1 & 1 \end{bmatrix} \begin{bmatrix} 1 \\ -2 \\ 0 \\ -3 \end{bmatrix} = \begin{bmatrix} -32 \\ -13 \\ -37 \\ 15 \end{bmatrix}$$

Solution: (−32, −13, −37, 15)

SECTION 9.4

The Determinant of a Matrix

- You should be able to determine the determinant of a matrix of order 2 or of order 3 by using the products of the diagonals.
- You should be able to use expansion by cofactors to find the determinant of a matrix of order 3 or greater.
- The determinant of a triangular matrix equals the product of the entries on the main diagonal.

Solutions to Selected Exercises

7. Find the determinant of

$$\begin{bmatrix} -7 & 6 \\ \frac{1}{2} & 3 \end{bmatrix}.$$

Solution:

$$\begin{vmatrix} -7 & 6 \\ \frac{1}{2} & 3 \end{vmatrix} = -7(3) - 6(\tfrac{1}{2}) = -24$$

11. Find the determinant of

$$\begin{bmatrix} 2 & -1 & 0 \\ 4 & 2 & 1 \\ 4 & 2 & 1 \end{bmatrix}.$$

Solution:

$$\left|\begin{matrix} 2 & -1 & 0 \\ 4 & 2 & 1 \\ 4 & 2 & 1 \end{matrix}\right|\begin{matrix} 2 & -1 \\ 4 & 2 \\ 4 & 2 \end{matrix} = 4 + (-4) + 0 - 4 - (-4) = 0$$

15. Find the determinant of

$$\begin{bmatrix} 1 & 4 & -2 \\ 3 & 6 & -6 \\ -2 & 1 & 4 \end{bmatrix}.$$

Solution:

$$\begin{vmatrix} 1 & 4 & -2 \\ 3 & 6 & -6 \\ -2 & 1 & 4 \end{vmatrix} \begin{matrix} 1 & 4 \\ 3 & 6 \\ -2 & 1 \end{matrix} = 24 + 48 + (-6) - 24 - (-6) - 48 = 0$$

19. Find the determinant of

$$\begin{bmatrix} x & y & 1 \\ -2 & -2 & 1 \\ 1 & 5 & 1 \end{bmatrix}.$$

Solution:

$$\begin{vmatrix} x & y & 1 \\ -2 & -2 & 1 \\ 1 & 5 & 1 \end{vmatrix} \begin{matrix} x & y \\ -2 & -2 \\ 1 & 5 \end{matrix} = -2x + y + (-10) - (-2) - 5x - (-2y) = -7x + 3y - 8$$

25. Find the determinant of the following matrix by the method of expansion by cofactors. Expand using (a) Row 1 and (b) Column 2.

$$\begin{bmatrix} -3 & 2 & 1 \\ 4 & 5 & 6 \\ 2 & -3 & 1 \end{bmatrix}$$

Solution:

(a) Expansion along the first row:

$$\begin{vmatrix} -3 & 2 & 1 \\ 4 & 5 & 6 \\ 2 & -3 & 1 \end{vmatrix} = -3C_{11} + 2C_{12} + 1C_{13} = -3(23) + 2(8) + 1(-22)$$
$$= -69 + 16 - 22$$
$$= -75$$

(b) Expansion along the second column:

$$\begin{vmatrix} -3 & 2 & 1 \\ 4 & 5 & 6 \\ 2 & -3 & 1 \end{vmatrix} = 2C_{12} + 5C_{22} - 3C_{32} = 2(8) + 5(-5) - 3(22)$$
$$= 16 - 25 - 66$$
$$= -75$$

31. Find the determinant of

$$\begin{bmatrix} 1 & 4 & -2 \\ 3 & 2 & 0 \\ -1 & 4 & 3 \end{bmatrix}.$$

Solution:
Expansion along the third column:

$$\begin{vmatrix} 1 & 4 & -2 \\ 3 & 2 & 0 \\ -1 & 4 & 3 \end{vmatrix} = -2\begin{vmatrix} 3 & 2 \\ -1 & 4 \end{vmatrix} + 0 + 3\begin{vmatrix} 1 & 4 \\ 3 & 2 \end{vmatrix}$$
$$= -2(14) + 3(-10)$$
$$= -28 - 30$$
$$= -58$$

35. Find the determinant of

$$\begin{bmatrix} 3 & 6 & -5 & 4 \\ -2 & 0 & 6 & 0 \\ 1 & 1 & 2 & 2 \\ 0 & 3 & -1 & -1 \end{bmatrix}.$$

Solution:
Expansion along the second row:

$$\begin{vmatrix} 3 & 6 & -5 & 4 \\ -2 & 0 & 6 & 0 \\ 1 & 1 & 2 & 2 \\ 0 & 3 & -1 & -1 \end{vmatrix} = 2\begin{vmatrix} 6 & -5 & 4 \\ 1 & 2 & 2 \\ 3 & -1 & -1 \end{vmatrix} - 6\begin{vmatrix} 3 & 6 & 4 \\ 1 & 1 & 2 \\ 0 & 3 & -1 \end{vmatrix}$$
$$= 2[6(0) - 1(9) + 3(-18)] - 6[3(-7) - (-18)]$$
$$= 2[-9 - 54] - 6[-21 + 18]$$
$$= -126 + 18$$
$$= -108$$

39. Find the determinant of

$$\begin{bmatrix} 3 & 2 & 4 & -1 & 5 \\ -2 & 0 & 1 & 3 & 2 \\ 1 & 0 & 0 & 4 & 0 \\ 6 & 0 & 2 & -1 & 0 \\ 3 & 0 & 5 & 1 & 0 \end{bmatrix}.$$

Solution:
Expansion along the second column:

$$\begin{vmatrix} 3 & 2 & 4 & -1 & 5 \\ -2 & 0 & 1 & 3 & 2 \\ 1 & 0 & 0 & 4 & 0 \\ 6 & 0 & 2 & -1 & 0 \\ 3 & 0 & 5 & 1 & 0 \end{vmatrix} = -2 \begin{vmatrix} -2 & 1 & 3 & 2 \\ 1 & 0 & 4 & 0 \\ 6 & 2 & -1 & 0 \\ 3 & 5 & 1 & 0 \end{vmatrix}$$

$$= -2(-2) \begin{vmatrix} 1 & 0 & 4 \\ 6 & 2 & -1 \\ 3 & 5 & 1 \end{vmatrix}$$

$$= 4[1(7) - 0 + 4(24)]$$

$$= 4[7 + 96]$$

$$= 412$$

43. Evaluate the determinant of

$$\begin{bmatrix} 4u & -1 \\ -1 & 2v \end{bmatrix}.$$

Determinants of this type occur in calculus.

Solution:

$$\begin{vmatrix} 4u & -1 \\ -1 & 2v \end{vmatrix} = 8uv - 1$$

47. Evaluate the determinant of

$$\begin{bmatrix} e^{2x} & e^{3x} \\ 2e^{2x} & 3e^{3x} \end{bmatrix}.$$

Determinants of this type occur in calculus.

Solution:

$$\begin{vmatrix} e^{2x} & e^{3x} \\ 2e^{2x} & 3e^{3x} \end{vmatrix} = 3e^{5x} - 2e^{5x} = e^{5x}$$

SECTION 9.5

Properties of Determinants

- You should know what effect each elementary row (column) operation has on the determinant of a matrix.
- You should know what conditions yield a determinant of zero.
- You should be able to use determinants to find inverses, if they exist, of matrices.

Solutions to Selected Exercises

5. State the property of determinants that verifies the equation.

$$\begin{vmatrix} 1 & 3 & 4 \\ -7 & 2 & -5 \\ 6 & 1 & 2 \end{vmatrix} = -\begin{vmatrix} 1 & 4 & 3 \\ -7 & -5 & 2 \\ 6 & 2 & 1 \end{vmatrix}$$

Solution:
Interchanging Columns 2 and 3 results in a change of sign of the determinant.

9. State the property of determinants that verifies the equation.

$$\begin{vmatrix} 5 & 0 & 10 \\ 25 & -30 & 40 \\ -15 & 5 & 20 \end{vmatrix} = 5^3 \begin{vmatrix} 1 & 0 & 2 \\ 5 & -6 & 8 \\ -3 & 1 & 4 \end{vmatrix}$$

Solution:
Each row was divided by 5, thus the resulting determinant is 5^3 times the original determinant.

13. State the property of determinants that verifies the equation.

$$\begin{vmatrix} 3 & 2 & 4 \\ -2 & 1 & 5 \\ 5 & -7 & -20 \end{vmatrix} = \begin{vmatrix} 7 & 2 & -6 \\ 0 & 1 & 0 \\ -9 & -7 & 15 \end{vmatrix}$$

Solution:
Adding multiples of Column 2 to Columns 1 and 3 leaves the determinant unchanged.

17. Use elementary row (or column) operations as aids for evaluating $\begin{vmatrix} 1 & 1 & 1 \\ 2 & -1 & -2 \\ 1 & -2 & -1 \end{vmatrix}$.

Solution:

$$\begin{vmatrix} 1 & 1 & 1 \\ 2 & -1 & -2 \\ 1 & -2 & -1 \end{vmatrix} = \begin{vmatrix} 1 & 1 & 1 \\ 0 & -3 & -4 \\ 0 & -3 & -2 \end{vmatrix} = -6$$

$-2R_1 + R_2$
$-R_1 + R_3$

21. Use elementary row (or column) operations as aids for evaluating $\begin{vmatrix} 3 & 8 & -7 \\ 0 & -5 & 4 \\ 8 & 1 & 6 \end{vmatrix}$.

Solution:

$$\begin{vmatrix} 3 & 8 & -7 \\ 0 & -5 & 4 \\ 8 & 1 & 6 \end{vmatrix} = 3(-34) + 8(-3) = -126$$

Expansion along
Column 1

25. Use elementary row (or column) operations as aids for evaluating $\begin{vmatrix} 9 & -4 & 2 & 5 \\ 2 & 7 & 6 & -5 \\ 4 & 1 & -2 & 0 \\ 7 & 3 & 4 & 10 \end{vmatrix}$.

Solution:

$$\begin{vmatrix} 9 & -4 & 2 & 5 \\ 2 & 7 & 6 & -5 \\ 4 & 1 & -2 & 0 \\ 7 & 3 & 4 & 10 \end{vmatrix} = \begin{vmatrix} 9 & -4 & 2 & 5 \\ 11 & 3 & 8 & 0 \\ 4 & 1 & -2 & 0 \\ 11 & 17 & 16 & 0 \end{vmatrix} \quad \begin{matrix} R_1 + R_2 \\ 2R_2 + R_4 \end{matrix}$$

$$= -5\begin{vmatrix} 11 & 3 & 8 \\ 4 & 1 & -2 \\ 11 & 17 & 16 \end{vmatrix} \quad \text{Expansion along Column 4}$$

$$= -5\begin{vmatrix} 11 & 3 & 8 \\ 4 & 1 & -2 \\ 0 & 14 & 8 \end{vmatrix} \quad -R_1 + R_3$$

$$= -5\begin{vmatrix} 11 & -11 & 0 \\ 4 & 1 & -2 \\ 0 & 14 & 8 \end{vmatrix} \quad -R_3 + R_1$$

$$= -5(11)\begin{vmatrix} 1 & -1 & 0 \\ 0 & 5 & -2 \\ 0 & 14 & 8 \end{vmatrix} \quad \begin{matrix} \frac{1}{11}R_1 \\ -4R_1 + R_2 \end{matrix}$$

$$= -55[40 + 28] = -3740$$

29. Use elementary row (or column) operations as aids for evaluating

$$\begin{vmatrix} 3 & -2 & 4 & 3 & 1 \\ -1 & 0 & 2 & 1 & 0 \\ 5 & -1 & 0 & 3 & 2 \\ 4 & 7 & -8 & 0 & 0 \\ 1 & 2 & 3 & 0 & 2 \end{vmatrix}.$$

Solution:

$$\begin{vmatrix} 3 & -2 & 4 & 3 & 1 \\ -1 & 0 & 2 & 1 & 0 \\ 5 & -1 & 0 & 3 & 2 \\ 4 & 7 & -8 & 0 & 0 \\ 1 & 2 & 3 & 0 & 2 \end{vmatrix} = \begin{vmatrix} 3 & -2 & 4 & 3 & 1 \\ -1 & 0 & 2 & 1 & 0 \\ -1 & 3 & -8 & -3 & 0 \\ 4 & 7 & -8 & 0 & 0 \\ -4 & 3 & 3 & -3 & 0 \end{vmatrix} \quad \begin{matrix} -2R_1 + R_3 \\ -R_3 + R_5 \end{matrix}$$

$$= \begin{vmatrix} -1 & 0 & 2 & 1 \\ -1 & 3 & -8 & -3 \\ 4 & 7 & -8 & 0 \\ -4 & 3 & 3 & -3 \end{vmatrix} \quad \text{Expansion along Column 5}$$

$$= \begin{vmatrix} -1 & 0 & 2 & 1 \\ -4 & 3 & -2 & 0 \\ 4 & 7 & -8 & 0 \\ -3 & 0 & 11 & 0 \end{vmatrix} \quad \begin{matrix} 3R_1 + R_3 \\ -R_2 + R_4 \end{matrix}$$

$$= -\begin{vmatrix} -4 & 3 & -2 \\ 4 & 7 & -8 \\ -3 & 0 & 11 \end{vmatrix} \quad \text{Expansion along Column 4}$$

$$= -\begin{vmatrix} -4 & 3 & -2 \\ 0 & 10 & -10 \\ -3 & 0 & 11 \end{vmatrix} \quad R_1 + R_2$$

$$= -[4(110) - 3(-10)]$$

$$= -[-440 + 30]$$

$$= 410$$

33. Use a determinant to determine whether the following matrix is invertible.

$$\begin{bmatrix} 14 & 7 & 0 \\ 2 & 3 & 0 \\ 1 & -5 & 2 \end{bmatrix}$$

Solution:

$$\begin{vmatrix} 14 & 7 & 0 \\ 2 & 3 & 0 \\ 1 & -5 & 2 \end{vmatrix} = 2\begin{vmatrix} 14 & 7 \\ 2 & 3 \end{vmatrix} = 2(42 - 14) = 56 \neq 0$$

The matrix *is* invertible.

37. Show that $\begin{vmatrix} 1 & x & x^2 \\ 1 & y & y^2 \\ 1 & z & z^2 \end{vmatrix} = (y-x)(z-x)(z-y)$.

Solution:

$$\begin{vmatrix} 1 & x & x^2 \\ 1 & y & y^2 \\ 1 & z & z^2 \end{vmatrix} = \begin{vmatrix} y & y^2 \\ z & z^2 \end{vmatrix} - \begin{vmatrix} x & x^2 \\ z & z^2 \end{vmatrix} + \begin{vmatrix} x & x^2 \\ y & y^2 \end{vmatrix}$$

$$\begin{aligned}
&= (yz^2 - y^2z) - (xz^2 - x^2z) + (xy^2 - x^2y) \\
&= yz^2 - xz^2 - y^2z + x^2z + xy(y-x) \\
&= z^2(y-x) - z(y^2 - x^2) + xy(y-x) \\
&= z^2(y-x) - z(y-x)(y+x) + xy(y-x) \\
&= (y-x)[z^2 - z(y+x) + xy] \\
&= (y-x)[z^2 - zy - zx + xy] \\
&= (y-x)[z^2 - zx - zy + xy] \\
&= (y-x)[z(z-x) - y(z-x)] \\
&= (y-x)(z-x)(z-y)
\end{aligned}$$

41. Verify the equation.

$$\begin{vmatrix} w & x \\ y & z \end{vmatrix} = -\begin{vmatrix} y & z \\ w & x \end{vmatrix}$$

Solution:

$$\begin{vmatrix} w & x \\ y & z \end{vmatrix} = wz - xy$$

$$-\begin{vmatrix} y & z \\ w & x \end{vmatrix} = -[xy - wz] = wz - xy$$

Therefore,

$$\begin{vmatrix} w & x \\ y & z \end{vmatrix} = -\begin{vmatrix} y & z \\ w & x \end{vmatrix}.$$

SECTION 9.6

Applications of Determinants and Matrices

- You should be able to use Cramer's Rule to solve a system of linear equations.
- Now you should be able to solve a system of linear equations by substitution, elimination, elementary row operations on an augmented matrix, using the inverse matrix, or Cramer's Rule.
- You should be able to find the area of a triangle in the xy-plane.

Solutions to Selected Exercises

5. Use Cramer's Rule to solve the system of equations.

$$\begin{aligned} 20x + 8y &= 11 \\ 12x - 24y &= 21 \end{aligned}$$

Solution:

$$x = \frac{\begin{vmatrix} 11 & 8 \\ 21 & -24 \end{vmatrix}}{\begin{vmatrix} 20 & 8 \\ 12 & -24 \end{vmatrix}} = \frac{-432}{-576} = \frac{3}{4}$$

$$y = \frac{\begin{vmatrix} 20 & 11 \\ 12 & 21 \end{vmatrix}}{\begin{vmatrix} 20 & 8 \\ 12 & -24 \end{vmatrix}} = \frac{288}{-576} = -\frac{1}{2}$$

Solution: $\left(\frac{3}{4}, -\frac{1}{2}\right)$

9. Use Cramer's Rule to solve the system of equations.

$$\begin{aligned} 3x + 6y &= 5 \\ 6x + 14y &= 11 \end{aligned}$$

Solution:

$$x = \frac{\begin{vmatrix} 5 & 6 \\ 11 & 14 \end{vmatrix}}{\begin{vmatrix} 3 & 6 \\ 6 & 14 \end{vmatrix}} = \frac{4}{6} = \frac{2}{3}$$

$$y = \frac{\begin{vmatrix} 3 & 5 \\ 6 & 11 \end{vmatrix}}{\begin{vmatrix} 3 & 6 \\ 6 & 14 \end{vmatrix}} = \frac{3}{6} = \frac{1}{2}$$

Solution: $\left(\frac{2}{3}, \frac{1}{2}\right)$

13. Use Cramer's Rule to solve the system of equations.

$$\begin{aligned} 3x + 4y + 4z &= 11 \\ 4x - 4y + 6z &= 11 \\ 6x - 6y \qquad &= \ 3 \end{aligned}$$

Solution:

$$x = \frac{\begin{vmatrix} 11 & 4 & 4 \\ 11 & -4 & 6 \\ 3 & -6 & 0 \end{vmatrix}}{\begin{vmatrix} 3 & 4 & 4 \\ 4 & -4 & 6 \\ 6 & -6 & 0 \end{vmatrix}} = \frac{252}{252} = 1$$

Substituting this value into the third equation gives us $y = \frac{1}{2}$. Then substituting these values back into either of the other equations gives us $z = \frac{3}{2}$.

Solution: $\left(1, \frac{1}{2}, \frac{3}{2}\right)$

19. Use Cramer's Rule to solve the system of equations.

$$\begin{aligned} 7x - 3y \qquad + 2w &= 41 \\ -2x + y \qquad - w &= -13 \\ 4x \qquad + z - 2w &= 12 \\ -x + y \qquad - w &= -8 \end{aligned}$$

Solution:

$$x = \frac{\begin{vmatrix} 41 & -3 & 0 & 2 \\ -13 & 1 & 0 & -1 \\ 12 & 0 & 1 & -2 \\ -8 & 1 & 0 & -1 \end{vmatrix}}{\begin{vmatrix} 7 & -3 & 0 & 2 \\ -2 & 1 & 0 & -1 \\ 4 & 0 & 1 & -2 \\ -1 & 1 & 0 & -1 \end{vmatrix}} = \frac{\begin{vmatrix} 41 & -3 & -2 \\ -13 & 1 & -1 \\ -8 & 1 & -1 \end{vmatrix}}{\begin{vmatrix} 7 & -3 & -2 \\ -2 & 1 & -1 \\ -1 & 1 & -1 \end{vmatrix}} = \frac{25}{5} = 5$$

$$y = \frac{\begin{vmatrix} 7 & 41 & 0 & 2 \\ -2 & -13 & 0 & -1 \\ 4 & 12 & 1 & -2 \\ -1 & -8 & 0 & -1 \end{vmatrix}}{\begin{vmatrix} 7 & -3 & 0 & 2 \\ -2 & 1 & 0 & -1 \\ 4 & 0 & 1 & -2 \\ -1 & 1 & 0 & -1 \end{vmatrix}} = \frac{\begin{vmatrix} 7 & 41 & 2 \\ -2 & -13 & -1 \\ -1 & -8 & -1 \end{vmatrix}}{\begin{vmatrix} 7 & -3 & -2 \\ -2 & 1 & -1 \\ -1 & 1 & -1 \end{vmatrix}} = \frac{0}{5} = 0$$

Substituting these values back into Equation 4 gives us $w = 3$. Then using these values in Equation 3 gives us $z = -2$.

Solution: $(5, 0, -2, 3)$

23. Use a determinant to find the area of the triangle with vertices $(-2, -3)$, $(2, -3)$, and $(0, 4)$.

Solution:

$$A = \frac{1}{2}\begin{vmatrix} -2 & -3 & 1 \\ 2 & -3 & 1 \\ 0 & 4 & 1 \end{vmatrix} = 14$$

29. Use a determinant to determine if the points $(2, -\frac{1}{2})$, $(-4, 4)$, and $(6, -3)$ are collinear.

Solution:

Since

$$\begin{vmatrix} 2 & -\frac{1}{2} & 1 \\ -4 & 4 & 1 \\ 6 & -3 & 1 \end{vmatrix} = -3 \neq 0,$$

the points are not collinear.

35. Use a determinant to find an equation of the line through the points $(-4, 3)$ and $(2, 1)$.

Solution:

$$\begin{vmatrix} x & y & 1 \\ -4 & 3 & 1 \\ 2 & 1 & 1 \end{vmatrix} = 0$$

$$2x + 6y - 10 = 0$$
$$x + 3y - 5 = 0$$

39. Write a cryptogram for LANDING SUCCESSFUL (see Example 6), using the matrix

$$A = \begin{bmatrix} 1 & 2 & 2 \\ 3 & 7 & 9 \\ -1 & -4 & -7 \end{bmatrix}.$$

Solution:

L A N D I N G – S U C C E S S F U L
[12 1 14] [4 9 14] [7 0 19] [21 3 3] [5 19 19] [6 21 12]

$$\begin{aligned}
[12 \quad 1 \quad 14]A &= [\ \ 1 \quad -25 \quad -65] \\
[\ 4 \quad 9 \quad 14]A &= [\ 17 \quad 15 \quad -9] \\
[\ 7 \quad 0 \quad 19]A &= [-12 \quad -62 \quad -119] \\
[21 \quad 3 \quad 3]A &= [\ 27 \quad 51 \quad 48] \\
[\ 5 \quad 19 \quad 19]A &= [\ 43 \quad 67 \quad 48] \\
[\ 6 \quad 21 \quad 12]A &= [\ 57 \quad 111 \quad 117]
\end{aligned}$$

Cryptogram:
1 −25 −65 17 15 −9 −12 −62 −119 27 51 48 43 67 48 57 111 117

43. Decode the cryptogram

20 17 −15 −12 −56 −104 1 −25 −65 62 143 181

using the inverse of the following matrix (see Example 7).

$$A = \begin{bmatrix} 1 & 2 & 2 \\ 3 & 7 & 9 \\ -1 & -4 & -7 \end{bmatrix}$$

Solution:
To find A^{-1}, use Gauss-Jordan elimination.

$$\begin{bmatrix} 1 & 2 & 2 & \vdots & 1 & 0 & 0 \\ 3 & 7 & 9 & \vdots & 0 & 1 & 0 \\ -1 & -4 & -7 & \vdots & 0 & 0 & 1 \end{bmatrix} \to \begin{bmatrix} 1 & 0 & 0 & \vdots & -13 & 6 & 4 \\ 0 & 1 & 0 & \vdots & 12 & -5 & -3 \\ 0 & 0 & 1 & \vdots & -5 & 2 & 1 \end{bmatrix}$$

$$A^{-1} = \begin{bmatrix} -13 & 6 & 4 \\ 12 & -5 & -3 \\ -5 & 2 & 1 \end{bmatrix}$$

$$\begin{aligned} [\;20 \quad 17 \quad -15]A^{-1} &= [19 \quad 5 \quad 14] \\ [-12 \quad -56 \quad -104]A^{-1} &= [\;4 \quad 0 \quad 16] \\ [\;1 \quad -25 \quad -65]A^{-1} &= [12 \quad 1 \quad 14] \\ [\;62 \quad 143 \quad 181]A^{-1} &= [\;5 \quad 19 \quad 0] \end{aligned}$$

19	5	14	4	0	16	12	1	14	5	19	0
S	E	N	D	–	P	L	A	N	E	S	–

REVIEW EXERCISES FOR CHAPTER 9

Solutions to Selected Exercises

3. Use matrices and elementary row operations to solve the system of equations.

$$\begin{aligned} 0.2x - 0.1y &= 0.07 \\ 0.4x - 0.5y &= -0.01 \end{aligned}$$

Solution:

$$\begin{bmatrix} 0.2 & -0.1 & \vdots & 0.07 \\ 0.4 & -0.5 & \vdots & -0.01 \end{bmatrix} \quad \begin{array}{r} 5R_1 \rightarrow \\ -2R_1 + R_2 \rightarrow \end{array} \begin{bmatrix} 1 & -0.5 & \vdots & 0.35 \\ 0 & -0.3 & \vdots & -0.15 \end{bmatrix}$$

$$\begin{array}{r} 0.5R_2 + R_1 \rightarrow \\ -\frac{1}{0.3}R_2 \rightarrow \end{array} \begin{bmatrix} 1 & 0 & \vdots & 0.6 \\ 0 & 1 & \vdots & 0.5 \end{bmatrix}$$

$x = 0.6$

$y = 0.5$

Solution: $(0.6, 0.5)$

7. Use matrices and elementary row operations to solve the system of equations.

$$\begin{aligned} 2x + 3y + 3z &= 3 \\ 6x + 6y + 12z &= 13 \\ 12x + 9y - z &= 2 \end{aligned}$$

Solution:

$$\begin{bmatrix} 2 & 3 & 3 & \vdots & 3 \\ 6 & 6 & 12 & \vdots & 13 \\ 12 & 9 & -1 & \vdots & 2 \end{bmatrix} \quad \begin{array}{r} \\ -3R_1 + R_2 \rightarrow \\ -2R_2 + R_3 \rightarrow \end{array} \begin{bmatrix} 2 & 3 & 3 & \vdots & 3 \\ 0 & -3 & 3 & \vdots & 4 \\ 0 & -3 & -25 & \vdots & -24 \end{bmatrix}$$

$$\begin{array}{r} R_2 + R_1 \rightarrow \\ \\ -R_2 + R_3 \rightarrow \end{array} \begin{bmatrix} 2 & 0 & 6 & \vdots & 7 \\ 0 & -3 & 3 & \vdots & 4 \\ 0 & 0 & -28 & \vdots & -28 \end{bmatrix}$$

$$\begin{array}{r} \frac{1}{2}R_1 \rightarrow \\ -\frac{1}{3}R_2 \rightarrow \\ -\frac{1}{28}R_3 \rightarrow \end{array} \begin{bmatrix} 1 & 0 & 3 & \vdots & \frac{7}{2} \\ 0 & 1 & -1 & \vdots & -\frac{4}{3} \\ 0 & 0 & 1 & \vdots & 1 \end{bmatrix}$$

$$z = 1$$
$$x - 3z = \frac{7}{2} \Rightarrow x = \frac{1}{2}$$
$$y - z = -\frac{4}{3} \Rightarrow y = -\frac{1}{3}$$

Solution: $\left(\frac{1}{2}, -\frac{1}{3}, 1\right)$

11. Use matrices and elementary row operations to solve the system of equations.

$$\begin{aligned} x + 2y + 6z &= 1 \\ 2x + 5y + 15z &= 4 \\ 3x + y + 3z &= -6 \end{aligned}$$

Solution:

$$\begin{bmatrix} 1 & 2 & 6 & \vdots & 1 \\ 2 & 5 & 15 & \vdots & 4 \\ 3 & 1 & 3 & \vdots & -6 \end{bmatrix} \quad \begin{matrix} \\ -2R_1 + R_2 \rightarrow \\ -3R_1 + R_3 \rightarrow \end{matrix} \begin{bmatrix} 1 & 2 & 6 & \vdots & 1 \\ 0 & 1 & 3 & \vdots & 2 \\ 0 & -5 & -15 & \vdots & -9 \end{bmatrix}$$

$$\begin{matrix} -2R_2 + R_1 \rightarrow \\ \\ 5R_2 + R_3 \rightarrow \end{matrix} \begin{bmatrix} 1 & 0 & 0 & \vdots & -3 \\ 0 & 1 & 3 & \vdots & 2 \\ 0 & 0 & 0 & \vdots & 1 \end{bmatrix}$$

$$\begin{aligned} x &= -3 \\ y + 3z &= 2 \\ 0 &= 1, \quad \text{Inconsistent} \end{aligned}$$

15. Perform the indicated matrix operation.

$$\begin{bmatrix} 1 & 2 \\ 5 & -4 \\ 6 & 0 \end{bmatrix} \begin{bmatrix} 6 & -2 & 8 \\ 4 & 0 & 0 \end{bmatrix}$$

Solution:

$$\begin{bmatrix} 1 & 2 \\ 5 & -4 \\ 6 & 0 \end{bmatrix} \begin{bmatrix} 6 & -2 & 8 \\ 4 & 0 & 0 \end{bmatrix} = \begin{bmatrix} 14 & -2 & 8 \\ 14 & -10 & 40 \\ 36 & -12 & 48 \end{bmatrix}$$

19. Perform the indicated matrix operation.

$$\begin{bmatrix} 1 & 3 & 2 \\ 0 & 2 & -4 \\ 0 & 0 & 3 \end{bmatrix} \begin{bmatrix} 4 & -3 & 2 \\ 0 & 3 & -1 \\ 0 & 0 & 2 \end{bmatrix}$$

Solution:

$$\begin{bmatrix} 1 & 3 & 2 \\ 0 & 2 & -4 \\ 0 & 0 & 3 \end{bmatrix} \begin{bmatrix} 4 & -3 & 2 \\ 0 & 3 & -1 \\ 0 & 0 & 2 \end{bmatrix} = \begin{bmatrix} 4 & 6 & 3 \\ 0 & 6 & -10 \\ 0 & 0 & 6 \end{bmatrix}$$

23. Solve for X in $3X + 2A = B$, given

$$A = \begin{bmatrix} -4 & 0 \\ 1 & -5 \\ -3 & 2 \end{bmatrix} \quad \text{and} \quad B = \begin{bmatrix} 1 & 2 \\ -2 & 1 \\ 4 & 4 \end{bmatrix}.$$

Solution:

$$\begin{aligned} X &= \frac{1}{3}[B - 2A] \\ &= \frac{1}{3}\left(\begin{bmatrix} 1 & 2 \\ -2 & 1 \\ 4 & 4 \end{bmatrix} - 2\begin{bmatrix} -4 & 0 \\ 1 & -5 \\ -3 & 2 \end{bmatrix}\right) \\ &= \frac{1}{3}\begin{bmatrix} 9 & 2 \\ -4 & 11 \\ 10 & 0 \end{bmatrix} \end{aligned}$$

27. Evaluate the determinant.

$$\begin{vmatrix} 3 & 0 & -4 & 0 \\ 0 & 8 & 1 & 2 \\ 6 & 1 & 8 & 2 \\ 0 & 3 & -4 & 1 \end{vmatrix}$$

Solution:

$$\begin{vmatrix} 3 & 0 & -4 & 0 \\ 0 & 8 & 1 & 2 \\ 6 & 1 & 8 & 2 \\ 0 & 3 & -4 & 1 \end{vmatrix} = 3\begin{vmatrix} 8 & 1 & 2 \\ 1 & 8 & 2 \\ 3 & -4 & 1 \end{vmatrix} - 4\begin{vmatrix} 0 & 8 & 2 \\ 6 & 1 & 2 \\ 0 & 3 & 1 \end{vmatrix} \quad \text{Expansion along Row 1}$$

$$\begin{aligned} &= 3[8(8 - (-8)) - 1(1 - 6) + 2(-4 - 24)] - 4[0 - 6(8 - 6) + 0] \\ &= 3[128 + 5 - 56] - 4[-12] \\ &= 279 \end{aligned}$$

31. Find the inverse of the following matrix.

$$\begin{bmatrix} 2 & 0 & 3 \\ -1 & 1 & 1 \\ 2 & -2 & 1 \end{bmatrix}$$

Solution:

$$\left[\begin{array}{rrr|rrr} 2 & 0 & 3 & 1 & 0 & 0 \\ -1 & 1 & 1 & 0 & 1 & 0 \\ 2 & -2 & 1 & 0 & 0 & 1 \end{array}\right] \quad \begin{array}{r} R_2 + R_1 \rightarrow \\ R_1 + R_2 \rightarrow \\ -2R_1 + R_3 \rightarrow \end{array} \left[\begin{array}{rrr|rrr} 1 & 1 & 4 & 1 & 1 & 0 \\ 0 & 2 & 5 & 1 & 2 & 0 \\ 0 & -4 & -7 & -2 & -2 & 1 \end{array}\right]$$

$$\begin{array}{r} -R_2 + R_1 \rightarrow \\ \frac{1}{2}R_2 \rightarrow \\ 4R_2 + R_3 \rightarrow \end{array} \left[\begin{array}{rrr|rrr} 1 & 0 & \frac{3}{2} & \frac{1}{2} & 0 & 0 \\ 0 & 1 & \frac{5}{2} & \frac{1}{2} & 1 & 0 \\ 0 & 0 & 3 & 0 & 2 & 1 \end{array}\right]$$

$$\begin{array}{r} -\frac{3}{2}R_3 + R_1 \rightarrow \\ -\frac{5}{2}R_3 + R_2 \rightarrow \\ \frac{1}{3}R_3 \rightarrow \end{array} \left[\begin{array}{rrr|rrr} 1 & 0 & 0 & \frac{1}{2} & -1 & -\frac{1}{2} \\ 0 & 1 & 0 & \frac{1}{2} & -\frac{2}{3} & -\frac{5}{6} \\ 0 & 0 & 1 & 0 & \frac{2}{3} & \frac{1}{3} \end{array}\right]$$

Inverse: $\begin{bmatrix} \frac{1}{2} & -1 & -\frac{1}{2} \\ \frac{1}{2} & -\frac{2}{3} & -\frac{5}{6} \\ 0 & \frac{2}{3} & \frac{1}{3} \end{bmatrix}$

33. Write the system of linear equations represented by the matrix equation.

$$\begin{bmatrix} 5 & 4 \\ -1 & 1 \end{bmatrix} \begin{bmatrix} x \\ y \end{bmatrix} = \begin{bmatrix} 2 \\ -22 \end{bmatrix}$$

Solution:

$$\begin{bmatrix} 5 & 4 \\ -1 & 1 \end{bmatrix} \begin{bmatrix} x \\ y \end{bmatrix} = \begin{bmatrix} 2 \\ -22 \end{bmatrix}$$

$$\begin{bmatrix} 5x + 4y \\ -x + y \end{bmatrix} = \begin{bmatrix} 2 \\ -22 \end{bmatrix}$$

$$\begin{aligned} 5x + 4y &= 2 \\ -x + y &= -22 \end{aligned}$$

37. Solve the system of linear equations using (a) the inverse of the coefficient matrix and (b) Cramer's Rule.

$$\begin{aligned} -3x - 3y - 4z &= 2 \\ y + z &= -1 \\ 4x + 3y + 4z &= -1 \end{aligned}$$

Solution:

(a)

$$\left[\begin{array}{ccc:ccc} -3 & -3 & -4 & 1 & 0 & 0 \\ 0 & 1 & 1 & 0 & 1 & 0 \\ 4 & 3 & 4 & 0 & 0 & 1 \end{array}\right] \begin{array}{r} R_3 + R_1 \rightarrow \\ \\ -4R_1 + R_3 \rightarrow \end{array} \left[\begin{array}{ccc:ccc} 1 & 0 & 0 & 1 & 0 & 1 \\ 0 & 1 & 1 & 0 & 1 & 0 \\ 0 & 3 & 4 & -4 & 0 & -3 \end{array}\right]$$

$$\begin{array}{r} \\ -R_3 + R_2 \rightarrow \\ -3R_2 + R_3 \rightarrow \end{array} \left[\begin{array}{ccc:ccc} 1 & 0 & 0 & 1 & 0 & 1 \\ 0 & 1 & 0 & 4 & 4 & 3 \\ 0 & 0 & 1 & -4 & -3 & -3 \end{array}\right]$$

$$\begin{bmatrix} x \\ y \\ z \end{bmatrix} = \begin{bmatrix} 1 & 0 & 1 \\ 4 & 4 & 3 \\ -4 & -3 & -3 \end{bmatrix} \begin{bmatrix} 2 \\ -1 \\ -1 \end{bmatrix} = \begin{bmatrix} 1 \\ 1 \\ -2 \end{bmatrix}$$

Solution: $(1, 1, -2)$

(b)

$$x = \frac{\begin{vmatrix} 2 & -3 & -4 \\ -1 & 1 & 1 \\ -1 & 3 & 4 \end{vmatrix}}{\begin{vmatrix} -3 & -3 & -4 \\ 0 & 1 & 1 \\ 4 & 3 & 4 \end{vmatrix}} = \frac{1}{1} = 1$$

$$y = \frac{\begin{vmatrix} -3 & 2 & -4 \\ 0 & -1 & 1 \\ 4 & -1 & 4 \end{vmatrix}}{\begin{vmatrix} -3 & -3 & -4 \\ 0 & 1 & 1 \\ 4 & 3 & 4 \end{vmatrix}} = \frac{1}{1} = 1$$

$$z = \frac{\begin{vmatrix} -3 & -3 & 2 \\ 0 & 1 & -1 \\ 4 & 3 & -1 \end{vmatrix}}{\begin{vmatrix} -3 & -3 & -4 \\ 0 & 1 & 1 \\ 4 & 3 & 4 \end{vmatrix}} = \frac{-2}{1} = -2$$

Solution: $(1, 1, -2)$

41. Use a determinant to find the area of the triangle with vertices (1, 0), (5, 0), and (5, 8).

Solution:

$$\text{Area} = \frac{1}{2}\begin{vmatrix} 1 & 0 & 1 \\ 5 & 0 & 1 \\ 5 & 8 & 1 \end{vmatrix} = \frac{1}{2}(32) = 16$$

45. Use a determinant to find the equation of the line through the points $(-4, 0)$ and $(4, 4)$.

Solution:

$$\begin{vmatrix} x & y & 1 \\ -4 & 0 & 1 \\ 4 & 4 & 1 \end{vmatrix} = 0$$

$$-4x + 8y - 16 = 0$$

$$x - 2y + 4 = 0$$

49. If A is a 3×3 matrix and $|A| = 2$, then what is the value of $|4A|$? Give the reason for your answer.

Solution:
$|4A| = 4^3(2) = 128$, since $4A$ means that each one of the three rows of A was multiplied by 4.

Practice Test for Chapter 9

1. Put the matrix in reduced echelon form.

$$\begin{bmatrix} 1 & -2 & 4 \\ 3 & -5 & 9 \end{bmatrix}$$

For Exercises 2–4, use matrices to solve the system of equations.

2. $\begin{aligned} 3x + 5y &= 3 \\ 2x - y &= -11 \end{aligned}$

3. $\begin{aligned} 2x + 3y &= -3 \\ 3x + 2y &= 8 \\ x + y &= 1 \end{aligned}$

4. $\begin{aligned} x + 3z &= -5 \\ 2x + y &= 0 \\ 3x + y - z &= 3 \end{aligned}$

5. Multiply $\begin{bmatrix} 1 & 4 & 5 \\ 2 & 0 & -3 \end{bmatrix} \begin{bmatrix} 1 & 6 \\ 0 & -7 \\ -1 & 2 \end{bmatrix}$.

6. Given $A = \begin{bmatrix} 9 & 1 \\ -4 & 8 \end{bmatrix}$ and $B = \begin{bmatrix} 6 & -2 \\ 3 & 5 \end{bmatrix}$, find $3A - 5B$.

7. Find $f(A)$:

$$f(x) = x^2 - 7x + 8, \quad A = \begin{bmatrix} 3 & 0 \\ 7 & 1 \end{bmatrix}.$$

8. True or false: $(A + B)(A + 3B) = A^2 + 4AB + 3B^2$ where A and B are matrices.

For Exercises 9–10, find the inverse of the matrix, if it exists.

9. $\begin{bmatrix} 1 & 2 \\ 3 & 5 \end{bmatrix}$

10. $\begin{bmatrix} 1 & 1 & 1 \\ 3 & 6 & 5 \\ 6 & 10 & 8 \end{bmatrix}$

11. Use an inverse matrix to solve the systems:

(a) $x + 2y = 4$
$3x + 5y = 1$

(b) $x + 2y = 3$
$3x + 5y = -2$

For Exercises 12–14, find the determinant of the matrix.

12. $\begin{bmatrix} 6 & -1 \\ 3 & 4 \end{bmatrix}$

13. $\begin{bmatrix} 1 & 3 & -1 \\ 5 & 9 & 0 \\ 6 & 2 & -5 \end{bmatrix}$

14. $\begin{bmatrix} 1 & 4 & 2 & 3 \\ 0 & 1 & -2 & 0 \\ 3 & 5 & -1 & 1 \\ 2 & 0 & 6 & 1 \end{bmatrix}$

15. True or false:

$$\begin{vmatrix} 3 & 0 & 0 \\ 0 & 3 & 0 \\ 0 & 0 & 3 \end{vmatrix} = -3^3 \begin{vmatrix} 1 & 0 & 0 \\ 0 & 0 & 1 \\ 0 & 1 & 0 \end{vmatrix}$$

16. Evaluate $\begin{bmatrix} 6 & 4 & 3 & 0 & 6 \\ 0 & 5 & 1 & 4 & 8 \\ 0 & 0 & 2 & 7 & 3 \\ 0 & 0 & 0 & 9 & 2 \\ 0 & 0 & 0 & 0 & 1 \end{bmatrix}$.

17. Use cofactors to find the inverse of $\begin{bmatrix} 1 & 3 & 0 \\ 0 & 4 & 5 \\ 0 & 1 & 2 \end{bmatrix}$.

For Exercises 18–20, use Cramer's Rule to find the indicated value.

18. $6x - 7y = 4$ Find x.
$2x + 5y = 11$

19. $3x + z = 1$ Find z.
$y + 4z = 3$
$x - y = 2$

20. $721.4x - 29.1y = 33.77$ Find y.
$45.9x + 105.6y = 19.85$

CHAPTER 10
Sequences, Series, and Probability

SECTION 10.1

Sequences and Summation Notation

- Given the general nth term in a sequence, you should be able to find, or list, the terms.
- You should be able to find an expression for the nth term of a sequence.
- You should be able to use sigma notation for a sum.

Solutions to Selected Exercises

5. Write the first five terms of the following sequence. (Assume n begins with 1.)

$$a_n = \frac{3^n}{n!}$$

Solution:

$$a_n = \frac{3^n}{n!}$$

$$a_1 = \frac{3^1}{1!} = 3$$

$$a_2 = \frac{3^2}{2!} = \frac{9}{2}$$

$$a_3 = \frac{3^3}{3!} = \frac{27}{6} = \frac{9}{2}$$

$$a_4 = \frac{3^4}{4!} = \frac{81}{24} = \frac{27}{8}$$

$$a_5 = \frac{3^5}{5!} = \frac{243}{120} = \frac{81}{40}$$

Solution: $3, \frac{9}{2}, \frac{9}{2}, \frac{27}{8}, \frac{81}{40}$

9. Write the first five terms of the following sequence. (Assume n begins with 1.)

$$a_n = \frac{n+1}{n}$$

Solution:

$$a_n = \frac{n+1}{n}$$
$$a_1 = \frac{2}{1} = 2$$
$$a_2 = \frac{3}{2}$$
$$a_3 = \frac{4}{3}$$
$$a_4 = \frac{5}{4}$$
$$a_5 = \frac{6}{5}$$

Solution: 2, $\frac{3}{2}$, $\frac{4}{3}$, $\frac{5}{4}$, $\frac{6}{5}$

13. Write the first five terms of the following sequence. (Assume n begins with 1.)

$$a_n = \frac{1+(-1)^n}{n}$$

Solution:

$$a_n = \frac{1+(-1)^n}{n}$$
$$a_1 = \frac{1+(-1)}{1} = \frac{0}{1} = 0$$
$$a_2 = \frac{1+(-1)^2}{2} = \frac{2}{2} = 1$$
$$a_3 = \frac{1+(-1)^3}{3} = \frac{0}{3} = 0$$
$$a_4 = \frac{1+(-1)^4}{4} = \frac{2}{4} = \frac{1}{2}$$
$$a_5 = \frac{1+(-1)^5}{5} = \frac{0}{5} = 0$$

Solution: 0, 1, 0, $\frac{1}{2}$, 0

19. Write the first five terms of the sequence $a_1 = 3$ and $a_{k+1} = 2(a_k - 1)$. (Assume n begins with 1.)

Solution:

$$a_1 = 3 \text{ and } a_{k+1} = 2(a_k - 1)$$
$$a_1 = 3$$
$$a_2 = 2(3 - 1) = 4$$
$$a_3 = 2(4 - 1) = 6$$
$$a_4 = 2(6 - 1) = 10$$
$$a_5 = 2(10 - 1) = 18$$

Solution: 3, 4, 6, 10, 18

21. Simplify the ratio

$$\frac{10!}{8!}.$$

Solution:

$$\frac{10!}{8!} = \frac{10 \cdot 9 \cdot 8!}{8!} = 10 \cdot 9 = 90$$

25. Simplify the ratio

$$\frac{(2n-1)!}{(2n+1)!}.$$

Solution:

$$\frac{(2n-1)!}{(2n+1)!} = \frac{(2n-1)!}{(2n+1)(2n)(2n-1)!} = \frac{1}{(2n+1)(2n)} = \frac{1}{2n(2n+1)}$$

29. Write an expression for the nth term of the sequence 0, 3, 8, 15, 24, ... (Assume n begins with 1.)

$$a_1 = 0 = 1^2 - 1$$
$$a_2 = 3 = 2^2 - 1$$
$$a_3 = 8 = 3^2 - 1$$
$$a_4 = 15 = 4^2 - 1$$
$$a_5 = 24 = 5^2 - 1$$

Therefore, $a_n = n^2 - 1$.

35. Write an expression for the nth term of the sequence $1+\frac{1}{1}, 1+\frac{1}{2}, 1+\frac{1}{3}, 1+\frac{1}{4}, 1+\frac{1}{5}, \ldots$. (Assume n begins with 1.)

Solution:

$$\begin{aligned} a_1 &= 1+\tfrac{1}{1} \\ a_2 &= 1+\tfrac{1}{2} \\ a_3 &= 1+\tfrac{1}{3} \\ a_4 &= 1+\tfrac{1}{4} \\ a_5 &= 1+\tfrac{1}{5} \end{aligned}$$

Therefore, $a_n = 1+\dfrac{1}{n}$.

39. Write an expression for the nth term of the sequence $1, -1, 1, -1, 1, \ldots$. (Assume n begins with 1.)

Solution:

$$\begin{aligned} a_1 &= 1 = (-1)^{1-1} &\text{or}\quad (-1)^{1+1} \\ a_2 &= -1 = (-1)^{2-1} &\text{or}\quad (-1)^{2+1} \\ a_3 &= 1 = (-1)^{3-1} &\text{or}\quad (-1)^{3+1} \\ a_4 &= -1 = (-1)^{4-1} &\text{or}\quad (-1)^{4+1} \\ a_5 &= 1 = (-1)^{5-1} &\text{or}\quad (-1)^{5+1} \end{aligned}$$

Therefore, $a_n = (-1)^{n-1}$ or $a_n = (-1)^{n+1}$.

43. Find the sum.

$$\sum_{k=0}^{3} \frac{1}{k^2+1}$$

Solution:

$$\begin{aligned} \sum_{k=0}^{3} \frac{1}{k^2+1} &= \frac{1}{(0)^2+1} + \frac{1}{(1)^2+1} + \frac{1}{(2)^2+1} + \frac{1}{(3)^2+1} \\ &= 1 + \frac{1}{2} + \frac{1}{5} + \frac{1}{10} \\ &= \frac{10+5+2+1}{10} \\ &= \frac{18}{10} \\ &= \frac{9}{5} \end{aligned}$$

49. Find the sum.

$$\sum_{i=1}^{4}(x^2+2i)$$

Solution:

$$\begin{aligned}\sum_{i=1}^{4}(x^2+2i) &= (x^2+2(1))+(x^2+2(2))+(x^2+2(3))+(x^2+2(4))\\ &= x^2+2+x^2+4+x^2+6+x^2+8\\ &= 4x^2+20\end{aligned}$$

51. Use sigma notation to write the sum

$$\frac{1}{3(1)}+\frac{1}{3(2)}+\frac{1}{3(3)}+\cdots+\frac{1}{3(9)}.$$

Solution:

$$\frac{1}{3(1)}+\frac{1}{3(2)}+\frac{1}{3(3)}+\cdots+\frac{1}{3(9)}=\sum_{i=1}^{9}\frac{1}{3i}$$

55. Use sigma notation to write the sum $3-9+27-81+243-729$.

Solution:

$$\begin{aligned}3-9+27-81+243-729 &= 3^1-3^2+3^3-3^4+3^5-3^6\\ &= \sum_{i=1}^{6}(-1)^{i+1}3^i\end{aligned}$$

59. Use sigma notation to write the sum

$$\frac{1}{4}+\frac{3}{8}+\frac{7}{16}+\frac{15}{32}+\frac{31}{64}.$$

Solution:

$$\begin{aligned}\frac{1}{4}+\frac{3}{8}+\frac{7}{16}+\frac{15}{32}+\frac{31}{64} &= \frac{2^1-1}{2^2}+\frac{2^2-1}{2^3}+\frac{2^3-1}{2^4}+\frac{2^4-1}{2^5}+\frac{2^5-1}{2^6}\\ &= \sum_{i=1}^{5}\frac{2^i-1}{2^{i+1}}\end{aligned}$$

61. Prove that

$$\sum_{i=1}^{n}(x_i - \overline{x}) = 0, \quad \text{where} \quad \overline{x} = \frac{1}{n}\sum_{i=1}^{n} x_i.$$

Solution:

$$\begin{aligned}
\sum_{i=1}^{n}(x_i - \overline{x}) &= \sum_{i=1}^{n} x_i - \sum_{i=1}^{n} \overline{x} \\
&= \sum_{i=1}^{n} x_i - n\overline{x} \\
&= \sum_{i=1}^{n} x_i - n\left(\frac{1}{n}\sum_{i=1}^{n} x_i\right) \\
&= 0
\end{aligned}$$

SECTION 10.2

Arithmetic Sequences

- You should be able to recognize an arithmetic sequence, find its common difference, and find its nth term.
- You should be able to find the nth partial sum of an arithmetic sequence with common difference d using the formula

$$S_n = \frac{n}{2}(a_1 + a_n) = \frac{n}{2}[2a_1 + (n-1)d].$$

Solutions to Selected Exercises

5. Determine whether the sequence $\frac{9}{4}$, 2, $\frac{7}{4}$, $\frac{3}{2}$, $\frac{5}{4}$, ... is arithmetic. If it is, find the common difference.

Solution:

$$\frac{9}{4}, 2, \frac{7}{4}, \frac{3}{2}, \frac{5}{4}, \ldots = \frac{9}{4}, \frac{8}{4}, \frac{7}{4}, \frac{6}{4}, \frac{5}{4}, \ldots$$

$$a_n = \frac{10}{4} - \frac{1}{4}n$$

Therefore, the sequence is arithmetic with $d = -\frac{1}{4}$.

9. Determine whether the sequence 5.3, 5.7, 6.1, 6.5, 6.9, ... is arithmetic. If it is, find the common difference.

Solution:

$$5.3, 5.7, 6.1, 6.5, 6.9, \ldots = 4.9 + 0.4,\ 4.9 + 2(0.4),\ 4.9 + 3(0.4),\ 4.9 + 4(0.4),\ 4.9 + 5(0.4), \ldots$$

$$a_n = 4.9 + 0.4n$$

Therefore, the sequence is arithmetic with $d = 0.4$.

13. Write the first five terms of the following sequence. Determine whether the sequence is arithmetic, and if it is, find the common difference.

$$a_n = \frac{1}{n+1}$$

Solution:

$$a_n = \frac{1}{n+1}$$
$$a_1 = \frac{1}{1+1} = \frac{1}{2}$$
$$a_2 = \frac{1}{2+1} = \frac{1}{3}$$
$$a_3 = \frac{1}{3+1} = \frac{1}{4}$$
$$a_4 = \frac{1}{4+1} = \frac{1}{5}$$
$$a_5 = \frac{1}{5+1} = \frac{1}{6}$$

The sequence is not arithmetic.

17. Write the first five terms of the sequence $a_1 = 1,\ a_2 = 1,\ a_n = a_{n-1} + a_{n-2},\ n \geq 3$. Determine whether the sequence is arithmetic, and if it is, find the common difference.

Solution:

$$a_1 = 1,\ a_2 = 1,\ a_n = a_{n-1} + a_{n-2},\ n \geq 3$$
$$a_1 = 1$$
$$a_2 = 1$$
$$a_3 = a_2 + a_1 = 1 + 1 = 2$$
$$a_4 = a_3 + a_2 = 2 + 1 = 3$$
$$a_5 = a_4 + a_3 = 3 + 2 = 5$$

The sequence 1, 1, 2, 3, 5, ... is not arithmetic.

21. Find a_n for the arithmetic sequence $a_1 = 100,\ d = -8,\ n = 8$.

Solution:

$$a_1 = 100,\ d = -8,\ n = 8$$
$$a_n = 100 + (n-1)(-8)$$
$$a_8 = 100 + (8-1)(-8) = 100 - 56 = 44$$

25. Find a_n for the arithmetic sequence $4,\ \frac{3}{2},\ -1,\ -\frac{7}{2},\ \ldots,\ n = 10$.

Solution:

$$4,\ \tfrac{3}{2},\ -1,\ -\tfrac{7}{2},\ \ldots,\ n = 10$$
$$a_n = 4 + (n-1)\left(-\tfrac{5}{2}\right)$$
$$a_{10} = 4 + (10-1)\left(-\tfrac{5}{2}\right) = 4 - \tfrac{45}{2} = -\tfrac{37}{2}$$

29. Write the first five terms of the arithmetic sequence $a_1 = -2.6,\ d = -0.4$.

Solution:

$$\begin{aligned} a_1 &= -2.6,\ d = -0.4 \\ a_2 &= -2.6 - 0.4, = -3 \\ a_3 &= -3 - 0.4 = -3.4 \\ a_4 &= -3.4 - 0.4 = -3.8 \\ a_5 &= -3.8 - 0.4 = -4.2 \end{aligned}$$

Solution: $-2.6,\ -3,\ -3.4,\ -3.8,\ -4.2$

35. Write the first five terms of the arithmetic sequence $a_8 = 26,\ a_{12} = 42$.

Solution:

$$\begin{aligned} a_8 &= 26,\ a_{12} = 42 \\ d &= \frac{42 - 26}{4} = 4 \\ a_n &= -6 + 4n \\ a_1 &= -2 \\ a_2 &= 2 \\ a_3 &= 6 \\ a_4 &= 10 \\ a_5 &= 14 \end{aligned}$$

Solution: $-2,\ 2,\ 6,\ 10,\ 14$

39. Find the nth partial sum of the arithmetic sequence $-6,\ -2,\ 2,\ 6,\ \ldots,\ n = 50$.

Solution:

$$\begin{aligned} &-6,\ -2,\ 2,\ 6,\ \ldots,\ n = 50 \\ &d = 4 \\ S_{50} &= \frac{50}{2}[2(-6) + (50 - 1)(4)] = 25[-12 + 196] = 25(184) = 4600 \end{aligned}$$

43. Find the nth partial sum of the arithmetic sequence $a_1 = 100,\ a_{25} = 220,\ n = 25$.

Solution:

$$\begin{aligned} a_1 &= 100,\ a_{25} = 220,\ n = 25 \\ S_{25} &= \frac{25}{2}(100 + 220) = \frac{25}{2}(320) = 4000 \end{aligned}$$

47. Find the sum.

$$\sum_{n=1}^{100} 5n$$

Solution:

$$a_n = 5n$$
$$a_1 = 5, \; a_{100} = 500$$
$$S_{100} = \tfrac{100}{2}(5 + 500) = 25,250$$

53. Find the sum.

$$\sum_{n=0}^{50} (1000 - 5n)$$

Solution:

$$\sum_{n=0}^{50} (1000 - 5n) = 1000 + \sum_{n=1}^{50} (1000 - 5n)$$
$$= 1000 + \frac{50}{2}(995 + 750) = 1000 + 43,625 = 44,625$$

59. Determine the seating capacity of an auditorium with 30 rows of seats if there are 20 seats in the first row, 24 seats in the second row, 28 seats in the third row, and so on.

Solution:

$$a_n = 16 + 4n$$
$$a_1 = 20, \; a_{30} = 136$$
$$S_{30} = \tfrac{30}{2}(20 + 136) = 2340 \text{ seats}$$

63. Insert three arithmetic means between the pair of numbers 3 and 6.

Solution:

$3, \; 6; \; k = 3$

$3, \; m_1, \; m_2, \; m_3, \; 6$

$$a_5 = 6 = 3 + 4d$$
$$d = \frac{3}{4}$$
$$m_1 = 3 + \frac{3}{4} = \frac{15}{4}$$
$$m_2 = \frac{15}{4} + \frac{3}{4} = \frac{18}{4} = \frac{9}{2}$$
$$m_3 = \frac{18}{4} + \frac{3}{4} = \frac{21}{4}$$

SECTION 10.3

Geometric Sequences and Series

- You should be able to identify a geometric sequence, find its common ratio, and find the nth term.
- You should be able to find the nth partial sum of a geometric sequence with common ratio r using the formula

$$S_n = \frac{a_1(1-r^n)}{1-r}, \quad r \neq 1.$$

- You should know that if $|r| < 1$, then

$$\sum_{n=0}^{\infty} a_1 r^n = \sum_{n=1}^{\infty} a_1 r^{n-1} = \frac{a_1}{1-r}.$$

- You should be able to write a repeating decimal as the ratio of two integers.

Solutions to Selected Exercises

3. Determine whether the sequence 3, 12, 21, 30, ... is geometric. If it is, find its common ratio.

Solution:

$$3,\ 12,\ 21,\ 30,\ \ldots$$
$$a_n = -6 + 9n$$

This an arithmetic sequence, not a geometric sequence.

7. Determine whether the sequence $\frac{1}{2}, \frac{2}{3}, \frac{3}{4}, \frac{4}{5}, \ldots$ is geometric. If it is, find its common ratio.

Solution:

$$\frac{1}{2}, \frac{2}{3}, \frac{3}{4}, \frac{4}{5}, \ldots$$
$$a_n = \frac{n}{n+1}$$

This is not a geometric series.

11. Write the first five terms of the geometric sequence $a_1 = 2,\ r = 3$.

Solution:

$$a_1 = 2, \ r = 3$$
$$a_2 = 2(3) = 6$$
$$a_3 = 2(3)^2 = 18$$
$$a_4 = 2(3)^3 = 54$$
$$a_5 = 2(3)^4 = 162$$

Solution: 2, 6, 18, 54, 162

15. Write the first five terms of the geometric sequence $a_1 = 5, \ r = -\frac{1}{10}$.

Solution:

$$a_1 = 5, \ r = -\frac{1}{10}$$
$$a_2 = 5\left(-\frac{1}{10}\right) = -\frac{1}{2}$$
$$a_3 = 5\left(-\frac{1}{10}\right)^2 = \frac{1}{20}$$
$$a_4 = 5\left(-\frac{1}{10}\right)^3 = -\frac{1}{200}$$
$$a_5 = 5\left(-\frac{1}{10}\right)^4 = \frac{1}{2000}$$

Solution: $5, \ -\frac{1}{2}, \ \frac{1}{20}, \ -\frac{1}{200}, \ \frac{1}{2000}, \ \ldots$

19. Find the nth term of the geometric sequence $a_1 = 4, \ r = \frac{1}{2}, \ n = 10$.

Solution:

$$a_1 = 4, \ r = \frac{1}{2}, \ n = 10$$
$$a_{10} = 4\left(\frac{1}{2}\right)^9 = \frac{1}{128} = \left(\frac{1}{2}\right)^7$$

23. Find the nth term of the geometric sequence $a_1 = 100, \ r = e^x, \ n = 9$.

Solution:

$$a_1 = 100, \ r = e^x, \ n = 9$$
$$a_9 = 100(e^x)^8 = 100e^{8x}$$

27. Find the nth term of the geometric sequence $a_1 = 16, \ a_4 = \frac{27}{4}, \ n = 3$.

Solution:

$a_1 = 16,\ a_4 = \dfrac{27}{4},\ n = 3$

$$a_4 = 16r^3 = \frac{27}{4}$$

$$r^3 = \frac{27}{64}$$

$$r = \frac{3}{4}$$

$$a_3 = 16\left(\frac{3}{4}\right)^2 = 16\left(\frac{9}{16}\right) = 9$$

31. A sum of \$1000 is invested at 10% interest. Find the amount after 10 years if the interest is compounded (a) annually, (b) semiannually, (c) quarterly, (d) monthly, and (e) daily.

Solution:

$$A = P\left(1+\frac{r}{n}\right)^{nt} = 1000\left(1+\frac{0.10}{n}\right)^{n(10)}$$

(a) $n = 1,\quad A = 1000(1+0.10)^{10} = \$2593.74$

(b) $n = 2,\quad A = 1000\left(1+\dfrac{0.10}{2}\right)^{2(10)} = \$2653.30$

(c) $n = 4,\quad A = 1000\left(1+\dfrac{0.10}{4}\right)^{4(10)} = \$2685.06$

(d) $n = 12,\quad A = 1000\left(1+\dfrac{0.10}{12}\right)^{12(10)} = \$2707.04$

(e) $n = 365,\quad A = 1000\left(1+\dfrac{0.10}{365}\right)^{365(10)} = \$2717.91$

35. Find the sum.

$$\sum_{n=0}^{20} 3\left(\frac{3}{2}\right)^n$$

Solution:

$$\sum_{n=0}^{20} 3\left(\frac{3}{2}\right)^n = \frac{3(1-(3/2)^{21})}{1-(3/2)} = 29,921.31$$

39. Find the sum.

$$\sum_{n=0}^{8} 2^n$$

Solution:

$$\sum_{n=0}^{8} 2^n = \frac{1-2^9}{1-2} = 511$$

43. A deposit of P dollars is made at the beginning of each month for T years in an account that pays R percent interest, compounded monthly. Let $N = 12T$ be the total number of deposits. The balance after T years is

$$A = P\left(1+\frac{R}{12}\right) + P\left(1+\frac{R}{12}\right)^2 + \cdots + P\left(1+\frac{R}{12}\right)^N.$$

Show that the balance is given by

$$A = P\left[\left(1+\frac{R}{12}\right)^N - 1\right]\left(1+\frac{12}{R}\right).$$

Solution:

$$\begin{aligned}
A &= P\left(1+\frac{R}{12}\right) + P\left(1+\frac{R}{12}\right)^2 + \cdots + P\left(1+\frac{R}{12}\right)^N \\
&= \left(1+\frac{R}{12}\right)\left[P + P\left(1+\frac{R}{12}\right) + \cdots + P\left(1+\frac{R}{12}\right)^{N-1}\right] \\
&= P\left(1+\frac{R}{12}\right)\sum_{n=1}^{N}\left(1+\frac{R}{12}\right)^{n-1} \\
&= P\left(1+\frac{R}{12}\right)\frac{1-\left(1+\frac{R}{12}\right)^N}{1-\left(1+\frac{R}{12}\right)} \\
&= P\left(1+\frac{R}{12}\right)\left(-\frac{12}{R}\right)\left[1-\left(1+\frac{R}{12}\right)^N\right] \\
&= P\left(\frac{12}{R}+1\right)\left[-1+\left(1+\frac{R}{12}\right)^N\right] \\
&= P\left[\left(1+\frac{R}{12}\right)^N - 1\right]\left(1+\frac{12}{R}\right)
\end{aligned}$$

47. Find the sum of the infinite geometric series .

$$\sum_{n=0}^{\infty}\left(\frac{1}{2}\right)^n = 1 + \frac{1}{2} + \frac{1}{4} + \frac{1}{8} + \cdots$$

Solution:

$$\sum_{n=0}^{\infty}\left(\frac{1}{2}\right)^n = 1+\frac{1}{2}+\frac{1}{4}+\frac{1}{8}+\cdots = \frac{1}{1-(1/2)} = 2$$

51. Find the sum of the infinite geometric series.

$$\sum_{n=0}^{\infty}4\left(\frac{1}{4}\right)^n = 4+1+\frac{1}{4}+\frac{1}{16}+\cdots$$

Solution:

$$\sum_{n=0}^{\infty}4\left(\frac{1}{4}\right)^n = 4+1+\frac{1}{4}+\frac{1}{16}+\cdots = \frac{4}{1-(1/4)} = \frac{16}{3}$$

55. Find the sum of the infinite geometric series $4-2+1-\frac{1}{2}+\cdots$.

Solution:

$$4-2+1-\frac{1}{2}+\cdots = \sum_{n=0}^{\infty}4\left(-\frac{1}{2}\right)^n = \frac{4}{1-(-1/2)} = \frac{8}{3}$$

59. A ball is dropped from a height of 16 feet. Each time it drops h feet, it rebounds $0.81h$ feet. Find the total distance traveled by the ball.

Solution:

$$\text{Total distance} = \left[\sum_{n=0}^{\infty}32(0.81)^n\right] - 16 = \frac{32}{1-0.81} - 16 = 152.42 \text{ ft}$$

63. Write 0.363636... as the ratio of two integers by considering it to be the sum of an infinite geometric series.

Solution:

$$0.363636\ldots = \sum_{n=0}^{\infty}0.36(0.01)^n = \frac{0.36}{1-0.01} = \frac{0.36}{0.99} = \frac{36}{99} = \frac{4}{11}$$

67. Write 1.363636... as the ratio of two integers by considering it to be the sum of an infinite geometric series.

Solution:

$$\begin{aligned}1.363636\ldots &= 1+\sum_{n=0}^{\infty}0.36(0.01)^n\\ &= 1+\frac{4}{11} \quad \text{From Exercise 63}\\ &= \frac{15}{11}\end{aligned}$$

SECTION 10.4

Mathematical Induction

- You should be sure that you understand the principle of mathematical induction. If P_n is a statement involving the positive integer n, where P_1 is true and the truth of P_k implies the truth of P_{k+1}, then P_n is true for all positive integers n.
- You should be able to use mathematical induction to find a formula for the nth term of a sequence.

Solutions to Selected Exercises

5. Find the following sum, using the formulas for the sums of powers of integers.

$$\sum_{n=1}^{6} n^4$$

Solution:

$$\sum_{n=1}^{N} n^4 = \frac{N(N+1)(2N+1)(3N^2+3N-1)}{30}$$

$$\sum_{n=1}^{6} n^4 = \frac{6(7)(13)(125)}{30} = 2275$$

9. Find S_{k+1} for

$$S_k = \frac{k^2(k+1)^2}{4}.$$

Solution:

$$S_k = \frac{k^2(k+1)^2}{4}$$

$$S_{k+1} = \frac{(k+1)^2((k+1)+1)^2}{4}$$

$$= \frac{(k+1)^2(k+2)^2}{4}$$

13. Use mathematical induction to prove the formula for every positive integer n.

$$2 + 7 + 12 + 17 + \cdots + (5n - 3) = \frac{n}{2}(5n - 1)$$

Solution:

When $n = 1$,

$$S_1 = 2 = \frac{1}{2}(5(1) - 1).$$

Assume that

$$S_k = 2 + 7 + 12 + 17 + \cdots + (5k - 3) = \frac{k}{2}(5k - 1).$$

Then,

$$\begin{aligned} S_{k+1} &= 2 + 7 + 12 + 17 + \cdots + (5k - 3) + [5(k + 1) - 3] \\ &= S_k + 5k + 5 - 3 \\ &= \frac{k}{2}(5k - 1) + 5k + 2 \\ &= \frac{5k^2 - k + 10k + 4}{2} \\ &= \frac{5k^2 + 9k + 4}{2} \\ &= \frac{(k + 1)(5k + 4)}{2} \\ &= \frac{(k + 1)}{2}[5(k + 1) - 1]. \end{aligned}$$

We conclude by mathematical induction that the formula is valid for all positive integer values of n.

17. Use mathematical induction to prove the formula for every positive integer n.

$$1 + 2 + 3 + 4 + \cdots + n = \frac{n(n + 1)}{2}$$

Solution:

When $n = 1$,

$$S_1 = 1 = \frac{1(1 + 1)}{2}.$$

Assume that

$$S_k = 1 + 2 + 3 + 4 + \cdots + k = \frac{k(k + 1)}{2}.$$

Then,

$$\begin{aligned} S_{k+1} &= 1+2+3+4+\cdots+k+k+1 \\ &= S_k+k+1 \\ &= \frac{k(k+1)}{2}+\frac{2(k+1)}{2} \\ &= \frac{(k+1)(k+2)}{2}. \end{aligned}$$

Therefore, we conclude that this formula holds for all positive integer values of n.

23. Use mathematical induction to prove the formula for every positive integer n.

$$\sum_{i=1}^{n} i(i+1) = \frac{n(n+1)(n+2)}{3}$$

Solution:
When $n = 1$,

$$S_1 = 2 = \frac{1(2)(3)}{3}.$$

Assume that

$$S_k = 1(2)+2(3)+3(4)+\cdots+k(k+1) = \frac{k(k+1)(k+2)}{3}.$$

Then,

$$\begin{aligned} S_{k+1} &= 1(2)+2(3)+3(4)+\cdots+k(k+1)+(k+1)(k+2) \\ &= S_k+(k+1)(k+2) \\ &= \frac{k(k+1)(k+2)}{3}+\frac{3(k+1)(k+2)}{3} \\ &= \frac{(k+1)(k+2)(k+3)}{3}. \end{aligned}$$

Thus, this formula is valid for all positive integer values of n.

27. Find a formula for the nth partial sum of the sequence.

$$1, \frac{9}{10}, \frac{81}{100}, \frac{729}{1000}, \ldots$$

Solution:

$$S_1 = 1$$
$$S_2 = 1 + \frac{9}{10} = 1 + \left(\frac{9}{10}\right)^1$$
$$S_3 = 1 + \frac{9}{10} + \frac{81}{100} = 1 + \left(\frac{9}{10}\right)^1 + \left(\frac{9}{10}\right)^2$$
$$S_4 = 1 + \frac{9}{10} + \frac{81}{100} + \frac{729}{1000} = 1 + \left(\frac{9}{10}\right)^1 + \left(\frac{9}{10}\right)^2 + \left(\frac{9}{10}\right)^3$$
$$\vdots$$
$$S_n = 1 + \left(\frac{9}{10}\right)^1 + \left(\frac{9}{10}\right)^2 + \left(\frac{9}{10}\right)^3 + \cdots + \left(\frac{9}{10}\right)^{n-1} = \sum_{i=1}^{n}\left(\frac{9}{10}\right)^{i-1}$$

Since this is a geometric series,

$$S_n = \frac{1 - \left(\frac{9}{10}\right)^n}{1 - \frac{9}{10}} = 10[1 - (0.9)^n].$$

31. Use mathematical induction to prove the inequality, $\left(\frac{4}{3}\right)^n > n,\ n \geq 7$.

Solution:

When $n = 7$,

$$\left(\frac{4}{3}\right)^7 \approx 7.4915 > 7.$$

Assume that

$$\left(\frac{4}{3}\right)^k > k,\ k > 7.$$

Then,

$$\left(\frac{4}{3}\right)^{k+1} = \left(\frac{4}{3}\right)^k \left(\frac{4}{3}\right) > k\left(\frac{4}{3}\right) = k + \frac{k}{3} > k + 1.$$

Thus,

$$\left(\frac{4}{3}\right)^{k+1} > k + 1.$$

Therefore,

$$\left(\frac{4}{3}\right)^n > n.$$

35. Use mathematical induction to prove the property $(ab)^n = a^n b^n$ for all positive integers n.

Solution:

When $n = 1$, $(ab)^1 = a^1 b^1 = ab$.

Assume that $(ab)^k = a^k b^k$.

Then,
$$\begin{aligned}(ab)^{k+1} &= (ab)^k(ab)\\ &= a^k b^k ab\\ &= a^{k+1} b^{k+1}.\end{aligned}$$

Thus, $(ab)^n = a^n b^n$.

39. Use mathematical induction to prove the Generalized Distributive Law: $x(y_1 + y_2 + \cdots + y_n) = xy_1 + xy_2 + \cdots + xy_n$.

Solution:

When $n = 1$, $x(y_1) = xy_1$.

Assume that $x(y_1 + y_2 + \cdots + y_k) = xy_1 + xy_2 + \cdots + xy_k$.

Then,
$$\begin{aligned}xy_1 + xy_2 + \cdots + xy_k + xy_{k+1} &= x(y_1 + y_2 + \cdots + y_k) + xy_{k+1}\\ &= x[(y_1 + y_2 + \cdots + y_k) + y_{k+1}]\\ &= x(y_1 + y_2 + \cdots + y_k + y_{k+1}).\end{aligned}$$

Hence, the formula holds.

43. Use mathematical induction to prove DeMoivre's Theorem: $[r(\cos\theta + i\sin\theta)]^n = r^n[\cos n\theta + i\sin n\theta]$.

Solution:

When $n = 1$, $[r(\cos\theta + i\sin\theta)]^1 = r^1[\cos\theta + i\sin\theta]$.

Assume that $[r(\cos\theta + i\sin\theta)]^k = r^k[\cos k\theta + i\sin k\theta]$.

Then,
$$\begin{aligned}[r(\cos\theta + i\sin\theta)]^{k+1} &= [r(\cos\theta + i\sin\theta)]^k[r(\cos\theta + i\sin\theta)]\\ &= r^k(\cos k\theta + i\sin k\theta)r(\cos\theta + i\sin\theta)\\ &= r^{k+1}[\cos(k+1)\theta + i\sin(k+1)\theta].\end{aligned}$$

Thus, DeMoivre's Theorem holds.

SECTION 10.5

The Binomial Theorem

- You should be able to use the formula

$$(x+y)^n = x^n + nx^{n-1}y + \frac{n(n-1)}{2!}x^{n-2}y^2 + \cdots + {}_nC_r{}^n x^{n-r}y^r + \cdots + y^n$$

where ${}_nC_r = \frac{n!}{(n-r)!r!}$, to expand $(x+y)^n$.

- You should be able to use Pascal's Triangle in binomial expansion.

Solutions to Selected Exercises

3. Evaluate ${}_{12}C_0$.

Solution:

$${}_{12}C_0 = \frac{12!}{(12-0)!0!} = \frac{12!}{(12!)(1)} = 1$$

7. Evaluate ${}_{100}C_{98}$.

Solution:

$${}_{100}C_{98} = \frac{100!}{(100-98)!98!} = \frac{100 \cdot 99 \cdot 98!}{2! \cdot 98!} = \frac{100 \cdot 99}{2} = 4950$$

11. Use the Binomial Theorem to expand $(x+y)^5$. Simplify your answer.

Solution:

$$\begin{aligned}(x+y)^5 &= x^5 + 5x^4y + {}_5C_2x^3y^2 + {}_5C_3x^2y^3 + 5xy^4 + y^5 \\ &= x^5 + 5x^4y + 10x^3y^2 + 10x^2y^3 + 5xy^4 + y^5\end{aligned}$$

15. Use the Binomial Theorem to expand $(r+3s)^6$. Simplify your answer.

Solution:

$$\begin{aligned}(r+3s)^6 &= r^6 + 6r^5(3s) + 15r^4(3s)^2 + 20r^3(3s)^3 + 15r^2(3s)^4 + 6r(3s)^5 + (3s)^6 \\ &= r^6 + 18r^5s + 135r^4s^2 + 540r^3s^3 + 1215r^2s^4 + 1458rs^5 + 729s^6\end{aligned}$$

19. Use the Binomial Theorem to expand $(1-2x)^3$. Simplify your answer.

Solution:

$$\begin{aligned}(1-2x)^3 &= [1+(-2x)]^3\\ &= 1^3 + 3(1)^2(-2x) + 3(1)(-2x)^2 + (-2x)^3\\ &= 1 - 6x + 12x^2 - 8x^3\end{aligned}$$

23. Use the Binomial Theorem to expand the following. Simplify your answer.

$$\left(\frac{1}{x}+y\right)^5$$

Solution:

$$\begin{aligned}\left(\frac{1}{x}+y\right)^5 &= \left(\frac{1}{x}\right)^5 + 5\left(\frac{1}{x}\right)^4 y + 10\left(\frac{1}{x}\right)^3 y^2 + 10\left(\frac{1}{x}\right)^2 y^3 + 5\left(\frac{1}{x}\right) y^4 + y^5\\ &= \frac{1}{x^5} + \frac{5y}{x^4} + \frac{10y^2}{x^3} + \frac{10y^3}{x^2} + \frac{5y^4}{x} + y^5\end{aligned}$$

27. Use the Binomial Theorem to expand $(2-3i)^6$. Simplify your answer by using the fact that $i^2 = -1$.

Solution:

$$\begin{aligned}(2-3i)^6 &= 2^6 - 6(2)^5(3i) + 15(2)^4(3i)^2 - 20(2)^3(3i)^3 + 15(2)^2(3i)^4 - 6(2)(3i)^5 + (3i)^6\\ &= 64 - 576i - 2160 + 4320i + 4860 - 2916i - 729\\ &= 2035 + 828i\end{aligned}$$

31. Expand $(2t-s)^5$, using Pascal's Triangle to determine the coefficients.

Solution:

$$\begin{array}{c} 1\\ 1 \quad 1\\ 1 \quad 2 \quad 1\\ 1 \quad 3 \quad 3 \quad 1\\ 1 \quad 4 \quad 6 \quad 4 \quad 1\\ 1 \quad 5 \quad 10 \quad 10 \quad 5 \quad 1\end{array}$$

$$\begin{aligned}(2t-s)^5 &= (2t)^5 - 5(2t)^4s + 10(2t)^3s^2 - 10(2t)^2s^3 + 5(2t)s^4 - s^5\\ &= 32t^5 - 80t^4s + 80t^3s^2 - 40t^2s^3 + 10ts^4 - s^5\end{aligned}$$

35. Find the term x^5 in the expansion of $(x+3)^{12}$.

Solution:

$$_{12}C_7 x^5 (3)^7 = \frac{12!3^7}{(12-7)!7!} = 1,732,104x^5$$

39. Find the term x^4y^{11} in the expansion of $(3x-2y)^{15}$.

Solution:

$$_{15}C_{11}(3x)^4(-2y)^{11} = \frac{15!3^4(-2)^{11}}{(15-11)!11!}x^4y^{11} = -226,437,120x^4y^{11}$$

45. Use the Binomial Theorem to expand $\left(\frac{1}{3}+\frac{2}{3}\right)^8$. In the study of probability, it is sometimes necessary to use the expansion of $(p+q)^n$, where $p+q=1$.

Solution:

$$\begin{aligned}\left(\frac{1}{3}+\frac{2}{3}\right)^8 &= \left(\frac{1}{3}\right)^8 + 8\left(\frac{1}{3}\right)^7\left(\frac{2}{3}\right) + 28\left(\frac{1}{3}\right)^6\left(\frac{2}{3}\right)^2 + 56\left(\frac{1}{3}\right)^5\left(\frac{2}{3}\right)^3 + 70\left(\frac{1}{3}\right)^4\left(\frac{2}{3}\right)^4 \\ &\quad + 56\left(\frac{1}{3}\right)^3\left(\frac{2}{3}\right)^5 + 28\left(\frac{1}{3}\right)^2\left(\frac{2}{3}\right)^6 + 8\left(\frac{1}{3}\right)\left(\frac{2}{3}\right)^7 + \left(\frac{2}{3}\right)^8 \\ &= \frac{1}{6561} + \frac{16}{6561} + \frac{112}{6561} + \frac{448}{6561} + \frac{1120}{6561} + \frac{1792}{6561} + \frac{1792}{6561} + \frac{1024}{6561} + \frac{256}{6561}\end{aligned}$$

51. Prove $_{n+1}C_m = {}_nC_m + {}_nC_{m-1}$ for all integers m and n, $0 \le m \le n$.

Solution:

$$\begin{aligned}{}_nC_m + {}_nC_{m-1} &= \frac{n!}{(n-m)!m!} + \frac{n!}{(n-m+1)!(m-1)!} \\ &= \frac{n!(n-m+1)!(m-1)! + n!(n-m)!m!}{(n-m)!m!(n-m+1)!(m-1)!} \\ &= \frac{n![(n-m+1)!(m-1)! + m!(n-m)!]}{(n-m)!m!(n-m+1)!(m-1)!} \\ &= \frac{n!\cancel{(m-1)}![(n-m+1)! + m(n-m)!]}{(n-m)!m!(n-m+1)!\cancel{(m-1)}!} \\ &= \frac{n!\cancel{(n-m)}![(n-m+1)+m]}{\cancel{(n-m)}!m!(n-m+1)!} \\ &= \frac{n![n+1]}{m!(n-m+1)!} \\ &= \frac{(n+1)!}{[(n+1)-m]!m!} \\ &= {}_{n+1}C_m\end{aligned}$$

SECTION 10.6

Counting Principles, Permutations, and Combinations

- You should know The Fundamental Principle of Counting.
- $_nP_r = \frac{n!}{(n-r)!}$ is the number of permutations of n elements taken r at a time.
- Given a set of n objects that has n_1 of one kind, n_2 of a second kind, and so on, the number of distinguishable permutations is

  $$\frac{n!}{n_1!n_2!\dots n_k!}.$$

- $_nC_r = \frac{n!}{(n-r)!r!}$ is the number of combinations of n elements taken r at a time.

Solutions to Selected Exercises

1. A small college needs two additional faculty members, a chemist and a statistician. In how many ways can these positions be filled if there are three applicants for the chemistry position and four for the position in statistics?

Solution:

$3 \cdot 4 = 12$ ways to fill the positions.

5. In a certain state the automobile license plates consist of two letters followed by a four-digit number. How many distinct license plate numbers can be formed?

Solution:

$26 \cdot 26 \cdot 10 \cdot 10 \cdot 10 \cdot 10 = 6,760,000$ distinct license plate numbers.

9. Three couples have reserved seats in a given row for a concert. In how many different ways can they be seated, given the following conditions?

(a) There are no seating restrictions.
(b) The two members of each couple wish to sit together.

Solution:

(a) $6! = 720$ different ways.
(b) $6 \cdot 4 \cdot 2 = 48$ different ways.

13. Evaluate ${}_8P_3$.

Solution:

$${}_8P_3 = \frac{8!}{(8-3)!} = \frac{8!}{5!} = 8 \cdot 7 \cdot 6 = 336$$

17. Evaluate ${}_{100}P_2$.

Solution:

$${}_{100}P_2 = \frac{100!}{(100-2)!} = \frac{100!}{98!} = 100 \cdot 99 = 9900$$

21. In how many ways can five children line up in one row to have their picture taken?

Solution:

$5! = 120$ ways.

25. In order to conduct a certain experiment, four students are randomly selected from a class of 20. How many different groups of four students are possible?

$${}_{20}C_4 = \frac{20!}{(20-4)!4!} = \frac{20!}{16!4!} = \frac{20 \cdot 19 \cdot 18 \cdot 17}{4 \cdot 3 \cdot 2} = 4845 \text{ different groups.}$$

29. Determine the number of three-digit numbers that can be formed from the ten digits 0, 1, 2, 3, 4, 5, 6, 7, 8, 9. (The leading digit cannot be zero.)

Solution:

$9 \cdot 10 \cdot 10 = 900$ different numbers.

35. An employer interviews eight people for four openings in the company. Three of the eight people are from a minority group. If all eight are qualified, in how many ways could the employer fill the four positions if (a) the selection is random and (b) exactly two are selected from the minority group?

Solution:

(a) $${}_8C_4 = \frac{8!}{(8-4)!4!} = \frac{8!}{4!4!} = \frac{8 \cdot 7 \cdot 6 \cdot 5}{4 \cdot 3 \cdot 2} = 70 \text{ ways.}$$

(b) $${}_3C_2 \cdot {}_5C_2 = \frac{3!}{(3-2)!2!} \cdot \frac{5!}{(5-2)!2!} = 3 \cdot 10 = 30 \text{ ways.}$$

37. Four people are to be selected at random from a group of four couples. In how many ways can this be done, given the following conditions?

(a) There are no restrictions.
(b) There is to be at least one couple in the group of four.
(c) The selection must include one member from each couple.

Solution:

(a) ${}_8C_4 = \frac{8!}{4!4!} = 70$ ways.

(b) There are 16 ways that a group of four can be formed without any couples in the group. Therefore, if at least one couple is to be in the group, there are $70 - 16 = 54$ ways that could occur.

(c) $2 \cdot 2 \cdot 2 \cdot 2 = 16$ ways.

41. Find the number of diagonals of an octagon.

Solution:

$$ {}_8C_2 - 8 = \frac{8!}{6!2!} - 8 = 28 - 8 = 20 \text{ diagonals.} $$

45. Find the number of distinguishable permutations of the letters A, A, Y, Y, Y, Y, X, X, X.

Solution:

$$ \frac{9!}{2!4!3!} = \frac{9 \cdot 8 \cdot 7 \cdot 6 \cdot 5}{2 \cdot 3 \cdot 2} = 1260 \text{ distinguishable permutations.} $$

49. Solve $14 \cdot {}_nP_3 = {}_{n+2}P_4$ for n.

Solution:

$$ \begin{aligned} 14 \cdot {}_nP_3 &= {}_{n+2}P_4 \\ \frac{14n!}{(n-3)!} &= \frac{(n+2)!}{((n+2)-4)!} \\ 14n(n-1)(n-2) &= (n+2)(n+1)(n)(n-1) \\ 0 &= (n+2)(n+1)(n)(n-1) - 14n(n-1)(n-2) \\ 0 &= n(n-1)[(n+2)(n+1) - 14(n-2)] \\ 0 &= n(n-1)[n^2 + 3n + 2 - 14n + 28] \\ 0 &= n(n-1)[n^2 - 11n + 30] \\ 0 &= n(n-1)(n-5)(n-6) \end{aligned} $$

$$ n = 0, \ n = 1, \ n = 5, \ n = 6 $$

Since $n \geq 3$, we have $n = 5$ or $n = 6$.

53. Prove ${}_nC_{n-1} = {}_nC_1$.

Solution:

$$ {}_nC_{n-1} = \frac{n!}{(n-(n-1))!(n-1)!} = \frac{n!}{(1)!(n-1)!} = \frac{n!}{(n-1)!1!} = {}_nC_1 $$

SECTION 10.7

Probability

You should know the following basic principles of probability.

- If an event A has $n(A)$ equally likely outcomes and its sample space has $n(S)$ equally likely outcomes, then the probability of event A is

$$P(A) = \frac{n(A)}{n(S)}.$$

- If A and B are mutually exclusive events, then $P(A \text{ or } B) = P(A) + P(B)$.
- If A and B are independent events, then the probability that both A and B will occur is $P(A)P(B)$.
- The complement of an event A is $P(A') = 1 - P(A)$.

Solutions to Selected Exercises

3. A coin is tossed three times. Find the probability of getting at least one head.

Solution:

$$S = \{HHH,\ HHT,\ HTH,\ HTT,\ THH,\ THT,\ TTH,\ TTT\}$$

$$P(TTT) = \frac{1}{8}$$

$$P(\text{at least one head}) = 1 - \frac{1}{8} = \frac{7}{8}$$

7. One card is selected from a standard deck of 52 playing cards. Find the probability of getting a black card that is not a face card.

Solution:

26 of the cards are black. Six of these are face cards (J, Q, K of clubs and spades). Therefore, there are 20 black cards that are not face cards.

$$P(A) = \frac{20}{52} = \frac{5}{13}$$

9. A six-sided die is tossed twice. Find the probability that the sum is 4.

Solution:

$$n(S) = 6(6) = 36$$
$$A = \{(1,\ 3),\ (2,\ 2),\ (3,\ 1)\}$$
$$P(A) = \frac{3}{36} = \frac{1}{12}$$

13. A six-sided die is tossed twice. Find the probability that the sum is odd and no more than 7.

Solution:

$$n(S) = 6 \cdot 6 = 36$$
$$A = \{(1,\ 2),\ (1,\ 4),\ (1,\ 6),\ (2,\ 1),\ (2,\ 3),\ (2,\ 5),\ (3,\ 2),\ (3,\ 4),\ (4,\ 1),\ (4,\ 3),\ (5,\ 2),\ (6,\ 1)\}$$
$$P(A) = \frac{12}{36} = \frac{1}{3}$$

17. Two marbles are drawn (the first is *not* replaced before the second is drawn) from a bag containing one green, two yellow, and three red marbles. Find the probability of drawing neither yellow marble.

Solution:

$$P(A) = \frac{{}_4C_2}{{}_6C_2} = \frac{\frac{4!}{2!2!}}{\frac{6!}{4!2!}} = \frac{6}{15} = \frac{2}{5}$$

23. Two integers (between 1 and 30 inclusive) are chosen by a random number generator on a computer. What is the probability that (a) the numbers are both even, (b) one number is even and one is odd, (c) both numbers are less than 10, and (d) the same number is chosen twice?

Solution:

(a) $P(EE) = \frac{15}{30} \cdot \frac{15}{30} = \frac{1}{4}$

(b) $P(EO \text{ or } OE) = 2\left(\frac{15}{30}\right)\left(\frac{15}{30}\right) = \frac{1}{2}$

(c) $P(N_1 < 10,\ N_2 < 10) = \frac{9}{30} \cdot \frac{9}{30} = \frac{9}{100}$

(d) $P(N_1 N_1) = \frac{30}{30} \cdot \frac{1}{30} = \frac{1}{30}$

27. Four letters and envelopes are addressed to four different people. If the letters are randomly inserted into the envelopes, what is the probability that (a) exactly one will be inserted in the correct envelope and (b) at least one will be inserted in the correct envelope?

Solution:

(a) $P((C, W, W, W), (W, C, W, W), (W, W, C, W), (W, W, W, C))$
$= 4\left(\frac{1}{4}\right)\left(\frac{4}{6}\right)\left(\frac{2}{4}\right)(1) = \frac{1}{3}$

(b) $1 - P(W, W, W, W) = 1 - \frac{3}{4}\left(\frac{3}{6}\right) = \frac{5}{8}$

31. A space vehicle has an independent back-up system for one of its communication networks. The probability that either system will function satisfactorily for the duration of a flight is 0.985. What is the probability that during a given flight (a) both systems function satisfactorily, (b) at least one system functions satisfactorily, and (c) both systems fail?

Solution:

(a) $P(SS) = (0.985)^2 = 0.9702$

(b) $P(S) = 1 - P(FF) = 1 - (0.015)^2 = 0.9998$

(c) $P(FF) = (0.015)^2 = 0.0002$

35. A particular binomial experiment has 10 trials. The probability of success is 10%.

(a) Find the probability that all 10 trials fail.
(b) Find the probability that all 10 trials succeed.
(c) Find the probability that at least one trial succeeds.
(d) Find the probability that exactly one trial succeeds.

Solution:

(a) $P(\text{all fail}) = (0.9)^{10}$

(b) $P(\text{all succeed}) = (0.1)^{10}$

(c) $P(\text{at least one success}) = 1 - P(\text{all fail}) = 1 - (0.9)^{10}$

(d) $P(\text{one success}) = {}_{10}C_1(0.1)^1(0.9)^9 = (0.9)^9$

REVIEW EXERCISES FOR CHAPTER 10

Solutions to Selected Exercises

3. Use sigma notation to write the sum $\frac{1}{2} + \frac{2}{3} + \frac{3}{4} + \cdots + \frac{9}{10}$.

Solution:

$$\frac{1}{2} + \frac{2}{3} + \frac{3}{4} + \cdots + \frac{9}{10} = \sum_{k=1}^{9} \frac{k}{k+1}$$

7. Find the sum.

$$\sum_{j=3}^{10} (2j - 3)$$

Solution:

$$\begin{aligned}\sum_{j=3}^{10} (2j - 3) &= [2(3) - 3] + [2(4) - 3] + [2(5) - 3] + [2(6) - 3] + [2(7) - 3] \\ &\quad + [2(8) - 3] + [2(9) - 3] + [2(10) - 3] \\ &= 3 + 5 + 7 + 9 + 11 + 13 + 15 + 17 \\ &= 80\end{aligned}$$

11. Find the sum.

$$\sum_{i=0}^{\infty} \left(\frac{7}{8}\right)^i$$

Solution:

$$\sum_{i=0}^{\infty} \left(\frac{7}{8}\right)^i = \frac{1}{1 - \frac{7}{8}} = 8$$

17. Find the sum.

$$\sum_{n=0}^{10} (n^2 + 3)$$

Solution:

$$\sum_{n=0}^{10}(n^2+3) = \sum_{n=0}^{10} n^2 + \sum_{n=0}^{10} 3 = \frac{10(11)(21)}{6} + 3(11) = 418$$

21. Write the first five terms of the geometric sequence, $a_1 = 4$, $r = -\frac{1}{4}$.

Solution:

$$a_1 = 4$$
$$a_2 = 4\left(-\frac{1}{4}\right) = -1$$
$$a_3 = -1\left(-\frac{1}{4}\right) = \frac{1}{4}$$
$$a_4 = \frac{1}{4}\left(-\frac{1}{4}\right) = -\frac{1}{16}$$
$$a_5 = -\frac{1}{16}\left(-\frac{1}{4}\right) = \frac{1}{64}$$

Solution: $4,\ -1,\ \frac{1}{4},\ -\frac{1}{16},\ \frac{1}{64}$

25. Write 0.454545... as the ratio of two integers by considering it to be the sum of an infinite geometric series.

Solution:

$$0.454545\ldots = \sum_{n=0}^{\infty} 0.45(0.01)^n = \frac{0.45}{1-0.01} = \frac{45}{99} = \frac{5}{11}$$

29. Write 0.01333... as the ratio of two integers by considering it to be the sum of an infinite geometric series.

Solution:

$$\begin{aligned} 0.01333\ldots &= 0.01 + \sum_{n=0}^{\infty} 0.003(0.1)^n \\ &= 0.01 + \frac{0.003}{1-0.1} \\ &= \frac{1}{100} + \frac{3}{900} \\ &= \frac{3}{300} + \frac{1}{300} \\ &= \frac{1}{75} \end{aligned}$$

31. Use mathematical induction to prove the formula $1 + 4 + \cdots + (3n - 2) = (n/2)(3n - 1)$ for every positive integer n.

Solution:

When $n = 1$,

$$1 = \frac{1}{2}[3(1) - 1] = 1.$$

Assume that

$$1 + 4 + 7 + \cdots + (3k - 2) = \frac{k}{2}(3k - 1).$$

Then,

$$\begin{aligned}
1 + 4 + 7 + \cdots + (3k - 2) + [3(k + 1) - 2] &= [1 + 4 + 7 + \cdots + (3k - 2)] + (3k + 1) \\
&= \frac{k}{2}(3k - 1) + (3k + 1) \\
&= \frac{k(3k - 1)}{2} + \frac{2(3k + 1)}{2} \\
&= \frac{3k^2 + 5k + 2}{2} \\
&= \frac{(k + 1)(3k + 2)}{2} \\
&= \frac{(k + 1)}{2}[3(k + 1) - 1]
\end{aligned}$$

Thus, the formula holds for all positive integers n.

37. Use the Binomial Theorem to expand the following binomial. Simplify your answer. [Remember that $i = \sqrt{-1}$.]

$$\left(\frac{2}{x} - 3x\right)^6$$

Solution:

$$\begin{aligned}
\left(\frac{2}{x} - 3x\right)^6 &= \left(\frac{2}{x}\right)^6 + 6\left(\frac{2}{x}\right)^5(-3x) + 15\left(\frac{2}{x}\right)^4(-3x)^2 + 20\left(\frac{2}{x}\right)^3(-3x)^3 \\
&\quad + 15\left(\frac{2}{x}\right)^2(-3x)^4 + 6\left(\frac{2}{x}\right)(-3x)^5 + (-3x)^6 \\
&= \frac{64}{x^6} - \frac{576}{x^4} + \frac{2160}{x^2} - 4320 + 4860x^2 - 2916x^4 + 729x^6
\end{aligned}$$

41. As a family increases in number, the number of different interpersonal relationships that exist increases at an even faster rate. Find the number of different interpersonal relationships that exist (between two people) if the number of members in the family is (a) 2, (b) 4, and (c) 6.

Solution:

(a) ${}_2C_2 = 1$
(b) ${}_4C_2 = 6$
(c) ${}_6C_2 = 15$

45. A man has five pairs of socks (no two pairs are the same color). If he randomly selects two socks from the drawer, what is the probability that he gets a matched pair?

Solution:

$$P(\text{pair}) = \frac{10}{10} \cdot \frac{1}{9} = \frac{1}{9}$$

49. Five cards are drawn from an ordinary deck of 52 playing cards. Find the probability of getting two pairs. (For example, the hand could be A-A-5-5-Q or 4-4-7-7-K.)

Solution:

$$P(2 \text{ pairs}) = \frac{(13)({}_4C_2)(12)({}_4C_2)(44)}{(2)({}_{52}C_5)} = 0.0475$$

Practice Test for Chapter 10

1. Write out the first five terms of the sequence $a_n = \dfrac{2n}{(n+2)!}$.

2. Write an expression for the nth term of the sequence $\{\frac{4}{3}, \frac{5}{9}, \frac{6}{27}, \frac{7}{81}, \frac{8}{243}, \ldots\}$.

3. Find the sum $\displaystyle\sum_{i=1}^{6}(2i-1)$.

4. Write out the first five terms of the arithmetic sequence where $a_1 = 23$ and $d = -2$.

5. Find a_n for the arithmetic sequence with $a_1 = 12$, $d = 3$, and $n = 50$.

6. Find the sum of the first 200 positive integers.

7. Write out the first five terms of the geometric sequence with $a_1 = 7$ and $r = 2$.

8. Evaluate $\displaystyle\sum_{n=0}^{9} 6\left(\frac{2}{3}\right)^n$.

9. Evaluate $\displaystyle\sum_{n=0}^{\infty}(0.03)^n$.

10. Use mathematical induction to prove that

$$1 + 2 + 3 + 4 + \cdots + n = \frac{n(n+1)}{2}.$$

11. Use mathematical induction to prove that $n! > 2^n$, $n \geq 4$.

12. Evaluate ${}_{13}C_4$.

13. Expand $(x+3)^5$.

14. Find the term involving x^7 in $(x-2)^{12}$.

15. Evaluate ${}_{30}P_4$.

16. How many ways can six people sit at a table with six chairs?

17. Twelve cars run in a race. How many different ways can they come in first, second, and third place? (Assume that there are no ties.)

18. Two six-sided dice are tossed. Find the probability that the total of the two dice is less than 5.

19. Two cards are selected at random from a deck of 52 playing cards. Find the probability that the first card is a King and the second card is a black ten.

20. A manufacturer has determined that for every 1000 units it produces, 3 will be faulty. What is the probability that an order of 50 units will have one or more faulty units?

CHAPTER 11

Some Topics in Analytic Geometry

SECTION 11.1

Introduction to Conics: Parabolas

- A *parabola* is the set of all points (x, y) that are equidistant from a fixed line (*directrix*) and a fixed point (*focus*) not on the line.
- The standard equation of a parabola with vertex (h, k) and:

  (a) Vertical axis $x = h$ and directrix $y = k - p$ is:

  $$(x - h)^2 = 4p(y - k)$$

  (b) Horizontal axis $y = k$ and directrix $x = h - p$ is:

  $$(y - k)^2 = 4p(x - h)$$

Solutions to Selected Exercises

5. Match $(y - 1)^2 = 4(x - 2)$ with the correct graph.

Solution:
The vertex is at $(2, 1)$ and the axis is horizontal.
Therefore, it matches graph (d).

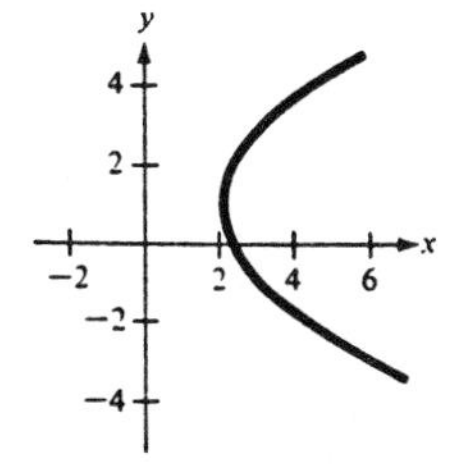

9. Find the vertex, focus, and directrix of the parabola $y^2 = -6x$.

Solution:

$$y^2 = -6x$$
$$(y - 0)^2 = 4\left(-\tfrac{3}{2}\right)(x - 0)$$

$h = 0,\ k = 0,\ p = -\frac{3}{2}$
Vertex: $(0, 0)$
Focus: $\left(-\frac{3}{2}, 0\right)$
Directrix: $x = 0 - \left(-\frac{3}{2}\right) = \frac{3}{2}$

13. Find the vertex, focus, and directrix of the parabola $(x-1)^2+8(y+2)=0$.

Solution:

$$\begin{aligned}(x-1)^2+8(y+2)&=0\\(x-1)^2&=-8(y+2)\\(x-1)^2&=4(-2)(y+2)\end{aligned}$$

$h=1,\ k=-2,\ p=-2$
Vertex: $(1,\ -2)$
Focus: $(1,\ -2+(-2))$ OR $(1,\ -4)$
Directrix: $y=-2-(-2)$ OR $y=0$

17. Find the vertex, focus, and directrix of the parabola $y=\frac{1}{4}(x^2-2x+5)$.

Solution:

$$\begin{aligned}y&=\tfrac{1}{4}(x^2-2x+5)\\4y&=x^2-2x+1-1+5\\4y&=(x-1)^2+4\\4y-4&=(x-1)^2\\(x-1)^2&=4(y-1)\\(x-1)^2&=4(1)(y-1)\end{aligned}$$

$h=1,\ k=1,\ p=1$
Vertex: $(1,\ 1)$
Focus: $(1,\ 1+1)$ OR $(1,\ 2)$
Directrix: $y=1-1$ OR $y=0$

21. Find the vertex, focus, and directrix of the parabola $y^2+6y+8x+25=0$.

Solution:

$$\begin{aligned}y^2+6y+8x+25&=0\\y^2+6y&=-8x-25\\y^2+6y+9&=-8x-25+9\\(y+3)^2&=-8x-16\\(y+3)^2&=-8(x+2)\\(y+3)^2&=4(-2)(x+2)\end{aligned}$$

$h=-2,\ k=-3,\ p=-2$
Vertex: $(-2,\ -3)$
Focus: $(-2+(-2),\ -3)$ OR $(-4,\ -3)$
Directrix: $x=-2-(-2)$ OR $x=0$

25. Find the vertex, focus, and directrix of the parabola $x^2 + 4x + 4y - 4 = 0$.

Solution:

$$
\begin{aligned}
x^2 + 4x + 4y - 4 &= 0 \\
x^2 + 4x &= -4y + 4 \\
x^2 + 4x + 4 &= -4y + 4 + 4 \\
(x+2)^2 &= -4y + 8 \\
(x+2)^2 &= -4(y-2) \\
(x+2)^2 &= 4(-1)(y-2)
\end{aligned}
$$

$h = -2,\ k = 2,\ p = -1$
Vertex: $(-2,\ 2)$
Focus: $(-2,\ 2 + (-1))$ OR $(-2,\ 1)$
Directrix: $y = 2 - (-1)$ OR $y = 3$

29. Find an equation of the specified parabola.

Vertex: $(0,\ 0)$
Focus: $(-2,\ 0)$

Solution:
The axis is horizontal with $p = -2$.

$$
\begin{aligned}
(y-0)^2 &= 4(-2)(x-0) \\
y^2 &= -8x
\end{aligned}
$$

33. Find an equation of the specified parabola.

Vertex: $(3,\ 2)$
Focus: $(1,\ 2)$

Solution:
The axis is horizontal with $p = -2$.

$$
\begin{aligned}
(y-2)^2 &= 4(-2)(x-3) \\
y^2 - 4y + 4 &= -8x + 24 \\
y^2 - 4y + 8x - 20 &= 0
\end{aligned}
$$

37. Find an equation of the specified parabola.

Focus: $(0,\ 0)$
Directrix: $y = 4$

Solution:
The vertex is $(0, 2)$. The axis is vertical with $p = -2$.

$$(x - 0)^2 = 4(-2)(y - 2)$$
$$x^2 = -8y + 16$$
$$x^2 + 8y - 16 = 0$$

41. Find an equation of the specified parabola.

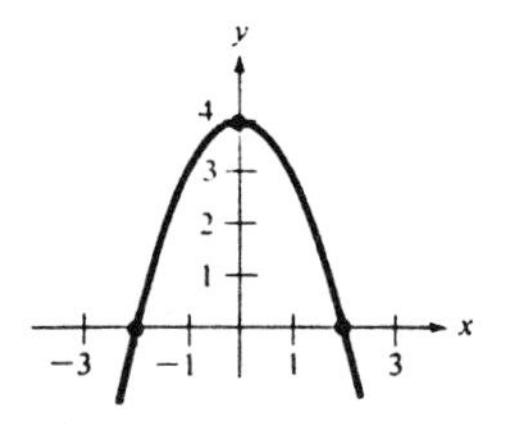

Solution:
The x-intercepts occur at $(\pm 2, 0)$ and the parabola opens downward.

$$y = -(x + 2)(x - 2)$$
$$y = -(x^2 - 4)$$
$$y = 4 - x^2$$

45. The receiver in a parabolic television dish antenna is 3 feet from the vertex and is located at the focus, as shown in the figure. Find an equation of a cross section of the reflector. (Assume the dish is directed upward and the vertex is at the origin.)

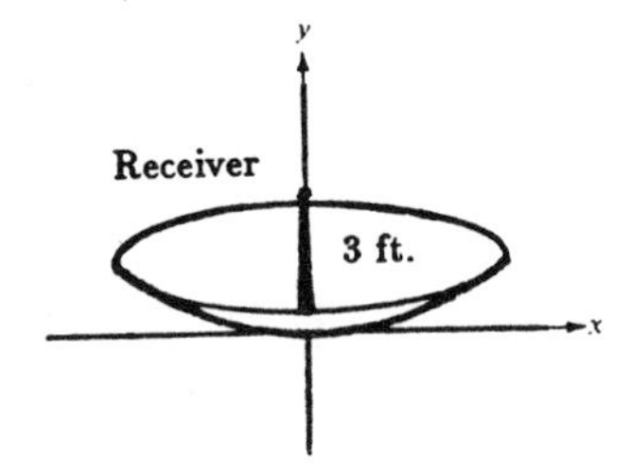

Solution:
The vertex is at $(0, 0)$ and the focus is at $(0, 3)$. Therefore, $p = 3$.

$$(x - 0)^2 = 4(3)(y - 0)$$
$$x^2 = 12y$$

49. Find the equation of the tangent line to the parabola $y = -2x^2$ at the point $(-1,\ -2)$.

Solution:

$$y = -2x^2$$

$$x^2 = -\frac{1}{2}y$$

$$x^2 = 4\left(-\frac{1}{8}\right)y \Rightarrow p = -\frac{1}{8}$$

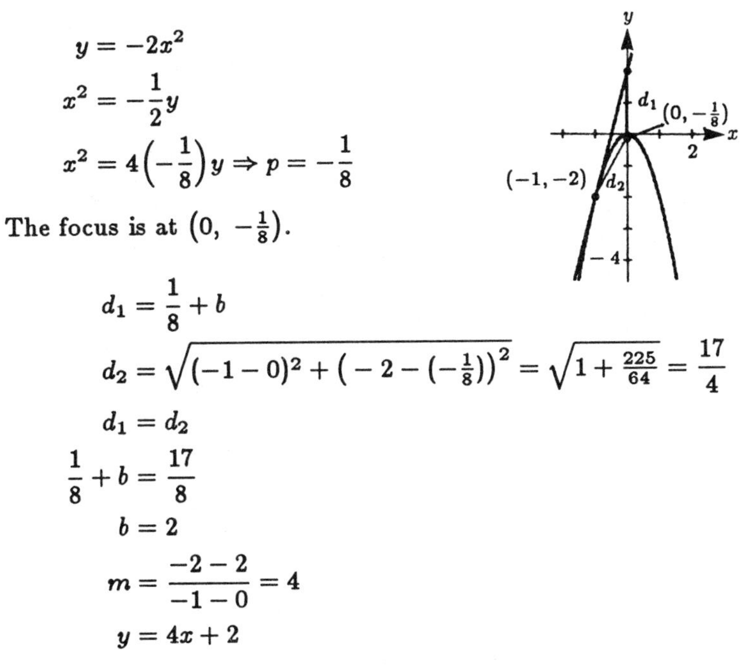

The focus is at $\left(0,\ -\frac{1}{8}\right)$.

$$d_1 = \frac{1}{8} + b$$

$$d_2 = \sqrt{(-1-0)^2 + \left(-2 - (-\tfrac{1}{8})\right)^2} = \sqrt{1 + \tfrac{225}{64}} = \frac{17}{4}$$

$$d_1 = d_2$$

$$\frac{1}{8} + b = \frac{17}{8}$$

$$b = 2$$

$$m = \frac{-2-2}{-1-0} = 4$$

$$y = 4x + 2$$

53. A ball is thrown horizontally from the top of a 75-foot tower with a velocity of 32 feet per second.

(a) Find the equation of the parabolic path.

(b) How far does the ball travel horizontally before striking the ground?

Solution:

(a)
$$y = -\frac{16}{v^2}x^2 + s$$

$$y = -\frac{16}{32^2}x^2 + 75$$

$$y = -\frac{1}{64}x^2 + 75 \text{ or}$$

$$x^2 + 64y = 4800$$

(b) When $y = 0$, we have

$$x^2 = 4800$$

$$x = \sqrt{4800} = 40\sqrt{3} \approx 69.28 \text{ feet away}$$

SECTION 11.2

Ellipses

- An *ellipse* is the set of all points (x, y) the sum of whose distances from two distinct points (*foci*) is constant.
- The standard equation of an ellipse with center (h, k) and major and minor axes of lengths $2a$ and $2b$ is:

  (a) $\dfrac{(x+h)^2}{a^2} + \dfrac{(y-k)^2}{b^2} = 1$ if the major axis is horizontal.

  (b) $\dfrac{(x-h)^2}{b^2} + \dfrac{(y-k)^2}{a^2} = 1$ if the major axis is vertical.
- $c^2 = a^2 - b^2$ where c is the distance from the center to a focus.
- The eccentricity of an ellipse is $e = \dfrac{c}{a}$.

Solutions to Selected Exercises

3. Match the following equation with the correct graph.

$$\frac{x^2}{9} + \frac{y^2}{4} = 1$$

Solution:
$a = 3,\ b = 2$
Center: $(0, 0)$
Major axis is horizontal
Therefore, it matches graph (c).

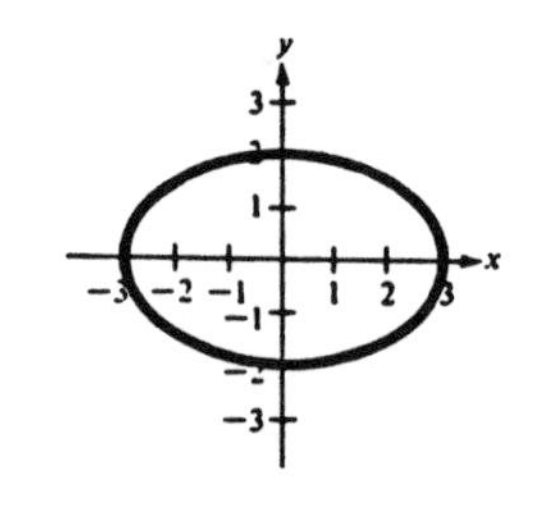

7. Find the center, foci, vertices, and eccentricity of the following ellipse and sketch its graph.

$$\frac{x^2}{25} + \frac{y^2}{16} = 1$$

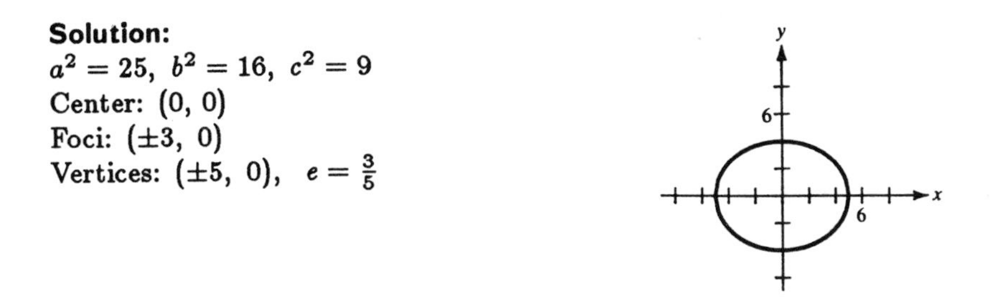

Solution:
$a^2 = 25,\ b^2 = 16,\ c^2 = 9$
Center: (0, 0)
Foci: $(\pm 3,\ 0)$
Vertices: $(\pm 5,\ 0)$, $e = \frac{3}{5}$

11. Find the center, foci, vertices, and eccentricity of the following ellipse and sketch its graph.

$$\frac{x^2}{9} + \frac{y^2}{5} = 1$$

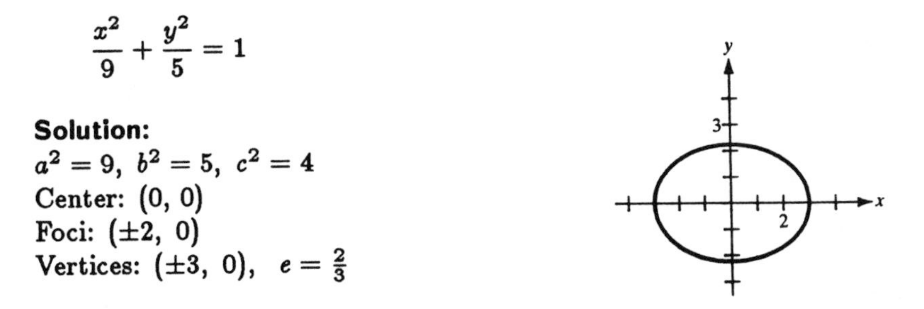

Solution:
$a^2 = 9,\ b^2 = 5,\ c^2 = 4$
Center: (0, 0)
Foci: $(\pm 2,\ 0)$
Vertices: $(\pm 3,\ 0)$, $e = \frac{2}{3}$

15. Find the center, foci, vertices, and eccentricity of the ellipse $3x^2 + 2y^2 = 6$ and sketch its graph.

Solution:

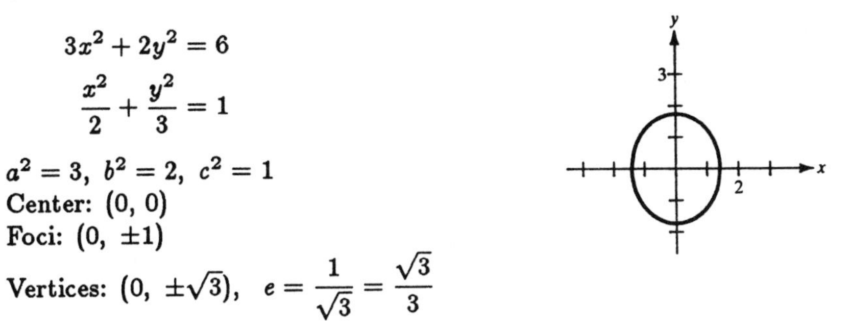

$$3x^2 + 2y^2 = 6$$
$$\frac{x^2}{2} + \frac{y^2}{3} = 1$$

$a^2 = 3,\ b^2 = 2,\ c^2 = 1$
Center: (0, 0)
Foci: $(0,\ \pm 1)$
Vertices: $(0,\ \pm\sqrt{3})$, $e = \frac{1}{\sqrt{3}} = \frac{\sqrt{3}}{3}$

19. Find the center, foci, vertices, and eccentricity of the following ellipse and sketch its graph.

$$\frac{(x-1)^2}{9} + \frac{(y-5)^2}{25} = 1$$

Solution:

$a^2 = 25,\ b^2 = 9,\ c^2 = 16$

Center: $(1, 5)$

Foci: $(1, 9), (1, 1)$

Vertices: $(1, 10), (1, 0),\ e = \frac{4}{5}$

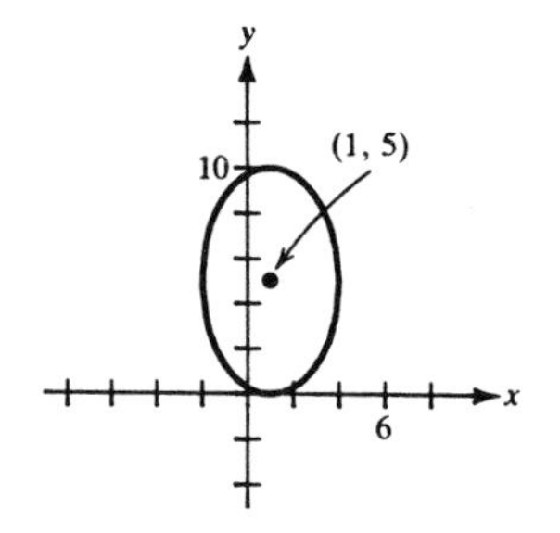

23. Find the center, foci, vertices, and eccentricity of the ellipse $16x^2 + 25y^2 - 32x + 50y + 16 = 0$ and sketch its graph.

Solution:

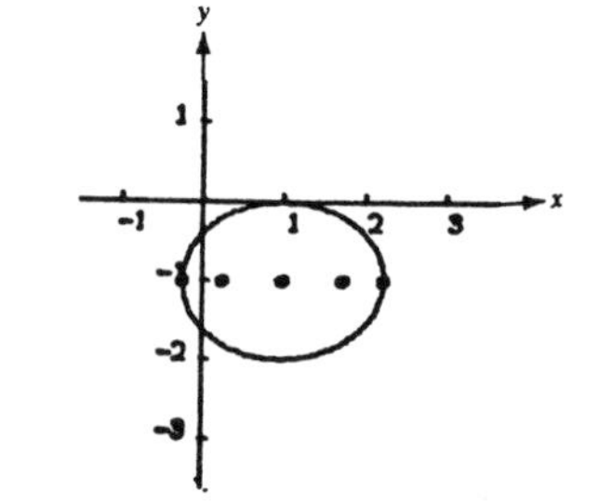

$$16x^2 + 25y^2 - 32x + 50y + 16 = 0$$

$$16(x^2 - 2x + 1) + 25(y^2 + 2y + 1) = -16 + 16 + 25$$

$$16(x-1)^2 + 25(y+1)^2 = 25$$

$$\frac{(x-1)^2}{25/16} + \frac{(y+1)^2}{1} = 1$$

$a^2 = \dfrac{25}{16},\ b^2 = 1,\ c^2 = \dfrac{9}{16}$

Center: $(1, -1)$

Foci: $\left(\frac{1}{4}, -1\right), \left(\frac{7}{4}, -1\right)$

Vertices: $\left(-\frac{1}{4}, -1\right), \left(\frac{9}{4}, -1\right),\ e = \frac{3}{5}$

27. Find an equation of the specified ellipse.

Vertices: $(\pm 6, 0)$

Foci: $(\pm 5, 0)$

Solution:

The major axis is horizontal with the center at $(0, 0)$.

$a = 6,\ c = 5$ implies $b = \sqrt{11}$.

$$\frac{(x-0)^2}{(6)^2} + \frac{(y-0)^2}{(\sqrt{11})^2} = 1$$

$$\frac{x^2}{36} + \frac{y^2}{11} = 1$$

31. Find an equation of the specified ellipse.

Vertices: $(0, \pm 2)$

Minor axis length 2

Solution:

The major axis is vertical with center at (0, 0).

$a = 2,\ 2b = 2 \Rightarrow b = 1$

$$\frac{(x-0)^2}{(1)^2} + \frac{(y-0)^2}{(2)^2} = 1$$

$$x^2 + \frac{y^2}{4} = 1$$

35. Find an equation of the specified ellipse.

Foci: (0, 0), (0, 8)
Major axis of length 16

Solution:

The major axis is vertical with center at (0, 4).

$2a = 16 \Rightarrow a = 8$

$c = 4 \Rightarrow b = \sqrt{48} = 4\sqrt{3}$

$$\frac{(x-0)^2}{(\sqrt{48})^2} + \frac{(y-4)^2}{(8)^2} = 1$$

$$\frac{x^2}{48} + \frac{(y-4)^2}{64} = 1$$

39. Find an equation of the specified ellipse.

Center: (3, 2), $a = 3c$
Foci: (1, 2), (5, 2)

Solution:

The major axis is horizontal with center at (3, 2).

$c = 2 \Rightarrow a = 3(2) = 6 \Rightarrow b = \sqrt{32}$

$$\frac{(x-3)^2}{(6)^2} + \frac{(y-2)^2}{(\sqrt{32})^2} = 1$$

$$\frac{(x-3)^2}{36} + \frac{(y-2)^2}{32} = 1$$

43. A fireplace arch is to be constructed in the shape of a semiellipse. The opening is to have a height of 2 feet at the center and a width of 5 feet along the base, as shown in the figure. The contractor will first draw the form of the ellipse by the method shown in Figure 11.12. Where should the tacks be placed and how long should the piece of string be?

Solution:

$a = \frac{5}{2},\ b = 2,\ c = \sqrt{\left(\frac{5}{2}\right)^2 - (2)^2} = \frac{3}{2}$

The tacks should be placed 1.5 feet from the center.

$d_1 + d_2 = 2a = 2\left(\frac{5}{2}\right) = 5$

The string should be 5 feet long.

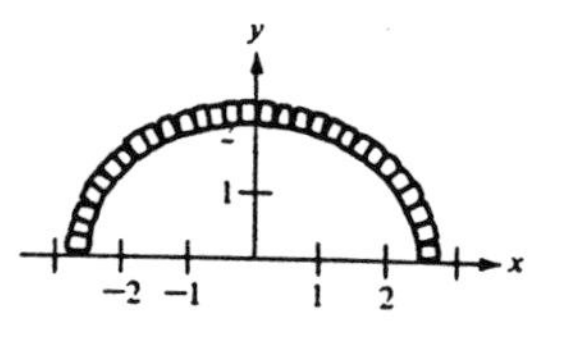

47. The earth moves in an elliptical orbit with the sun at one of the foci. The length of half the major axis is 93 million miles and the eccentricity is 0.017. Find the least and greatest distances between the earth and the sun.

Solution:

$$e = \frac{c}{a},\ a = 93,000,000,\ e = 0.017,\ 0.017 = \frac{c}{93,000,000},\ c \approx 1,581,000$$

Least distance: $a - c \approx 91,419,000$ miles

Greatest distance: $a + c \approx 94,581,000$ miles

51. Show that the equation of an ellipse can be written

$$\frac{(x-h)^2}{a^2} + \frac{(y-k)^2}{a^2(1-e^2)} = 1.$$

Note that as e approaches zero, with a remaining fixed, the ellipse approaches a circle of radius a.

Solution:

$$\frac{x^2}{a^2} + \frac{y^2}{b^2} = 1$$

$$\frac{x^2}{a^2} + \frac{y^2}{a^2(b^2/a^2)} = 1$$

$$\frac{x^2}{a^2} + \frac{y^2}{a^2(a^2-c^2)/a^2} = 1$$

$$\frac{x^2}{a^2} + \frac{y^2}{a^2(1-e^2)} = 1$$

As $e \Rightarrow 0,\ 1 - e^2 \Rightarrow 1$ and we have $\frac{x^2}{a^2} + \frac{y^2}{a^2} = 1$ or the circle $x^2 + y^2 = a^2$.

SECTION 11.3

Hyperbolas

- A *hyperbola* is the set of all points (x, y) the difference of whose distances from two fixed points (*foci*) is constant.
- The standard equation of a hyperbola with center (h, k) and transverse and conjugate axes of lengths $2a$ and $2b$ is:

  (a) $\dfrac{(x-h)^2}{a^2} - \dfrac{(y-k)^2}{b^2} = 1$ if the transverse axis is horizontal.

  (b) $\dfrac{(y-k)^2}{a^2} - \dfrac{(x-h)^2}{b^2} = 1$ if the transverse axis is vertical.
- $c^2 = a^2 + b^2$ where c is the distance from the center to a focus.
- The asymptotes of a hyperbola are:

  (a) $y = k \pm \dfrac{b}{a}(x-h)$ if the transverse axis is horizontal.

  (b) $y = k \pm \dfrac{a}{b}(x-h)$ if the transverse axis is vertical.
- The eccentricity of a hyperbola is $e = \dfrac{c}{a}$.
- To classify a nondegenerate conic from its general equation $Ax^2+Cy^2+Dx+Ey+F=0$:

  (a) If $A = C$ $(A \neq 0,\ C \neq 0)$, then it is a circle.
  (b) If $AC = 0$ $(A = 0$ or $C = 0$, but not both), then it is a parabola.
  (c) If $AC > 0$, then it is an ellipse.
  (d) If $AC < 0$, then it is a hyperbola.

Solutions to Selected Exercises

5. Match the following equation with the correct graph.

$$\frac{(x-2)^2}{9} - \frac{y^2}{4} = 1$$

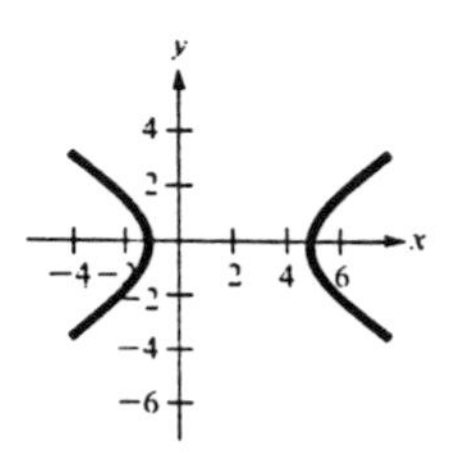

Solution:
$a = 3,\ b = 2$
Center: $(2, 0)$
Horizontal transverse axis
Matches graph (d)

9. Find the center, vertices, and foci of the following hyperbola and sketch its graph, using asymptotes as an aid.

$$\frac{y^2}{1} - \frac{x^2}{4} = 1$$

Solution:

$a = 1,\ b = 2,\ c = \sqrt{5}$
Center: $(0,\ 0)$
Vertices: $(0,\ \pm 1)$
Foci: $(0,\ \pm\sqrt{5})$
Asymptotes: $y = \pm\frac{1}{2}x$

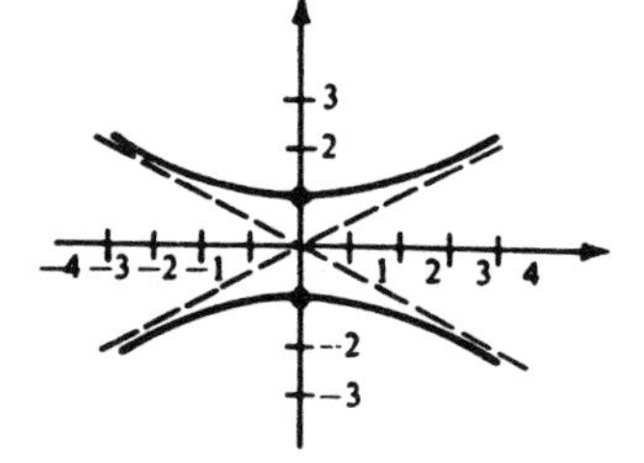

13. Find the center, vertices, and foci of the hyperbola $2x^2 - 3y^2 = 6$ and sketch its graph, using asymptotes as an aid.

Solution:

$$2x^2 - 3y^2 = 6$$

$$\frac{x^2}{3} - \frac{y^2}{2} = 1$$

$a = \sqrt{3},\ b = \sqrt{2},\ c = \sqrt{5}$
Center: $(0,\ 0)$
Vertices: $(\pm\sqrt{3},\ 0)$
Foci: $(\pm\sqrt{5},\ 0)$
Asymptotes: $y = \pm\sqrt{\frac{2}{3}}\,x$

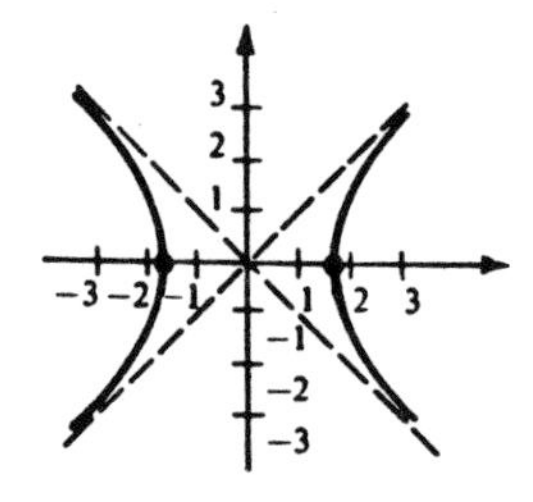

19. Find the center, vertices, and foci of the hyperbola $(y+6)^2 - (x-2)^2 = 1$ and sketch its graph, using asymptotes as an aid.

Solution:

$$\frac{(y+6)^2}{1} - \frac{(x-2)^2}{1} = 1$$

$a = 1,\ b = 1,\ c = \sqrt{2}$
Center: $(2,\ -6)$
Vertices: $(2,\ -5),\ (2,\ -7)$
Foci: $(2,\ -6 \pm \sqrt{2})$
Asymptotes: $y = -6 \pm (x - 2)$

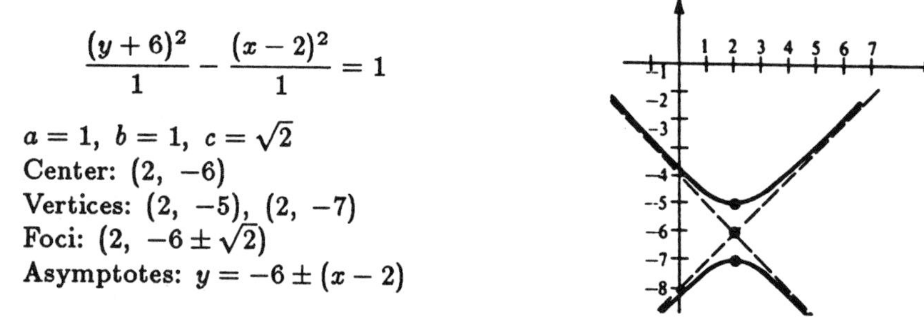

23. Find the center, vertices, and foci of the hyperbola $9y^2 - x^2 + 2x + 54y + 62 = 0$.

Solution:

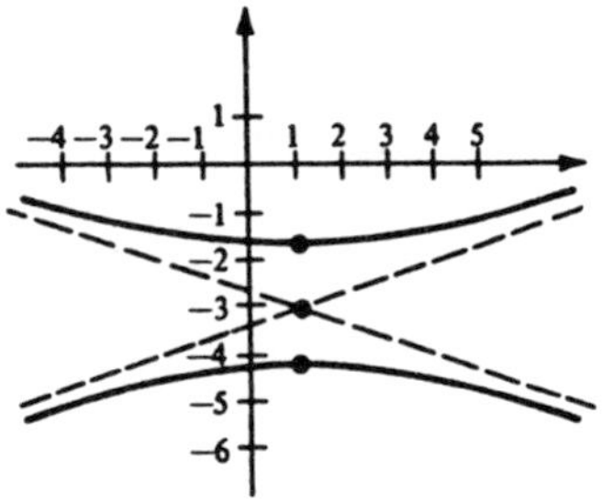

$$9y^2 - x^2 + 2x + 54y + 62 = 0$$
$$9(y^2 + 6y + 9) - (x^2 - 2x + 1) = -62 - 1 + 81$$
$$\frac{(y+3)^2}{2} - \frac{(x-1)^2}{18} = 1$$

$a = \sqrt{2}$, $b = 3\sqrt{2}$, $c = 2\sqrt{5}$

Center: $(1, -3)$

Vertices: $(1, -3 \pm \sqrt{2})$

Foci: $(1, -3 \pm 2\sqrt{5})$

Asymptotes: $y = -3 \pm \frac{1}{3}(x-1)$

29. Find an equation of the hyperbola with vertices at $(\pm 1, 0)$, and asymptotes $y = \pm 3x$.

Solution:

$a = 1$, $b/a = 3 \Rightarrow b = 3$

Center: $(0, 0)$

The transverse axis is horizontal.

$$\frac{(x-0)^2}{(1)^2} - \frac{(y-0)^2}{(3)^2} = 1$$
$$\frac{x^2}{1} - \frac{y^2}{9} = 1$$

33. Find an equation of the hyperbola with vertices at $(4, 1)$, $(4, 9)$ and foci at $(4, 0)$, $(4, 10)$.

Solution:

$a = 4$, $c = 5 \Rightarrow b = 3$

Center: $(4, 5)$

The transverse axis is vertical.

$$\frac{(y-5)^2}{(4)^2} - \frac{(x-4)^2}{(3)^2} = 1$$
$$\frac{(y-5)^2}{16} - \frac{(x-4)^2}{9} = 1$$

37. Find an equation of the hyperbola with vertices at $(-2, 1)$, $(2, 1)$ and solution point at $(4, 3)$.

Solution:

$a = 2$

Center: $(0, 1)$

The transverse axis is horizontal.

$$\frac{(x-0)^2}{(2)^2} - \frac{(y-1)^2}{b^2} = 1$$

$$\frac{x^2}{4} - \frac{(y-1)^2}{b^2} = 1$$

$$\frac{(4)^2}{4} - \frac{(3-1)^2}{b^2} = 1$$

$$4 - \frac{4}{b^2} = 1$$

$$4 = 3b^2$$

$$b^2 = \frac{4}{3}$$

$$\frac{x^2}{4} - \frac{(y-1)^2}{4/3} = 1$$

$$\frac{x^2}{4} - \frac{3(y-1)^2}{4} = 1$$

41. Three listening stations located at $(4400, 0)$, $(4400, 1100)$, and $(-4400, 0)$ hear an explosion. If the latter two stations heard the sound 1 second and 5 seconds after the first, respectively, where did the explosion occur? Assume that the coordinate system is measured in feet and that sound travels at the rate of 1100 feet per second.

Solution:

The listening stations are located at $A(4400, 0)$, $B(4400, 1100)$, and $C(-4400, 0)$. Since B heard the explosion 1 second after A heard it, the explosion must have occurred at a point that is 1100 feet farther from B than from A. Since A and B are 1100 feet apart, we conclude that the explosion occurred below A on the line $x = 4400$. Furthermore, since C heard the explosion 5 seconds after A, the explosion must have occurred 5500 feet farther away from C than from A. The collection of all such points is a hyperbola with foci at A and C, center at $(0, 0)$ and $2a = 5500$. Thus,

$$2a = 5500 \quad \text{or} \quad a = 2750$$

$$2c = 8800 \quad \text{or} \quad c = 4400$$

$$b = \sqrt{(4400)^2 - (2750)^2} = 550\sqrt{39}$$

and we have

$$\frac{x^2}{(2750)^2} - \frac{y^2}{(550\sqrt{39})^2} = 1.$$

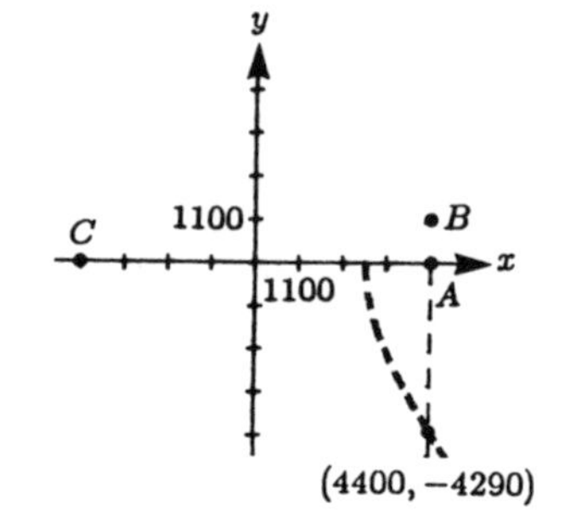

Finally, if $x = 4400$ and $y < 0$, we have

$$\frac{y^2}{(550\sqrt{39})^2} = \frac{(4400)^2}{(2750)^2} - 1$$

$$y = -550\sqrt{39}\sqrt{\frac{64-25}{25}} = -110\sqrt{39}\sqrt{39} = -4290$$

45. Classify the graph of the equation $4x^2 - y^2 - 4x - 3 = 0$ as a circle, a parabola, an ellipse, or a hyperbola.

Solution:
$4x^2 - y^2 - 4x - 3 = 0$
$A = 4, \ C = -1$
$AC = (4)(-1) = -4 < 0$
Therefore, the graph is a hyperbola.

49. Classify the graph of the equation $25x^2 - 10x - 200y - 119 = 0$ as a circle, a parabola, an ellipse, or a hyperbola.

Solution:
$25x^2 - 10x - 200y - 119 = 0$
$A = 25, \ C = 0$
$AC = (25)(0) = 0$
Therefore, the graph is a parabola.

SECTION 11.4

Rotation and the General Second-Degree Equation

- The general second-degree equation $Ax^2+Bxy+Cy^2+Dx+Ey+F=0$ can be rewritten as $A'(x')^2+C'(y')^2+D'x'+E'y'+F'=0$ by rotating the coordinate axes through the angle θ, where $\cot 2\theta = (A-C)/B$.
- The coefficients of the new system are:

  $A' = A\cos^2\theta + B\cos\theta\sin\theta + C\sin^2\theta$
  $C' = A\sin^2\theta - B\cos\theta\sin\theta + C\cos^2\theta$
  $D' = D\cos\theta + E\sin\theta$
  $E' = -D\sin\theta + E\cos\theta$
  $F' = F$
- The graph of the nondegenerate equation $Ax^2+Bxy+Cy^2+Dx+Ey+F=0$ is:

  (a) An ellipse or circle if $B^2-4AC<0$.
  (b) A parabola if $B^2-4AC=0$.
  (c) A hyperbola if $B^2-4AC>0$.

Solutions to Selected Exercises

3. Rotate the axes to eliminate the xy-term in the equation $9x^2+24xy+16y^2+90x-130y=0$. Sketch the graph of the resulting equation, showing both sets of axes.

Solution:

$A=9,\ B=24,\ C=16,\ D=90,\ E=-130,\ F=0$

$$\cot 2\theta = \frac{A-C}{B} = -\frac{7}{24} \quad \text{or} \quad \theta = \frac{1}{2}\operatorname{arccot}\left(-\frac{7}{24}\right) \approx 53.13^\circ$$

Since $\cot 2\theta = -\dfrac{7}{24}$, $\cos 2\theta = -\dfrac{7}{25}$.

$$\sin\theta = \frac{\sqrt{1+(7/25)}}{\sqrt{2}} = \frac{4}{5}$$

$$\cos\theta = \frac{\sqrt{1-(7/25)}}{\sqrt{2}} = \frac{3}{5}$$

$$A' = 9\left(\frac{9}{25}\right) + 24\left(\frac{4}{5}\right)\left(\frac{3}{5}\right) + 16\left(\frac{16}{25}\right) = 25$$

$$C' = 9\left(\frac{16}{25}\right) - 24\left(\frac{4}{5}\right)\left(\frac{3}{5}\right) + 16\left(\frac{9}{25}\right) = 0$$

$$D' = 90\left(\frac{3}{5}\right) - 130\left(\frac{4}{5}\right) = -50$$

$$E' = -90\left(\frac{4}{5}\right) - 130\left(\frac{3}{5}\right) = -150$$

$$F' = 0$$

$$25x'^2 - 50x' - 150y' = 0$$

$$y' = \frac{x'^2}{6} - \frac{x'}{3}$$

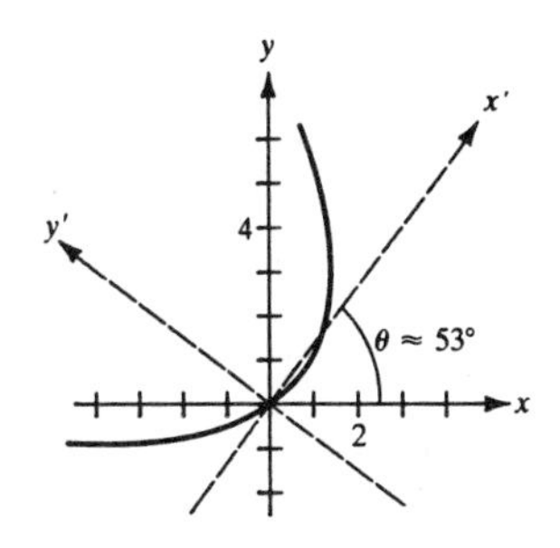

7. Rotate the axes to eliminate the xy-term in the equation $xy - 2y - 4x = 0$. Sketch the graph of the resulting equation, showing both sets of axes.

Solution:

$A = 0,\ B = 1,\ C = 0,\ D = -4,\ E = -2,\ F = 0$

$$\cot 2\theta = 0,\ 2\theta = \frac{\pi}{2},\ \theta = \frac{\pi}{4}$$

$$\sin\theta = \cos\theta = \frac{\sqrt{2}}{2}$$

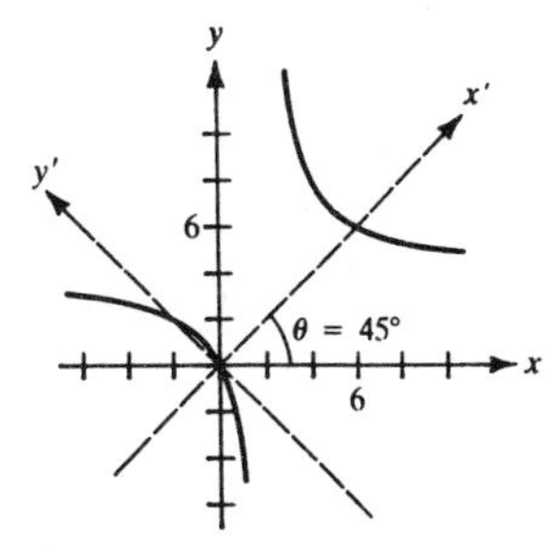

$$A' = 0 + \frac{1}{2} + 0 = \frac{1}{2}$$

$$C' = 0 - \frac{1}{2} + 0 = -\frac{1}{2}$$

$$D' = -4\left(\frac{\sqrt{2}}{2}\right) - 2\left(\frac{\sqrt{2}}{2}\right) = -3\sqrt{2}$$

$$E' = 4\left(\frac{\sqrt{2}}{2}\right) - 2\left(\frac{\sqrt{2}}{2}\right) = \sqrt{2}$$

$$F' = 0$$

$$\frac{1}{2}x'^2 - \frac{1}{2}y'^2 - 3\sqrt{2}\,x' + \sqrt{2}\,y' = 0$$

$$\frac{(x' - 3\sqrt{2})^2}{16} - \frac{(y' - \sqrt{2})^2}{16} = 1$$

11. Rotate the axes to eliminate the xy-term in the equation $3x^2 - 2\sqrt{3}xy + y^2 + 2x + 2\sqrt{3}y = 0$. Sketch the graph of the resulting equation, showing both sets of axes.

Solution:

$A = 3,\ B = -2\sqrt{3},\ C = 1,\ D = 2,\ E = 2\sqrt{3},\ F = 0$

$$\cot 2\theta = -\frac{1}{\sqrt{3}},\quad \theta = 60°$$

$$\sin\theta = \frac{\sqrt{3}}{2},\quad \cos\theta = \frac{1}{2}$$

$$A' = 3\left(\frac{1}{4}\right) - 2\sqrt{3}\left(\frac{\sqrt{3}}{2}\right)\left(\frac{1}{2}\right) + \left(\frac{3}{4}\right) = 0$$

$$C' = 3\left(\frac{3}{4}\right) + 2\sqrt{3}\left(\frac{\sqrt{3}}{2}\right)\left(\frac{1}{2}\right) + \left(\frac{1}{4}\right) = 4$$

$$D' = 2\left(\frac{1}{2}\right) + 2\sqrt{3}\left(\frac{\sqrt{3}}{2}\right) = 4$$

$$E' = -2\left(\frac{\sqrt{3}}{2}\right) + 2\sqrt{3}\left(\frac{1}{2}\right) = 0$$

$$F' = 0$$

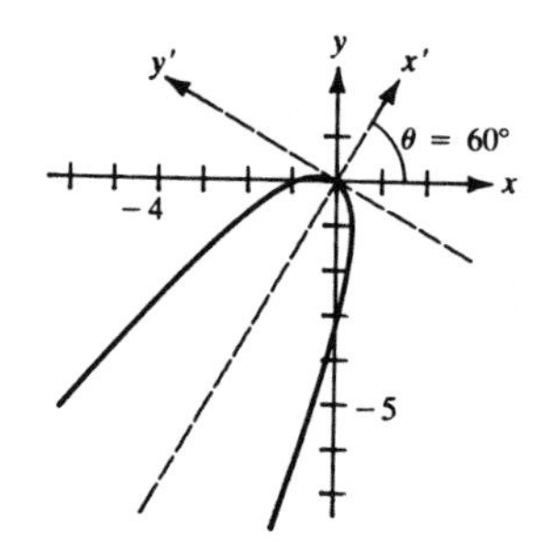

$$4y'^2 + 4x' = 0$$

$$x' = -(y')^2$$

15. Rotate the axes to eliminate the xy-term in the equation $32x^2 + 50xy + 7y^2 = 52$. Sketch the graph of the resulting equation, showing both sets of axes.

Solution:

$A = 32,\ B = 50,\ C = 7,\ D = 0,\ E = 0,\ F = -52$

$\cot 2\theta = \frac{1}{2},\ \cos 2\theta = \frac{1}{\sqrt{5}},\ \theta \approx 31.72°$

$$\sin\theta = \frac{\sqrt{1 - (1/\sqrt{5})}}{\sqrt{2}} = \sqrt{\frac{\sqrt{5}-1}{2\sqrt{5}}}$$

$$\cos\theta = \frac{\sqrt{1 + (1/\sqrt{5})}}{\sqrt{2}} = \sqrt{\frac{\sqrt{5}+1}{2\sqrt{5}}}$$

$$A' = 32\left(\frac{\sqrt{5}+1}{2\sqrt{5}}\right) + 50\sqrt{\frac{\sqrt{5}-1}{2\sqrt{5}}}\left(\sqrt{\frac{\sqrt{5}+1}{2\sqrt{5}}}\right) + 7\left(\frac{\sqrt{5}-1}{2\sqrt{5}}\right) = \frac{39 + 25\sqrt{5}}{2} \approx 47.451$$

$$C' = 32\left(\frac{\sqrt{5}-1}{2\sqrt{5}}\right) - 50\sqrt{\frac{\sqrt{5}-1}{2\sqrt{5}}}\left(\sqrt{\frac{\sqrt{5}+1}{2\sqrt{5}}}\right) + 7\left(\frac{\sqrt{5}+1}{2\sqrt{5}}\right) = \frac{39 - 25\sqrt{5}}{2} \approx -8.451$$

$$D' = E' = 0$$

$$F' = -52$$

$$47.451x'^2 - 8.451y'^2 - 52 = 0$$

$$\frac{x'^2}{1.096} - \frac{y'^2}{6.153} = 1$$

19. Use the discriminant to determine whether the graph of $13x^2 - 8xy + 7y^2 - 45 = 0$ is a parabola, an ellipse, or a hyperbola.

Solution:
$A = 13,\ B = -8,\ C = 7$
$B^2 - 4AC = (-8)^2 - 4(13)(7) = -300 < 0$
Therefore, the graph is an ellipse or a circle.

23. Use the discriminant to determine whether the graph of $x^2 + 4xy + 4y^2 - 5x - y - 3 = 0$ is a parabola, an ellipse, or a hyperbola.

Solution:
$A = 1,\ B = 4,\ C = 4$
$B^2 - 4AC = (4)^2 - 4(1)(4) = 0$
Therefore, the graph is a parabola.

25. Show that the equation $x^2 + y^2 = r^2$ is invariant under rotation of axes.

Solution:

$$\begin{aligned}(x')^2 + (y')^2 &= [x\cos\theta + y\sin\theta]^2 + [y\cos\theta - x\sin\theta]^2 \\ &= x^2\cos^2\theta + 2xy\cos\theta\sin\theta + y^2\sin^2\theta + y^2\cos^2\theta - 2xy\cos\theta\sin\theta + x^2\sin^2\theta \\ &= x^2(\cos^2\theta + \sin^2\theta) + y^2(\sin^2\theta + \cos^2\theta) \\ &= x^2 + y^2 = r^2\end{aligned}$$

SECTION 11.5

Polar Coordinates

- In polar coordinates you do not have unique representation of points. The point (r, θ) can be represented by $(r, \theta + 2n\pi)$ or by $(-r, \theta + (2n+1)\pi)$ where n is any integer. The pole is represented by $(0, \theta)$ where θ is any angle.
- To convert from polar coordinates to rectangular coordinates, use the following relationships.

  $x = r\cos\theta$
  $y = r\sin\theta$

- To convert from rectangular coordinates to polar coordinates, use the following relationships.

  $r = \pm\sqrt{x^2 + y^2}$
  $\tan\theta = y/x$

  If θ is in the same quadrant as the point (x, y), then r is positive. If θ is in the opposite quadrant as the point (x, y), then r is negative.

- You should be able to convert rectangular equations to polar form and vice versa.

Solutions to Selected Exercises

3. Plot the polar point $(-1, 5\pi/4)$ and find the corresponding rectangular coordinates.

Solution:

$r = -1,\ \theta = \dfrac{5\pi}{4}$

$$x = (-1)\cos\frac{5\pi}{4} = (-1)\left(-\frac{\sqrt{2}}{2}\right) = \frac{\sqrt{2}}{2}$$

$$y = (-1)\sin\frac{5\pi}{4} = (-1)\left(-\frac{\sqrt{2}}{2}\right) = \frac{\sqrt{2}}{2}$$

$\left(-1, \dfrac{5\pi}{4}\right)$ corresponds to $\left(\dfrac{\sqrt{2}}{2}, \dfrac{\sqrt{2}}{2}\right)$.

y
1
(√2/2, √2/2)
1
x

7. Plot the polar point $(\sqrt{2}, 2.36)$ and find the corresponding rectangular coordinates.

Solution:
$r = \sqrt{2}$, $\theta = 2.36$ (in radians)

$$x = \sqrt{2}\cos 2.36 \approx -1.004$$
$$y = \sqrt{2}\sin 2.36 \approx 0.996$$

$(\sqrt{2},\ 2.36)$ corresponds to $(-1.004,\ 0.996)$.

(−1.004, 0.996)

11. Find two sets of polar coordinates for the rectangular point $(-3,\ 4)$, using $0 \le \theta < 2\pi$.

Solution:
$x = -3,\ y = 4$
$r = \pm\sqrt{(-3)^2 + (4)^2} = \pm 5$
$\tan\theta = -\frac{4}{3},\ \theta \approx 2.214$
The point $(-3,\ 4)$ is in Quadrant II as is the angle $\theta = 2.214$. Thus, one polar representation is $(5,\ 2.214)$. Another representation is $(-5,\ 2.214 + \pi) \approx (-5,\ 5.356)$.

15. Find two sets of polar coordinates for the rectangular point $(4,\ 6)$, using $0 \le \theta < 2\pi$.

Solution:
$x = 4,\ y = 6$
$r = \pm\sqrt{(4)^2 + (6)^2} = \pm 2\sqrt{13}$
$\tan\theta = \frac{6}{4},\ \theta \approx 0.983$
Since $(4,\ 6)$ is in Quadrant I and $\theta = 0.983$ is in Quadrant I, one representation in polar coordinates is $(2\sqrt{13},\ 0.983)$. Another representation is $(-2\sqrt{13},\ 0.983 + \pi) \approx (-2\sqrt{13},\ 4.124)$.

19. Convert the equation $x^2 + y^2 - 2ax = 0$ to polar form.

Solution:

$$x^2 + y^2 - 2ax = 0$$
$$r^2 - 2ar\cos\theta = 0$$
$$r(r - 2a\cos\theta) = 0$$
$$r = 0 \quad \text{OR} \quad r = 2a\cos\theta$$

Since $r = 0$ is the pole and is also on the graph of $r = 2a\cos\theta$, we just have $r = 2a\cos\theta$.

23. Convert the equation $x = 10$ to polar form.

Solution:

$$x = 10$$
$$r\cos\theta = 10$$
$$r = \frac{10}{\cos\theta}$$
$$r = 10\sec\theta$$

27. Convert the equation $xy = 4$ to polar form.

Solution:

$$\begin{aligned} xy &= 4 \\ (r\cos\theta)(r\sin\theta) &= 4 \\ r^2 &= \frac{4}{\cos\theta\sin\theta} \\ r^2 &= \frac{4}{\frac{1}{2}\sin 2\theta} \\ r^2 &= 8\csc 2\theta \end{aligned}$$

31. Convert the equation $r = 4\sin\theta$ to rectangular form.

Solution:

$$\begin{aligned} r &= 4\sin\theta \\ r^2 &= 4r\sin\theta \\ x^2 + y^2 &= 4y \\ x^2 + y^2 - 4y &= 0 \end{aligned}$$

35. Convert the equation $r = 2\csc\theta$ to rectangular form.

Solution:

$$\begin{aligned} r &= 2\csc\theta \\ r &= \frac{2}{\sin\theta} \\ r\sin\theta &= 2 \\ y &= 2 \end{aligned}$$

39. Convert the equation $r = \dfrac{6}{2 - 3\sin\theta}$ to rectangular form.

Solution:

$$\begin{aligned} r &= \frac{6}{2 - 3\sin\theta} \\ r(2 - 3\sin\theta) &= 6 \\ 2r - 3r\sin\theta &= 6 \\ 2r &= 6 + 3r\sin\theta \\ 2(\pm\sqrt{x^2 + y^2}) &= 6 + 3y \\ 4(x^2 + y^2) &= (6 + 3y)^2 \\ 4x^2 + 4y^2 &= 36 + 36y + 9y^2 \\ 4x^2 - 5y^2 - 36y - 36 &= 0 \end{aligned}$$

41. Show that the distance between $(r_1,\ \theta_1)$ and $(r_2,\ \theta_2)$ is $\sqrt{{r_1}^2 + {r_2}^2 - 2r_1r_2\cos(\theta_1 - \theta_2)}$.

Solution:

$(r_1,\ \theta_1)$ corresponds to the point $(r_1\cos\theta_1,\ r_1\sin\theta_1)$ in rectangular coordinates. Likewise, $(r_2,\ \theta_2)$ corresponds to the point $(r_2\cos\theta_2,\ r_2\sin\theta_2)$. In rectangular coordinates we use the distance formula, $d = \sqrt{(x_2 - x_1)^2 + (y_2 - y_1)^2}$, to find the distance between two points.

$$
\begin{aligned}
d &= \sqrt{(r_2\cos\theta_2 - r_1\cos\theta_1)^2 + (r_2\sin\theta_2 - r_1\sin\theta_1)^2} \\
&= \sqrt{{r_2}^2\cos^2\theta_2 - 2r_1r_2\cos\theta_1\cos\theta_2 + {r_1}^2\cos^2\theta_1 + {r_2}^2\sin^2\theta_2 - 2r_1r_2\sin\theta_1\sin\theta_2 + {r_1}^2\sin^2\theta_1} \\
&= \sqrt{{r_1}^2(\cos^2\theta_1 + \sin^2\theta_1) + {r_2}^2(\cos^2\theta_2 + \sin^2\theta_2) - 2r_1r_2(\cos\theta_1\cos\theta_2 + \sin\theta_1\sin\theta_2)} \\
&= \sqrt{{r_1}^2(1) + {r_2}^2(1) - 2r_1r_2\cos(\theta_1 - \theta_2)} \\
&= \sqrt{{r_1}^2 + {r_2}^2 - 2r_1r_2\cos(\theta_1 - \theta_2)}
\end{aligned}
$$

SECTION 11.6

Graphs of Polar Equations

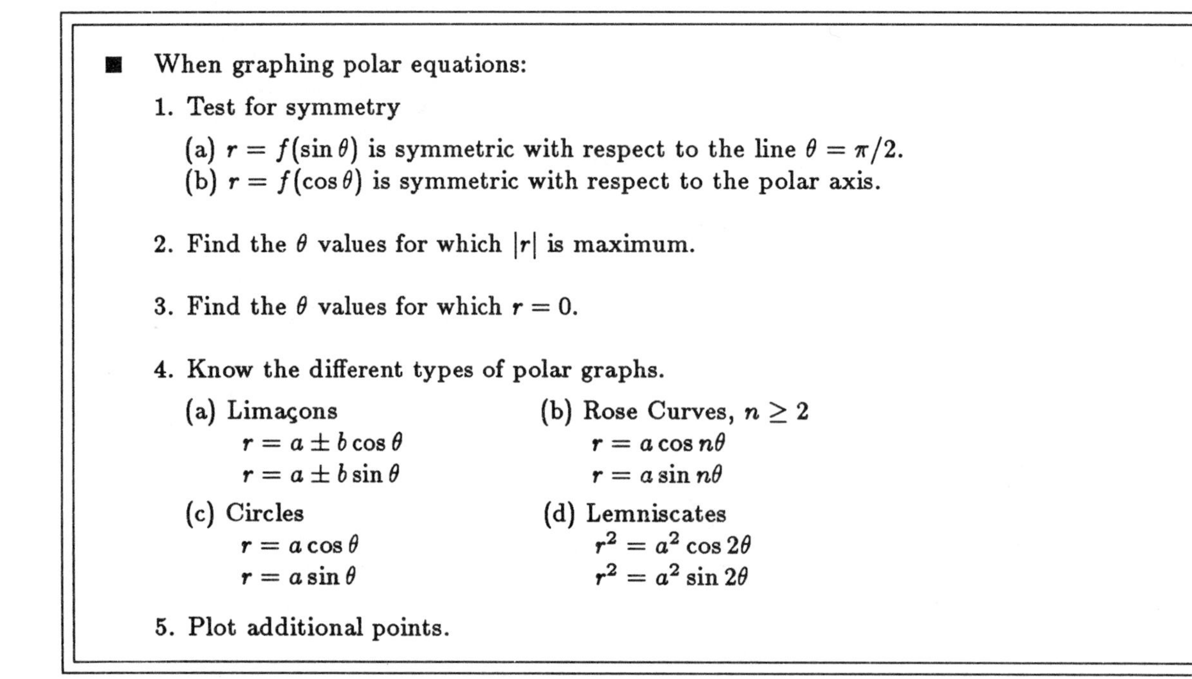

- When graphing polar equations:
  1. Test for symmetry
     (a) $r = f(\sin\theta)$ is symmetric with respect to the line $\theta = \pi/2$.
     (b) $r = f(\cos\theta)$ is symmetric with respect to the polar axis.
  2. Find the θ values for which $|r|$ is maximum.
  3. Find the θ values for which $r = 0$.
  4. Know the different types of polar graphs.
     (a) Limaçons
     $r = a \pm b\cos\theta$
     $r = a \pm b\sin\theta$
     (b) Rose Curves, $n \geq 2$
     $r = a\cos n\theta$
     $r = a\sin n\theta$
     (c) Circles
     $r = a\cos\theta$
     $r = a\sin\theta$
     (d) Lemniscates
     $r^2 = a^2\cos 2\theta$
     $r^2 = a^2\sin 2\theta$
  5. Plot additional points.

Solutions to Selected Exercises

3. Sketch the graph of $\theta = \pi/6$.

Solution:

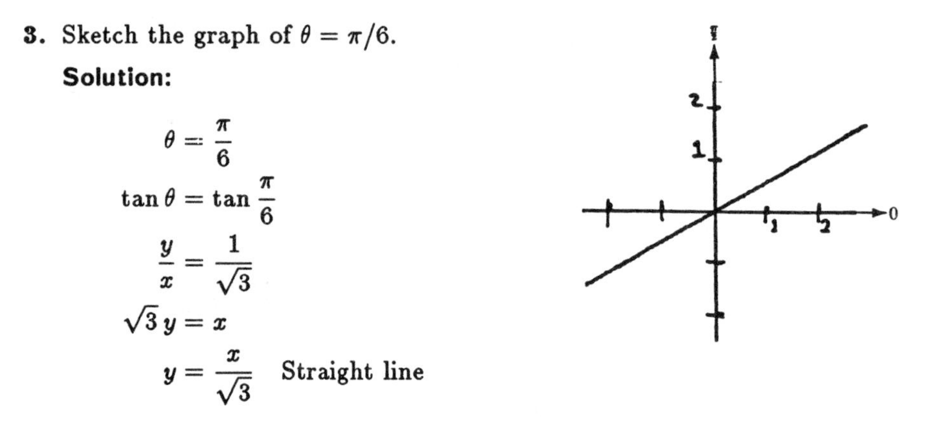

$$\theta = \frac{\pi}{6}$$

$$\tan\theta = \tan\frac{\pi}{6}$$

$$\frac{y}{x} = \frac{1}{\sqrt{3}}$$

$$\sqrt{3}\,y = x$$

$$y = \frac{x}{\sqrt{3}} \quad \text{Straight line}$$

7. Sketch the graph of $r = 3(1 - \cos\theta)$.

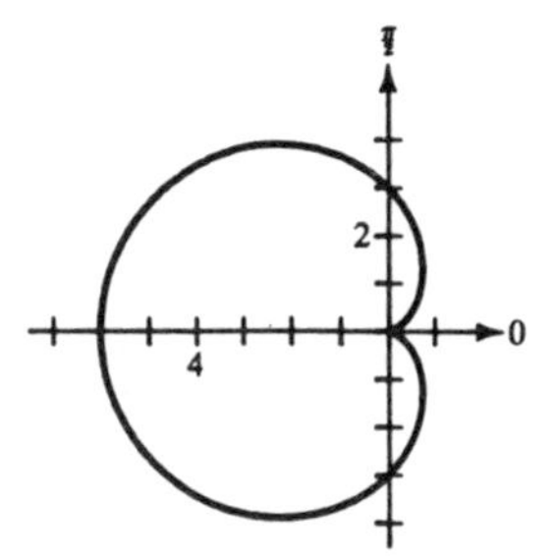

Solution:

$a/b = 1$, so the graph is a cardioid.
Symmetric to the polar axis since r is a function of $\cos\theta$.
Maximum value of $|r|$ is 6 and occurs when $\theta = \pi$.
The zero of r occurs when $\theta = 0$.

θ	0	$\frac{\pi}{3}$	$\frac{\pi}{2}$	$\frac{2\pi}{3}$	π
r	0	$\frac{3}{2}$	3	$\frac{9}{2}$	6

11. Sketch the graph of $r = 2 + 3\sin\theta$.

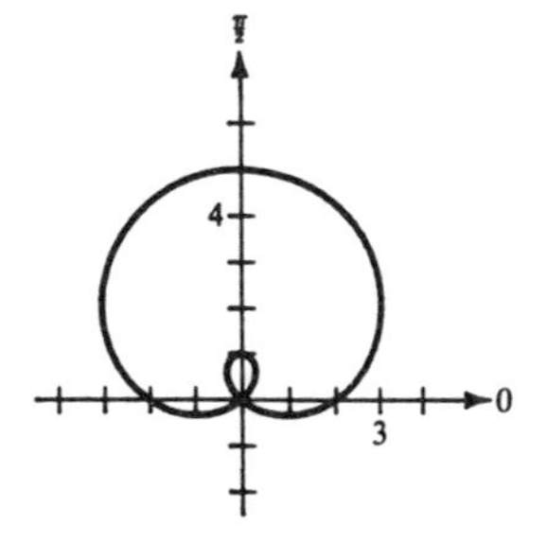

Solution:

$a/b = 2/3 < 1$, so the graph is a limaçon with inner loop. Symmetric to $\theta = \pi/2$ since r is a function of $\sin\theta$. Maximum value of $|r|$ is 5 and occurs when $\theta = \pi/2$. The zero of r occurs when $\theta \approx 2.412$.

θ	$-\pi$	$-\frac{\pi}{6}$	0	$\frac{\pi}{6}$	$\frac{\pi}{2}$	π	2.412	$\frac{3\pi}{2}$
r	2	$\frac{1}{2}$	2	$\frac{7}{2}$	5	2	0	-1

15. Sketch the graph of $r = 3 - 2\cos\theta$.

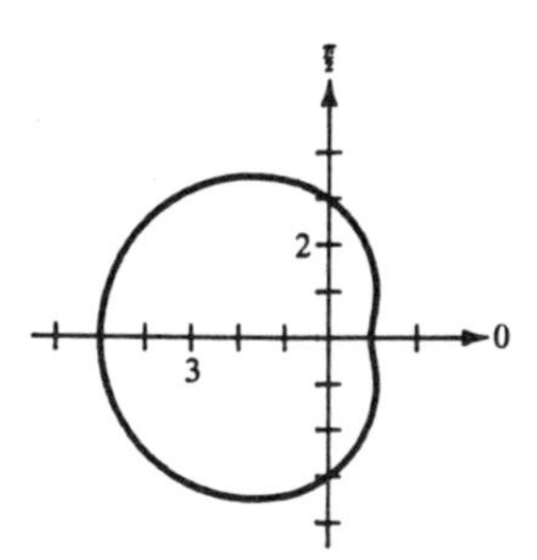

Solution:

$a/b = 3/2 > 1$, so the graph is a dimpled limaçon. Symmetric to the polar axis since r is a function of $\cos\theta$. Maximum value of $|r|$ is 5 and occurs when $\theta = \pi$. There are no zeros of r.

θ	0	$\frac{\pi}{3}$	$\frac{\pi}{2}$	$\frac{2\pi}{3}$	π
r	1	2	3	4	5

19. Sketch the graph of $r = 2\cos 3\theta$.

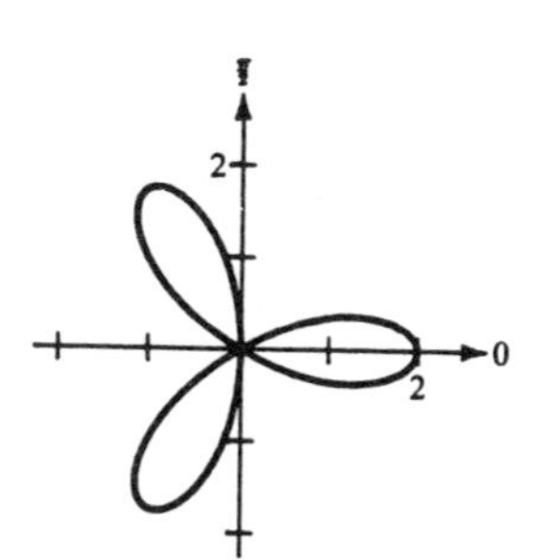

Solution:

The graph is a rose curve with three petals. Symmetric to the polar axis. Maximum value of $|r|$ is 2 and occurs when $\theta = 0$, $\theta = \pi/3$, and $\theta = 2\pi/3$. Zeros of r occur when $\theta = \pi/6$, $\theta = \pi/2$, and $\theta = 5\pi/6$.

θ	0	$\frac{\pi}{2}$	$\frac{\pi}{6}$	$\frac{\pi}{4}$	$\frac{\pi}{3}$	$\frac{5\pi}{12}$	$\frac{\pi}{2}$
r	2	$\sqrt{2}$	0	$-\sqrt{2}$	-2	$-\sqrt{2}$	0

23. Sketch the graph of $r = 2\sec\theta$.

Solution:

$$r = 2\sec\theta$$
$$r = \frac{2}{\cos\theta}$$
$$r\cos\theta = 2$$
$$x = 2 \text{ is a vertical line.}$$

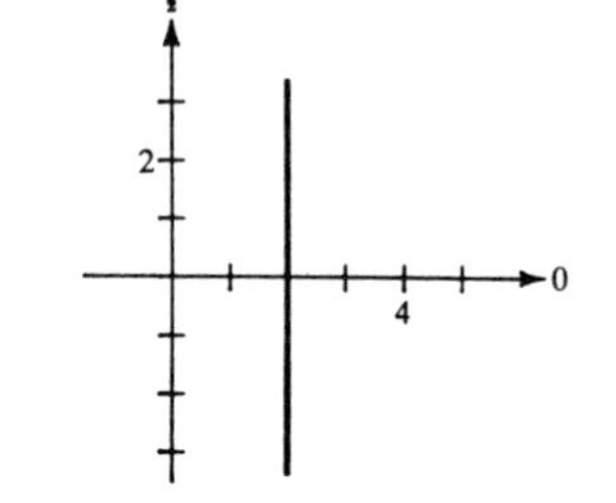

27. Sketch the graph $r^2 = 4\cos 2\theta$.

Solution:
The graph is a lemniscate. Symmetric to the polar axis, the line $\theta = \pi/2$, and the pole. Maximum value of $|r|$ is 2 and occurs when $\theta = 0$. The zeros of r occur when $\theta = \pi/4$ and $\theta = 3\pi/4$.

θ	0	$\frac{\pi}{6}$	$\frac{\pi}{4}$
r	± 2	$\pm\sqrt{2}$	0

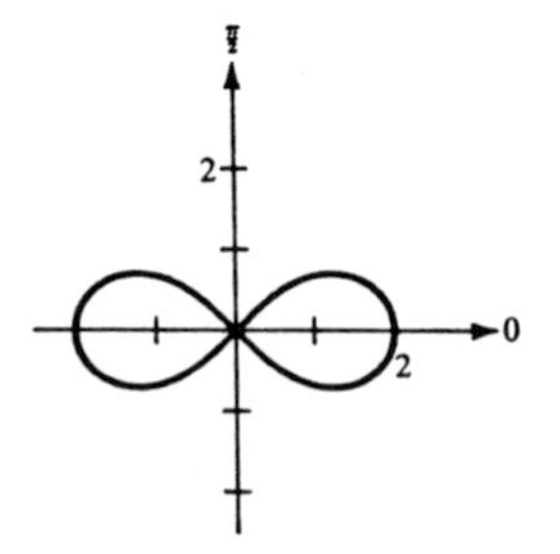

33. Write the equation for the limaçon $r = 2 - \sin\theta$ after it has been rotated by the given amount.

(a) $\frac{\pi}{4}$ (b) $\frac{\pi}{2}$ (c) π (d) $\frac{3\pi}{2}$

Solution:
Refer to Exercises 31 and 32.

(a) $r = 2 - \sin\left(\theta - \frac{\pi}{4}\right)$
$$= 2 - \frac{\sqrt{2}}{2}(\sin\theta - \cos\theta)$$

(b) $r = 2 - (-\cos\theta) = 2 + \cos\theta$

(c) $r = 2 - (-\sin\theta) = 2 + \sin\theta$

(d) $r = 2 - \cos\theta$

35. Sketch the graphs of the equations.

(a) $r = 1 - \sin\theta$ (b) $r = 1 - \sin\left(\theta - \frac{\pi}{4}\right)$

Solution:

(a) Cardioid

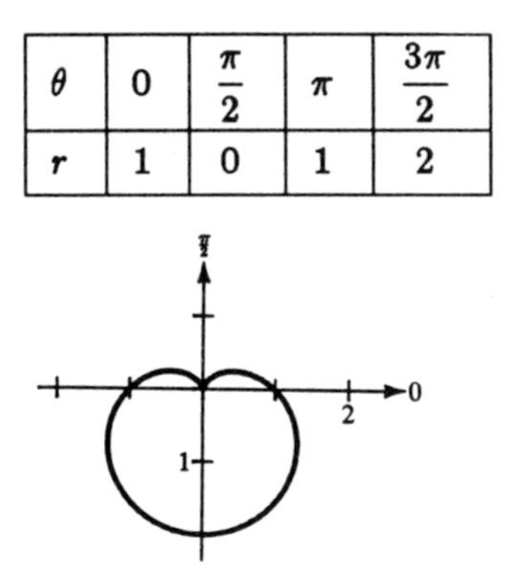

θ	0	$\frac{\pi}{2}$	π	$\frac{3\pi}{2}$
r	1	0	1	2

(b) Rotate the graph of $r = 1 - \sin\theta$ through the angle $\pi/4$.

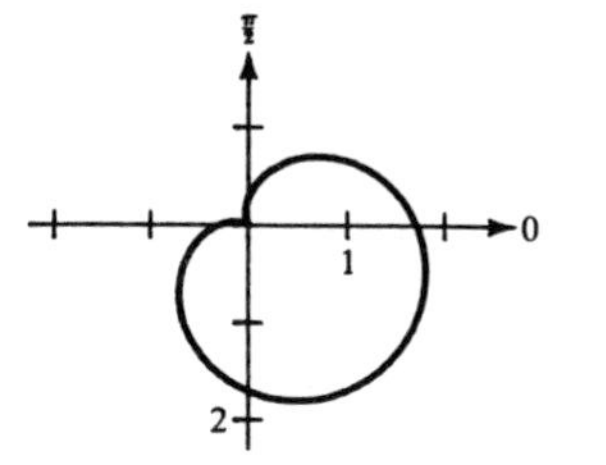

SECTION 11.7

Polar Equations of Conics

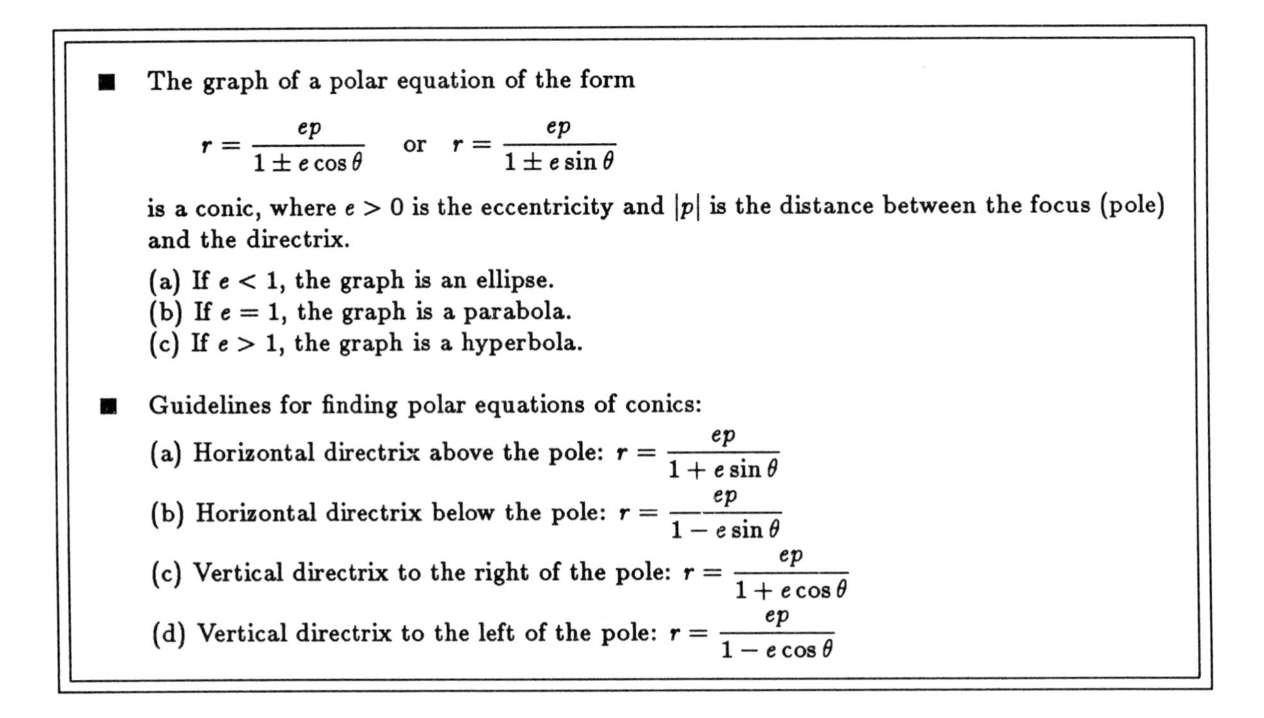

- The graph of a polar equation of the form

$$r = \frac{ep}{1 \pm e\cos\theta} \quad \text{or} \quad r = \frac{ep}{1 \pm e\sin\theta}$$

is a conic, where $e > 0$ is the eccentricity and $|p|$ is the distance between the focus (pole) and the directrix.

(a) If $e < 1$, the graph is an ellipse.
(b) If $e = 1$, the graph is a parabola.
(c) If $e > 1$, the graph is a hyperbola.

- Guidelines for finding polar equations of conics:

(a) Horizontal directrix above the pole: $r = \dfrac{ep}{1 + e\sin\theta}$

(b) Horizontal directrix below the pole: $r = \dfrac{ep}{1 - e\sin\theta}$

(c) Vertical directrix to the right of the pole: $r = \dfrac{ep}{1 + e\cos\theta}$

(d) Vertical directrix to the left of the pole: $r = \dfrac{ep}{1 - e\cos\theta}$

Solutions to Selected Exercises

1. Match the following with one of the given graphs.

$$r = \frac{6}{1 - \cos\theta}$$

Solution:
$e = 1$, so the graph is a parabola.
Vertex: $(3, \pi)$
Matches graph (c)

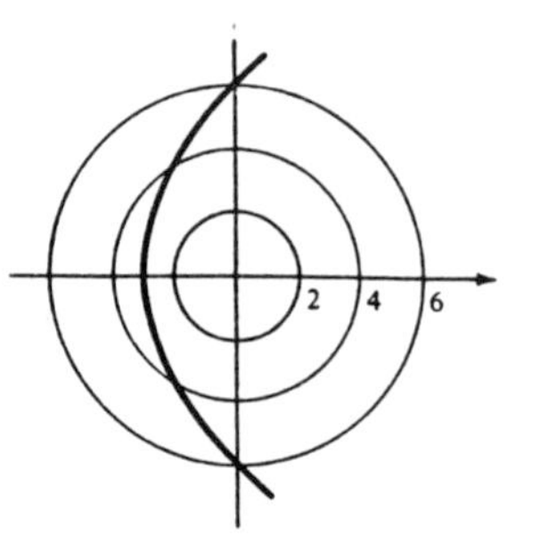

7. Identify and sketch the graph of $r = \dfrac{2}{1 - \cos\theta}$.

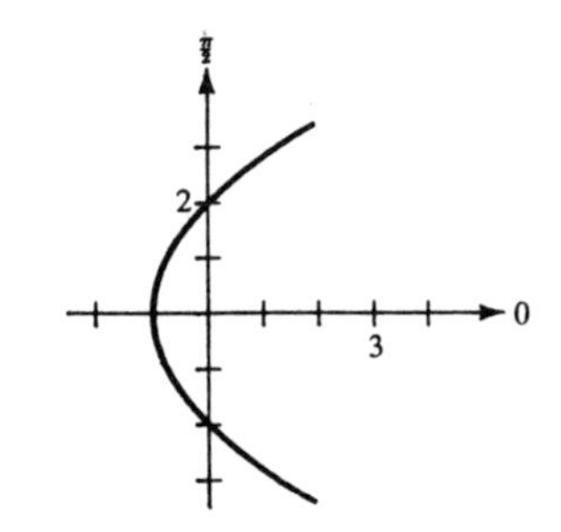

Solution:
$e = 1$, so the graph is a parabola.
Vertex: $(1, \pi)$

θ	$\frac{\pi}{2}$	π	$\frac{3\pi}{2}$
r	2	1	2

11. Identify and sketch the graph of $r = \dfrac{2}{2 + \cos\theta}$.

Solution:

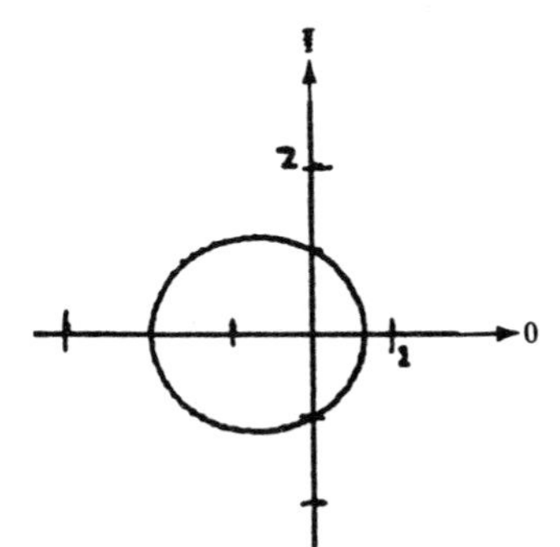

$$r = \frac{2}{2 + \cos\theta}$$
$$r = \frac{1}{1 + \frac{1}{2}\cos\theta}$$

$e = \frac{1}{2} < 1$, so the graph is an ellipse.

θ	0	$\frac{\pi}{2}$	π	$\frac{3\pi}{2}$
r	$\frac{2}{3}$	1	2	1

15. Identify and sketch the graph of $r = \dfrac{5}{1 + 2\cos\theta}$.

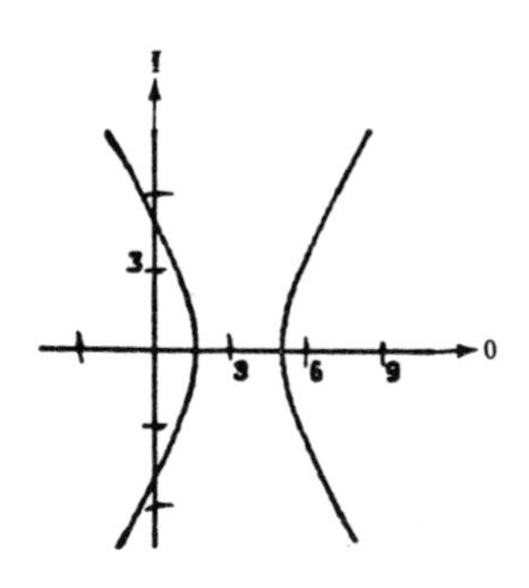

Solution:
$e = 2 > 1$, so the graph is a hyperbola.

θ	0	$\frac{\pi}{2}$	π	$\frac{3\pi}{2}$
r	$\frac{5}{3}$	5	-5	5

19. Identify and sketch the graph of $r = \dfrac{3}{2 - 6\cos\theta}$.

Solution:

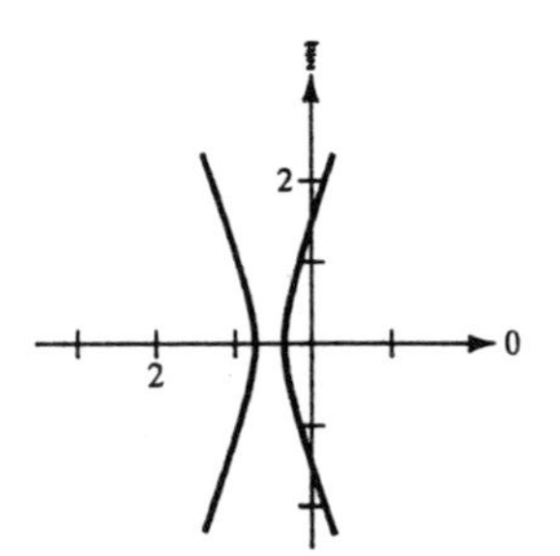

$$r = \frac{3}{2 - 6\cos\theta}$$
$$r = \frac{3/2}{1 - 3\cos\theta}$$

$e = 3 > 1$, so the graph is a hyperbola.

θ	0	$\frac{\pi}{2}$	π	$\frac{3\pi}{2}$
r	$-\frac{3}{4}$	$\frac{3}{2}$	$\frac{3}{8}$	$\frac{3}{2}$

23. Find a polar equation of the ellipse with focus at $(0,\ 0)$, $e = \frac{1}{2}$, and directrix $y = 1$.

Solution:

$e = \frac{1}{2},\ y = 1,\ p = 1$

$$r = \frac{ep}{1 + e\sin\theta}$$

$$r = \frac{\frac{1}{2}}{1 + \frac{1}{2}\sin\theta}$$

$$r = \frac{1}{2 + \sin\theta}$$

27. Find a polar equation of the parabola with focus at $(0,\ 0)$ and vertex at $(1,\ -\pi/2)$.

Solution:

$e = 1,\ p = 2$

$$r = \frac{ep}{1 - e\sin\theta}$$

$$r = \frac{2}{1 - \sin\theta}$$

31. Find a polar equation of the hyperbola with focus at $(0,\ 0)$ and vertices at $(1,\ 3\pi/2)$, $(9,\ 3\pi/2)$.

Solution:

Center: $\left(5,\ \frac{3\pi}{2}\right)$; $c = 5,\ a = 4,\ e = \frac{c}{a} = \frac{5}{4}$

$$r = \frac{ep}{1 - e\sin\theta}$$

$$r = \frac{\frac{5}{4}p}{1 - \frac{5}{4}\sin\theta}$$

$$r = \frac{5p}{4 - 5\sin\theta}$$

$$1 = \frac{5p}{4 - 5\sin\frac{3\pi}{2}}$$

$$1 = \frac{5p}{9}$$

$$p = \frac{9}{5}$$

$$r = \frac{5\left(\frac{9}{5}\right)}{4 - 5\sin\theta}$$

$$r = \frac{9}{4 - 5\sin\theta}$$

35. Find a polar equation of the parabola with focus at $(0, 0)$ and vertex at $(5, \pi)$.

Solution:

Directrix: $x = -10, \quad e = 1, \quad p = 10$

$$r = \frac{ep}{1 - e\cos\theta}$$

$$r = \frac{10}{1 - \cos\theta}$$

39. Use the results of Exercises 37 and 38 to write the polar form of

$$\frac{x^2}{169} + \frac{y^2}{144} = 1.$$

Solution:

$a = 13, \quad b = 12, \quad c = 5, \quad e = \frac{5}{13}$

$$r^2 = \frac{b^2}{1 - e^2\cos^2\theta}$$

$$r^2 = \frac{144}{1 - \left(\frac{25}{169}\right)\cos^2\theta}$$

43. Use the results of Exercises 37 and 38 to write the polar form of the hyperbola with one focus at $(5,\ \pi/2)$ and vertices at $(4,\ \pi/2)$, $(4,\ -\pi/2)$.

Solution:

Center: $(0,\ 0)$, $a = 4$, $c = 5$, $b = 3$, $e = \frac{5}{4}$

$$\frac{x^2}{16} - \frac{y^2}{9} = 1$$

$$r^2 = \frac{-b^2}{1 - e^2 \cos^2 \theta}$$

$$r^2 = \frac{-9}{1 - \left(\frac{25}{16}\right) \cos^2 \theta}$$

47. An earth satellite in a 100-mile-high circular orbit around the earth has a velocity of approximately 17,500 miles per hour. If this velocity is multiplied by $\sqrt{2}$, then the satellite will have the minimum velocity necessary to escape the earth's gravity and it will follow a parabolic path with the center of the earth as the focus. Find a polar equation of the parabolic path of the satellite (assume the radius of the earth is 4000 miles).

Solution:

Directrix: $y = 8200$, $e = 1$, $p = 8200$

$$r = \frac{ep}{1 + e \sin \theta}$$

$$r = \frac{8200}{1 + \sin \theta}$$

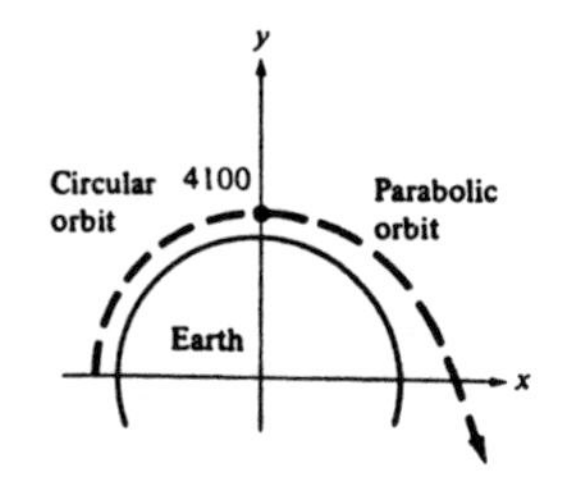

SECTION 11.8

Plane Curves and Parametric Equations

- If f and g are continuous functions of t on an interval I, then the set of ordered pairs $(f(t),\ g(t))$ is called a *plane curve* C. The equations $x = f(t)$ and $y = g(t)$ are called *parametric equations* for C and t is called the *parameter*.
- To eliminate the parameter:

  (a) Solve for t in one equation and substitute into the second equation.
  (b) Use trigonometric identities.
- You should be able to find the parametric equations for a graph.

Solutions to Selected Exercises

5. Sketch the curve represented by the parametric equations, $x = \sqrt{t}$ and $y = 1 - t$, and write the corresponding rectangular equation by eliminating the parameter.

Solution:
Since $x = \sqrt{t}$, we know that $x \geq 0$. $x^2 = t$ substituted into $y = 1 - t$ yields the equation $y = 1 - x^2$, $x \geq 0$.

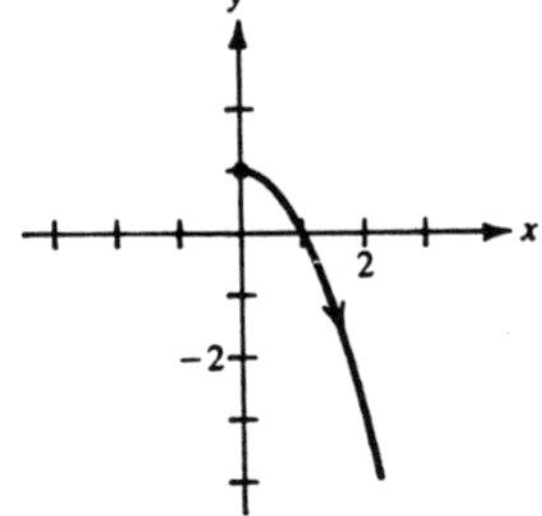

9. Sketch the curve represented by the parametric equations, $x = t - 1$ and $y = t/(t-1)$, and write the corresponding rectangular equation by eliminating the parameter.

Solution:

$$x = t - 1 \Rightarrow t = x + 1$$

$$y = \frac{t}{t-1} = \frac{x+1}{(x+1)-1} = \frac{x+1}{x}$$

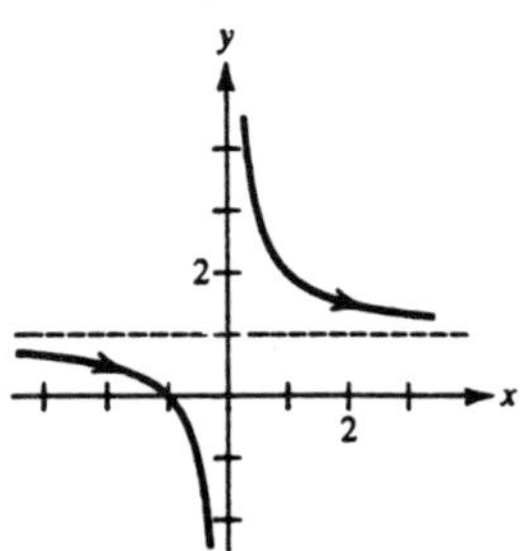

13. Sketch the curve represented by the parametric equations, $x = 4\sin 2\theta$ and $y = 2\cos 2\theta$, and write the corresponding rectangular equation by eliminating the parameter.

Solution:

$$x = 4\sin 2\theta \Rightarrow \sin 2\theta = \frac{x}{4}$$

$$y = 2\cos 2\theta \Rightarrow \cos 2\theta = \frac{y}{2}$$

$$\sin^2 2\theta + \cos^2 2\theta = 1$$

$$\left(\frac{x}{4}\right)^2 + \left(\frac{y}{2}\right)^2 = 1$$

$$\frac{x^2}{16} + \frac{y^2}{4} = 1$$

17. Sketch the curve represented by the parametric equations, $x = \sec\theta$ and $y = \cos\theta$, and write the corresponding rectangular equation by eliminating the parameter.

Solution:

Since $x = \sec\theta$, we have the restriction $x \le -1$ or $x \ge 1$.
Since $y = \cos\theta$, we have the restriction $-1 \le y \le 1$.

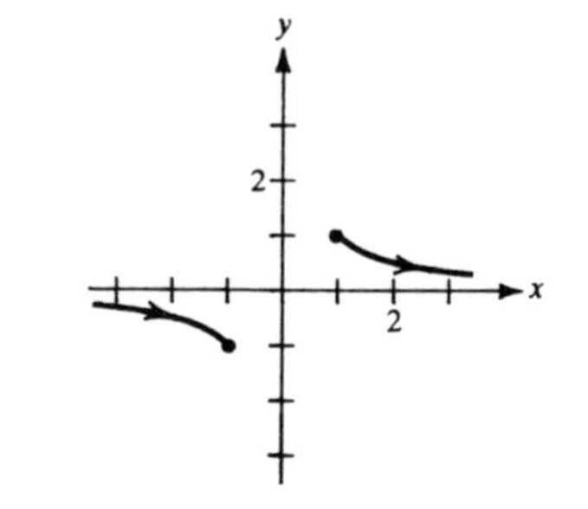

$$x = \sec\theta \Rightarrow \frac{1}{x} = \cos\theta$$

$$y = \cos\theta \Rightarrow y = \frac{1}{x}, \quad x \le -1 \text{ OR } x \ge 1,\ -1 \le y \le 1$$

21. Sketch the curve represented by the parametric equations, $x = 4 + 2\cos\theta$ and $y = -1 + 4\sin\theta$, and write the corresponding rectangular equation by eliminating the parameter.

Solution:

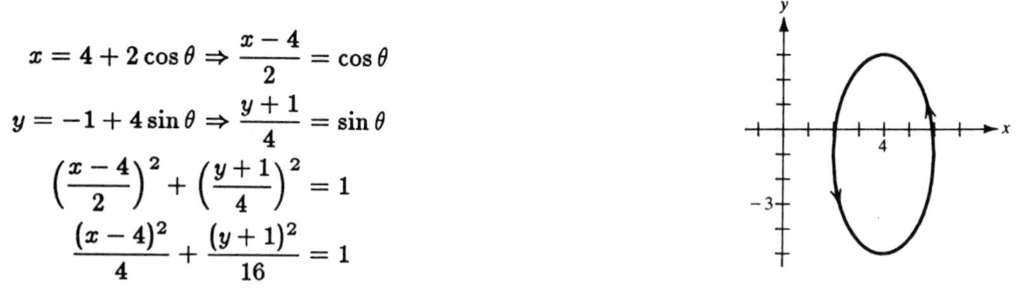

$$x = 4 + 2\cos\theta \Rightarrow \frac{x-4}{2} = \cos\theta$$

$$y = -1 + 4\sin\theta \Rightarrow \frac{y+1}{4} = \sin\theta$$

$$\left(\frac{x-4}{2}\right)^2 + \left(\frac{y+1}{4}\right)^2 = 1$$

$$\frac{(x-4)^2}{4} + \frac{(y+1)^2}{16} = 1$$

25. Sketch the curve represented by the parametric equations, $x = e^{-t}$ and $y = e^{3t}$, and write the corresponding rectangular equation by eliminating the parameter.

Solution:
Since $x = e^{-t}$ and $y = e^{3t}$, we have $x > 0$ and $y > 0$.

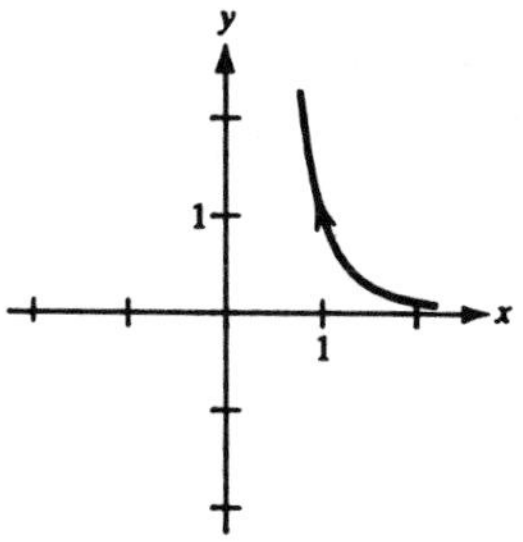

$$x = e^{-t} \Rightarrow \frac{1}{x} = e^{t}$$
$$y = e^{3t} = (e^{t})^3 = \left(\frac{1}{x}\right)^3 = \frac{1}{x^3}$$

where $x > 0$ and $y > 0$.

29. Eliminate the parameter and obtain the standard form of the rectangular equation of the curve.

Ellipse: $x = h + a\cos\theta,\ y = k + b\sin\theta$

Solution:

$$x = h + a\cos\theta \Rightarrow \frac{x-h}{a} = \cos\theta$$
$$y = k + b\sin\theta \Rightarrow \frac{y-k}{b} = \sin\theta$$
$$\left(\frac{x-h}{a}\right)^2 + \left(\frac{y-k}{b}\right)^2 = 1$$
$$\frac{(x-h)^2}{a^2} + \frac{(y-k)^2}{b^2} = 1$$

33. Find a set of parametric equations for the circle with center at $(2, 1)$ and radius 4.

Solution:
From Exercise 28 we have $x = h + r\cos\theta,\ y = k + r\sin\theta$. Using $h = 2,\ k = 1$ and $r = 4$, we have $x = 2 + 4\cos\theta,\ y = 1 + 4\sin\theta$. This solution is not unique.

37. Find a set of parametric equations for the hyperbola with vertices at $(\pm 4,\ 0)$ and foci at $(\pm 5,\ 0)$.

Solution:
From Exercise 30 we have $x = h + a\sec\theta,\ y = k + b\tan\theta$. Using $(h,\ k) = (0,\ 0),\ a = 4,\ c = 5,$ and $b = 3$, we have $x = 4\sec\theta,\ y = 3\tan\theta$. This solution is not unique.

39. Find two different sets of parametric equations for the rectangular equation $y = x^3$.

Solution:

Examples

$x = t,\qquad y = t^3$

$x = \sqrt[3]{t},\qquad y = t$

$x = \tan t,\qquad y = \tan^3 t$

$x = t - 4,\qquad y = (t-4)^3$

and so on.

43. Sketch the curve represented by the parametric equations.

Witch of Agnesi: $x = 2\cot\theta,\ y = 2\sin^2\theta$

Solution:

$$x = 2\cot\theta \Rightarrow \theta = \operatorname{arccot}\frac{x}{2}$$

$$y = 2\sin^2\theta \Rightarrow y = 2\sin^2\left(\operatorname{arccot}\frac{x}{2}\right)$$

$$y = 2\left(\frac{2}{\sqrt{x^2+4}}\right)^2$$

$$y = \frac{8}{x^2+4}$$

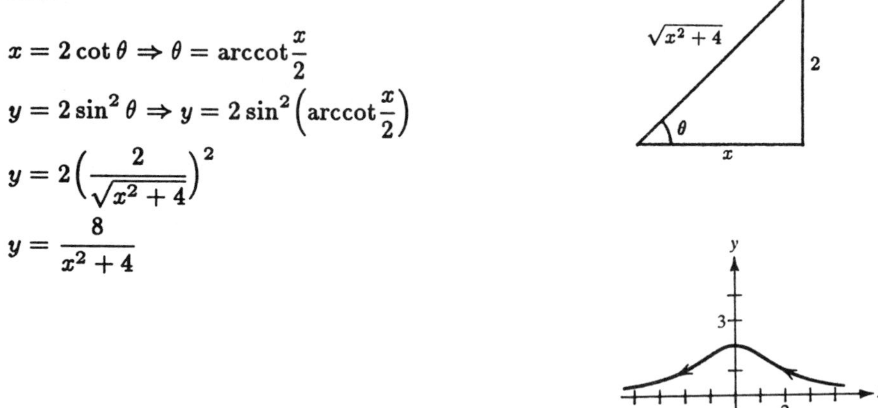

47. A wheel of radius a rolls along a straight line without slipping, as shown in the figure. Find the parametric equation for the curve described by a point P that is b units from the center of the wheel. This curve is called a *curtate cycloid* when $b < a$.

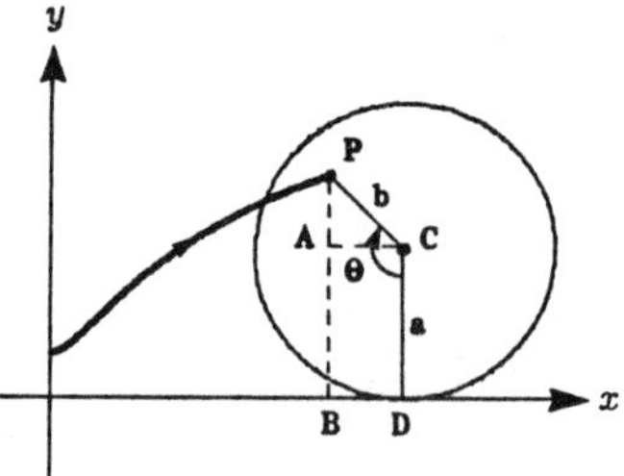

Solution:

When the circle has rolled θ radians, we know that the center is at $(a\theta,\ a)$.

$$\sin\theta = \sin(180^\circ - \theta) = \frac{|AC|}{b} = \frac{|BD|}{b} \quad \text{or} \quad |BD| = b\sin\theta$$

$$\cos\theta = -\cos(180^\circ - \theta) = \frac{|AP|}{-b} \quad \text{or} \quad |AP| = -b\cos\theta$$

Therefore, $x = a\theta - b\sin\theta$ and $y = a - b\cos\theta$.

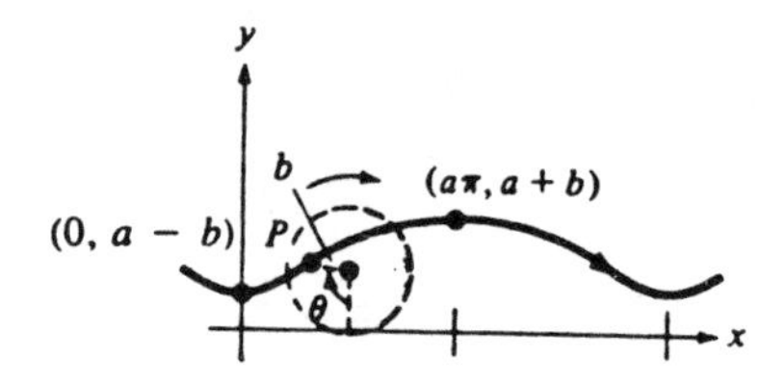

REVIEW EXERCISES FOR CHAPTER 11

5. Identify and sketch the graph of the rectangular equation $3x^2 + 2y^2 - 12x + 12y + 29 = 0$.

Solution:

Since $AC = 3(2) = 6 > 0$, the graph is an ellipse.

$$3x^2 + 2y^2 - 12x + 12y + 29 = 0$$
$$3(x^2 - 4x + 4) + 2(y^2 + 6y + 9) = -29 + 12 + 18$$
$$3(x-2)^2 + 2(y+3)^2 = 1$$
$$\frac{(x-2)^2}{1/3} + \frac{(y+3)^2}{1/2} = 1$$

Center: $(2,\ -3)$

Vertices: $\left(2,\ -3 \pm \frac{\sqrt{2}}{2}\right)$

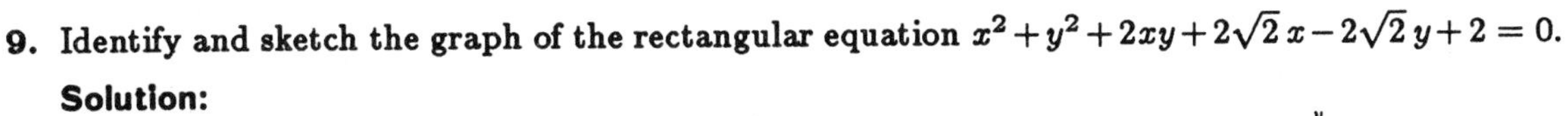

9. Identify and sketch the graph of the rectangular equation $x^2 + y^2 + 2xy + 2\sqrt{2}\,x - 2\sqrt{2}\,y + 2 = 0$.

Solution:

Since $B^2 - 4AC = 0$, the graph is a parabola.

$A = 1,\ B = 2,\ C = 1,\ D = 2\sqrt{2},\ E = -2\sqrt{2},\ F = 2$

$\cot 2\theta = 0,\ 2\theta = \frac{\pi}{2},\ \theta = \frac{\pi}{4},\ \sin\theta = \frac{\sqrt{2}}{2},\ \cos\theta = \frac{\sqrt{2}}{2}$

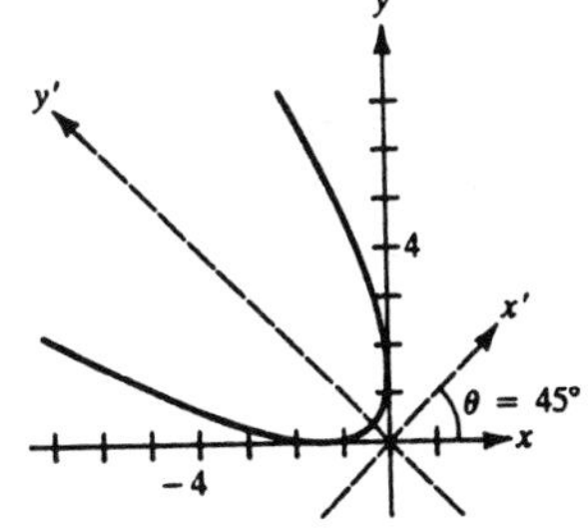

$$A' = \left(\frac{\sqrt{2}}{2}\right)^2 + 2\left(\frac{\sqrt{2}}{2}\right)^2 + \left(\frac{\sqrt{2}}{2}\right)^2 = 2$$
$$C' = \left(\frac{\sqrt{2}}{2}\right)^2 - 2\left(\frac{\sqrt{2}}{2}\right)^2 + \left(\frac{\sqrt{2}}{2}\right)^2 = 0$$
$$D' = 2\sqrt{2}\left(\frac{\sqrt{2}}{2}\right) - 2\sqrt{2}\left(\frac{\sqrt{2}}{2}\right) = 0$$
$$E' = -2\sqrt{2}\left(\frac{\sqrt{2}}{2}\right) - 2\sqrt{2}\left(\frac{\sqrt{2}}{2}\right) = -4$$
$$F' = 2$$

$$2x'^2 - 4y' + 2 = 0$$
$$x'^2 = 4\left(\frac{1}{2}\right)\left(y' - \frac{1}{2}\right)$$

Vertex: $(x',\ y') = \left(0,\ \frac{1}{2}\right)$ or $(x,\ y) = \left(\frac{-1}{\sqrt{8}},\ \frac{1}{\sqrt{8}}\right)$

13. Find a rectangular equation for the parabola with vertex at $(0, 2)$ and directrix $x = -3$.

Solution:

$p = 3, \ (h, k) = (0, 2)$

$$(y - k)^2 = 4p(x - h)$$
$$(y - 2)^2 = 4(3)(x - 0)$$
$$y^2 - 4y + 4 = 12x$$
$$y^2 - 12x - 4y + 4 = 0$$

17. Find a rectangular equation for the ellipse with vertices at $(0, \pm 6)$ and passes through the point $(2, 2)$.

Solution:

$(h, k) = (0, 0)$

Vertical major axis with $a = 6$

$\frac{x^2}{b^2} + \frac{y^2}{36} = 1$

Since the graph passes through the points $(2, 2)$, we have

$$\frac{4}{b^2} + \frac{4}{36} = 1$$
$$36 + b^2 = 9b^2$$
$$36 = 8b^2$$
$$\frac{36}{8} = b^2$$
$$\frac{9}{2} = b^2$$
$$\frac{x^2}{9/2} + \frac{y^2}{36} = 1$$
$$\frac{2x^2}{9} + \frac{y^2}{36} = 1$$

21. Find a rectangular equation of the hyperbola with foci at $(0, 0)$, $(8, 0)$ and asymptotes $y = \pm 2(x - 4)$.

Solution:

$(h, k) = (4, 0)$

The transverse axis is horizontal with $c = 4$. Also from the slopes of the asymptotes $\pm b/a = \pm 2$ or $\pm b = \pm 2a$. Now $c^2 = a^2 + b^2 = a^2 + (2a)^2 = 16$. Therefore,

$$a^2 = \frac{16}{5}, \ b^2 = \frac{64}{5} \text{ and } \frac{(x-4)^2}{16/5} - \frac{y^2}{64/5} = 1, \ \frac{5(x-4)^2}{16} - \frac{5y^2}{64} = 1.$$

25. Find an equation of the tangent line to

$$\frac{x^2}{100} + \frac{y^2}{25} = 1$$

at the point $(-8,\ 3)$. The tangent line to the conic

$$\frac{x^2}{a^2} \pm \frac{y^2}{b^2} = 1$$

at the point $(x_0,\ y_0)$ is given by

$$\frac{x_0 x}{a^2} \pm \frac{y_0 y}{b^2} = 1.$$

Solution:

$(x_0,\ y_0) = (-8,\ 3)$
$a^2 = 100,\ b^2 = 25$

$$\frac{-8x}{100} + \frac{3y}{25} = 1$$
$$-8x + 12y = 100$$
$$2x - 3y = -25$$

29. Identify and sketch the graph of the polar equation $r = 4$.

Solution:

$r = 4$ is a circle centered at the pole of radius 4.

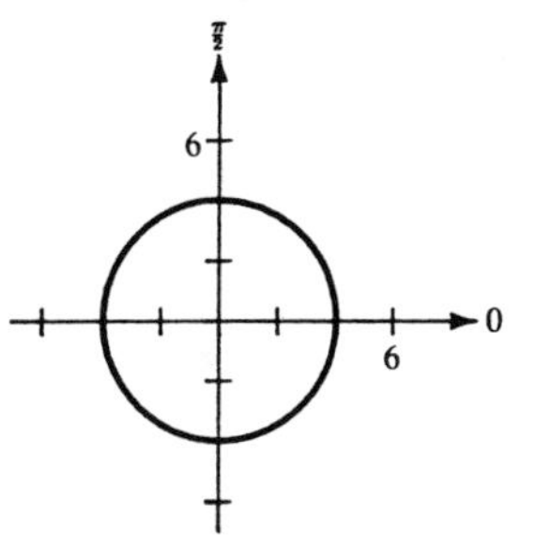

33. Identify and sketch the graph of the polar equation $r = -2(1 + \cos\theta)$.

Solution:

$r = -2(1 + \cos\theta)$ is a cardioid; symmetric to the polar axis.

θ	0	$\frac{\pi}{3}$	$\frac{\pi}{2}$	$\frac{2\pi}{3}$	π
r	-4	-3	-2	-1	0

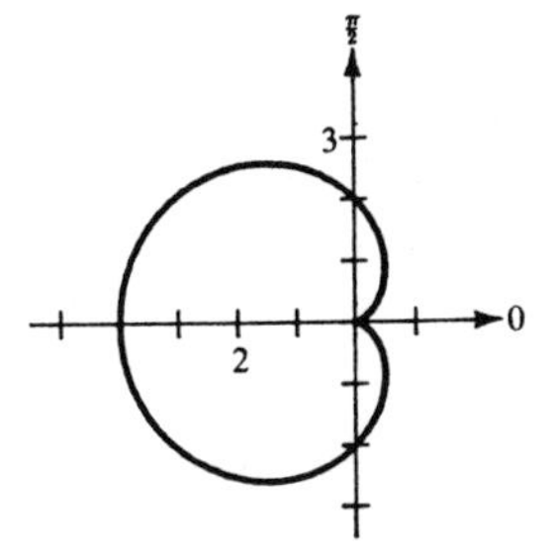

37. Identify and sketch the graph of the polar equation $r = -3\cos 2\theta$.

Solution:

$r = -3\cos 2\theta$ is a rose curve with four petals; symmetric to the polar axis, $\theta = \pi/2$, and the pole. Maximum value of $|r|$ is 3.

$(-3,\ 0),\ \left(3,\ \frac{\pi}{2}\right),\ (-3,\ \pi),\ \left(3,\ \frac{3\pi}{2}\right)$

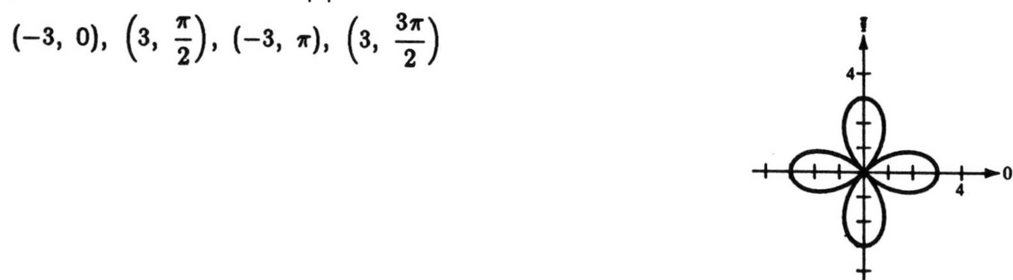

41. Identify and sketch the graph of the polar equation $r = \dfrac{3}{\cos(\theta - (\pi/4))}$.

Solution:

$$\begin{aligned}
r &= \frac{3}{\cos(\theta - (\pi/4))}\\
r\cos\left(\theta - \frac{\pi}{4}\right) &= 3\\
r\left[\cos\theta\cos\frac{\pi}{4} + \sin\theta\sin\frac{\pi}{4}\right] &= 3\\
r\left[\frac{\sqrt{2}}{2}\cos\theta + \frac{\sqrt{2}}{2}\sin\theta\right] &= 3\\
\frac{\sqrt{2}}{2}r\cos\theta + \frac{\sqrt{2}}{2}r\sin\theta &= 3\\
r\cos\theta + r\sin\theta &= 3\sqrt{2}\\
x + y &= 3\sqrt{2}
\end{aligned}$$

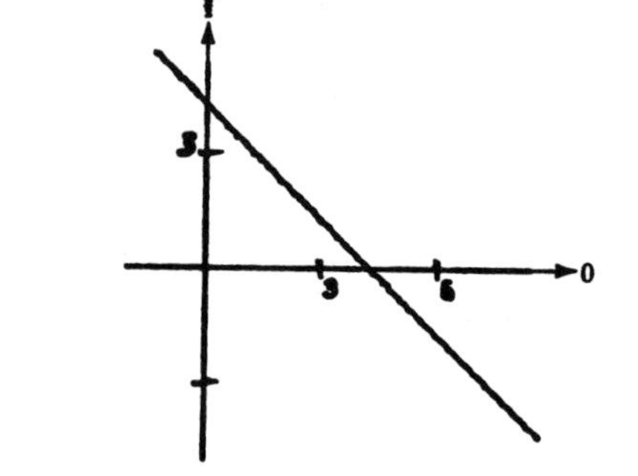

45. Convert $r = 3\cos\theta$ to rectangular form.

Solution:

$$\begin{aligned}
r &= 3\cos\theta\\
r^2 &= 3r\cos\theta\\
x^2 + y^2 &= 3x\\
x^2 + y^2 - 3x &= 0
\end{aligned}$$

49. Convert $r^2 = \cos 2\theta$ to rectangular form.

Solution:

$$\begin{aligned} r^2 &= \cos 2\theta \\ r^2 &= 2\cos^2\theta - 1 \\ x^2 + y^2 &= 2\left(\frac{x^2}{x^2+y^2}\right) - 1 \\ (x^2+y^2)^2 &= 2x^2 - (x^2+y^2) \\ x^4 + 2x^2y^2 + y^4 &= x^2 - y^2 \\ x^4 + 2x^2y^2 + y^4 - x^2 + y^2 &= 0 \end{aligned}$$

53. Find a polar equation for a parabola with vertex at $(2, \pi)$ and focus at $(0, 0)$.

Solution:

$e = 1, \ p = 4$

$$r = \frac{ep}{1 - e\cos\theta}$$

$$r = \frac{4}{1 - \cos\theta}$$

57. Find a polar equation for a circle with center at $(0, 5)$ and passes through $(0, 0)$.

Solution:

The radius is 5.

$$\begin{aligned} x^2 + (y-5)^2 &= 25 \\ x^2 + y^2 - 10y &= 0 \\ r^2 - 10r\sin\theta &= 0 \\ r(r - 10\sin\theta) &= 0 \\ r &= 10\sin\theta \end{aligned}$$

61. Sketch the curve represented by the parametric equations, $x = 1+4t$, $y = 2-3t$, and if possible, write the corresponding rectangular equation by eliminating the parameter.

Solution:

$$x = 1 + 4t \Rightarrow t = \frac{x-1}{4}$$

$$y = 2 - 3t \Rightarrow y = 2 - 3\left(\frac{x-1}{4}\right)$$

$$y = -\frac{3}{4}x + \frac{11}{4}$$

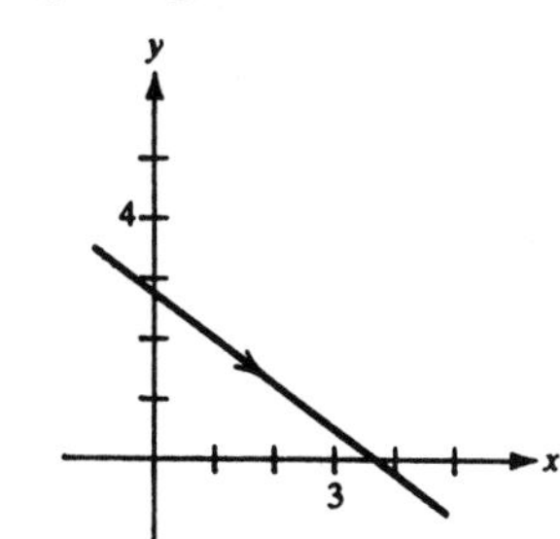

65. Sketch the curve represented by the parametric equations, $x = 6\cos\theta$, $y = 6\sin\theta$, and if possible, write the corresponding rectangular equation by eliminating the parameter.

Solution:

$$x = 6\cos\theta \Rightarrow \cos\theta = \frac{x}{6}$$

$$y = 6\sin\theta \Rightarrow \sin\theta = \frac{y}{6}$$

$$\left(\frac{x}{6}\right)^2 + \left(\frac{y}{6}\right)^2 = 1$$

$$x^2 + y^2 = 36$$

69. Sketch the curve represented by the parametric equations, $x = \cos^3\theta$, $y = 4\sin^3\theta$, and if possible, write the corresponding rectangular equation by eliminating the parameter.

Solution:

$$x = \cos^3\theta \Rightarrow \cos\theta = x^{1/3}$$

$$y = 4\sin^3\theta \Rightarrow \sin\theta = \left(\frac{y}{4}\right)^{1/3}$$

$$(x^{1/3})^2 + \left[\left(\frac{y}{4}\right)^{1/3}\right]^2 = 1$$

$$x^{2/3} + \left(\frac{y}{4}\right)^{2/3} = 1$$

Practice Test for Chapter 11

1. Find the vertex, focus, and directrix of the parabola $x^2 - 6x - 4y + 1 = 0$.

2. Find an equation of the parabola with its vertex at $(2, -5)$ and focus at $(2, -6)$.

3. Find the center, foci, vertices, and eccentricity of the ellipse $x^2 + 4y^2 - 2x + 32y + 61 = 0$.

4. Find an equation of the ellipse with vertices $(0, \pm 6)$ and eccentricity $e = \frac{1}{2}$.

5. Find the center, vertices, foci, and asymptotes of the hyperbola $16y^2 - x^2 - 6x - 128y + 231 = 0$.

6. Find an equation of the hyperbola with vertices at $(\pm 3, 2)$ and foci at $(\pm 5, 2)$.

7. Rotate the axes to eliminate the xy-term. Sketch the graph of the resulting equation, showing both sets of axes. $5x^2 + 2xy + 5y^2 - 10 = 0$

8. Use the discriminant to determine whether the graph of the equation is a parabola, ellipse, or hyperbola.

(a) $6x^2 - 2xy + y^2 = 0$

(b) $x^2 + 4xy + 4y^2 - x - y + 17 = 0$

9. Convert the polar point $\left(\sqrt{2}, \dfrac{3\pi}{4}\right)$ to rectangular coordinates.

10. Convert the rectangular point $(\sqrt{3}, -1)$ to polar coordinates.

11. Convert the rectangular equation $4x - 3y = 12$ to polar form.

12. Convert the polar equation $r = 5\cos\theta$ to rectangular form.

13. Sketch the graph of $r = 1 - \cos\theta$.

14. Sketch the graph of $\theta = \dfrac{3\pi}{4}$.

15. Sketch the graph of $r = 5\sin 2\theta$.

16. Sketch the graph of $r = \dfrac{3}{6 - \cos\theta}$.

17. Find a polar equation of the parabola with its vertex at $(6, \pi/2)$ and focus at $(0, 0)$.

For 18–20, eliminate the parameter and write the corresponding rectangular equation.

18. $x = \sqrt{t+1}, \quad y = 3 + t$

19. $x = 3 - 2\sin\theta, \quad y = 1 + 5\cos\theta$

20. $x = e^{2t}, \quad y = e^{4t}$

CHAPTER 1

Practice Test Solutions

1. $4 + 3(18 - 11) = 4 + 3(7) = 4 + 21 = 25$

2. $\left(\frac{4}{15} \div 2\right) - \left(5 \times \frac{8}{15}\right) = \frac{4}{15} \cdot \frac{1}{2} - \frac{5}{1} \cdot \frac{8}{15} = \frac{2}{15} - \frac{8}{3} = \frac{2}{15} - \frac{40}{15} = -\frac{38}{15}$

3. $|x - (-6)| < 4$

$|x + 6| < 4$

4. $0.0000439 = 4.39 \times 10^{-5}$

5. $(3x^2y^{-1})^2(4x^{-2}y)^{-1} = 9x^4y^{-2}4^{-1}x^2y^{-1} = \dfrac{9x^6}{4y^3}$

6. $\sqrt[3]{81x^5y^6} = \sqrt[3]{27x^3y^6 3x^2} = 3xy^2\sqrt[3]{3x^2}$

7. $\dfrac{4}{\sqrt[3]{2}} = \dfrac{4}{\sqrt[3]{2}} \cdot \dfrac{\sqrt[3]{2}}{\sqrt[3]{2}} \cdot \dfrac{\sqrt[3]{2}}{\sqrt[3]{2}} = \dfrac{4\sqrt[3]{4}}{\sqrt[3]{8}} = \dfrac{4\sqrt[3]{4}}{2} = 2\sqrt[3]{4}$

8.
$$\begin{aligned}(x+3)(x^2 - 4x - 7) &= (x+3)(x^2) + (x+3)(-4x) + (x+3)(-7)\\ &= x^3 + 3x^2 - 4x^2 - 12x - 7x - 21\\ &= x^3 - x^2 - 19x - 21\end{aligned}$$

9. $x^4 - 81 = (x^2 + 9)(x^2 - 9) = (x^2 + 9)(x + 3)(x - 3)$

10.
$$\begin{aligned}x^5 - 4x^3 - x^2 + 4 &= x^3(x^2 - 4) - (x^2 - 4)\\ &= (x^3 - 1)(x^2 - 4)\\ &= (x - 1)(x^2 + x + 1)(x - 2)(x + 2)\end{aligned}$$

11. $8x^2 + 6x - 9 = (2x + 3)(4x - 3)$

12. $\dfrac{8x^3 + 8x^2y}{x^2y + xy^2} = \dfrac{8x^2(x + y)}{xy(x + y)} = \dfrac{8x}{y}$

13.
$$\begin{aligned}\frac{3x}{x^2 - x - 6} - \frac{2}{x - 3} &= \frac{3x}{(x+2)(x-3)} - \frac{2}{x-3} \cdot \frac{x+2}{x+2}\\ &= \frac{3x - 2(x+2)}{(x+2)(x-3)} = \frac{3x - 2x - 4}{(x+2)(x-3)} = \frac{x - 4}{(x+2)(x-3)}\end{aligned}$$

14.
$$\frac{\dfrac{1}{x+1} - \dfrac{1}{x}}{\dfrac{1}{x^2 + x}} = \frac{\dfrac{x - (x+1)}{x(x+1)}}{\dfrac{1}{x(x+1)}} = \frac{-1}{x(x+1)} \cdot \frac{x(x+1)}{1} = -1$$

15.
$$\frac{x^2 - 49}{x^2 + 6x - 7} \cdot \frac{x^2 - 1}{x} = \frac{(x+7)(x-7)}{(x+7)(x-1)} \cdot \frac{(x+1)(x-1)}{x} = \frac{(x-7)(x+1)}{x}$$

16.

$$\frac{1}{x+2} - \frac{3}{x-4} = \frac{5}{x^2-2x-8}$$

$$(x+2)(x-4)\left[\frac{1}{x+2} - \frac{3}{x-4}\right] = (x+2)(x-4)\left[\frac{5}{(x+2)(x-4)}\right]$$

$$(x-4) - 3(x+2) = 5$$

$$x - 4 - 3x - 6 = 5$$

$$-2x - 10 = 5$$

$$-2x = 15$$

$$x = -\frac{15}{2}$$

17. $(x+12)^2 = 20$

$$x + 12 = \pm\sqrt{20}$$

$$x + 12 = \pm 2\sqrt{5}$$

$$x = -12 \pm 2\sqrt{5}$$

18. $3x^2 + 6x + 2 = 0$

$$x^2 + 2x + \frac{2}{3} = 0$$

$$x^2 + 2x + 1 = -\frac{2}{3} + 1$$

$$(x+1)^2 = \frac{1}{3}$$

$$x + 1 = \pm\sqrt{\frac{1}{3}}$$

$$x + 1 = \pm\frac{\sqrt{3}}{3}$$

$$x = -1 \pm \frac{\sqrt{3}}{3} = \frac{-3 \pm \sqrt{3}}{3}$$

19. $2x^2 - 3x - 5 = 0$

$$a = 2,\ b = -3,\ c = -5$$

$$x = \frac{-(-3) \pm \sqrt{(-3)^2 - 4(2)(-5)}}{2(2)}$$

$$= \frac{3 \pm \sqrt{9+40}}{4} = \frac{3 \pm \sqrt{49}}{4} = \frac{3 \pm 7}{4}$$

$$x = \frac{3+7}{4} = \frac{10}{4} = \frac{5}{2}$$

$$x = \frac{3-7}{4} = \frac{-4}{4} = -1$$

20.

$$-3 \le \frac{4-x}{2} \le 5$$

$$-6 \le 4 - x \le 10$$

$$-10 \le -x \le 6$$

$$10 \ge x \ge -6 \text{ OR } -6 \le x \le 10$$

21. $|x - 15| \ge 10$

$$x - 15 \le -10 \quad \text{OR} \quad x - 15 \ge 10$$

$$x \le 5 \quad \text{OR} \quad x \ge 25$$

22.

$$x^3 - 9x \le 0$$
$$x(x^2 - 9) \le 0$$
$$x(x + 3)(x - 3) \le 0$$

Critical numbers: $x = 0, \ x = -3, \ x = 3$

Test intervals: $(-\infty, \ -3), \ (-3, \ 0), \ (0, \ 3), \ (3, \ \infty)$

$(-)(-)(-) < 0$	$(-)(+)(-) > 0$	$(+)(+)(-) < 0$	$(+)(+)(+) > 0$
YES	NO	YES	NO

Number line: -3, 0, 3

Solution: $x \le -3$ or $0 \le x \le 3$

23.

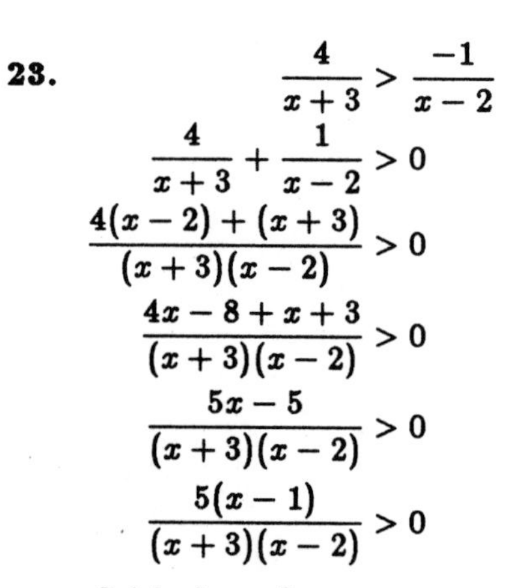

$$\frac{4}{x+3} > \frac{-1}{x-2}$$
$$\frac{4}{x+3} + \frac{1}{x-2} > 0$$
$$\frac{4(x-2) + (x+3)}{(x+3)(x-2)} > 0$$
$$\frac{4x - 8 + x + 3}{(x+3)(x-2)} > 0$$
$$\frac{5x - 5}{(x+3)(x-2)} > 0$$
$$\frac{5(x-1)}{(x+3)(x-2)} > 0$$

Critical numbers: $x = 1, \ x = -3, \ x = 2$

$\frac{(-)}{(-)(-)} < 0$	$\frac{(-)}{(+)(-)} > 0$	$\frac{(+)}{(+)(-)} < 0$	$\frac{(+)}{(+)(+)} > 0$
NO	YES	NO	YES

Number line: -3, -2, -1, 0, 1, 2

Solution: $-3 < x < 1$ or $x > 2$

24. $x^2(1 - 2x)^{4/3} - 3x(1 - 2x)^{1/3} = x(1 - 2x)^{1/3}[x(1 - 2x) - 3]$

$= x(1 - 2x)^{1/3}(x - 2x^2 - 3) = x(1 - 2x)^{1/3}(-2x^2 + x - 3)$

25. $\sqrt{36 + x^2} = 6 + x$ is false.

$\sqrt{36 + x^2}$ cannot be reduced any further.

$\sqrt{36 + 12x + x^2} = \sqrt{(6 + x)^2} = 6 + x$

CHAPTER 2

Practice Test Solutions

1. $d = \sqrt{(4-0)^2 + (-1-3)^2}$
$= \sqrt{16+16}$
$= \sqrt{32}$
$= 4\sqrt{2}$

2. Midpoint: $\left(\frac{4+0}{2}, \frac{-1+3}{2}\right) = (2,\ 1)$

3. $6 = \sqrt{(x-0)^2 + (-2-0)^2}$
$6 = \sqrt{x^2+4}$
$36 = x^2 + 4$
$x^2 = 32$
$x = \pm\sqrt{32}$
$x = \pm 4\sqrt{2}$

4. x-intercept: Let $y = 0$; $0 = \frac{x-2}{x+3}$
$0 = x - 2$
$x = 2 \quad (2,\ 0)$

y-intercept: Let $x = 0$; $y = \frac{0-2}{0+3}$
$y = -\frac{2}{3} \quad \left(0,\ -\frac{2}{3}\right)$

5. $xy^2 = 6$
$x(-y)^2 = 6 \Rightarrow xy^2 = 6$ x-axis symmetry
$(-x)y^2 = 6 \Rightarrow xy^2 = -6$ No y-axis symmetry
$(-x)(-y)^2 = 6 \Rightarrow xy^2 = -6$ No origin symmetry

6. x-intercepts: $(0,\ 0)$, $(2,\ 0)$, $(-2,\ 0)$
Origin symmetry:

x	0	1	−1	2	−2	3
y	0	−3	3	0	0	15

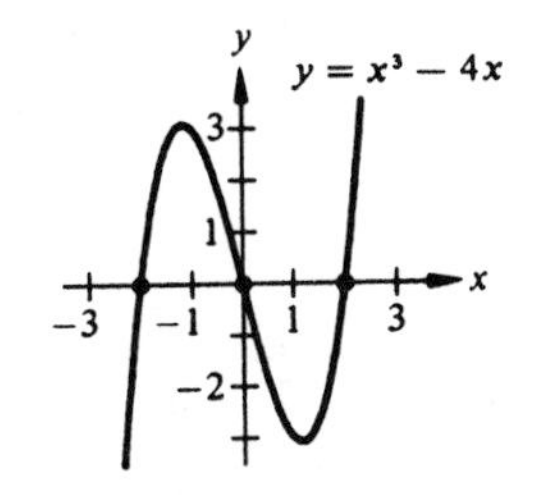

7. $x^2 + y^2 - 6x + 2y + 6 = 0$
$x^2 - 6x + \underline{9} + y^2 + 2y + \underline{1} = -6 + 9 + 1$
$(x-3)^2 + (y-1)^2 = 4$

Center: $(3,\ -1)$
Radius: 2

8. $f(x-3) = (x-3)^2 - 2(x-3) + 1$
$= x^2 - 6x + 9 - 2x + 6 + 1$
$= x^2 - 8x + 16$

9. $$f(3) = 12 - 11 = 1$$
$$\frac{f(x) - f(3)}{x - 3} = \frac{(4x - 11) - 1}{x - 3} = \frac{4x - 12}{x - 3} = \frac{4(x - 3)}{x - 3} = 4$$

10. $f(x) = \sqrt{36 - x^2} = \sqrt{(6 + x)(6 - x)}$

Domain: $[-6, 6]$

Range: $[0, 6]$

11. (a) $6x - 5y + 4 = 0$

$$y = \frac{6x + 4}{5} \quad \text{function}$$

(b) $x^2 + y^2 = 9$

$$y = \pm\sqrt{9 - x^2} \quad \text{not a function}$$

(c) $y^3 = x^2 + 6$

$$y = \sqrt[3]{x^2 + 6} \quad \text{function}$$

12. Parabola: Vertex $(0, -5)$

Intercepts: $(0, -5)$, $(\pm\sqrt{5}, 0)$

y-axis symmetry

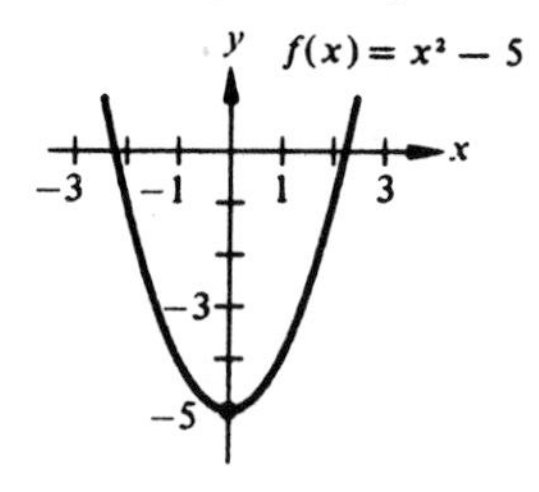

13. Intercepts: $(0, 3)$, $(-3, 0)$

x	0	1	−1	2	−2	−3	−4
y	3	4	2	5	1	0	1

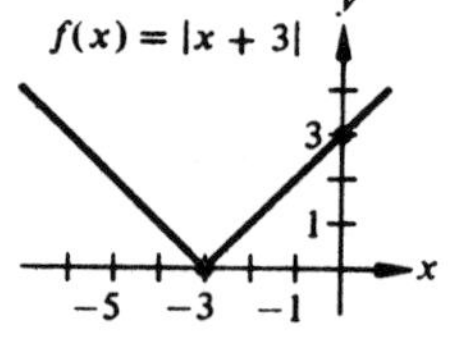

14.

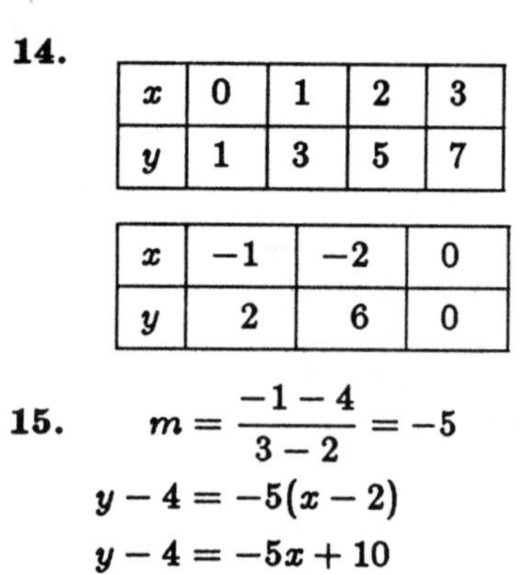

x	0	1	2	3
y	1	3	5	7

x	−1	−2	0
y	2	6	0

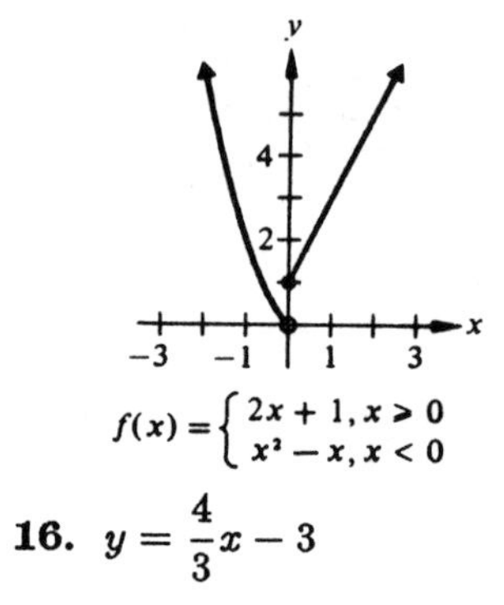

15. $$m = \frac{-1 - 4}{3 - 2} = -5$$
$$y - 4 = -5(x - 2)$$
$$y - 4 = -5x + 10$$
$$y = -5x + 14$$

16. $y = \frac{4}{3}x - 3$

17. $2x + 3y = 0$

$$y = -\tfrac{2}{3}x$$
$$m_1 = -\tfrac{2}{3}$$
$$\perp m_2 = \tfrac{3}{2} \text{ through } (4, 1)$$
$$y - 1 = \tfrac{3}{2}(x - 4)$$
$$y - 1 = \tfrac{3}{2}x - 6$$
$$y = \tfrac{3}{2}x - 5$$

18. $(5, 32)$ and $(9, 44)$

$$m = \frac{44 - 32}{9 - 5} = \frac{12}{4} = 3$$
$$y - 32 = 3(x - 5)$$
$$y - 32 = 3x - 15$$
$$y = 3x + 17$$

When $x = 20$, $y = 3(20) + 17$

$$y = \$77$$

19. $f(g(x)) = f(2x + 3)$

$$= (2x + 3)^2 - 2(2x + 3) + 16$$
$$= 4x^2 + 12x + 9 - 4x - 6 + 16$$
$$= 4x^2 + 8x + 19$$

20.

$$f(x) = x^3 + 7$$
$$y = x^3 + 7$$
$$x = \sqrt[3]{y - 7}$$
$$f^{-1}(x) = \sqrt[3]{x - 7}$$

21. (a) $f(x) = |x - 6|$ is not one-to-one.
For example, $f(0) = 6$ and $f(12) = 6$.

(b) $f(x) = ax + b$, $a \neq 0$ is one-to-one.

(c) $f(x) = x^3 - 19$ is one-to-one.

22.

$$f(x) = \sqrt{\frac{3 - x}{x}}, \quad 0 < x \leq 3$$
$$y = \sqrt{\frac{3 - x}{x}}$$
$$y^2 = \frac{3 - x}{x}$$
$$xy^2 = 3 - x$$
$$xy^2 + x = 3$$
$$x(y^2 + 1) = 3$$
$$x = \frac{3}{y^2 + 1}$$
$$f^{-1}(x) = \frac{3}{x^2 + 1}$$

23. $y = kx$ and $y = 30$ when $x = 5$

$$30 = k(5)$$
$$6 = k$$
$$y = 6x$$

24. $y = k/x$ and $y = 0.5$ when $x = 14$

$$0.5 = k/14$$
$$7 = k$$
$$y = 7/x$$

25. $z = \dfrac{kx^2}{y}$ and $z = 3$ when $x = 3$, $y = -6$

$$3 = \frac{k(3)^2}{-6}$$
$$-18 = 9k$$
$$-2 = k$$
$$z = \frac{-2x^2}{y}$$

CHAPTER 3

Practice Test Solutions

1. $f(x) = x^2 - 9$

Vertex: $(0, -9)$

x-intercepts: $(3, 0)$, $(-3, 0)$

y-intercept: $(0, -9)$

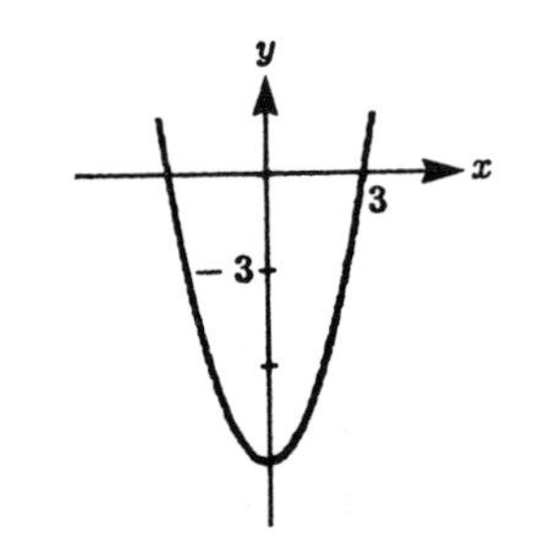

2. $g(x) = 3x^2 - 5x - 28$

Vertex: $-\dfrac{b}{2a} = \dfrac{5}{6}$, $g\left(\dfrac{5}{6}\right) = -\dfrac{361}{12}$

$\left(\dfrac{5}{6}, -\dfrac{361}{12}\right)$

x-intercepts: $0 = 3x^2 - 5x - 28$

$0 = (3x + 7)(x - 4)$

$x = -\dfrac{7}{3}$, $x = 4$

$\left(-\dfrac{7}{3}, 0\right)$, $(4, 0)$

y-intercept: $(0, -28)$

3. Vertex: $(-2, 1)$

Point: $(3, 4)$

$f(x) = a(x + 2)^2 + 1$

$4 = a(3 + 2)^2 + 1$

$3 = 25a \Rightarrow a = \frac{3}{25}$

$f(x) = \frac{3}{25}(x + 2)^2 + 1$

4.

$$f(x) = x^4 - 5x^2 + 4$$
$$x^4 - 5x^2 + 4 = 0$$
$$(x^2 - 1)(x^2 - 4) = 0$$
$$(x + 1)(x - 1)(x + 2)(x - 2) = 0$$

Zeros: ± 1, ± 2

5. Zeros: -1, $\sqrt{3}$ and $-\sqrt{3}$

$$f(x) = (x - (-1))(x - \sqrt{3})(x - (-\sqrt{3})) = (x + 1)(x - \sqrt{3})(x + \sqrt{3})$$
$$= (x + 1)(x^2 - 3) = x^3 + x^2 - 3x - 3$$

6. $f(x) = x^3 - 9x = x(x+3)(x-3)$

Intercepts: (0, 0), (3, 0), (−3, 0)

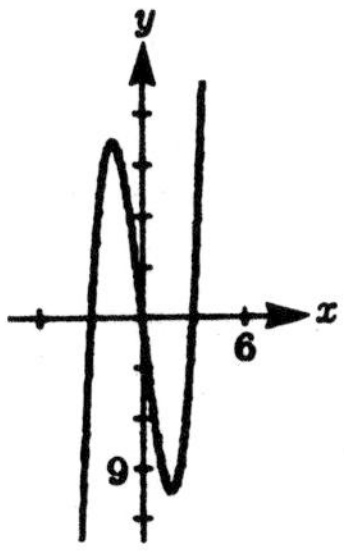

7.

$$\begin{array}{r}
6x \\
x^2 + 0x - 2 \enclose{longdiv}{6x^3 + 0x^2 - 5x + 1} \\
-(6x^3 + 0x^2 - 12x) \\ \hline
7x + 1
\end{array}$$

$$\frac{6x^3 - 5x + 1}{x^2 - 2} = 6x + \frac{7x + 1}{x^2 - 2}$$

8.

$$\begin{array}{r|rrrrr}
-3 & 1 & 3 & -9 & 0 & 1 \\
 & & -3 & 0 & 27 & -81 \\ \hline
 & 1 & 0 & -9 & 27 & -80
\end{array}$$

$$\frac{x^4 + 3x^3 - 9x^2 + 1}{x + 3} = x^3 - 9x + 27 - \frac{80}{x + 3}$$

9.

$$\begin{array}{r|rrrrrrr}
-2 & 1 & 0 & 0 & -3 & 0 & 1 & -5 \\
 & & -2 & 4 & -8 & 22 & -44 & 86 \\ \hline
 & 1 & -2 & 4 & -11 & 22 & -43 & 81
\end{array}$$

$f(-2) = 81$

10. $0 = x^3 - 2x^2 - 5x + 6$

Possible rational zeros: $\pm 1, \pm 2, \pm 3, \pm 6$

$$\begin{array}{r|rrrr}
1 & 1 & -2 & -5 & 6 \\
 & & 1 & -1 & -6 \\ \hline
 & 1 & -1 & -6 & 0
\end{array}$$

$$\begin{aligned} x^3 - 2x^2 - 5x + 6 &= (x-1)(x^2 - x - 6) \\ 0 &= (x-1)(x-3)(x+2) \end{aligned}$$

Zeros: 1, 3, −2

11. $0 = x^3 + x^2 - 5x - 6$

Possible rational zeros: $\pm 1, \pm 2, \pm 3, \pm 6$

$$\begin{array}{r|rrrr}
-2 & 1 & 1 & -5 & -6 \\
 & & -2 & 2 & 6 \\ \hline
 & 1 & -1 & -3 & 0
\end{array}$$

$x^3 + x^2 - 5x - 6 = (x+2)(x^2 - x - 3)$

Zeros: $x = -2$

$$x = \frac{-(-1) \pm \sqrt{(-1)^2 - 4(1)(-3)}}{2(1)} = \frac{1 \pm \sqrt{13}}{2}$$

12. $p(x) = 4x^3 - 7x^2 + x - 20$

Possible rational zeros: $\pm 1, \pm 2, \pm 4, \pm 5, \pm 10, \pm 20, \pm \frac{1}{2}, \pm \frac{1}{4}, \pm \frac{5}{2}, \pm \frac{5}{4}$

13. $5i^3 - (\sqrt{-9})^2 = -5i - (-9) = 9 - 5i$

14. $(6+5i)(-2+9i) = -12 + 54i - 10i + 45i^2$
$= -12 - 45 + (54 - 10)i$
$= -57 + 44i$

15. $\dfrac{3+2i}{4-7i} = \dfrac{3+2i}{4-7i} \cdot \dfrac{4+7i}{4+7i}$
$= \dfrac{12 + 21i + 8i + 14i^2}{16 + 49}$
$= \dfrac{-2 + 29i}{65} = -\dfrac{2}{65} + \dfrac{29}{65}i$

16. $f(x) = x^2 - 3x + 5$
$x = \dfrac{-(-3) \pm \sqrt{(-3)^2 - 4(1)(5)}}{2(1)}$
$= \dfrac{3 \pm \sqrt{9 - 20}}{2} = \dfrac{3 \pm \sqrt{11}\,i}{2}$

17. $p(x) = x^5 + x^3 - 8x^2 - 8$
$0 = x^3(x^2 + 1) - 8(x^2 + 1)$
$0 = (x^3 - 8)(x^2 + 1)$
$0 = (x - 2)(x^2 + 2x + 4)(x^2 + 1)$
$x = 2$
$x = \dfrac{-2 \pm \sqrt{(2)^2 - 4(1)(4)}}{2(1)}$
$= \dfrac{-2 \pm \sqrt{-12}}{2}$
$= \dfrac{-2 \pm 2\sqrt{3}\,i}{2} = -1 \pm \sqrt{3}\,i$
$x = \pm i$

Zeros: $x = 2$, $x = -1 \pm \sqrt{3}\,i$, $x = \pm i$

18. Zeros: 0, 4, $2 + i$
Since $2 + i$ is a zero, so is $2 - i$.
$p(x) = (x - 0)(x - 4)[x - (2 + i)][x - (2 - i)]$
$= x(x - 4)(x - 2 - i)(x - 2 + i)$
$= (x^2 - 4x)(x^2 - 4x + 5)$
$= x^4 - 8x^3 + 21x^2 - 20x$

19. $h(x) = \dfrac{x^2 + 1}{x^2 - 16}$
Vertical asymptotes: $x = 4$, $x = -4$
Horizontal asymptote: $y = 1$

20. $f(x) = \dfrac{x + 3}{x - 2}$
x-intercept: $(-3,\ 0)$
y-intercept: $(0,\ -\frac{3}{2})$
Vertical asymptote: $x = 2$
Horizontal asymptote: $y = 1$

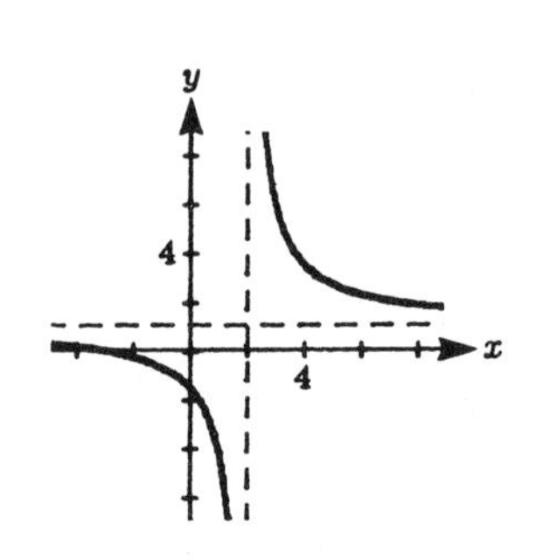

21. $g(x) = \dfrac{x}{x^2 - 1}$

x-intercept: $(0, 0)$

Vertical asymptote: $x = 1,\ x = -1$

Horizontal asymptote: $y = 0$

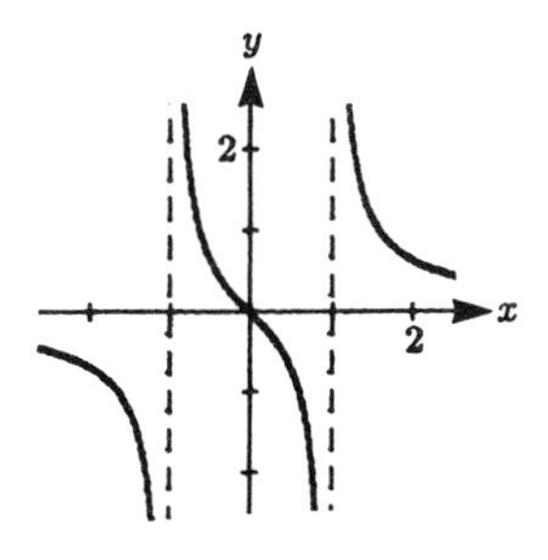

22. $h(x) = \dfrac{x^2 + 6x - 7}{x + 3}$

$$\begin{array}{r|l} & x + 3 \\ \hline x+3 & x^2 + 6x - 7 \\ & -(x^2 + 3x) \\ \hline & 3x - 7 \\ & -(3x + 9) \\ \hline & -16 \end{array}$$

$$\frac{x^2 + 6x - 7}{x + 3} = x + 3 - \frac{16}{x + 3}$$

Slant asymptote: $y = x + 3$

23. $\dfrac{8x - 2}{x^2 - 2x - 8} = \dfrac{8x - 2}{(x + 2)(x - 4)} = \dfrac{A}{x + 2} + \dfrac{B}{x - 4}$

$8x - 2 = A(x - 4) + B(x + 2)$

Let $x = -2, \quad -18 = -6A, \quad A = 3$

Let $x = 4, \quad 30 = 6B, \quad B = 5$

$$\frac{8x - 2}{x^2 - 2x - 8} = \frac{3}{x + 2} + \frac{5}{x - 4}$$

24.

$$\frac{3x^2 + 15}{x^3 + 3x} = \frac{3x^2 + 15}{x(x^2 + 3)} = \frac{A}{x} + \frac{Bx + C}{x^2 + 3}$$

$$3x^2 + 15 = A(x^2 + 3) + (Bx + C)x$$

$$3x^2 + 15 = Ax^2 + 3A + Bx^2 + Cx$$

$$3x^2 + 0x + 15 = (A + B)x^2 + Cx + 3A$$

$$3 = A + B$$

$$0 = C$$

$$15 = 3A \Rightarrow A = 5, \quad B = -2$$

$$\frac{3x^2 + 15}{x^3 + 3x} = \frac{5}{x} + \frac{-2x + 0}{x^2 + 3} = \frac{5}{x} - \frac{2x}{x^2 + 3}$$

25.

$$\begin{array}{r|l} & \quad 2x \\ x^2+4x+4 & \quad 2x^3+8x^2+9x-1 \\ & -(2x^3+8x^2+8x) \\ & \qquad\qquad x-1 \end{array}$$

$$\frac{2x^3+8x^2+9x-1}{x^2+4x+4} = 2x + \frac{x-1}{x^2+4x+4}$$

$$\frac{x-1}{(x+2)^2} = \frac{A}{x+2} + \frac{B}{(x+2)^2}$$

$$x-1 = A(x+2)+B$$

$$x-1 = Ax+2A+B$$

$$1 = A$$

$$-1 = 2A+B, \quad B=-3$$

$$\frac{2x^3+8x^2+9x-1}{x^2+4x+4} = 2x + \frac{1}{x+2} - \frac{3}{(x+2)^2}$$

CHAPTER 4.

Practice Test Solutions

1. $x^{3/5} = 8$

$$x = 8^{5/3} = (\sqrt[3]{8})^5 = 2^5 = 32$$

2. $3^{x-1} = \frac{1}{81}$

$$3^{x-1} = 3^{-4}$$
$$x - 1 = -4$$
$$x = -3$$

3. $f(x) = 2^{-x} = \left(\frac{1}{2}\right)^x$

x	-2	-1	0	1	2
$f(x)$	4	2	1	$\frac{1}{2}$	$\frac{1}{4}$

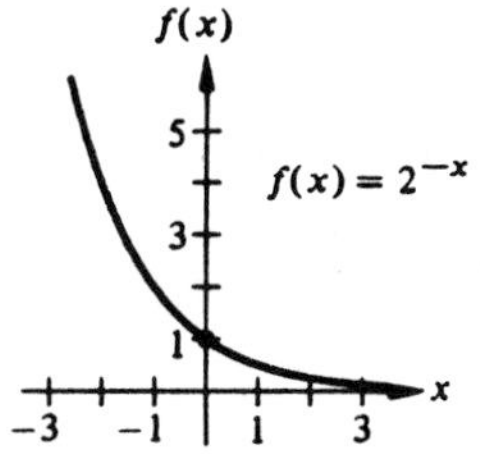

$f(x)$

$f(x) = 2^{-x}$

4. $g(x) = e^x + 1$

x	-2	-1	0	1	2
$g(x)$	1.14	1.37	2	3.72	8.39

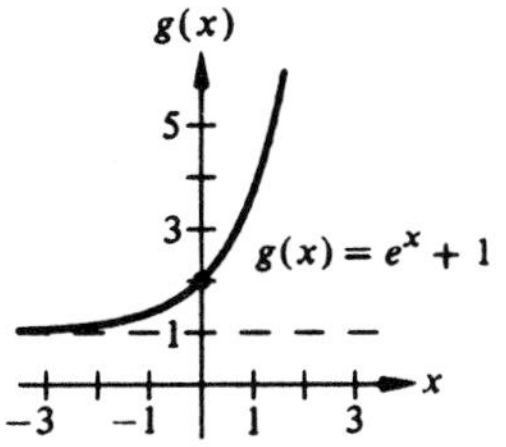

$g(x)$

$g(x) = e^x + 1$

5. $A = P\left(1 + \dfrac{r}{n}\right)^{nt}$

(a) $A = 5000\left(1 + \dfrac{0.09}{12}\right)^{12(3)} \approx \$6543.23$

(b) $A = 5000\left(1 + \dfrac{0.09}{4}\right)^{4(3)} \approx \$6530.25$

(c) $A = 5000e^{(0.09)(3)} \approx \$6549.82$

6. $7^{-2} = \frac{1}{49}$

$$\log_7 \tfrac{1}{49} = -2$$

7. $x - 4 = \log_2 \frac{1}{64}$

$$2^{x-4} = \tfrac{1}{64}$$
$$2^{x-4} = 2^{-6}$$
$$x - 4 = -6$$
$$x = -2$$

8. $\log_b \sqrt[4]{8/25} = \frac{1}{4}\log_b \frac{8}{25}$

$$= \tfrac{1}{4}[\log_b 8 - \log_b 25]$$
$$= \tfrac{1}{4}[\log_b 2^3 - \log_b 5^2]$$
$$= \tfrac{1}{4}[3\log_b 2 - 2\log_b 5]$$
$$= \tfrac{1}{4}[3(0.3562) - 2(0.8271)]$$
$$= -0.1464$$

9. $5\ln x - \dfrac{1}{2}\ln y + 6\ln z = \ln x^5 - \ln\sqrt{y} + \ln z^6$

$$= \ln\left(\frac{x^5 z^6}{\sqrt{y}}\right)$$

10. $\log_9 28 = \dfrac{\log 28}{\log 9} \approx 1.5166$

11. $\log N = 0.6646$

$N = 10^{0.6646} \approx 4.62$

12.

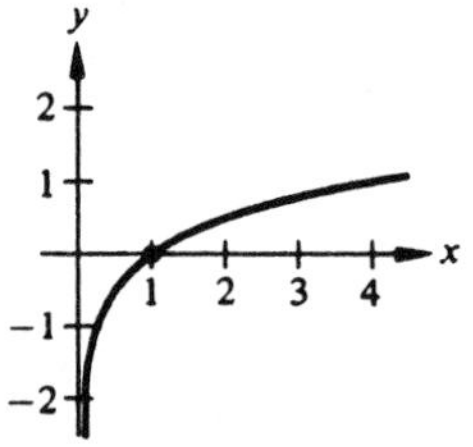

13. Domain:

$$x^2 - 9 > 0$$
$$(x+3)(x-3) > 0$$
$$x < -3 \text{ or } x > 3$$

14. $y = \ln(x-2)$

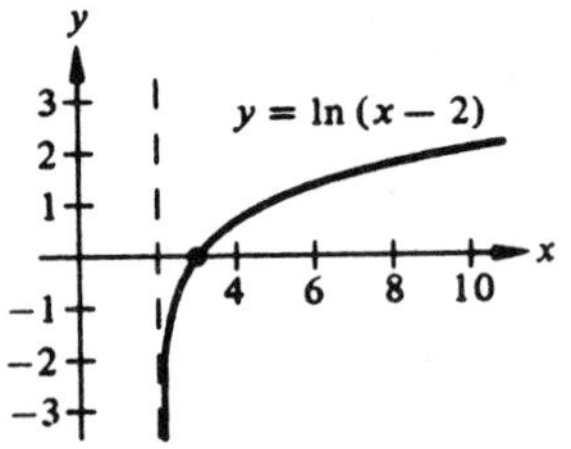

15. $\dfrac{\ln x}{\ln y} \neq \ln(x - y)$ since $\dfrac{\ln x}{\ln y} = \log_y x$

16. $5^x = 41$

$$x = \log_5 41 = \frac{\ln 41}{\ln 5} \approx 2.3074$$

17. $x - x^2 = \log_5 \frac{1}{25}$

$$5^{x-x^2} = \tfrac{1}{25}$$
$$5^{x-x^2} = 5^{-2}$$
$$x - x^2 = -2$$
$$0 = x^2 - x - 2$$
$$0 = (x+1)(x-2)$$
$$x = -1 \text{ or } x = 2$$

18. $\log_2 x + \log_2(x-3) = 2$

$$\log_2[x(x-3)] = 2$$
$$x(x-3) = 2^2$$
$$x^2 - 3x = 4$$
$$x^2 - 3x - 4 = 0$$
$$(x+1)(x-4) = 0$$
$$x = 4$$
$$x = -1 \text{ (extraneous solution)}$$

19. $\dfrac{e^x + e^{-x}}{3} = 4$

$$e^x(e^x + e^{-x}) = 12e^x$$
$$e^{2x} + 1 = 12e^x$$
$$e^{2x} - 12e^x + 1 = 0$$
$$e^x = \frac{12 \pm \sqrt{144 - 4}}{2}$$

$e^x = 11.9161$ or $e^x = 0.0839$

$x = \ln 11.9161$ $\quad x = \ln 0.0839$

$x \approx 2.4779$ $\quad x \approx -2.4779$

20. $A = Pe^{rt}$

$$12,000 = 6,000e^{0.13t}$$
$$2 = e^{0.13t}$$
$$0.13t = \ln 2$$
$$t = \frac{\ln 2}{0.13}$$

$t \approx 5.3319$ yrs or 5 yrs 4 months

CHAPTER 5

Practice Test Solutions

1. (a) $350^\circ = 350\left(\frac{\pi}{180}\right) = \frac{35\pi}{18}$

(b) $\frac{5\pi}{9} = \frac{5\pi}{9} \cdot \frac{180}{\pi} = 100^\circ$

2. (a) $135^\circ 14' 12'' = 135 + \frac{14}{60} + \frac{12}{3600}$

$\approx 135.2367^\circ$

(b) $-22.569^\circ = -(22 + 0.569(60)')$

$= -22^\circ 34.14'$

$= -(22^\circ 34' + 0.14(60)'')$

$\approx -22^\circ 34' 8''$

3. (a) $\frac{5\pi}{6}$ corresponds to the point $\left(-\frac{\sqrt{3}}{2}, \frac{1}{2}\right)$

$\sin \frac{5\pi}{6} = y = \frac{1}{2}$

(b) $\frac{5\pi}{4}$ corresponds to the point $\left(-\frac{\sqrt{2}}{2}, -\frac{\sqrt{2}}{2}\right)$

$\tan \frac{5\pi}{4} = \frac{y}{x} = 1$

4. (a) $\sin 7\pi = \sin(6\pi + \pi) = \sin \pi = 0$

(b) $\cos\left(-\frac{13\pi}{3}\right) = \cos\left(-4\pi - \frac{\pi}{3}\right)$

$= \cos\left(-\frac{\pi}{3}\right)$

$= \cos \frac{\pi}{3} = \frac{1}{2}$

5. $\cos \theta = \frac{2}{3}$

$x = 2,\ r = 3,\ y = \sqrt{9-4} = \sqrt{5}$

$\tan \theta = \frac{y}{x} = \frac{\sqrt{5}}{2}$

6. $\sin \theta = 0.9063$

$\theta = \arcsin(0.9063)$

$\theta \approx 65^\circ$ or $\frac{13\pi}{36}$

7. $\tan 20^\circ = \frac{35}{x}$

$x = \frac{35}{\tan 20^\circ} \approx 96.1617$

8. $\theta = \frac{6\pi}{5}$, θ is in Quadrant III.

Reference angle: $\frac{6\pi}{5} - \pi = \frac{\pi}{5}$ or 36°

9. $\csc 3.92 = \frac{1}{\sin 3.92} \approx -1.4242$

10. $\tan \theta = 6 = \frac{6}{1}$, θ lies in Quadrant III.

$y = -6,\ x = -1,\ r = \sqrt{36+1} = \sqrt{37}$,

so $\sec \theta = \frac{\sqrt{37}}{-1} \approx -6.0828$.

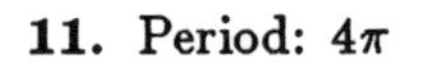

11. Period: 4π

Amplitude: 3

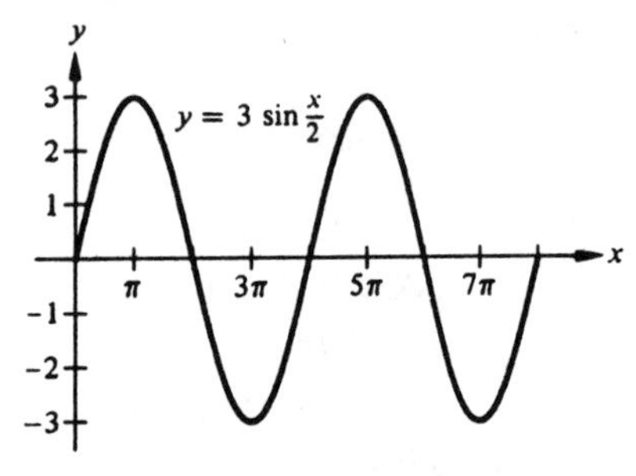

12. Period: 2π

Amplitude: 2

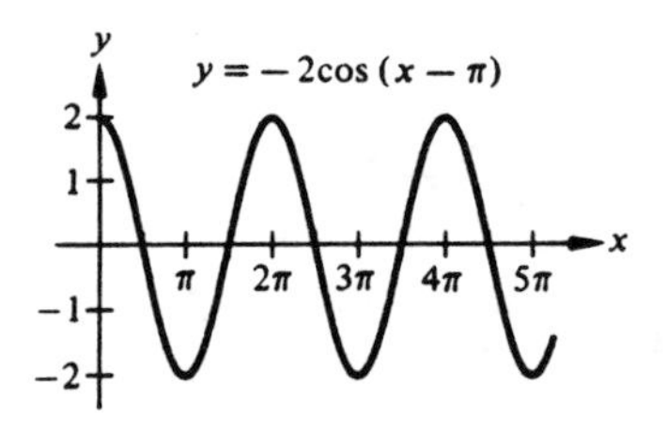

13. Period: $\dfrac{\pi}{2}$

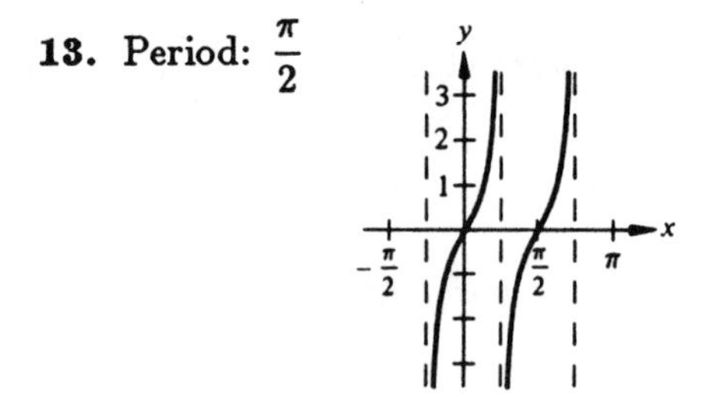

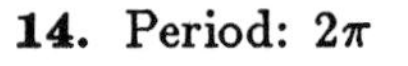

14. Period: 2π

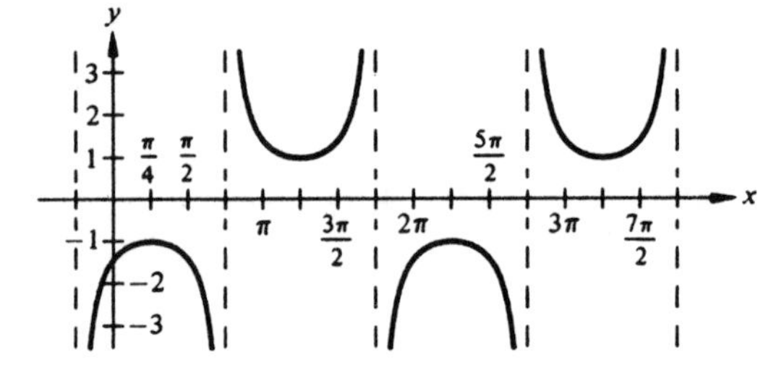

15.

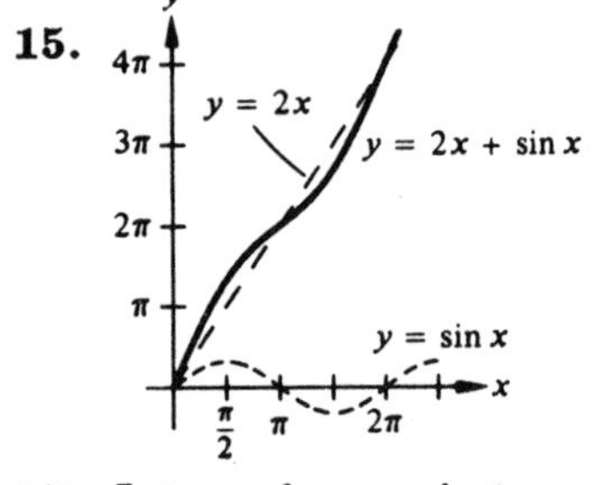

16.

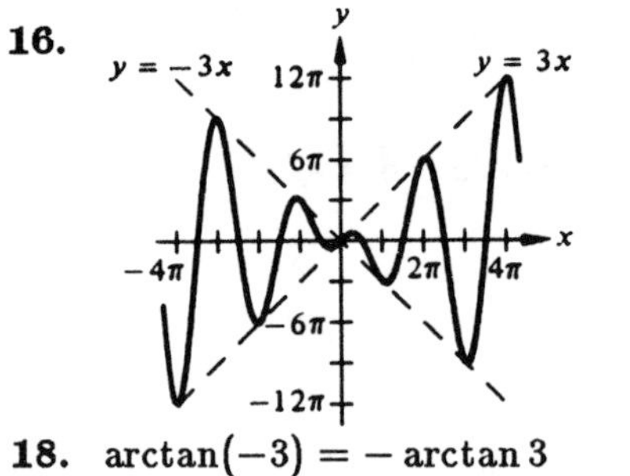

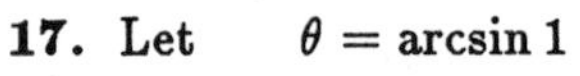

17. Let $\theta = \arcsin 1$

$$\sin\theta = 1$$

$$\theta = \frac{\pi}{2}$$

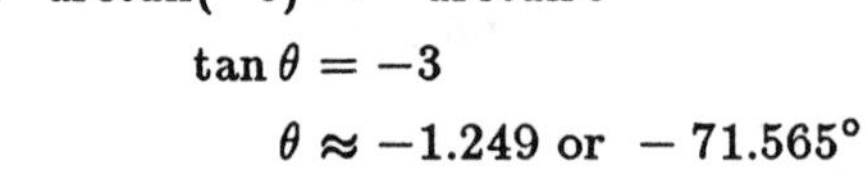

18. $\arctan(-3) = -\arctan 3$

$$\tan\theta = -3$$

$$\theta \approx -1.249 \text{ or } -71.565^\circ$$

19. $\sin\left(\arccos\dfrac{4}{\sqrt{35}}\right)$

$$\sin\theta = \frac{\sqrt{19}}{\sqrt{35}} \approx 0.7368$$

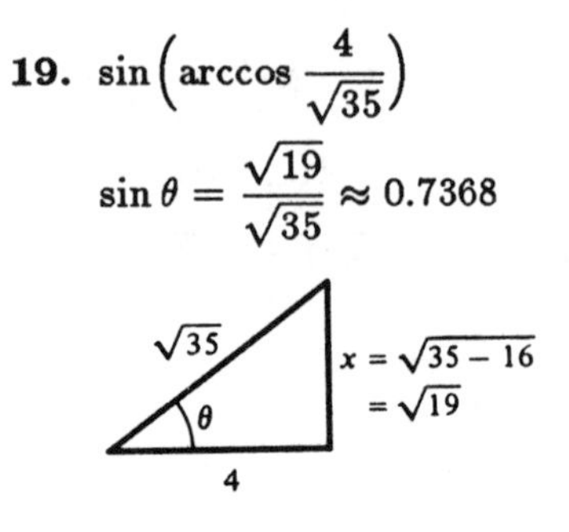

20. $\cos\left(\arcsin\dfrac{x}{4}\right)$

$$\cos\theta = \frac{\sqrt{16 - x^2}}{4}$$

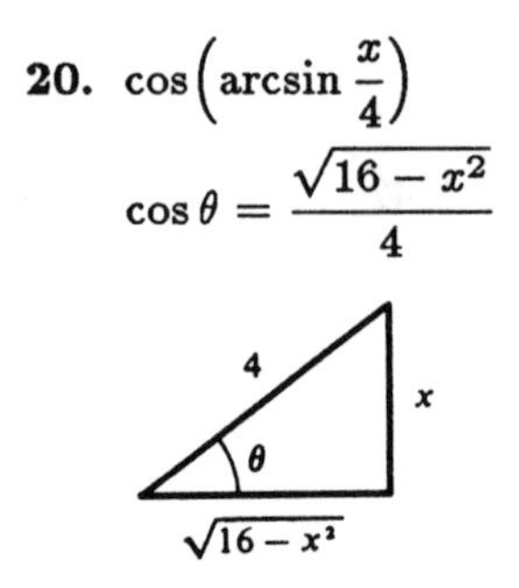

21. Given $A = 40°,\ c = 12$

$B = 90° - 40° = 50°$

$\sin 40° = \dfrac{a}{12}$

$a = 12 \sin 40° \approx 7.713$

$\cos 40° = \dfrac{b}{12}$

$b = 12 \cos 40° \approx 9.192$

22. Given $B = 6.84°,\ a = 21.3$

$A = 90° - 6.84° = 83.16°$

$\sin 83.16° = \dfrac{21.3}{c}$

$c = \dfrac{21.3}{\sin 83.16°} \approx 21.453$

$\tan 83.16° = \dfrac{21.3}{b}$

$b = \dfrac{21.3}{\tan 83.16°} \approx 2.555$

23. Given $a = 5,\ b = 9$

$c = \sqrt{25 + 81} = \sqrt{106} \approx 10.296$

$\tan A = \frac{5}{9}$

$A = \arctan \frac{5}{9} \approx 29.055°$

$B = 90° - 29.055° = 60.945°$

24. $\sin 67° = \dfrac{x}{20}$

$x = 20 \sin 67° \approx 18.41'$

25. $\tan 5° = \dfrac{250}{x}$

$x = \dfrac{250}{\tan 5°}$

$\approx 2857.513'$

≈ 0.541 mi

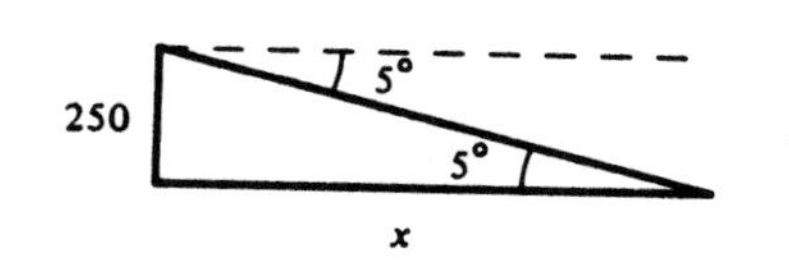

CHAPTER 6

Practice Test Solutions

1. $\tan x = \dfrac{4}{11}$, $\sec x < 0 \Rightarrow x$ is in Quadrant III.

$y = -4,\ x = -11,\ r = \sqrt{16+121} = \sqrt{137}$

$$\sin x = -\frac{4}{\sqrt{137}} = -\frac{4\sqrt{137}}{137} \qquad \csc x = -\frac{\sqrt{137}}{4}$$

$$\cos x = -\frac{11}{\sqrt{137}} = -\frac{11\sqrt{137}}{137} \qquad \sec x = -\frac{\sqrt{137}}{11}$$

$$\tan x = \frac{4}{11} \qquad \cot x = \frac{11}{4}$$

2. $$\frac{\sec^2 x + \csc^2 x}{\csc^2 x(1+\tan^2 x)} = \frac{\sec^2 x + \csc^2 x}{\csc^2 x + (\csc^2 x)\tan^2 x} = \frac{\sec^2 x + \csc^2 x}{\csc^2 x + \dfrac{1}{\sin^2 x}\cdot\dfrac{\sin^2 x}{\cos^2 x}}$$

$$= \frac{\sec^2 x + \csc^2 x}{\csc^2 x + \dfrac{1}{\cos^2 x}} = \frac{\sec^2 x + \csc^2 x}{\csc^2 x + \sec^2 x} = 1$$

3. $$\ln|\tan\theta| - \ln|\cot\theta| = \ln\frac{|\tan\theta|}{|\cot\theta|} = \ln\left|\frac{\sin\theta/\cos\theta}{\cos\theta/\sin\theta}\right| = \ln\left|\frac{\sin^2\theta}{\cos^2\theta}\right| = \ln|\tan^2\theta| = 2\ln|\tan\theta|$$

4. $\cos\left(\dfrac{\pi}{2} - x\right) = \dfrac{1}{\csc x}$ is true since $\cos\left(\dfrac{\pi}{2} - x\right) = \sin x = \dfrac{1}{\csc x}$.

5. $\sin^4 x + (\sin^2 x)\cos^2 x = \sin^2 x(\sin^2 x + \cos^2 x) = \sin^2 x(1) = \sin^2 x$

6. $(\csc x + 1)(\csc x - 1) = \csc^2 x - 1 = \cot^2 x$

7. $$\frac{\cos^2 x}{1-\sin x}\cdot\frac{1+\sin x}{1+\sin x} = \frac{\cos^2 x(1+\sin x)}{1-\sin^2 x} = \frac{\cos^2 x(1+\sin x)}{\cos^2 x} = 1 + \sin x$$

8. $$\frac{1+\cos\theta}{\sin\theta} + \frac{\sin\theta}{1+\cos\theta} = \frac{(1+\cos\theta)^2 + \sin^2\theta}{\sin\theta(1+\cos\theta)}$$

$$= \frac{1 + 2\cos\theta + \cos^2\theta + \sin^2\theta}{\sin\theta(1+\cos\theta)} = \frac{2+2\cos\theta}{\sin\theta(1+\cos\theta)} = \frac{2}{\sin\theta} = 2\csc\theta$$

9. $\tan^4 x + 2\tan^2 x + 1 = (\tan^2 x + 1)^2 = (\sec^2 x)^2 = \sec^4 x$

10. (a) $\sin 105° = \sin(60° + 45°) = \sin 60° \cos 45° + \cos 60° \sin 45°$

$$= \frac{\sqrt{3}}{2} \cdot \frac{\sqrt{2}}{2} + \frac{1}{2} \cdot \frac{\sqrt{2}}{2} = \frac{\sqrt{2}}{4}(\sqrt{3} + 1)$$

(b) $\tan 15° = \tan(60° - 45°) = \dfrac{\tan 60° - \tan 45°}{1 + \tan 60° \tan 45°}$

$$= \frac{\sqrt{3} - 1}{1 + \sqrt{3}} \cdot \frac{1 - \sqrt{3}}{1 - \sqrt{3}} = \frac{2\sqrt{3} - 1 - 3}{1 - 3} = \frac{2\sqrt{3} - 4}{-2} = 2 - \sqrt{3}$$

11. $(\sin 42°) \cos 38° - (\cos 42°) \sin 38° = \sin(42° - 38°) = \sin 4°$

12. $\tan\left(\theta + \dfrac{\pi}{4}\right) = \dfrac{\tan\theta + \tan(\pi/4)}{1 - (\tan\theta)\tan(\pi/4)} = \dfrac{\tan\theta + 1}{1 - \tan\theta(1)} = \dfrac{1 + \tan\theta}{1 - \tan\theta}$

13. $\sin(\arcsin x - \arccos x) = \sin(\arcsin x)\cos(\arccos x) - \cos(\arcsin x)\sin(\arccos x)$

$$= (x)(x) - (\sqrt{1 - x^2})(\sqrt{1 - x^2}) = x^2 - (1 - x^2) = 2x^2 - 1$$

14. (a) $\cos(120°) = \cos[2(60°)] = 2\cos^2 60° - 1 = 2\left(\dfrac{1}{2}\right)^2 - 1 = -\dfrac{1}{2}$

(b) $\tan(300°) = \tan[2(150°)] = \dfrac{2\tan 150°}{1 - \tan^2 150°} = \dfrac{2\sqrt{3}/3}{1 - (1/3)} = \sqrt{3}$

15. (a) $\sin 22.5° = \sin \dfrac{45°}{2} = \sqrt{\dfrac{1 - \cos 45°}{2}} = \sqrt{\dfrac{1 - \sqrt{2}/2}{2}} = \dfrac{\sqrt{2 - \sqrt{2}}}{2}$

(b) $\tan \dfrac{\pi}{12} = \tan \dfrac{\pi/6}{2} = \dfrac{\sin(\pi/6)}{1 + \cos(\pi/6)} = \dfrac{1/2}{1 + \sqrt{3}/2} = \dfrac{1}{2 + \sqrt{3}} = 2 - \sqrt{3}$

16. $\sin\theta = \dfrac{4}{5}$, θ lies in Quadrant II $\Rightarrow \cos\theta = -\dfrac{3}{5}$.

$$\cos\frac{\theta}{2} = \sqrt{\frac{1 + \cos\theta}{2}} = \sqrt{\frac{1 - 3/5}{2}} = \sqrt{\frac{2}{10}} = \frac{1}{\sqrt{5}} = \frac{\sqrt{5}}{5}$$

17. $(\sin^2 x)\cos^2 x = \dfrac{1 - \cos 2x}{2} \cdot \dfrac{1 + \cos 2x}{2} = \dfrac{1}{4}[1 - \cos^2 2x] = \dfrac{1}{4}\left[1 - \dfrac{1 + \cos 4x}{2}\right]$

$$= \frac{1}{8}[2 - (1 + \cos 4x)] = \frac{1}{8}[1 - \cos 4x]$$

18. $6(\sin 5\theta)\cos 2\theta = 6\{\tfrac{1}{2}[\sin(5\theta + 2\theta) + \sin(5\theta - 2\theta)]\} = 3[\sin 7\theta + \sin 3\theta]$

19. $\sin(x + \pi) + \sin(x - \pi) = 2\left(\sin \dfrac{[(x + \pi) + (x - \pi)]}{2}\right) \cos \dfrac{[(x + \pi) - (x - \pi)]}{2}$

$$= 2(\sin x)\cos\pi = -2\sin x$$

20. $\dfrac{\sin 9x + \sin 5x}{\cos 9x - \cos 5x} = \dfrac{2\sin 7x \cos 2x}{-2\sin 7x \sin 2x} = -\dfrac{\cos 2x}{\sin 2x} = -\cot 2x$

21. $\frac{1}{2}[\sin(u+v) - \sin(u-v)] = \frac{1}{2}\{(\sin u)\cos v + (\cos u)\sin v - [(\sin u)\cos v - (\cos u)\sin v]\}$
$= \frac{1}{2}[2(\cos u)\sin v] = (\cos u)\sin v$

22. $4\sin^2 x = 1$

$\sin^2 x = \dfrac{1}{4}$

$\sin x = \pm\dfrac{1}{2}$

$\sin x = \dfrac{1}{2}$ or $\sin x = -\dfrac{1}{2}$

$x = \dfrac{\pi}{6}$ or $\dfrac{5\pi}{6}$ $\quad$ $x = \dfrac{7\pi}{6}$ or $\dfrac{11\pi}{6}$

23. $\tan^2\theta + (\sqrt{3} - 1)\tan\theta - \sqrt{3} = 0$

$(\tan\theta - 1)(\tan\theta + \sqrt{3}) = 0$

$\tan\theta = 1$ or $\tan\theta = -\sqrt{3}$

$\theta = \dfrac{\pi}{4}$ or $\dfrac{5\pi}{4}$ $\quad$ $\theta = \dfrac{2\pi}{3}$ or $\dfrac{5\pi}{3}$

24. $\sin 2x = \cos x$

$2(\sin x)\cos x - \cos x = 0$

$\cos x(2\sin x - 1) = 0$

$\cos x = 0$ or $\sin x = \dfrac{1}{2}$

$x = \dfrac{\pi}{2}$ or $\dfrac{3\pi}{2}$ $\quad$ $x = \dfrac{\pi}{6}$ or $\dfrac{5\pi}{6}$

25. $\tan^2 x - 6\tan x + 4 = 0$

$\tan x = \dfrac{-(-6) \pm \sqrt{(-6)^2 - 4(1)(4)}}{2(1)}$

$\tan x = \dfrac{6 \pm \sqrt{20}}{2} = 3 \pm \sqrt{5}$

$\tan x = 3 + \sqrt{5}$ or $\tan x = 3 - \sqrt{5}$

$x = 1.3821$ or 4.5237 $\quad$ $x = 0.6524$ or 3.7940

CHAPTER 7

Practice Test Solutions

1. $C = 180° - (40° + 12°) = 128°$

$a = \sin 40° \left(\frac{100}{\sin 12°}\right) \approx 309.164$

$c = \sin 128° \left(\frac{100}{\sin 12°}\right) \approx 379.012$

2. $\sin A = 5\left(\frac{\sin 150°}{20}\right) = 0.1250$

$A \approx 7.181°$

$B \approx 180° - (150° + 7.181°) = 22.819°$

$b = \sin 22.819° \left(\frac{20}{\sin 150°}\right) \approx 15.513$

3. $\text{Area} = \frac{1}{2}ab \sin C$

$= \frac{1}{2}(3)(5) \sin 130°$

≈ 5.745 square units

4. $h = b \sin A,\ a = 10$

$= 35 \sin 22.5$

≈ 13.394

Since $a < h$ and A is acute, the triangle has no solution.

5. $\cos A = \frac{(53)^2 + (38)^2 - (49)^2}{2(53)(38)} \approx 0.4598$

$A \approx 62.627°$

$\cos B = \frac{(49)^2 + (38)^2 - (53)^2}{2(49)(38)} \approx 0.2782$

$B \approx 73.847°$

$C = 180° - (62.627° + 73.847°) = 43.526°$

6. $c^2 = (100)^2 + (300)^2 - 2(100)(300) \cos 29°$

≈ 47522.8176

$c \approx 218$

$\cos A = \frac{(300)^2 + (218)^2 - (100)^2}{2(300)(218)} \approx 0.9750$

$A \approx 12.85°$

$B = 180° - (12.85° + 29°) = 138.15°$

7. $s = \frac{a + b + c}{2} = \frac{4.1 + 6.8 + 5.5}{2} = 8.2$

$\text{Area} = \sqrt{s(s-a)(s-b)(s-c)}$

$= \sqrt{8.2(8.2 - 4.1)(8.2 - 6.8)(8.2 - 5.5)}$

≈ 11.273 square units

8. $x^2 = (40)^2 + (70)^2 - 2(40)(70) \cos 168°$

≈ 11977.6266

$x \approx 109.442$ miles

9. $\mathbf{w} = 4(3\mathbf{i} + \mathbf{j}) - 7(-\mathbf{i} + 2\mathbf{j})$

$= 19\mathbf{i} - 10\mathbf{j}$

10. $\frac{\mathbf{v}}{\|\mathbf{v}\|} = \frac{5\mathbf{i} - 3\mathbf{j}}{\sqrt{25 + 9}} = \frac{5}{\sqrt{34}}\mathbf{i} - \frac{3}{\sqrt{34}}\mathbf{j}$

$= \frac{5\sqrt{34}}{34}\mathbf{i} - \frac{3\sqrt{34}}{34}\mathbf{j}$

11. $\mathbf{u} = 6\mathbf{i} + 5\mathbf{j} \quad \mathbf{v} = 2\mathbf{i} - 3\mathbf{j} \quad \mathbf{w} = -4\mathbf{i} - 8\mathbf{j}$

$|\mathbf{u}| = \sqrt{61} \quad |\mathbf{v}| = \sqrt{13} \quad |\mathbf{w}| = \sqrt{80}$

$$\cos\theta = \frac{61 + 13 - 80}{2\sqrt{61}\sqrt{13}}$$

$$\theta \approx 96.116°$$

12. $\tan 30° = \frac{y}{x} = \frac{1}{\sqrt{3}}$

$\mathbf{u} = \sqrt{3}\,\mathbf{i} + \mathbf{j}$ but $|\mathbf{u}| = 2$

$\mathbf{v} = 2\mathbf{u} = 2\sqrt{3}\,\mathbf{i} + 2\mathbf{j}$

13. $r = \sqrt{25 + 25} = \sqrt{50} = 5\sqrt{2}$

$\tan\theta = \frac{-5}{5} = -1$

Since z is in Quadrant IV,

$\theta = 315°$

$z = 5\sqrt{2}(\cos 315° + i\sin 315°)$

14. $\cos 225° = -\frac{\sqrt{2}}{2} \quad \sin 225° = -\frac{\sqrt{2}}{2}$

$$z = 6\left(-\frac{\sqrt{2}}{2} - i\frac{\sqrt{2}}{2}\right) = -3\sqrt{2} - 3\sqrt{2}i$$

15. $$[7(\cos 23° + i\sin 23°)][4(\cos 7° + i\sin 7°)] = 7(4)[\cos(23° + 7°) + i\sin(23° + 7°)] = 28(\cos 30° + i\sin 30°)$$

16. $$\frac{9\left(\cos\frac{5\pi}{4} + i\sin\frac{5\pi}{4}\right)}{3(\cos\pi + i\sin\pi)} = \frac{9}{3}\left[\cos\left(\frac{5\pi}{4} - \pi\right) + i\sin\left(\frac{5\pi}{4} - \pi\right)\right] = 3\left(\cos\frac{\pi}{4} + i\sin\frac{\pi}{4}\right)$$

17. $$(2 + 2i)^8 = [2\sqrt{2}(\cos 45° + i\sin 45°)]^8 = (2\sqrt{2})^8[\cos(8)(45°) + i\sin(8)(45°)] = 4096[\cos 360° + i\sin 360°] = 4096$$

18. $z = 8\left(\cos\frac{\pi}{3} + i\sin\frac{\pi}{3}\right), \quad n = 3$

The cube roots of z are:

For $K = 0$, $\sqrt[3]{8}\left[\cos\frac{\pi/3}{3} + i\sin\frac{\pi/3}{3}\right] = 2\left(\cos\frac{\pi}{9} + i\sin\frac{\pi}{9}\right)$

For $K = 1$, $\sqrt[3]{8}\left[\cos\frac{\pi/3 + 2\pi}{3} + i\sin\frac{\pi/3 + 2\pi}{3}\right] = 2\left(\cos\frac{7\pi}{9} + i\sin\frac{7\pi}{9}\right)$

For $K = 2$, $\sqrt[3]{8}\left[\cos\frac{\pi/3 + 4\pi}{3} + i\sin\frac{\pi/3 + 4\pi}{3}\right] = 2\left(\cos\frac{13\pi}{9} + i\sin\frac{13\pi}{9}\right)$

19. $x^3 = -125 = 125(\cos \pi + i \sin \pi)$

For $K = 0$, $\sqrt[3]{125}\left(\cos \frac{\pi}{3} + i \sin \frac{\pi}{3}\right) = 5\left(\cos \frac{\pi}{3} + i \sin \frac{\pi}{3}\right)$

For $K = 1$, $\sqrt[3]{125}\left(\cos \frac{\pi + 2\pi}{3} + i \sin \frac{\pi + 2\pi}{3}\right) = -5$

For $K = 2$, $\sqrt[3]{125}\left(\cos \frac{\pi + 4\pi}{3} + i \sin \frac{\pi + 4\pi}{3}\right) = 5\left(\cos \frac{5\pi}{3} + i \sin \frac{5\pi}{3}\right)$

20. $x^4 = -i = 1\left(\cos \frac{3\pi}{2} + i \sin \frac{3\pi}{2}\right)$

For $K = 0$, $\cos \frac{3\pi/2}{4} + i \sin \frac{3\pi/2}{4} = \cos \frac{3\pi}{8} + i \sin \frac{3\pi}{8}$

For $K = 1$, $\cos \frac{3\pi/2 + 2\pi}{4} + i \sin \frac{3\pi/2 + 2\pi}{4} = \cos \frac{7\pi}{8} + i \sin \frac{7\pi}{8}$

For $K = 2$, $\cos \frac{3\pi/2 + 4\pi}{4} + i \sin \frac{3\pi/2 + 4\pi}{4} = \cos \frac{11\pi}{8} + i \sin \frac{11\pi}{8}$

For $K = 3$, $\cos \frac{3\pi/2 + 6\pi}{4} + i \sin \frac{3\pi/2 + 6\pi}{4} = \cos \frac{15\pi}{8} + i \sin \frac{15\pi}{8}$

CHAPTER 8

Practice Test Solutions

1. $x + y = 1$

$3x - y = 15 \Rightarrow y = 3x - 15$

$$x + (3x - 15) = 1$$
$$4x = 16$$
$$x = 4$$
$$y = -3$$

2. $x - 3y = -3 \Rightarrow x = 3y - 3$

$x^2 + 6y = 5$

$$(3y - 3)^2 + 6y = 5$$
$$9y^2 - 18y + 9 + 6y = 5$$
$$9y^2 - 12y + 4 = 0$$
$$(3y - 2)^2 = 0$$
$$y = \tfrac{2}{3}$$
$$x = -1$$

3. $x + y + z = 6 \Rightarrow z = 6 - x - y$

$2x - y + 3z = 0 \qquad 2x - y + 3(6 - x - y) = 0 \Rightarrow -x - 4y = -18$

$5x + 2y - z = -3 \qquad 5x + 2y - (6 - x - y) = -3 \Rightarrow 6x + 3y = 3$

$$x = 18 - 4y$$
$$6(18 - 4y) + 3y = 3$$
$$-21y = -105$$
$$y = 5$$
$$x = 18 - 4y = -2$$
$$z = 6 - x - y = 3$$

4. $x + y = 110 \Rightarrow y = 110 - x$

$xy = 2800$

$$x(110 - x) = 2800$$
$$0 = x^2 - 110x + 2800$$
$$0 = (x - 40)(x - 70)$$

$x = 40$ or $x = 70$

$y = 70 \qquad y = 40$

5. $2x + 2y = 170 \Rightarrow y = \dfrac{170 - 2x}{2} = 85 - x$

$xy = 2800$

$$\begin{aligned} x(85 - x) &= 2800 \\ 0 &= x^2 - 85x + 2800 \\ 0 &= (x - 25)(x - 60) \end{aligned}$$

$x = 25$ or $x = 60$

$y = 60$ $\quad$ $y = 25$

Dimensions: $60' \times 25'$

6.
$$\begin{aligned} 2x + 15y &= 4 \Rightarrow 2x + 15y = 4 \\ x - 3y &= 23 \Rightarrow 5x - 15y = 115 \\ & \qquad 7x = 119 \\ & \qquad x = 17 \\ & \qquad y = \frac{x - 23}{3} \\ & \qquad = -2 \end{aligned}$$

7.
$$\begin{aligned} x + y &= 2 \Rightarrow 19x + 19y = 38 \\ 38x - 19y &= 7 \Rightarrow 38x - 19y = 7 \\ & \qquad 57x = 45 \end{aligned}$$

$$x = \frac{45}{57} = \frac{15}{19}$$

$$y = 2 - x = \frac{38}{19} - \frac{15}{19} = \frac{23}{19}$$

8.
$$\begin{aligned} 0.4x + 0.5y &= 0.112 \Rightarrow 0.28x + 0.35y = 0.0784 \\ 0.3x - 0.7y &= -0.131 \Rightarrow 0.15x - 0.35y = -0.0655 \\ & \qquad 0.43x = 0.0129 \end{aligned}$$

$$x = \frac{0.0129}{0.43} = 0.03$$

$$y = \frac{0.112 - 0.4x}{0.5} = 0.20$$

9. Let x = amount in 11% fund and y = amount in 13% fund.

$x + y = 17000 \Rightarrow y = 17000 - x$

$0.11x + 0.13y = 2080$

$$\begin{aligned} 0.11x + 0.13(17000 - x) &= 2080 \\ -0.02x &= -130 \\ x &= \$6500 \\ y &= \$10,500 \end{aligned}$$

10. $(4, 3), (1, 1), (-1, -2), (-2, -1)$

$$n = 4, \sum_{i=1}^{4} x_i = 2, \sum_{i=1}^{4} y_i = 1, \sum_{i=1}^{4} x_i^2 = 22, \sum_{i=1}^{4} x_i y_i = 17$$

$$\begin{aligned} 4b + 2a &= 1 &\Rightarrow\quad 4b + 2a &= 1 \\ 2b + 22a &= 17 &\Rightarrow\quad -4b - 44a &= -34 \\ & & -42a &= -33 \end{aligned}$$

$$a = \frac{33}{42} = \frac{11}{14}$$

$$b = \frac{1}{4}\left(1 - 2\left(\frac{33}{42}\right)\right) = -\frac{1}{7}$$

$$y = ax + b = \frac{11}{14}x - \frac{1}{7}$$

11.

$$\begin{aligned} x + y &= -2 \\ 2x - y + z &= 11 \\ 4y - 3z &= -20 \end{aligned} \quad\Rightarrow\quad \begin{aligned} -2x - 2y &= 4 \\ 2x - y + z &= 11 \\ \hline -3y + z &= 15 \end{aligned} \qquad \begin{aligned} -9y + 3z &= 45 \\ 4y - 3z &= -20 \\ \hline -5y &= 25 \\ y &= -5 \\ x &= 3 \\ z &= 0 \end{aligned}$$

12.

$$\begin{aligned} 4x - y + 5z &= 4 &\Rightarrow\quad 4x - y + 5z &= 4 \\ 2x + y - z &= 0 &\Rightarrow\quad -4x - 2y + 2z &= 0 \\ 2x + 4y + 8z &= 0 & -3y + 7z &= 4 \end{aligned}$$

$$\begin{aligned} 2x + 4y + 8z &= 0 \\ -2x - y + z &= 0 \\ \hline 3y + 9z &= 0 \\ -3y + 7z &= 4 \\ \hline 16z &= 4 \\ z &= \tfrac{1}{4} \\ y &= -\tfrac{3}{4} \\ x &= \tfrac{1}{2} \end{aligned}$$

13. $3x + 2y - z = 5 \Rightarrow 6x + 4y - 2z = 10$

$6x - y + 5z = 2 \Rightarrow -6x + y - 5z = -2$

$$5y - 7z = 8$$

$$y = \frac{8 + 7z}{5}$$

$$3x + 2y - z = 5$$

$$12x - 2y + 10z = 4$$

$$15x + 9z = 9$$

$$x = \frac{9 - 9z}{15} = \frac{3 - 3z}{5}$$

Let $z = a$, then $x = \dfrac{3 - 3a}{5}$ and $y = \dfrac{8 + 7a}{5}$.

14. $y = ax^2 + bx + c$ passes through $(0, -1)$, $(1, 4)$, and $(2, 13)$.

At $(0, -1)$, $-1 = a(0)^2 + b(0) + c \Rightarrow c = -1$

At $(1, 4)$, $4 = a(1)^2 + b(1) - 1 \Rightarrow 5 = a + b \Rightarrow 5 = a + b$

At $(2, 13)$, $13 = a(2)^2 + b(2) - 1 \Rightarrow 14 = 4a + 2b \Rightarrow -7 = -2a - b$

$$-2 = -a$$

$$a = 2$$

$$b = 3$$

Thus, $y = 2x^2 + 3x - 1$.

15. $s = \frac{1}{2}at^2 + v_0t + s_0$ passes through $(1, 12)$, $(2, 5)$, and $(3, 4)$.

At $(1, 12)$, $12 = \frac{1}{2}a + v_0 + s_0 \Rightarrow 24 = a + 2v_0 + 2s_0$

At $(2, 5)$, $5 = 2a + 2v_0 + s_0 \Rightarrow -5 = -2a - 2v_0 - s_0$

At $(3, 4)$, $4 = \frac{9}{2}a + 3v_0 + s_0$ $\quad 19 = -a + s_0$

$$15 = 6a + 6v_0 + 3s_0$$

$$-8 = -9a - 6v_0 - 2s_0$$

$$7 = -3a + s_0$$

$$-19 = a - s_0$$

$$-12 = -2a$$

$$a = 6$$

$$s_0 = 25$$

$$v_0 = -16$$

Thus, $s = \frac{1}{2}(6)t^2 - 16t + 25 = 3t^2 - 16t + 25$.

16. $x^2 + y^2 \geq 9$

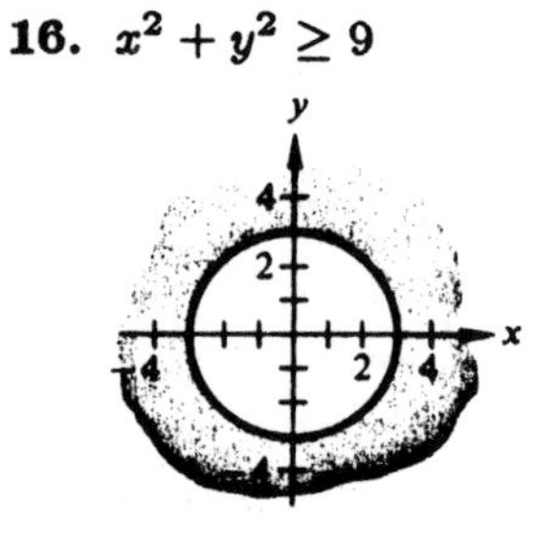

17. $x + y \leq 6$

$x \geq 2$

$y \geq 0$

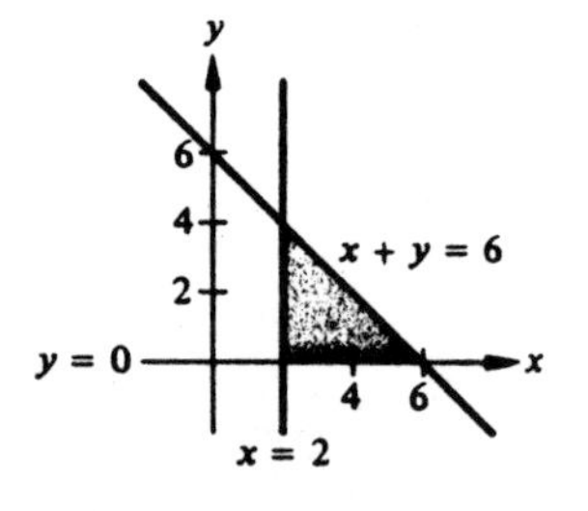

18. Line through (0, 0) and (0, 7):

$x = 0$

Line through (0, 0) and (2, 3):

$y = \frac{3}{2}x$ or $3x - 2y = 0$

Line through (0, 7) and (2, 3):

$y = -2x + 7$ or $2x + y = 7$

Inequalities: $x \geq 0$

$3x - 2y \leq 0$

$2x + y \leq 7$

19. Vertices: (0, 0), (0, 7), (6, 0), (3, 5)

$C = 30x + 26y$

At (0, 0), $C = 0$

At (0, 7), $C = 182$

At (6, 0), $C = 180$

At (3, 5), $C = 220$

The maximum value of C is 220.

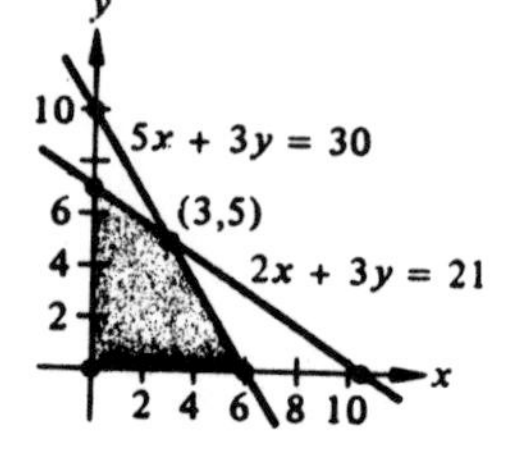

20. $x^2 + y^2 \leq 4$

$(x-2)^2 + y^2 \geq 4$

CHAPTER 9

Practice Test Solutions

1. $\begin{bmatrix} 1 & -2 & 4 \\ 3 & -5 & 9 \end{bmatrix} -3R_1 + R_2 \rightarrow \begin{bmatrix} 1 & -2 & 4 \\ 0 & 1 & -3 \end{bmatrix} 2R_2 + R_1 \rightarrow \begin{bmatrix} 1 & 0 & -2 \\ 0 & 1 & -3 \end{bmatrix}$

2. $\begin{aligned} 3x + 5y &= 3 \\ 2x - y &= -11 \end{aligned}$ $\begin{bmatrix} 3 & 5 & \vdots & 3 \\ 2 & -1 & \vdots & -11 \end{bmatrix}$ $-R_2 + R_1 \rightarrow \begin{bmatrix} 1 & 6 & \vdots & 14 \\ 2 & -1 & \vdots & -11 \end{bmatrix}$

$$-2R_1 + R_2 \rightarrow \begin{bmatrix} 1 & 6 & \vdots & 14 \\ 0 & -13 & \vdots & -39 \end{bmatrix}$$

$$-\tfrac{1}{13}R_2 \rightarrow \begin{bmatrix} 1 & 6 & \vdots & 14 \\ 0 & 1 & \vdots & 3 \end{bmatrix}$$

$$-6R_2 + R_1 \rightarrow \begin{bmatrix} 1 & 0 & \vdots & -4 \\ 0 & 1 & \vdots & 3 \end{bmatrix}$$

Solution: $x = -4, \ y = 3$

3. $\begin{aligned} 2x + 3y &= -3 \\ 3x + 2y &= 8 \\ x + y &= 1 \end{aligned}$ $\begin{bmatrix} 2 & 3 & \vdots & -3 \\ 3 & 2 & \vdots & 8 \\ 1 & 1 & \vdots & 1 \end{bmatrix}$ $R_3 \leftrightarrow R_1$ $\begin{bmatrix} 1 & 1 & \vdots & 1 \\ 3 & 2 & \vdots & 8 \\ 2 & 3 & \vdots & -3 \end{bmatrix}$

$$\begin{matrix} \\ -3R_1 + R_2 \rightarrow \\ 2R_1 + R_3 \rightarrow \end{matrix} \begin{bmatrix} 1 & 1 & \vdots & 1 \\ 0 & -1 & \vdots & 5 \\ 0 & -1 & \vdots & 5 \end{bmatrix}$$

$$\begin{matrix} R_2 + R_1 \rightarrow \\ -R_2 \rightarrow \\ -R_2 + R_3 \rightarrow \end{matrix} \begin{bmatrix} 1 & 0 & \vdots & 6 \\ 0 & 1 & \vdots & -5 \\ 0 & 0 & \vdots & 0 \end{bmatrix}$$

Solution: $x = 6, \ y = -5$

4. $\begin{aligned} x \quad\;\; + 3z &= -5 \\ 2x + y \quad\;\; &= 0 \\ 3x + y - z &= 3 \end{aligned}$ $\left[\begin{array}{rrr:r} 1 & 0 & 3 & -5 \\ 2 & 1 & 0 & 0 \\ 3 & 1 & -1 & 3 \end{array}\right]$ $\begin{array}{l} \\ -2R_1 + R_2 \rightarrow \\ -3R_1 + R_3 \rightarrow \end{array}$ $\left[\begin{array}{rrr:r} 1 & 0 & 3 & -5 \\ 0 & 1 & -6 & 10 \\ 0 & 1 & -10 & 18 \end{array}\right]$

$$\begin{array}{r} \\ \\ -R_2 + R_3 \rightarrow \end{array} \left[\begin{array}{rrr:r} 1 & 0 & 3 & -5 \\ 0 & 1 & -6 & 10 \\ 0 & 0 & -4 & 8 \end{array}\right]$$

$$\begin{array}{r} -3R_3 + R_1 \rightarrow \\ 6R_3 + R_2 \rightarrow \\ -\frac{1}{4}R_4 \rightarrow \end{array} \left[\begin{array}{rrr:r} 1 & 0 & 0 & 1 \\ 0 & 1 & 0 & -2 \\ 0 & 0 & 1 & -2 \end{array}\right]$$

Solution: $x = 1,\ y = -2,\ z = -2$

5. $\begin{bmatrix} 1 & 4 & 5 \\ 2 & 0 & -3 \end{bmatrix} \begin{bmatrix} 1 & 6 \\ 0 & -7 \\ -1 & 2 \end{bmatrix} = \begin{bmatrix} -4 & -12 \\ 5 & 6 \end{bmatrix}$

6. $$\begin{aligned} 3A - 5B &= 3\begin{bmatrix} 9 & 1 \\ -4 & 8 \end{bmatrix} - 5\begin{bmatrix} 6 & -2 \\ 3 & 5 \end{bmatrix} \\ &= \begin{bmatrix} 27 & 3 \\ -12 & 24 \end{bmatrix} - \begin{bmatrix} 30 & -10 \\ 15 & 25 \end{bmatrix} \\ &= \begin{bmatrix} -3 & 13 \\ -27 & -1 \end{bmatrix} \end{aligned}$$

7. $$\begin{aligned} f(A) &= \begin{bmatrix} 3 & 0 \\ 7 & 1 \end{bmatrix}^2 - 7\begin{bmatrix} 3 & 0 \\ 7 & 1 \end{bmatrix} + 8\begin{bmatrix} 1 & 0 \\ 0 & 1 \end{bmatrix} \\ &= \begin{bmatrix} 3 & 0 \\ 7 & 1 \end{bmatrix}\begin{bmatrix} 3 & 0 \\ 7 & 1 \end{bmatrix} - \begin{bmatrix} 21 & 0 \\ 49 & 7 \end{bmatrix} + \begin{bmatrix} 8 & 0 \\ 0 & 8 \end{bmatrix} \\ &= \begin{bmatrix} 9 & 0 \\ 28 & 1 \end{bmatrix} - \begin{bmatrix} 21 & 0 \\ 49 & 7 \end{bmatrix} + \begin{bmatrix} 8 & 0 \\ 0 & 8 \end{bmatrix} \\ &= \begin{bmatrix} -4 & 0 \\ -21 & 2 \end{bmatrix} \end{aligned}$$

8. False since
$$\begin{aligned} (A + B)(A + 3B) &= A(A + 3B) + B(A + 3B) \\ &= A^2 + 3AB + BA + 3B^2 \end{aligned}$$

9. $\left[\begin{array}{rr:rr} 1 & 2 & 1 & 0 \\ 3 & 5 & 0 & 1 \end{array}\right]$ $\begin{array}{l} \\ -3R_1 + R_2 \rightarrow \end{array}$ $\left[\begin{array}{rr:rr} 1 & 2 & 1 & 0 \\ 0 & -1 & -3 & 1 \end{array}\right]$

$$\begin{array}{r} 2R_2 + R_1 \rightarrow \\ -R_2 \rightarrow \end{array} \left[\begin{array}{rr:rr} 1 & 0 & -5 & 2 \\ 0 & 1 & 3 & -1 \end{array}\right]$$

$$A^{-1} = \begin{bmatrix} -5 & 2 \\ 3 & -1 \end{bmatrix}$$

10. $\begin{bmatrix} 1 & 1 & 1 & \vdots & 1 & 0 & 0 \\ 3 & 6 & 5 & \vdots & 0 & 1 & 0 \\ 6 & 10 & 8 & \vdots & 0 & 0 & 1 \end{bmatrix}$ $\begin{matrix} \\ -3R_1 + R_2 \rightarrow \\ -6R_1 + R_3 \rightarrow \end{matrix}$ $\begin{bmatrix} 1 & 1 & 1 & \vdots & 1 & 0 & 0 \\ 0 & 3 & 2 & \vdots & -3 & 1 & 0 \\ 0 & 4 & 2 & \vdots & -6 & 0 & 1 \end{bmatrix}$

$$\begin{matrix} -R_2 + R_1 \rightarrow \\ \frac{1}{3}R_2 \rightarrow \\ -4R_2 + R_3 \rightarrow \end{matrix} \begin{bmatrix} 1 & 0 & \frac{1}{3} & \vdots & 2 & -\frac{1}{3} & 0 \\ 0 & 1 & \frac{2}{3} & \vdots & -1 & \frac{1}{3} & 0 \\ 0 & 0 & -\frac{2}{3} & \vdots & -2 & -\frac{4}{3} & 1 \end{bmatrix}$$

$$\begin{matrix} \frac{1}{2}R_3 + R_1 \rightarrow \\ R_3 + R_2 \rightarrow \\ -\frac{3}{2}R_3 \rightarrow \end{matrix} \begin{bmatrix} 1 & 0 & 0 & \vdots & 1 & -1 & \frac{1}{2} \\ 0 & 1 & 0 & \vdots & -3 & -1 & 1 \\ 0 & 0 & 1 & \vdots & 3 & 2 & -\frac{3}{2} \end{bmatrix}$$

$$A^{-1} = \begin{bmatrix} 1 & -1 & \frac{1}{2} \\ -3 & -1 & 1 \\ 3 & 2 & -\frac{3}{2} \end{bmatrix}$$

11. (a) $\begin{aligned} x + 2y &= 4 \\ 3x + 5y &= 1 \end{aligned}$ $\begin{bmatrix} 1 & 2 & \vdots & 1 & 0 \\ 3 & 5 & \vdots & 0 & 1 \end{bmatrix}$ $\begin{matrix} \\ -3R_1 + R_2 \rightarrow \end{matrix}$ $\begin{bmatrix} 1 & 2 & \vdots & 1 & 0 \\ 0 & -1 & \vdots & -3 & 1 \end{bmatrix}$

$$\begin{matrix} -2R_2 + R_1 \rightarrow \\ -R_2 \rightarrow \end{matrix} \begin{bmatrix} 1 & 0 & \vdots & -5 & 2 \\ 0 & 1 & \vdots & 3 & -1 \end{bmatrix}$$

$$X = A^{-1}B = \begin{bmatrix} -5 & 2 \\ 3 & -1 \end{bmatrix} \begin{bmatrix} 4 \\ 1 \end{bmatrix} = \begin{bmatrix} -18 \\ 11 \end{bmatrix}$$

$x = -18, \ y = 11$

(b) $\begin{aligned} x + 2y &= 3 \\ 3x + 5y &= -2 \end{aligned}$

$$X = A^{-1}B = \begin{bmatrix} -5 & 2 \\ 3 & -1 \end{bmatrix} \begin{bmatrix} 3 \\ -2 \end{bmatrix} = \begin{bmatrix} -19 \\ 11 \end{bmatrix}$$

$x = -19, \ y = 11$

12. $\begin{vmatrix} 6 & -1 \\ 3 & 4 \end{vmatrix} = 24 - (-3) = 27$

13. $\begin{vmatrix} 1 & 3 & -1 \\ 5 & 9 & 0 \\ 6 & 2 & -5 \end{vmatrix} \begin{matrix} 1 & 3 \\ 5 & 9 \\ 6 & 2 \end{matrix} = (-45 + 0 - 10) - (-54 + 0 - 75) = 74$

14. $\begin{vmatrix} 1 & 4 & 2 & 3 \\ 0 & 1 & -2 & 0 \\ 3 & 5 & -1 & 1 \\ 2 & 0 & 6 & 1 \end{vmatrix} = \begin{vmatrix} 1 & 2 & 3 \\ 3 & -1 & 1 \\ 2 & 6 & 1 \end{vmatrix} + 2\begin{vmatrix} 1 & 4 & 3 \\ 3 & 5 & 1 \\ 2 & 0 & 1 \end{vmatrix} = 51 + 2(-29) = -7$ Expansion along Row 2.

15. $\begin{vmatrix} 3 & 0 & 0 \\ 0 & 3 & 0 \\ 0 & 0 & 3 \end{vmatrix} = 3(3)(3)\begin{vmatrix} 1 & 0 & 0 \\ 0 & 1 & 0 \\ 0 & 0 & 1 \end{vmatrix} = -3^3\begin{vmatrix} 1 & 0 & 0 \\ 0 & 0 & 1 \\ 0 & 1 & 0 \end{vmatrix}$

True

16. $\begin{vmatrix} 6 & 4 & 3 & 0 & 6 \\ 0 & 5 & 1 & 4 & 8 \\ 0 & 0 & 2 & 7 & 3 \\ 0 & 0 & 0 & 9 & 2 \\ 0 & 0 & 0 & 0 & 1 \end{vmatrix} = 6(5)(2)(9)(1) = 540$

17. $A = \begin{bmatrix} 1 & 3 & 0 \\ 0 & 4 & 5 \\ 0 & 1 & 2 \end{bmatrix} \quad |A| = 3 \quad A^t = \begin{bmatrix} 1 & 0 & 0 \\ 3 & 4 & 1 \\ 0 & 5 & 2 \end{bmatrix}$

$$A^{-1} = \frac{1}{3}\begin{bmatrix} \begin{vmatrix} 4 & 1 \\ 5 & 2 \end{vmatrix} & -\begin{vmatrix} 3 & 1 \\ 0 & 2 \end{vmatrix} & \begin{vmatrix} 3 & 4 \\ 0 & 5 \end{vmatrix} \\ -\begin{vmatrix} 0 & 0 \\ 5 & 2 \end{vmatrix} & \begin{vmatrix} 1 & 0 \\ 0 & 2 \end{vmatrix} & -\begin{vmatrix} 1 & 0 \\ 0 & 5 \end{vmatrix} \\ \begin{vmatrix} 0 & 0 \\ 4 & 1 \end{vmatrix} & -\begin{vmatrix} 1 & 0 \\ 3 & 1 \end{vmatrix} & \begin{vmatrix} 1 & 0 \\ 3 & 4 \end{vmatrix} \end{bmatrix} = \frac{1}{3}\begin{bmatrix} 3 & -6 & 15 \\ 0 & 2 & -5 \\ 0 & -1 & 4 \end{bmatrix} = \begin{bmatrix} 1 & -2 & 5 \\ 0 & \frac{2}{3} & -\frac{5}{3} \\ 0 & -\frac{1}{3} & \frac{4}{3} \end{bmatrix}$$

18. $x = \dfrac{\begin{vmatrix} 4 & -7 \\ 11 & 5 \end{vmatrix}}{\begin{vmatrix} 6 & -7 \\ 2 & 5 \end{vmatrix}} = \dfrac{97}{44}$

19. $z = \dfrac{\begin{vmatrix} 3 & 0 & 1 \\ 0 & 1 & 3 \\ 1 & -1 & 2 \end{vmatrix}}{\begin{vmatrix} 3 & 0 & 1 \\ 0 & 1 & 4 \\ 1 & -1 & 0 \end{vmatrix}} = \dfrac{14}{11}$

20. $y = \dfrac{\begin{vmatrix} 721.4 & 33.77 \\ 45.9 & 19.85 \end{vmatrix}}{\begin{vmatrix} 721.4 & -29.1 \\ 45.9 & 105.6 \end{vmatrix}} = \dfrac{12{,}769.747}{77{,}515.530} \approx 0.1647$

CHAPTER 10

Practice Test Solutions

1. $a_n = \dfrac{2n}{(n+2)!}$

$a_1 = \dfrac{2(1)}{3!} = \dfrac{2}{6} = \dfrac{1}{3}$

$a_2 = \dfrac{2(2)}{4!} = \dfrac{4}{24} = \dfrac{1}{6}$

$a_3 = \dfrac{2(3)}{5!} = \dfrac{6}{120} = \dfrac{1}{20}$

$a_4 = \dfrac{2(4)}{6!} = \dfrac{8}{720} = \dfrac{1}{90}$

$a_5 = \dfrac{2(5)}{7!} = \dfrac{10}{5040} = \dfrac{1}{504}$

$\left\{\dfrac{1}{3}, \dfrac{1}{6}, \dfrac{1}{20}, \dfrac{1}{90}, \dfrac{1}{504}, \ldots\right\}$

2. $a_n = \dfrac{n+3}{3^n}$

3. $\displaystyle\sum_{i=1}^{6}(2i-1) = 1+3+5+7+9+11 = 36$

4. $a_1 = 23, \ d = -2$

$a_2 = a_1 + d = 21$

$a_3 = a_2 + d = 19$

$a_4 = a_3 + d = 17$

$a_5 = a_4 + d = 15$

$\{23, 21, 19, 17, 15, \ldots\}$

5. $a_1 = 12, \ d = 3, \ n = 50$

$a_n = a_1 + (n-1)d$

$a_{50} = 12 + (50-1)3 = 159$

6. $a_1 = 1$

$a_{200} = 200$

$S_n = \dfrac{n}{2}(a_1 + a_n)$

$S_{200} = \dfrac{200}{2}(1 + 200) = 20,100$

7. $a_1 = 7, \ r = 2$

$a_2 = a_1 r = 14$

$a_3 = a_2 r = 28$

$a_4 = a_3 r = 56$

$a_5 = a_4 r = 112$

$\{7, 14, 28, 56, 112, \ldots\}$

8. $\displaystyle\sum_{n=0}^{9} 6\left(\frac{2}{3}\right)^n, \ a_1 = 6, \ r = \frac{2}{3}, \ n = 9$

$S_n = \dfrac{a_1(1-r^n)}{1-r}$

$= \dfrac{6(1-(\frac{2}{3})^9)}{1-\frac{2}{3}} \approx 17.5318$

9. $\sum_{n=0}^{\infty}(0.03)^n$, $a_1 = 1$, $r = 0.03$

$$S_n = \frac{a_1}{1-r} = \frac{1}{1-0.03} = \frac{1}{0.97} = \frac{100}{97} \approx 1.0309$$

10. For $n = 1$, $1 = \frac{1(1+1)}{2}$.

Assume that $1 + 2 + 3 + 4 + \cdots + k = \frac{k(k+1)}{2}$.

Now for $n = k + 1$,

$$\begin{aligned} 1 + 2 + 3 + 4 + \cdots + k + (k+1) &= \frac{k(k+1)}{2} + k + 1 \\ &= \frac{k(k+1)}{2} + \frac{2(k+1)}{2} \\ &= \frac{(k+1)(k+2)}{2}. \end{aligned}$$

Thus, $1 + 2 + 3 + 4 + \cdots + n = \frac{n(n+1)}{2}$ for all integers $n \geq 1$.

11. For $n = 4$, $4! > 2^4$.
Assume that $k! > 2^k$.
Then $(k+1)! = (k+1)(k!) > (k+1)2^k > 2 \cdot 2^k = 2^{k+1}$.
Thus, $n! > 2^n$ for all integers $n \geq 4$.

12. ${}_{13}C_4 = \frac{13!}{(13-4)!4!} = 715$

13. $$\begin{aligned} (x+3)^5 &= x^5 + 5x^4(3) + 10x^3(3)^2 + 10x^2(3)^3 + 5x(3)^4 + (3)^5 \\ &= x^5 + 15x^4 + 90x^3 + 270x^2 + 405x + 243 \end{aligned}$$

14. ${}_{12}C_5 x^7(-2)^5 = -25,344x^7$

15. ${}_{30}P_4 = \frac{30!}{(30-4)!} = 657,720$

16. $6! = 720$ ways

17. ${}_{12}P_3 = 1320$

18. $$\begin{aligned} P(2) + P(3) + P(4) &= \frac{1}{36} + \frac{2}{36} + \frac{3}{36} \\ &= \frac{6}{36} = \frac{1}{6} \end{aligned}$$

19. $P(K,\ B10) = \frac{4}{52} \cdot \frac{2}{51} = \frac{2}{663}$

20. Let A = probability of no faulty units.

$$P(A) = \left(\frac{997}{1000}\right)^{50} \approx 0.8605$$

$$P(A') = 1 - P(A) \approx 0.1395$$

CHAPTER 11

Practice Test Solutions

1. $x^2 - 6x - 4y + 1 = 0$

$$x^2 - 6x + 9 = 4y - 1 + 9$$

$$(x-3)^2 = 4y + 8$$

$$(x-3)^2 = 4(1)(y+2) \Rightarrow p = 1$$

Vertex: $(3,\ -2)$

Focus: $(3,\ -1)$

Directrix: $y = -3$

2. Vertex: $(2,\ -5)$

Focus: $(2,\ -6)$

Vertical axis;

opens downward with $p = -1$

$$(x-h)^2 = 4p(y-k)$$

$$(x-2)^2 = 4(-1)(y+5)$$

$$x^2 - 4x + 4 = -4y - 20$$

$$x^2 - 4x + 4y + 24 = 0$$

3. $x^2 + 4y^2 - 2x + 32y + 61 = 0$

$$(x^2 - 2x + 1) + 4(y^2 + 8y + 16) = -61 + 1 + 64$$

$$(x-1)^2 + 4(y+4)^2 = 4$$

$$\frac{(x-1)^2}{4} + \frac{(y+4)^2}{1} = 1$$

$a = 2,\ b = 1,\ c = \sqrt{3}$

Horizontal major axis

Center: $(1,\ -4)$

Foci: $(1 \pm \sqrt{3},\ -4)$

Vertices: $(3,\ -4),\ (-1,\ -4)$

Eccentricity: $e = \sqrt{3}/2$

4. Vertices: $(0,\ \pm 6)$

Eccentricity: $e = 1/2$

Center: $(0,\ 0)$

Vertical major axis

$a = 6,\ e = \frac{c}{a} = \frac{c}{6} = \frac{1}{2} \Rightarrow c = 3,$

$b^2 = (6)^2 - (3)^2 = 27$

$$\frac{x^2}{27} + \frac{y^2}{36} = 1$$

5. $16y^2 - x^2 - 6x - 128y + 231 = 0$

$$16(y^2 - 8y + 16) - (x^2 + 6x + 9) = -231 + 256 - 9$$
$$16(y-4)^2 - (x+3)^2 = 16$$
$$\frac{(y-4)^2}{1} - \frac{(x+3)^2}{16} = 1$$

$a = 1,\ b = 4,\ c = \sqrt{17}$

Center: $(-3,\ 4)$; vertical transverse axis

Vertices: $(-3,\ 5),\ (-3,\ 3)$

Foci: $(-3,\ 4 \pm \sqrt{17})$

Asymptotes: $y = 4 \pm \frac{1}{4}(x+3)$

6. Vertices: $(\pm 3,\ 2)$

Foci: $(\pm 5,\ 2)$

Center: $(0,\ 2)$; horizontal transverse axis

$a = 3,\ c = 5,\ b = 4$

$$\frac{(x-0)^2}{9} - \frac{(y-2)^2}{16} = 1$$
$$\frac{x^2}{9} - \frac{(y-2)^2}{16} = 1$$

7. $5x^2 + 2xy + 5y^2 - 10 = 0$

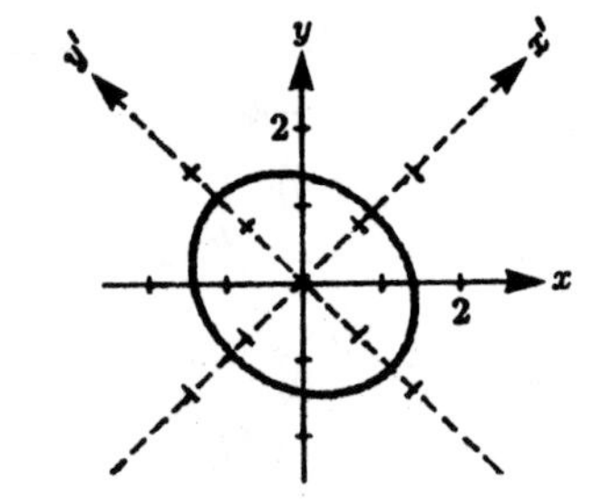

$A = 5,\ B = 2,\ C = 5,\ D = 0,\ E = 0,\ F = -10$

$$\cot 2\theta = \frac{5-5}{2} = 0$$
$$2\theta = \frac{\pi}{2} \Rightarrow \theta = \frac{\pi}{4}$$

$$A' = 5\cos^2\frac{\pi}{4} + 2\cos\frac{\pi}{4}\sin\frac{\pi}{4} + 5\sin^2\frac{\pi}{4} = 6$$
$$C' = 5\sin^2\frac{\pi}{4} - 2\cos\frac{\pi}{4}\sin\frac{\pi}{4} + 5\cos^2\frac{\pi}{4} = 4$$
$$D' = E' = 0$$
$$F' = -10$$

$$6(x')^2 + 4(y')^2 - 10 = 0$$
$$\frac{3(x')^2}{5} + \frac{2(y')^2}{5} = 1$$
$$\frac{(x')^2}{5/3} + \frac{(y')^2}{5/2} = 1$$

Ellipse centered at the origin

8. (a) $6x^2 - 2xy + y^2 = 0$

$A = 6,\ B = -2,\ C = 1$

$B^2 - 4AC = (-2)^2 - 4(6)(1) = -20 < 0$ Ellipse

(b) $x^2 + 4xy + 4y^2 - x - y + 17 = 0$

$A = 1,\ B = 4,\ C = 4$

$B^2 - 4AC = (4)^2 - 4(1)(4) = 0$ Parabola

9. Polar: $\left(\sqrt{2},\ \frac{3\pi}{4}\right)$

$$x = \sqrt{2}\cos\frac{3\pi}{4} = \sqrt{2}\left(-\frac{1}{\sqrt{2}}\right) = -1$$

$$y = \sqrt{2}\sin\frac{3\pi}{4} = \sqrt{2}\left(\frac{1}{\sqrt{2}}\right) = 1$$

Rectangular: $(-1,\ 1)$

10. Rectangular: $(\sqrt{3},\ -1)$

$$r = \pm\sqrt{(\sqrt{3})^2 + (-1)^2} = \pm 2$$

$$\tan\theta = \frac{\sqrt{3}}{-1} = -\sqrt{3}$$

$$\theta = \frac{2\pi}{3} \quad \text{or} \quad \theta = \frac{5\pi}{3}$$

Polar: $\left(-2,\ \frac{2\pi}{3}\right)$ or $\left(2,\ \frac{5\pi}{3}\right)$

11. Rectangular: $4x - 3y = 12$

Polar: $4r\cos\theta - 3r\sin\theta = 12$

$$r(4\cos\theta - 3\sin\theta) = 12$$

$$r = \frac{12}{4\cos\theta - 3\sin\theta}$$

12. Polar: $r = 5\cos\theta$

$$r^2 = 5r\cos\theta$$

Rectangular: $x^2 + y^2 = 5x$

$$x^2 + y^2 - 5x = 0$$

13. $r = 1 - \cos\theta$

Cardioid

Symmetry: Polar axis

Maximum value of $|r|$: $r = 2$ when $\theta = \pi$

Zero of r : $r = 0$ when $\theta = 0$

θ	0	$\frac{\pi}{2}$	π	$\frac{3\pi}{2}$
r	0	1	2	1

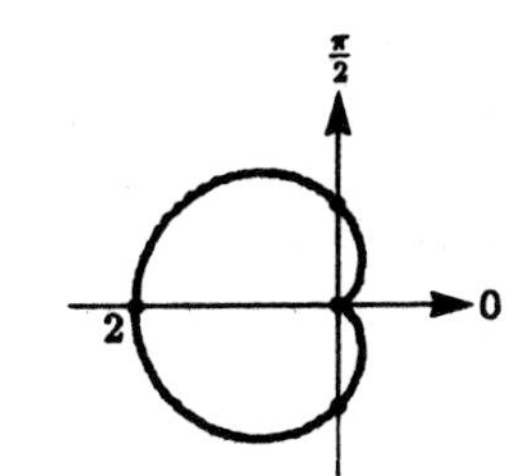

14. $\theta = \dfrac{3\pi}{4}$

Line

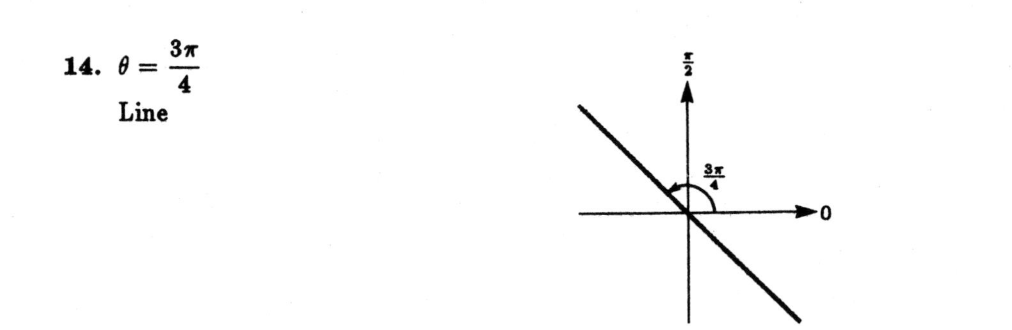

15. $r = 5 \sin 2\theta$

Rose curve with four petals

Symmetry: Polar axis, $\theta = \dfrac{\pi}{2}$, and pole

Maximum value of $|r|$: $|r| = 5$ when $\theta = \dfrac{\pi}{4}, \dfrac{3\pi}{4}, \dfrac{5\pi}{4}, \dfrac{7\pi}{4}$

Zeros of r: $r = 0$ when $\theta = 0, \dfrac{\pi}{2}, \pi, \dfrac{3\pi}{2}$

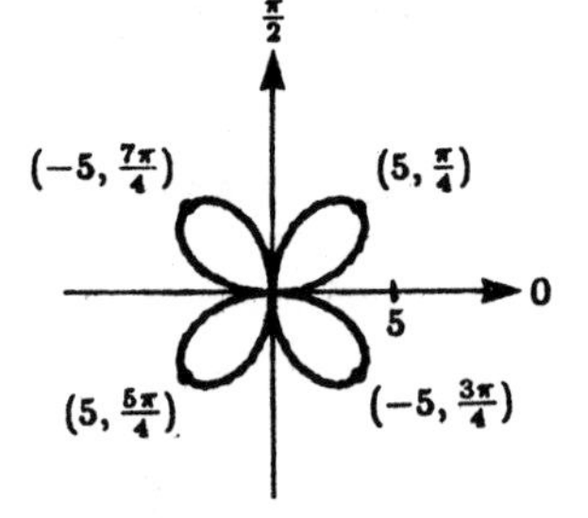

16. $r = \dfrac{3}{6 - \cos\theta}$

$r = \dfrac{\frac{1}{2}}{1 - \frac{1}{6}\cos\theta}$

$e = \frac{1}{6} < 1$, so the graph is an ellipse.

θ	0	$\frac{\pi}{2}$	π	$\frac{3\pi}{2}$
r	$\frac{3}{5}$	$\frac{1}{2}$	$\frac{3}{7}$	$\frac{1}{2}$

17. Parabola

Vertex: $\left(6, \frac{\pi}{2}\right)$

Focus: $(0, 0)$

$e = 1$

$$r = \frac{ep}{1 + e\sin\theta}$$

$$r = \frac{p}{1 + \sin\theta}$$

$$6 = \frac{p}{1 + \sin(\pi/2)}$$

$$6 = \frac{p}{2}$$

$$12 = p$$

$$r = \frac{12}{1 + \sin\theta}$$

18. $x = \sqrt{t+1}, \quad y = 3 + t$

$x \geq 0, \quad t = x^2 - 1$

$y = 3 + (x^2 - 1)$

$y = 2 + x^2, \quad x \geq 0$

19. $x = 3 - 2\sin\theta, \quad y = 1 + 5\cos\theta$

$$\frac{x-3}{-2} = \sin\theta, \quad \frac{y-1}{5} = \cos\theta$$

$$\left(\frac{x-3}{-2}\right)^2 + \left(\frac{y-1}{5}\right)^2 = 1$$

$$\frac{(x-3)^2}{4} + \frac{(y-1)^2}{25} = 1$$

20. $x = e^{2t}, \quad y = e^{4t}$

$x > 0, \quad y > 0$

$x = e^{2t} \Rightarrow \ln x = 2t \Rightarrow t = \frac{1}{2}\ln x$

$y = e^{4t} = e^{4(1/2 \ln x)} = e^{2\ln x} = e^{\ln x^2} = x^2$

$y = x^2, \quad x > 0, \quad y > 0$

Alternate solution:

$y = e^{4t} = (e^{2t})^2 = x^2, \quad x > 0, \quad y > 0$